Sustainable Supply Chain Management and Optimization

Sustainable Supply Chain Management and Optimization

Editors

Shaojian Qu
Qingguo Bai
Ying Ji
Congjun Rao

MDPI • Basel • Beijing • Wuhan • Barcelona • Belgrade • Manchester • Tokyo • Cluj • Tianjin

Editors
Shaojian Qu
School of Management
Science and Technology
Nanjing University of
Information Science and
Technology
Nanjing City
China

Qingguo Bai
School of Management
Qufu Normal University
Qufu City
China

Ying Ji
School of Management
Shanghai University
Shanghai City
China

Congjun Rao
School of Science
Wuhan University of
Technology
Wuhan City
China

Editorial Office
MDPI
St. Alban-Anlage 66
4052 Basel, Switzerland

This is a reprint of articles from the Special Issue published online in the open access journal *Sustainability* (ISSN 2071-1050) (available at: www.mdpi.com/journal/sustainability/special_issues/SSCMO).

For citation purposes, cite each article independently as indicated on the article page online and as indicated below:

LastName, A.A.; LastName, B.B.; LastName, C.C. Article Title. *Journal Name* **Year**, *Volume Number*, Page Range.

ISBN 978-3-0365-7581-0 (Hbk)
ISBN 978-3-0365-7580-3 (PDF)

Contents

About the Editors

Shaojian Qu

Shaojian Qu is a professor at the School of Management Engineering, Nanjing University of Information Science and Technology. He is a distinguished professor of Longshan Scholars, a doctoral supervisor, a researcher of Asia-Pacific Logistics Institute and Business School of National University of Singapore, and a Shanghai distinguished professor of Oriental Scholars. He works in the areas of big data analysis, modeling theory, robust optimization, decision theory, supply chain management, investment portfolios, etc. He has published more than 80 papers as the first or corresponding author, among which more than 70 SCI articles have been published in international journals, including 4 *EJOR* articles and 1 article each in *JOTA* and *IEEE T Ind Inform*. He has published two monographs and served as a reviewer for *EJOR, Informs J Comput, OMEGA, Inform Sci, IEEE T Ind Inform, Systems Engineering Theory and Practice, Operations Research and Management*, the *Management Journal, China Management Science*. He has presided over more than 10 provincial- and ministerial-level projects. Currently, he is a member of the editorial Board of the *Operations Research and Management Journal, Computer Modeling in Engineering & Sciences*, and the *SCI Journal*. He is also a guest editor of the *Journal of Mathematics and Mathematical Problems in Engineering*, an expert of the National Natural Science Foundation of China, an expert of the National Social Science Foundation of China, an expert of the National Talent Program (Changjiang Scholars and Young Changjiang Scholars), an executive Director of the Chinese Society of Economic Mathematics and Management Mathematics, an executive director of the Game Theory Branch of the Chinese Society of Operations Research, an executive Director of the Sustainable Operation and Management System Branch of the Chinese Society of Systems Engineering, and an executive director of the Intelligent Decision and Game Branch of the Chinese Society of Systems Engineering.

Qingguo Bai

Qingguo Bai is a professor at the School of Management, Qufu Normal University. He is the deputy dean of the School of Management, a doctoral supervisor, a Doctor of Management at Huazhong University of Science and Technology, a young expert of Shandong Taishan Scholars, and a leader of the Youth Innovation Team of Shandong Colleges and universities. His research interests include sustainable supply chain management, and operational research theory and application. He has published more than 60 papers in *Systems Engineering Theory and Practice, Chinese Management Science*, the *Journal of Management Engineering*, the *Journal of Management*, the *European Journal of Operational Research*, the {International Journal of Production Economics, the *European Journal of Operational Research*, the *International Journal of Production Economics, Transportation Research Part E: Logistics and Transportation Review, Annals of Operations Research*, and *Computers & Industrial Engineering*. He presided over two (completed) projects for the National Natural Science Foundation and five projects at the provincial and ministerial levels. As the first author, he won second prize and third prize for the Shandong Province Social Science Excellent Achievements award, and first prize for the Shandong Province Soft Science Excellent Achievements award. As a key member, he won first prize for the Shandong Province Social Science Excellent Achievements award and third prize for the Hubei Natural Science award. He has been to the Department of Systems Engineering and Engineering Management of the Chinese University of Hong Kong and the Industrial Engineering Department of Concordia University in Canada for academic exchanges and visits.

Ying Ji

Ying Ji is a professor at the School of Management, Shanghai University; a doctoral supervisor; a Shanghai Young Oriental Scholar; a researcher at the Business School of the National University of Singapore; a former associate professor at the Harbin Institute of Technology; a former principal of Information Management and Information System at the University of Shanghai for Science and Technology; and a deputy secretary general of the Shanghai Excellent Young Female Teachers Association of Education System. In recent years, he has published more than 60 papers in SCI or SSCI international journals as the first author or corresponding author, including *EJOR, FGCS, IEEE TII, JOTA, INFORM FUSION*, and other internationally renowned journals. He has presided over three National Natural Science Foundation projects, participated in one key project of the National Natural Science Foundation, and participated in a major project of the Ministry of Industry and Information Technology.

Congjun Rao

Congjun Rao is a professor at the School of Science, Wuhan University of Technology. His research interests cover statistical prediction and decision, system control and optimization, applied mathematics and mechanics, etc. He has published more than 90 academic papers, including nearly 40 papers indexed by SCI and SSCI and more than 50 papers indexed by EI and has published 1 monograph in *Science Press*. He has presided over two National Natural Science Foundation projects (one general project and one youth project each), two China Postdoctoral Science Foundation projects (one specially funded project and one general project each), two Natural Science Foundation projects for Hubei Province (general), one Hubei Technology Innovation Special (soft science) research project, and five Hubei Provincial Education Department projects (two key projects); has been awarded one Outstanding young and middle-aged people award; and has participated in five National Natural Science Foundation projects as a core member.

Preface to "Sustainable Supply Chain Management and Optimization"

In modern society, the concept of sustainable development is more and more popular, and supply chain management has been continuously incorporating sustainable development criteria into its decision making. Sustainable development is an economic growth model that focuses on long-term development. It presents a challenge to meet the needs of the present generation without compromising the needs of future generations while protecting the environment. To address this challenge, we encourage sustainable supply chain management and optimization.

Supply chain management is an integrated management idea and method that carries out the planning and control of logistics from the supplier to the end user in a supply chain. Supply chain optimization refers to "optimizing the scheme under constrained conditions or limited resources". Therefore, we study the optimization of supply chain management under sustainable development and put forward an optimized supply chain scheme under the constraints of multiple factors such as internal resources, costs, corporate responsibility, external environment, external supervision, and externality of enterprises.

To realize the sustainable development of supply chain management, the first step is to realize economic sustainability of the supply chain and its node enterprises, which can continuously meet the actual needs of certain customers and form a long-term economic growth model. The second is to achieve sustainable responsibility; every company and the entire supply chain should shoulder corporate social responsibility. Third is to achieve environmental sustainability and to make full use of and integrate social resources.

A sustainable supply chain is a complex, comprehensive, and dynamic multi-disciplinary system, and emphasis should be placed on research objectives such as coordinating economic benefits and environmental impacts. Its normative research paradigm, evaluation model, and performance indicators, as well as strategic sustainable governance, are research hotspots in the field of supply chain management, and its theoretical system and practice need more development and exploration. Sustainable supply chain management addresses the sustainability of supply chains from the economic, environmental, corporate, and social perspectives. Its application area is the entire value chain and life cycle of a product/service, from development to the end of the life cycle. This book focuses on how to make supply chains more sustainable through optimization.

Shaojian Qu, Qingguo Bai, Ying Ji, and Congjun Rao
Editors

sustainability

Editorial

Sustainable Supply Chain Management and Optimization

Shaojian Qu [1,*] and Ying Ji [2]

1 School of Management Science and Engineering, Nanjing University of Information Science & Technology, Nanjing 210044, China
2 School of Management, Shanghai University, Shanghai 200444, China
* Correspondence: qushaojian@usst.edu.cn

Citation: Qu, S.; Ji, Y. Sustainable Supply Chain Management and Optimization. *Sustainability* **2023**, *15*, 3844. https://doi.org/10.3390/su15043844

Received: 15 February 2023
Accepted: 17 February 2023
Published: 20 February 2023

Due to the complex and changing global trade environment and the intensification of economic and trade conflicts, enterprises have become more cautious about economic development. Therefore, an increasing number of enterprises are paying increasing attention to the construction of and investment in the supply chains for sustainable development. For example, in 1996, The Manufacturing Research Council (MRC) at Michigan State University, with a grant of USD 400,000 from the National Science Foundation (NSF), conducted a study on "Environmentally Responsible Manufacturing (ERM)", which introduced the concept of green supply chains and emphasized their importance. Green supply chain integrates the environment into the whole supply chain, from the raw materials provided by suppliers to the disposal of goods after consumption by customers, the whole process should consider the comprehensive use of resources and environmental protection, reduce production activities that cause harm to human beings and the environment, and ultimately optimize economic and environmental benefits.

In 1987, as the awareness of environmental protection gradually became more prevalent, the WCED introduced the concept of sustainable development, for which the fundamental ideas were to protect the environment, ensure the sustainability of resources, and then encourage economic and social developments [1]. Since this concept was introduced, sustainable development has spread to many industries. In the 1990s, scholars began to study sustainable supply chain management for environmental protection, as well as contributing to social development in the process of supply chain management. Sustainable supply chain development is already reflected in all aspects of supply chain management: development and design [2], production and packaging, marketing and distribution, consumption and recycling, etc. Currently, a safe, stable, and sustainable supply chain is not only positioned as the core aspect of enterprise development but is also gradually becoming integral to the sustainable development of the industrial chain. Modern society is facing very serious environmental and resource problems. In this context, sustainable supply chain management is a sustainable development model used in modern enterprises that takes these two issues into account, in order to achieve good economic and social benefits. In the process of implementing sustainable supply chain management, there are still many problems faced by enterprises, which need to be studied and continuously improved.

Studies that address all the challenges and possible influences of sustainable supply chain management are welcomed for this Special Issue, entitled "Sustainable Supply Chain Management and Optimization". Such studies will help to analyze and develop solutions in the field of sustainable supply chain management and develop effective methods for future research. The main focus of this Special Issue is to identify management factors for sustainable supply chain development—such as carbon footprint and emissions, waste, air pollution, big data, cost management, agricultural supply chain, and supply chain finance—to promote the innovative development of sustainable supply chain systems [3,4].

We wish to thank all authors for submitting their papers to this Special Issue, and acknowledge all the reviewers for their careful and timely verdicts, which helped to improve the quality of this Special Issue.

Funding: This research received no external funding.

Acknowledgments: We would thank the Editors and reviewers for the time and the efforts spent in handling and reviewing this paper. Their valuable comments allowed me to improve the research. We also gratefully acknowledge the support of the National Natural Science Foundation of China (Nos. 72171149, 72171123).

Conflicts of Interest: The authors declare no conflict of interest.

References

1. Mo, Z.; Fu, L.; Wen, X.; Liu, H.; Zhou, Q. The influence of government subsidy policies on sustainable supply chain decision-making. *Syst. Eng.* **2020**, *38*, 85–92.
2. Yang, F.; Wu, Z.; Teng, X.; Qu, S. Robust Counterpart Models for Fresh Agricultural Product Routing Planning Considering Carbon Emissions and Uncertainty. *Sustainability* **2022**, *14*, 14992. [CrossRef]
3. Qu, S.; Xu, Y.; Ji, Y.; Feng, C.; Wei, J.; Jiang, S. Data-Driven Robust Data Envelopment Analysis for Evaluating the Carbon Emissions Efficiency of Provinces in China. *Sustainability* **2022**, *14*, 13318. [CrossRef]
4. Guo, L.; Shang, Y. Decision-Making of Cross-Border E-Commerce Platform Supply Cvhains Considering Information Sharing and Free Shipping. *Sustainability* **2023**, *15*, 3350. [CrossRef]

sustainability

MDPI

Article

Optimal Pricing Strategy of New Products and Remanufactured Products Considering Consumers' Switching Purchase Behavior

Hao Li [1,2], Qing Xiao [1,*] and Ting Peng [3]

1 School of Economics and Management, Chongqing Jiaotong University, Chongqing 400074, China
2 Western China Transportation-Economy-Society Development Study Center, Chongqing 400074, China
3 School of Economics and Management, University of Electronic Science and Technology of China, Chengdu 611731, China
* Correspondence: xqingcq@126.com

Abstract: Due to income constraints, increased awareness of environmental protection and preference for new products, consumers generate switching purchases between new and remanufactured products, which often lead to a "cannibalization effect" in the market, and make sellers fall into a vicious circle of price reduction. Considering consumers' switching purchase behavior, this study examines the pricing problem of new products and remanufactured products in the competitive market environment. Based on two-period duopoly asymmetric price game models, there has been less research on the effectiveness of the price matching strategy and the traditional dynamic pricing strategy, which is the issue that this paper is dedicated to discussing. This study analyzes the equilibrium profits and their influencing factors under the dynamic pricing and price matching strategies of sellers, and discusses the simplified solution of the model. The results show that consumer learning costs, initial consumers and product differences can affect the sellers' pricing decisions. Consumers' learning costs of products reduces the equilibrium profit of the manufacturer and increases that of the remanufacturer. Initial consumers are not always advantageous for sellers' profitability. Product differences affect the determination of the seller's equilibrium strategy. In the optimal strategy, the remanufacturer should insist on price matching, while the manufacturer should choose dynamic pricing or price matching according to the product differences. This study provides sellers with insights to choose appropriate and custom pricing strategies to maximize profit as well as prevent the majority of consumers switching purchase.

Keywords: remanufactured products; price matching; duopoly; switching purchase behavior

Citation: Li, H.; Xiao, Q.; Peng, T. Optimal Pricing Strategy of New Products and Remanufactured Products Considering Consumers' Switching Purchase Behavior. *Sustainability* **2023**, *15*, 5246. https://doi.org/10.3390/su15065246

Academic Editor: Riccardo Testa

Received: 31 January 2023
Revised: 22 February 2023
Accepted: 13 March 2023
Published: 15 March 2023

1. Introduction

With the rapid development of science and technology, the speed of product renewal has become faster. According to the data analysis of the Global New Products Database (GNPD), products that have been launched in Europe, the Middle East and Africa accelerate significantly in the first half of 2022.The industries involved extend from emerging electronic products, cosmetics and fashion to traditional fields such as home furnishing and automobiles. At the same time, with the advent of the global era of green environmental protection, the recycling and remanufacturing of old products has received more and more attention. Remanufactured products are obtained by renovating or remanufacturing recycled old products (or parts), which are not significantly different from new products in function and appearance. Apple's official website shows that the products sold in the online store include manufactured mobile phones, and that each Apple-certified remanufactured product has undergone a rigorous refurbishment process, including comprehensive testing according to the same rigorous functional standards as new Apple products.

In order to meet the needs of consumers at different levels, manufacturers need to produce and sell new products and remanufactured products at the same time, which leads to the generation of consumer switching purchase behavior. For example, consumers who prefer remanufactured products at the beginning may delay their purchase and opt for the new product because of a strong desire to try new functions or new materials. However, due to income constraints, increased awareness of environmental protection and so on, consumers who want to buy new products will choose to compromise and finally buy remanufactured products. Considering these switching purchase behavior of consumers and the competition between new products and remanufactured products, manufacturers and remanufacturers face great challenges in product sales. First of all, new products and remanufactured products often lead to a "cannibalization effect" in the market, which greatly reduces the success rate of new products on the market. Secondly, consumers' switching purchase between new products and remanufactured products makes sellers fall into a vicious circle of price reduction in order to retain the original demand. The current research has generally concluded that consumers' switching purchase behavior has a negative impact on sellers' profits. Research shows that if the manufacturer or remanufacturer ignores some of the above-mentioned strategic behavior in the pricing decision, both of them may incur about 20% less profits [1]. Li et al. [2] have only analyzed equilibrium profits under dynamic pricing, and found that dynamic pricing leads to a decline in corporate equilibrium profits and consumer rationality when firms sharply discount prices. When introducing the price matching strategy to optimize the seller's profit, scholars have drawn different conclusions. For example, Anne and Shaffer [3] show that when both asymmetric product substitutability and shelf space availability are considered, the price under PMG may even fall below the competitive level, while Chen et al. [4] argue that price-match guarantees can generate a competition-enhancing effect. Additionally, in the actual business process, the Chairman of Best Buy, has pointed out that consumer switching purchase behavior made the company "face very difficult challenges, and the only effective way to break the bottom line of sales profit of products is to re-plan the pricing strategy of the sales market".

Some manufacturers try to apply dynamic pricing as a reactive response to dispose of products, and they dynamically adjust prices over time. However, one of the drawbacks of dynamic pricing strategy is that the manufacturers and remanufacturers cannot segment consumers completely because the strategy often prompts consumers to postpone and switch purchases between new products and remanufactured products. In general, scholars find it difficult to eliminate or weaken the negative impact of consumer behavior by using dynamic pricing. Given that consumers' perception of quality is positively correlated with product price, some manufacturers use price matching guarantee (PMG) to discourage consumers from postponing and switching purchases. Price matching guarantee (PMG) is a policy and practice that quickly matches another seller's quotation for the same or similar goods, or refunds the difference in price within a short time after the sale. This policy and practice has attracted increasing interest from the academics and practitioners [5]. McWilliams and Gerstner prove that the PMG strategy can be used to prevent consumers from switching purchases because it reduces the dissatisfaction of consumers who find a lower price elsewhere after purchase [6]. In fact, manufacturers and remanufacturers are trying to adopt price matching. For example, Huawei relaunched the P40 Pro 5G mobile phone in 2022 and supported the sale of new and remanufactured products at the same price. Additionally, some scholars have questioned whether when there are too many consumers, PMGs are unprofitable, and they believe that sellers' equilibrium price matching strategies depend on the relative importance of the demand-expansion and competition-intensifying effects [7]. Based on these points, this paper will analyze the effectiveness of price matching. We mainly study the competitive equilibrium of dynamic pricing and matching pricing when new products and remanufactured products are sold at the same time, and solve the following problems:

(1) Under the dynamic pricing strategy, does consumers' switching purchase behavior between new products and manufactured products have a negative impact on manufacturers and remanufacturers? How should this impact be described?

(2) Can the price matching strategy mitigate the effects of consumers' switching purchase behavior when new products and remanufactured products are sold at the same time?

To address these questions, we first present the traditional dynamic pricing strategy. Then, we introduce three types of price matching strategies. These strategies are as follows: first, only the manufacturer implements price matching and uses the price of the remanufactured product in the next period as their own pricing (BMP). Second, only the remanufacturer implements price matching and uses the price of the new product in the next period as their own pricing (OMP). Third, the manufacturer and the remanufacturer match the pricing of a product from their respective competitors in the next period, which suggests that the seller wants to retain their consumers and encourage them to buy products in the current period (RMP). The third matching strategy combines the two previous matching strategies to make the seller's sales exclusive, with little possibility for consumers to shift. Finally, we obtain the relevant factors that affect the profits of manufacturers and remanufacturers and the applicable conditions of the price matching strategy through the equilibrium analysis.

On the basis of the existing research, this paper discusses the following three innovative points: (1) The form of consumers' switching buying behavior; (2) The simplification of the model; (3) The comparison of optimal strategies. First, we introduce consumers' switching purchase behavior into the pricing decision of sellers' products in two periods. Specifically, we consider that consumers' switching behavior occurs between new products and remanufactured products in different periods and divide them into two categories. One is that consumers who view remanufactured products at the beginning may delay their purchase and turn to buy new products; the other is that consumers who view new products will switch to buying remanufactured products. Second, we simplify the solution process of the equilibrium profits of sellers under different strategies, which reduces the space for optimal decision-making of sellers. Third, we analyze the effectiveness of the price matching strategy compared with the traditional dynamic pricing strategy. The price matching strategy takes a competitor's future price as a seller's current price, which helps to explain the connotation of price matching. Considering consumers' switching purchase behavior, the only choice for the remanufacturer is to take the future price of the competitor's new products as the current product price, while the manufacturer can judge whether to match the future price of competitors according to the differences between new products and remanufactured products. This provides some guidance for the pricing of sellers.

The remainder of this paper is organized as follows. Section 2 briefly reviews the literature, while Section 3 introduces the basic model in detail, including the model framework, assumptions, and method. Section 4 analyses the dynamic pricing strategy equilibrium and the price matching strategies equilibrium. Section 5 conducts a numerical analysis on the impact of the related factors influencing the sellers' profits and their optimal strategy. Section 6 presents the conclusions and directions for future research. All relevant proofs appear in Appendix A.

2. Literature Review

This study is related to two research streams: the pricing strategy in the case of simultaneous sales of multiple products, and the pricing strategy selection problem considering consumers' switching purchase behavior.

Revenue management originates from the US airline industry in the end of 1970s, for a long time, dynamic pricing had been widely used by sellers to sell perishable products such as air tickets, whose product prices change with time. The extant literature has reviewed dynamic pricing from different perspectives. Levin and McGill [8] found that in a multi-period dynamic game model, sellers can increase revenue by implementing dynamic pricing strategies when consumers have price comparison behavior. Dasu and Tong [9] and Chew et al. [10] reach the same conclusion. Prasad et al. [11] find that the number of non-comparable consumers at a certain threshold can improve sellers' revenues based on a two-stage dynamic pricing model. Liu and Zhang [12] highlight that companies with low product quality have greater profit losses. Li et al. [2] show that dynamic pricing leads to a decline in corporate equilibrium profits and consumer rationality when firms sharply discount prices. The new products and remanufactured products we studied have similar characteristics to perishable products such as air tickets. In the analysis of the dynamic pricing model, scholars have mentioned the strategic behavior of consumers and their preference for products, but there is no detailed description of consumers' switching purchase behavior. In the era of advocating for green development and consumer personalized demand, it is particularly critical to choose the perspective from which to analyze the reasonable pricing of new products and remanufactured products. Our problem is essentially a more complex dynamic program that considers the competitive environment and consumers' behavior.

Different products are sold at the same time, which means fierce competition for profits among the sellers behind their respective products. Therefore, sellers have to grasp the reaction of competitors when setting their pricing strategies. The phenomenon of the duopoly competition has been a topic of discussion since the seminal papers of early researchers. Moorthy [13] examines how consumer preferences, costs, and price competition affect a firm's competitive product strategy under a duopoly competition. Li et al. [14] demonstrate that on the competitive pricing of duopoly, increasing the differentiation between sellers could reverse the unfavorable situation in the competition. The price competition between the new products and remanufactured products discussed in this paper is essentially the competitive problem of duopoly and multi-products. Actually, the competition between new manufactured products and remanufactured products within and across competitive channels significantly affects the pricing decisions of manufacturers. Wen et al. [15] found that when there is new product competition between the remanufacturer and the manufacturer, adding direct channels can improve the remanufacturer's ability to resist the uncertainty of consumer behavior. Similarly, Liu et al. [16] proposed that if manufacturers decide to produce remanufactured products, they can also sell remanufactured products in their online direct channels. Niu et al. [17] considered sellers who sell new products and remanufactured products at the same time, and found that retailers can eliminate their yield uncertainty by producing remanufactured products themselves or relying on third-party remanufacturers. Qiao and Su [18] built two-period game models to address the choice issues of the original equipment manufacturer's licensing strategy and the independent remanufacturer's distribution channel, and found that the original equipment manufacturer's licensing strategy choice depends on the sizes of the fixed licensing fee and the obtained willingness-to-pay. For the analysis of consumer behavior, Ma et al. [19] found that when only the reference quality effect behavior is considered, higher unit remanufacturing cost, lower remanufacturing rates, and higher customers' discount rates can offset some of the negative impacts of reference quality effects. Baghdadi et al. [20] believe that customers are often skeptical about the quality and durability of remanufactured products, they develop a Stackelberg game model to optimize the pricing decisions for remanufactured products. Cai et al. [21] used reverse induction to obtain the equilibrium profits under the strategy the supplier actively shares to demand information from the retailer. Through the Stackelberg game analysis, Han et al. [22] found that under decentralized decision-making, the optimal profit of construction and demolition waste resources with government subsidies using the supply chain is higher. Yang et al. [23] showed that when the consumers' valuation difference for remanufactured products is small, the monopolistic original equipment manufacturer can increase their profit by reminding consumers to consider the effects of anticipated regret. Among many consumer behaviors, transfer purchase behavior is one of the most common behaviors [24]. Therefore, a large number of scholars have discussed the product pricing decisions under the transfer purchase behavior of consumers. Consumer switching purchase behavior is a multi-party price comparison behavior based on utility maximization, which often occurs between different products or different channels of the same product. Shin [25] confirmed that differentiated services will ease the price competition between sellers, resulting in higher profits for both sides of the competition.

Liu et al. [26] found that prices and profits of sellers declined when consumers switch channels to buy products. Different from the above research literature, we mainly focus on the problem of duopoly competition between the manufacturer and the remanufacturer, meanwhile, analyzing consumers' switching purchase behavior based on a two-period dynamic game model.

Previous studies have shown that the emergence of consumer switch purchase behavior intensifies market competition, so scholars have proposed many pricing optimization methods. Marketing researchers have paid close attention to PMGs and documented their impact on price competition. The price matching strategy requires a seller to quickly match the price of the same or similar goods of another seller, or refund the price difference within a short time after the sale, so as to achieve the goal of consistent pricing of products in different sales channels. Hviid and Shaffer [27] found that an infinitesimally small hassle cost could attenuate the ability of PMG to mitigate price competition. Anne and Shaffer [3] showed that when both asymmetric product substitutability and shelf space availability are considered, the price under PMG may even fall below the competitive level. Hess and Gerstner [28] also agree with the above views. In contrast, Chen et al. [4] argued that price-match guarantees can generate a competition-enhancing effect, considering that price-match encourages consumers' price search behavior, and thus, exaggerates price competition. In addition, when consumers are heterogeneous, PMG can be used to price discriminate one consumer type over the others. Janssen and Parakhonyak [29] found that PMG increases consumers' option value of purchase and raises the prices of products. Xing and Liu [30] point out that the establishment of selective refund contracts based on price matching can improve the service level of retailers, even if some consumers switch to purchase. Chen and Chen [31] found that the seller with experience advantages can implement price matching strategies according to the shopping costs of consumers, which can help reduce the negative impact of consumer switching purchase on the seller's income. Mehra et al. [32] proposed price matching as a short-term strategy and an exclusive product assortment as a long-term strategy. Zhao et al. [33] found that a price matching strategy is more effective than discrete-time dynamic pricing in the presence of strategic purchasing behavior. According to the definition of price matching strategy in the existing research, we discuss the competitive equilibrium between a manufacturer and a remanufacturer when the current price of a seller is equal to the future price of the competitor. Moreover, scholars have not reached an agreement on the effectiveness of the price matching strategy, such as Anne and Shaffer [3], who have obtained that the price matching strategy would have a positive impact on the seller, while Chen et al. [4] argued that price-match guarantees can generate a negative impact. Therefore, it is necessary for us to further discuss the effectiveness of the sellers' price matching strategy, especially in the competitive environment of manufacturers and remanufactures.

To highlight the differences between new products and remanufactured products, we reflect the product difference and consumer learning cost into the utility function of consumers. To the best of our knowledge, few studies have simultaneously analyzed the impact of different consumer switching purchase behaviors on the competition between new and remanufactured products, and so we consider different consumers in a sales cycle: some consumers may first visit the manufacturer but then switch to purchasing remanufactured products; some consumers may first visit the remanufacturer but switch to purchasing new products. Then, we try to judge the effectiveness of the price matching strategy through equilibrium analysis between new products and remanufactured products.

3. Hypotheses and Methods

In a common market, a manufacturer (denoted with the subscript r) and a remanufacturer (denoted with the subscript o) compete to sell new products and remanufactured products to consumers. Each seller releases N products to the spot market. At the end of the sales period, the residual value of the product is zero. We refer to the manufacturer or the remanufacturer who encounters the consumer in the first period as the seller $i(i = r, o)$; if no purchase has occurred before the second period, the consumer visits the seller j, $j = r, o$ and $j \neq i$. When a seller encounters the consumer in the second period, they will be aware that the consumer has visited the competitor in the previous period. In period t, the price of a seller posted is denoted by p_{ti}, where $t = 1, 2$, $i = r, o$. We assume that both sellers are risk-neutral, aiming at the maximization of profits and denoting the seller i's expected profit-to-go function in period t as π_{ti}.

Hypothesis 1: *In the two periods, the manufacturer only sells new products and the remanufacturer only sells remanufactured products.*

Hypothesis 2: *Manufacturers and remanufacturers price products for the purpose of maximizing profits.*

There are N units similar consumers, whose valuations for the product are drawn independently from a uniform distribution on [0,1]. Additionally, N_i consumers visit the seller $i(i = r, o)$ in the first period, with $N_r + N_o = N$. In the second period, it is possible for each consumer to visit another seller, which means that consumers either buy from the manufacturer or the remanufacturer. The number of consumers buying from seller i in period t is denoted as $n_{ti}(t = 1, 2, i = r, o)$. Each consumer's reservation price for an ideal new product purchased from the manufacturer is v, and for the remanufactured product sold by the remanufacturer is θv. $\theta \in (0, 1]$ represents the difference between new products and remanufactured products, which is reflected in the differences in raw materials, manufacturing processes, functions, etc. Unlike purchasing remanufactured products, consumers can obtain the freshness and real experience of product updates through new marketing forms, such as a new product release and free trial service, which helps to improve the valuation of new products. $f(v)$ denotes the density of v and the distribution function is $F(v)$, which are well-known to sellers. Since new products have just entered the market, consumers do not know their functional performance, and need to spend time and energy to learn. In contrast, consumers are familiar with remanufactured products because they have been used or sold in the market before. We assume that the learning cost of each consumer for an ideal new product from the manufacturer is s ($s \in [0, 1]$). Moreover, the product cost information and use report of remanufactured products in the market are relatively complete; we ignore the consumer learning cost for an ideal remanufactured product purchased from the remanufacturer. In daily life, fast fashion clothing brands, such as Everlane, HoneyBy and Story Mfg, will actively disclose their production costs to consumers. Research institutions, such as IHS Markit, will release various product analysis reports, so s represents the timeliness of consumers' access to authoritative information and the difficulty of the learning product. The utility function of each consumer who purchases the new product is $U_{tr} = v - p_{tr} - s$; the utility function of each consumer who purchases from the remanufactured product is $U_{to} = \theta v - p_{to}$ and $t = (1, 2)$. When $U_{2i} > U_{1j}$ $(i = r, o,\ j = r, o,\ i \neq j)$, a consumer's switching purchase behavior occurs. Thus, let $q_{(n_{1i}, N_i)}$ denote the probability of N_i consumers who visit seller i to buy n_{1i} products in the first period, where $n_{1i} = 0.1 \ldots N_i$, $i = r, o$. Let $q_{(n_{2j}, N_i - n_{1i})}$ denote the probability of $N_i - n_{1i}$ consumers visiting the seller j to buy in the second period, where $n_{2j} = 0.1 \ldots (N_i - n_{1i})$, $j = r, o, i \neq j$.

Hypothesis 3: *A large number of consumers visit sellers at random, and each consumer can only know the pricing of their corresponding products after visiting the seller.*

Hypothesis 4: *Consumers make purchase decisions with the goal of maximizing their personal utility. Some consumers visit the manufacturer first to learn about the information of new products. Consumers who have purchased new products withdraw from the market and consumers who have not purchased products will turn to visit the remanufacturer to make purchase decisions. The other consumers visit the remanufacturer first to learn about the information of remanufactured products.*

Consumers who have purchased remanufactured products withdraw from the market and consumers who have not purchased products will turn to visit the remanufacturer to make purchase decisions.

The parameters involved in the paper and their meanings are shown in Table 1.

Table 1. The parameters involved in the paper and their meanings.

Symbol	Meanings
r	The subscript of the manufacturer
o	The subscript of the remanufacturer
t	Sales period
i	A seller in the first period
j	A seller in the second period
N	Number of consumers initially entering the market
n	Number of consumers purchasing products
q	The probability of consumers purchasing products
v	Each consumer's reservation price for an ideal new product purchased from the manufacturer
θ	Difference between new products and remanufactured products
s	Consumer learning costs
U	Consumer utility
p	The product pricing
π	The profits of a seller

This study first presents the traditional dynamic pricing strategy (TDP) as a benchmark, and then introduces and analyses three price matching strategies (BMP, OMP and RMP). Among them, the three price matching situations are as follows: the situation when only the manufacturer implements price matching is called the BMP strategy; the situation when only the remanufacturer implement price matching is called the OMP strategy; the situation when the manufacturer and the remanufacturer match the pricing of a product from their respective competitors is called RMP. From the above analysis, the decision-making process of two sellers can be obtained, as shown in Figure 1.

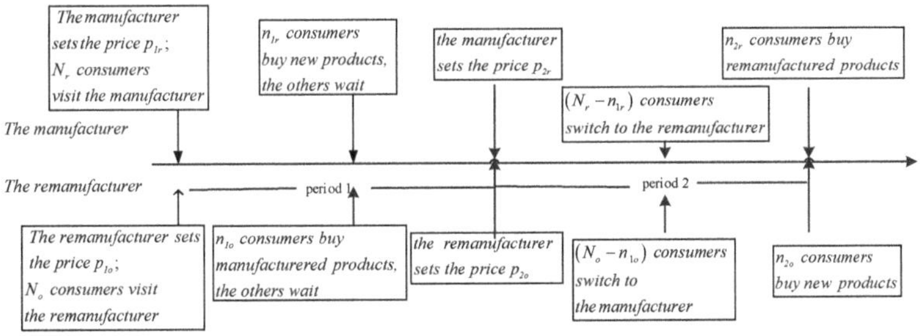

Figure 1. The decision-making process.

The decision-making process of the traditional dynamic pricing strategy is as follows: Step 1, in the first period, N_i, consumers visits the two sellers, and each seller announces the product pricing, $p_{1i}, i = r, o$; Step 2, consumers decide whether to buy or not; Step 3, in the second period, $(N_i - n_{1i})$, consumers switch to another seller for comparison; Step 4, each seller announces the product pricing, $p_{2i}, i = r, o$; Step 5, consumers who have not purchased products before make purchase decisions in the second period. If the BMP or OMP strategy is implemented, the seller will announce p_{1i}^{BMP} (p_{1i}^{OMP}) in Step 1, and $p_{1i}^{BMP} = p_{2j}^{BMP}$ ($p_{1i}^{OMP} = p_{2j}^{OMP}$), $i, j = r, o, i \neq j$. If the RMP strategy is implemented, the sellers will announce p_{1r}^{RMP} and p_{1o}^{RMP} in Step 1, then set $p_{1r}^{RMP} = p_{2o}^{RMP}$ and $p_{1o}^{RMP} = p_{2r}^{RMP}$. Next, we will first analyze the TDP strategy and then establish three models of price matching strategies and discuss them in combination with practical cases.

4. Model Analysis and Results

4.1. Traditional Dynamic Pricing Strategy Equilibrium Analysis

TDP strategies are widely used by sellers. Consider the switching buying behavior of consumers across periods and products; if $U_{2o} > U_{1r}$, that is $\theta v - p_{2o}^{TDP} > v - p_{1r}^{TDP} - s$, the probability that a consumer visits the manufacturer and then buys the remanufactured product is $F\left(\frac{p_{1r}^{TDP}-p_{2o}^{TDP}+s}{1-\theta}\right)$, and the probability of N_r consumers buying n_{1r} products from the manufacturer in the first period is $q_{(n_{1r},N_r)}^{TDP} = \frac{N_r!}{n_{1r}!(N_r-n_{1r})!}\left[1-F\left(\frac{p_{1r}^{TDP}-p_{2o}^{TDP}+s}{1-\theta}\right)\right]^{n_{1r}} F\left(\frac{p_{1r}^{TDP}-p_{2o}^{TDP}+s}{1-\theta}\right)^{N_r-n_{1r}}$. If $U_{2r} > U_{1o}$, that is $v - p_{2r} - s > \theta v - p_{1o}$, the probability that a consumer visits the remanufacturer and then buys the new products is $1-F\left(\frac{p_{2r}^{TDP}-p_{1o}^{TDP}+s}{1-\theta}\right)$, and the probability of N_o consumers buying n_{1o} products from the remanufacturer in the first period is $q_{(n_{1o},N_o)}^{TDP} = \frac{N_o!}{n_{1o}!(N_o-n_{1o})!} F\left(\frac{p_{2r}^{TDP}-p_{1o}^{TDP}+s}{1-\theta}\right)^{n_{1o}}\left[1-F\left(\frac{p_{2r}^{TDP}-p_{1o}^{TDP}+s}{1-\theta}\right)\right]^{N_o-n_{1o}}$. Then, we analyze the probability of consumers switching in the second period. Consumers waiting for the second period will buy the new products when $U_{2i} > 0$ $(i = r, o)$, and so the probability of $N_r - n_{1r}$ consumers buying n_{2o} remanufactured products is $q_{(n_{2o},N_r-n_{1r})}^{TDP} = \frac{(N_r-n_{1r})!}{n_{2o}!(N_r-n_{1r}-n_{2o})!}\left[1-F\left(\frac{p_{2o}^{TDP}}{\theta}\right)\right]^{n_{2o}} F\left(\frac{p_{2o}^{TDP}}{\theta}\right)^{N_r-n_{1r}-n_{2o}}$. Similarly, the probability of $N_o - n_{1o}$ consumers buying n_{2r} new products is $q_{(n_{2r},N_o-n_{1o})}^{TDP} = \frac{(N_o-n_{1o})!}{n_{2r}!(N_o-n_{1o}-n_{2r})!}\left[1-F(p_{2r}^{TDP}+s)\right]^{n_{2r}} F(p_{2r}^{TDP}+s)^{N_o-n_{1o}-n_{2r}}$.

$$\pi_r^{TDP}(N_r, N_o) = \underset{p_{1r}^{TDP}}{Max} \sum_{n_{1r}=0}^{N_r} q_{(n_{1r},N_r)}^{TDP}\left[p_{1r}^{TDP}\cdot n_{1r} + \sum_{n_{1o}=0}^{N_o} q_{(n_{1o},N_o)}^{TDP}\cdot p_{2r}^{TDP}\cdot \sum_{n_{2r}=0}^{N_o-n_{1o}}\left(n_{2r}\cdot q_{(n_{2r},N_o-n_{1o})}^{TDP}\right)\right] \tag{1}$$

$$\pi_o^{TDP}(N_r, N_o) = \underset{p_{1o}^{TDP}}{Max} \sum_{n_{1o}=0}^{N_o} q_{(n_{1o},N_o)}^{TDP}\left[p_{1o}^{TDP}\cdot n_{1r} + \sum_{n_{1r}=0}^{N_r} q_{(n_{1r},N_r)}^{TDP}\cdot p_{2o}^{TDP}\cdot \sum_{n_{2o}=0}^{N_r-n_{1r}}\left(n_{2o}\cdot q_{(n_{2o},N_r-n_{1r})}^{TDP}\right)\right] \tag{2}$$

From Equations (1) and (2), the quantity of products sold in each period is uncertain, and it is difficult to determine the equilibrium price and profits of the manufacturer and the remanufacturer. Therefore, we consider simplifying the solution process and discuss the equilibrium decisions of the two sellers when $N = 1$.

When a consumer first visits the manufacturer, the probability that he/she makes a purchase in the first period is $1-F\left(\frac{p_{1r}^{TDP}-p_{2o}^{TDP}+s}{1-\theta}\right)$, and the expected profits of the manufacturer in the first period can be simplified as

$$\begin{gathered}\pi_{1r}^{TDP}(1,0) = \max_{p_{1r}^{TDP}} p_{1r}^{TDP}\cdot\left[1-F\left(\frac{p_{1r}^{TDP}-p_{2o}^{TDP}+s}{1-\theta}\right)\right] \\ s.t.\ v - p_{1r}^{TDP} - s \geq 0\end{gathered} \tag{3}$$

The expected profits of the remanufacturer in the second period can be simplified as

$$\begin{gathered}\pi_{2o}^{TDP}(1,0) = \max_{p_{2O}^{TDP}} p_{2o}^{TDP}\cdot F\left(\frac{p_{1r}^{TDP}-p_{2o}^{TDP}+s}{1-\theta}\right)\cdot\left[1-\left(\frac{p_{2o}^{TDP}}{\theta}\right)\right] \\ s.t.\ \theta v - p_{2o}^{TDP} > 0\end{gathered} \tag{4}$$

When a consumer first visits the remanufacturer, the expected profits of the remanufacturer in the first period can be simplified as

$$\begin{gathered}\pi_{1o}^{TDP}(0,1) = \max_{p_{1o}^{TDP}} p_{1o}^{TDP}\cdot F\left(\frac{p_{2r}^{TDP}-p_{1o}^{TDP}+s}{1-\theta}\right) \\ s.t.\ \theta v - p_{1o}^{TDP} > 0\end{gathered} \tag{5}$$

The expected profits of the manufacturer in the second period can be simplified as

$$\pi_{2r}^{TDP}(0,1) = \max_{p_{2r}^{TDP}} p_{2r}^{TDP}\left[1 - F\left(\frac{p_{2r}^{TDP} - p_{1o}^{TDP} + s}{1-\theta}\right)\right] \cdot \left[1 - F\left(p_{2r}^{TDP} + s\right)\right]$$
$$s.t\ v - p_{2_r}^{TDP} - s > 0 \tag{6}$$

Based on the analysis above, the two sellers' equilibrium profits can be calculated as shown in Theorem 1. To simplify the presentation, we define

$$\lambda = \sqrt{4s^2 - 12s\theta + 17\theta^2 + 8s - 12\theta + 4}$$
$$\mu = \sqrt{9s^2 + 8s\theta + 16\theta^2 - 22s - 24\theta + 17} \tag{7}$$

Theorem 1. *Under the TDP strategy, when there is only one consumer in the market, the optimal expected profits of sellers are as follows:*

$$\pi_{1r}^{TDP^*}(1,0) = \frac{(10-6s-7\theta-\lambda)^2}{256(1-\theta)},\ \pi_{2o}^{TDP^*}(1,0) = \frac{(3s-4\theta+7-\mu)^2}{256(1-\theta)},$$

$$\pi_{2r}^{TDP^*}(0,1) = \frac{(7-5s+4\theta-\mu)(9-3s-12\theta+\mu)(1-3s+4\theta+\mu)}{1024\theta(1-\theta)},$$

$$\pi_{1o}^{TDP^*}(0,1) = \frac{(2+\theta+2s-\lambda)(6+6s-9\theta+\lambda)(7\theta-2s-2+\lambda)}{1024\theta(1-\theta)}.$$

The proof of Theorem 1 is provided in Appendix A. Theorem 1 shows that when there is only one consumer in the market, the sellers' equilibrium profits are affected by the service difference between two sellers and the learning cost. According to Equations (1) and (2), we obtain Theorem 2 after a series of calculations.

Theorem 2. *Under the TDP strategy, when there are N consumers in the market, the expected profits of the two sellers are as follows:*

$$\pi_r^{TDP^*}(N_r, N_o) = N_r \cdot \pi_{1r}^{TDP^*}(1,0) + N_o \cdot \pi_{2r}^{TDP^*}(0,1) \tag{8}$$

$$\pi_o^{TDP^*}(N_r, N_o) = N_o \cdot \pi_{1o}^{TDP^*}(1,0) + N_r \cdot \pi_{2o}^{TDP^*}(0,1) \tag{9}$$

The proof of Theorem 2 is provided in Appendix A. According to Theorem 2, when there are multiple consumers, the sellers' optimal profits can be transformed into the expected profits of the sellers when there is only one consumer in the market multiplied by the number of consumers who initially arrive at the corresponding seller. This helps us analyze the equilibrium profit of sellers. According to the above simplified procedure, $\theta \in (0,1]$, in order to ensure that the product price and the seller's income are not negative, we randomly take the value of s, and assume that $N_R = N_O = 50$. Taking "$s = 0.1$" as an example, the changes in sellers' profits with θ are shown in Figure 2. In fact, taking other values of s can result in similar changes in sellers' profits.

As shown in Figure 2, with the increase in θ, the profits of the manufacturer decreases, and the profits of the remanufacturer increases first and then decreases. It can be seen that it is difficult for sellers to eliminate the adverse effects of consumers' switching purchase behavior, which leads to a decrease in profits. Therefore, this study will introduce some new pricing strategies to improve sellers' profits.

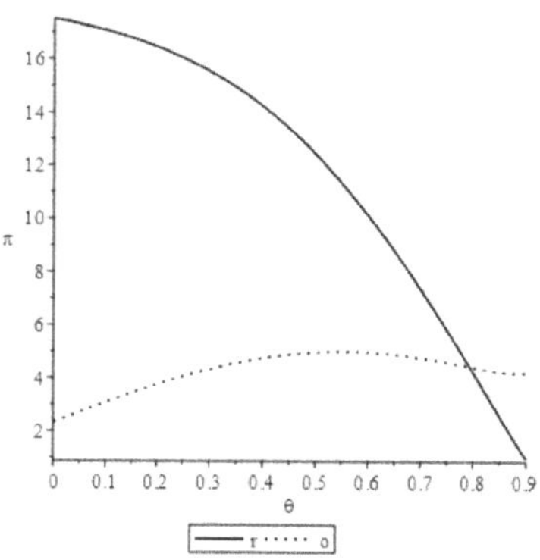

Figure 2. Changes in the two sellers' profits with θ ($s = 0.1$, $N_R = N_O = 50$).

4.2. Equilibrium Analysis for Price Matching Strategies

4.2.1. BMP Strategy Equilibrium Analysis

Under the BMP strategy, the manufacturer first wants to increase their profits by increasing the number of targeted consumers who prefer the new products when $p_{1r}^{BMP} = p_{2o}^{BMP}$, so that the consumer will not be tempted by the low price of the remanufacturer and switch to them. The remanufacturer, meanwhile, does not implement matching pricing but sets a new price for each period. Similar to the TDP strategy analysis, the profits of sellers are as follows:

$$\pi_r^{BMP}(N_r, N_o) = \underset{p_{tr}^{BMP}}{Max} \sum_{n_{1r}=0}^{N_r} q_{(n_{1r},N_r)}^{BMP} \left[p_{1r}^{BMP} \cdot n_{1r} + \sum_{n_{1o}=0}^{N_o} q_{(n_{1o},N_o)}^{BMP} \cdot p_{2r}^{BMP} \cdot \sum_{n_{2r}=0}^{N_o-n_{1o}} \left(n_{2r} \cdot q_{(n_{2r},N_o-n_{1o})}^{BMP} \right) \right] \tag{10}$$

$$\pi_o^{BMP}(N_r, N_o) = \underset{p_{to}^{BMP}}{Max} \sum_{n_{1o}=0}^{N_o} q_{(n_{1o},N_o)}^{BMP} \left[p_{1_o}^{BMP} \cdot n_{1o} + \sum_{n_{1r}=0}^{N_r} q_{(n_{1r},N_r)}^{BMP} \cdot p_{2O}^{BMP} \cdot \sum_{n_{2o}=0}^{N_r-n_{1r}} \left(n_{2o} \cdot q_{(n_{2o},N_r-n_{1r})}^{BMP} \right) \right] \tag{11}$$

where the probability of N_r consumers buying n_{1r} products from the manufacturer is $q_{(n_{1r},N_r)}^{BMP} = \frac{N_r!}{n_{1r}!(N_r-n_{1r})!} \left[1 - F\left(\frac{s}{1-\theta}\right)\right]^{n_{1r}} F\left(\frac{p_{1r}^{BMP} - p_{2o}^{BMP} + s}{1-\theta}\right)^{N_r-n_{1r}}$; the probability of N_o consumers buying n_{1o} products from the remanufacturer is $q_{(n_{1o},N_o)}^{BMP} = \frac{N_o!}{n_{1o}!(N_o-n_{1o})!} \left[F\left(\frac{p_{1o}^{BMP} - p_{2r}^{BMP} - s}{1-\theta}\right)\right]^{n_{1o}}$ $\left[1 - F\left(\frac{p_{1o}^{BMP} - p_{2r}^{BMP} - s}{1-\theta}\right)\right]^{N_o-n_{1o}}$; the probability of $N_r - n_{1r}$ consumers buying n_{2o} products from the remanufacturer is $q_{(n_{2o},N_r-n_{1r})}^{BMP} = \frac{(N_r-n_{1r})!}{n_{2o}!(N_r-n_{1r}-n_{2o})!} \left[1 - F\left(\frac{p_{2o}^{BMP}}{\theta}\right)\right]^{n_{2o}}$ $F\left(\frac{p_{2o}^{BMP}}{\theta}\right)^{N_r-n_{1r}-n_{2o}}$; and the probability of $N_o - n_{1o}$ consumers buying n_{2r} products from the remanufacturer is $q_{(n_{2r},N_o-n_{1o})}^{BMP} = \frac{(N_o-n_{1o})!}{n_{2r}!(N_o-n_{1o}-n_{2r})!} \left[1 - F\left(p_{2r}^{BMP} + s\right)\right]^{n_{2r}} \left[F\left(p_{2r}^{BMP} + s\right)\right]^{N_o-n_{1o}-n_{2r}}$.

When $N = 1$, the expected profits of the manufacturer in the first period can be written as

$$\pi_{1r}^{BMP}(1,0) = p_{1r}^{BMP} \cdot \left[1 - F\left(\frac{s}{1-\theta}\right)\right] = max p_{2o}^{BMP} \cdot \left[1 - F\left(\frac{s}{1-\theta}\right)\right] \\ s.t.\ v - p_{1r}^{BMP} - s > 0 \tag{12}$$

The expected profits of the remanufacturer in the second period can be written as

$$\pi_{2o}^{BMP}(1,0) = \max_{p_{2o}^{BMP}} p_{2o}^{BMP} \cdot F\left(\frac{s}{1-\theta}\right) \cdot \left[1 - \left(\frac{p_{2o}^{BMP}}{\theta}\right)\right] \quad (13)$$

$$s.t.\ \theta v - p_{2o}^{BMP} > 0$$

The expected profits of the remanufacturer in the first period are as follows:

$$\pi_{1o}^{BMP}(0,1) = \max_{p_{1o}^{BMP}} p_{1o}^{BMP} \cdot F\left(\frac{p_{2r}^{BMP} - p_{1o}^{BMP} + s}{1-\theta}\right) \quad (14)$$

$$s.t.\ \theta v - p_{1o}^{BMP} > 0$$

The expected profits of the manufacturer in the second period are as follows:

$$\pi_{2r}^{BMP}(0,1) = \max_{p_{2r}^{BMP}} p_{2r}^{BMP}\left[1 - F\left(\frac{p_{2r}^{BMP} - p_{1o}^{BMP} + s}{1-\theta}\right)\right] \cdot \left[1 - F\left(p_{2r}^{BMP} + s\right)\right] \quad (15)$$

$$s.t.\ v - p_{2r}^{BMP} - s > 0$$

Theorem 3. *Under the BMP strategy, when there is only one consumer in the market, the optimal profits of the two sellers are as follows:*

$$\pi_{1r}^{BMP^*}(1,0) = \frac{\theta(1-\theta-s)}{2(1-\theta)},\ \pi_{2o}^{BMP^*}(1,0) = \frac{\theta s}{4(1-\theta)},\ \pi_{1o}^{BMP^*}(0,1) = \frac{(7+3s-4\theta+\mu(\theta))^2}{256(1-\theta)},$$

$$\pi_{2r}^{BMP^*}(0,1) = \frac{(7-5s-4\theta+\mu(\theta))(3s+12\theta-9+\mu(\theta))(3s-1-4\theta+\mu(\theta))}{1024(1-\theta)}$$

The proof of Theorem 3 is in Appendix A.

Theorem 4. *Under the BMP strategy, when there are N consumers in the market, the expected profits of the two sellers are as follows:*

$$\pi_r^{BMP^*}(N_r, N_o) = N_r \cdot \pi_{1r}^{BMP^*}(1,0) + N_o \cdot \pi_{2r}^{BMP^*}(0,1) \quad (16)$$

$$\pi_o^{BMP^*}(N_r, N_o) = N_o \cdot \pi_{1o}^{BMP^*}(1,0) + N_r \cdot \pi_{2o}^{BMP^*}(0,1) \quad (17)$$

The proof of Theorem 4 is provided in Appendix A. According to Theorem 4, only the manufacturer can match the price, and the two sellers' optimal profit functions can also be simplified.

4.2.2. OMP Strategy Equilibrium Analysis

Using OMP strategy, the remanufacturer first wants to increase their profits by increasing the number of targeted consumers who prefer purchasing remanufactured products. When $p_{1o}^{OMP} = p_{2r}^{OMP}$, the remanufacturer can improve the probability of target consumers to buy products in the current period. Similar to Section 4.2.1, the profits of the manufacturer (π_r^{OMP}) and the remanufacturer (π_o^{OMP}) are

$$\pi_r^{OMP}(N_r, N_o) = \underset{p_{tr}^{OMP}}{Max} \sum_{n_{1r}=0}^{N_r} q_{(n_{1r},N_r)}^{OMP}\left[p_{1r}^{OMP} \cdot n_{1r} + \sum_{n_{1o}=0}^{N_o} q_{(n_{1o},N_o)}^{OMP} \cdot p_{2r}^{OMP} \cdot \sum_{n_{2r}=0}^{N_o-n_{1o}}\left(n_{2r} \cdot q_{(n_{2r},N_o-n_{1o})}^{OMP}\right)\right] \quad (18)$$

$$\pi_o^{OMP}(N_r, N_o) = \underset{p_{to}^{OMP}}{Max} \sum_{n_{1o}=0}^{N_o} q_{(n_{1o},N_o)}^{OMP} \left[p_{1o}^{OMP} \cdot n_{1o} + \sum_{n_{1r}=0}^{N_r} q_{(n_{1r},N_r)}^{OMP} \cdot p_{2o}^{OMP} \cdot \sum_{n_{2o}=0}^{N_r-n_{1r}} \left(n_{2o} \cdot q_{(n_{2o},N_r-n_{1r})}^{OMP} \right) \right] \tag{19}$$

where the probability of consumers buying products from the remanufacturer in the second period is $q_{(n_{2o},N_r-n_{1r})}^{OMP} = \frac{(N_r-n_{1r})!}{n_{2o}!(N_r-n_{1r}-n_{2o})!}\left[1-F\left(\frac{p_{2o}^{OMP}}{\theta}\right)\right]^{n_{2o}} F\left(\frac{p_{2o}^{OMP}}{\theta}\right)^{N_r-n_{1r}-n_{2o}}$; the probability of consumers buying products from the manufacturer in the first period is $q_{(n_{1r},N_r)}^{OMP} = \frac{N_r!}{n_{1r}!(N_r-n_{1r})!}\left[1-F\left(\frac{p_{1r}^{OMP}-p_{2o}^{OMP}+s}{1-\theta}\right)\right]^{n_{1r}} F\left(\frac{p_{1r}^{OMP}-p_{2o}^{OMP}+s}{1-\theta}\right)^{N_r-n_{1r}}$ the probability of consumers buying products from the remanufacturer in the first period is $q_{(n_{1o},N_o)}^{OMP} = \frac{N_o!}{n_{1o}!(N_o-n_{1o})!}\left[F\left(\frac{s}{1-\theta}\right)\right]^{n_{1o}}\left[1-F\left(\frac{s}{1-\theta}\right)\right]^{N_o-n_{1o}}$ and the probability of consumers buying products from the manufacturer in the second period is $q_{(n_{2r},N_o-n_{1o})}^{OMP} = \frac{(N_o-n_{1o})!}{n_{2r}!(N_o-n_{1o}-n_{2r})!}\left[1-F(p_{2r}^{OMP}+s)\right]^{n_{2r}}\left[F(p_{2r}^{OMP}+s)\right]^{N_o-n_{1o}-n_{2r}}$.

Similar to the previous analysis, when there is only one consumer who visits the manufacturer in the market, the expected profits of the two sellers are simplified as

$$\begin{gathered} \pi_{1r}^{OMP}(1,0) = \underset{p_{1r}^{OMP}}{max}\, p_{1r}^{OMP} \cdot \left[1-F\left(\frac{p_{1r}^{OMP}-p_{2o}^{OMP}+s}{1-\theta}\right)\right] \\ s.t.\ v - p_{1r}^{OMP} - s > 0 \end{gathered} \tag{20}$$

$$\begin{gathered} \pi_{2o}^{OMP}(1,0) = \underset{p_{2o}^{OMP}}{max}\, p_{2o}^{OMP} \cdot F\left(\frac{p_{1r}^{OMP}-p_{2o}^{OMP}+s}{1-\theta}\right) \cdot \left[1-\left(\frac{p_{2o}^{OMP}}{\theta}\right)\right] \\ s.t.\ \theta v - p_{2o}^{OMP} > 0 \end{gathered} \tag{21}$$

When there is only one consumer who visits the remanufacturer in the market, the expected profits of the two sellers are

$$\begin{gathered} \pi_{1o}^{OMP}(0,1) = max\, p_{1o}^{OMP} \cdot F\left(\frac{s}{1-\theta}\right) = p_{2r}^{OMP} \cdot F\left(\frac{s}{1-\theta}\right) \\ s.t.\ \theta v - p_{1o}^{OMP} > 0 \end{gathered} \tag{22}$$

$$\begin{gathered} \pi_{2r}^{OMP}(0,1) = \underset{p_{2r}^{OMP}}{max}\, p_{2r}^{OMP}\left[1-F\left(\frac{s}{1-\theta}\right)\right] \cdot \left[1-F(p_{2r}^{OMP}+s)\right] \\ s.t.\ v - p_{2r}^{OMP} - s > 0 \end{gathered} \tag{23}$$

The equilibrium prices of the OMP strategy are shown in Theorem 5.

Theorem 5. *Under the OMP strategy, when there is only one consumer in the market, the optimal profits of the two sellers are as follows:*

$$\pi_{1r}^{OMP^*}(1,0) = \frac{(10-6s-7\theta+\lambda(\theta))^2}{256(1-\theta)},$$

$$\pi_{2o}^{OMP^*}(1,0) = \frac{(2s+\theta+2+\lambda(\theta))(9\theta-6s-6+\lambda(\theta))(2s+2-7\theta+\lambda(\theta))}{1024\theta(1-\theta)},$$

$$\pi_{2r}^{OMP^*}(0,1) = \frac{(1-s)^2(1-\theta-s)}{4(1-\theta)},\ \pi_{1o}^{OMP^*}(0,1) = \frac{(1-s)s}{2(1-\theta)}.$$

The proof of Theorem 5 is in Appendix A.

Theorem 6. *Under the OMP strategy, when there are N consumers in the market, the expected profits of the two sellers are as follows:*

$$\pi_r^{OMP^*}(N_r, N_o) = N_r \cdot \pi_{1r}^{OMP^*}(1,0) + N_o \cdot \pi_{2r}^{OMP^*}(0,1) \tag{24}$$

$$\pi_o^{OMP^*}(N_r, N_o) = N_o \cdot \pi_{1o}^{OMP^*}(1,0) + N_r \cdot \pi_{2o}^{OMP^*}(0,1) \tag{25}$$

The proof of Theorem 6 is provided in Appendix A. According to Theorem 6, only the remanufacturer can match the price and the sellers' optimal profits can also be simplified.

4.2.3. RMP Strategy Equilibrium Analysis

Under the RMP strategy, both sellers in the first period match competitors' second prices, which means $p_{1r}^{RMP} = p_{2o}^{RMP}$ and $p_{1r}^{RMP} = p_{2o}^{RMP}$; then, the probability of N_r consumers buying n_{1r} products from the manufacturer in the first period is $q_{(n_{1r},N_r)}^{RMP} = \frac{N_r!}{n_{1r}!(N_r-n_{1r})!}\left[1-F\left(\frac{s}{1-\theta}\right)\right]^{n_{1r}} F\left(\frac{s}{1-\theta}\right)^{N_r-n_{1r}}$ and the probability of N_o consumers buying n_{1o} products from the remanufacturer in the first period is $q_{(n_{1o},N_o)}^{RMP} = \frac{N_o!}{n_{1o}!(N_o-n_{1o})!} F\left(\frac{s}{1-\theta}\right)^{n_{1o}}\left[1-F\left(\frac{s}{1-\theta}\right)\right]^{N_o-n_{1o}}$. Similar to Section 4.2.1, the profits of the sellers are as follows:

$$\pi_r^{RMP}(N_r, N_o) = \underset{p_{tr}^{RMP}}{Max} \sum_{n_{1r}=0}^{N_r} q_{(n_{1r},N_r)}^{RMP}\left[p_{1r}^{RMP} \cdot n_{1r} + \sum_{n_{1o}=0}^{N_o} q_{(n_{1o},N_o)}^{RMP} \cdot p_{2r}^{RMP} \cdot \sum_{n_{2r}=0}^{N_o-n_{1o}}\left(n_{2r} \cdot q_{(n_{2r},N_o-n_{1o})}^{RMP}\right)\right] \tag{26}$$

$$\pi_o^{RMP}(N_r, N_o) = \underset{p_{to}^{RMP}}{Max} \sum_{n_{1o}=0}^{N_o} q_{(n_{1o},N_o)}^{RMP}\left[p_{1_o}^{RMP} \cdot n_{1o} + \sum_{n_{1r}=0}^{N_r} q_{(n_{1r},N_r)}^{RMP} \cdot p_{2O}^{RMP} \cdot \sum_{n_{2o}=0}^{N_r-n_{1r}}\left(n_{2o} \cdot q_{(n_{2o},N_r-n_{1r})}^{RMP}\right)\right] \tag{27}$$

where the probability of consumers buying products from the manufacturer in the first period is $q_{(n_{1r},N_r)}^{RMP} = \frac{N_r!}{n_{1r}!(N_r-n_{1r})!}\left[1-F\left(\frac{s}{1-\theta}\right)\right]^{n_{1r}} F\left(\frac{s}{1-\theta}\right)^{N_r-n_{1r}}$; the probability of consumers buying products from the remanufacturer in the second period is $q_{(n_{2o},N_r-n_{1r})}^{RMP} = \frac{(N_r-n_{1r})!}{n_{2o}!(N_r-n_{1r}-n_{2o})!}\left[1-F\left(\frac{p_{2o}^{RMP}}{\theta}\right)\right]^{n_{2o}} F\left(\frac{p_{2o}^{RMP}}{\theta}\right)^{N_r-n_{1r}-n_{2o}}$; the probability of consumers buying products from the remanufacturer in the first period is $q_{(n_{1o},N_o)}^{RMP} = \frac{N_o!}{n_{1o}!(N_o-n_{1o})!}\left[F\left(\frac{s}{1-\theta}\right)\right]^{n_{1o}}\left[1-F\left(\frac{s}{1-\theta}\right)\right]^{N_o-n_{1o}}$ and the probability of consumers buying products from the manufacturer in the second period is $q_{(n_{2r},N_o-n_{1o})}^{RMP} = \frac{(N_o-n_{1o})!}{n_{2r}!(N_o-n_{1o}-n_{2r})!}\left[1-F\left(p_{2r}^{RMP}+s\right)\right]^{n_{2r}}\left[F\left(p_{2r}^{RMP}+s\right)\right]^{N_o-n_{1o}-n_{2r}}$.

We also discuss the equilibrium decision of the two sellers when $N = 1$. When there is only one consumer in the market, the expected profits of the two sellers are:

$$\begin{gathered}\pi_{1r}^{RMP}(1,0) = max\ p_{1r}^{RMP} \cdot \left[1-F\left(\frac{s}{1-\theta}\right)\right] = p_{2o}^{RMP} \cdot \left[1-F\left(\frac{s}{1-\theta}\right)\right] \\ s.t.\ v - p_{1r}^{RMP} - s > 0\end{gathered} \tag{28}$$

$$\begin{gathered}\pi_{2o}^{RMP}(1,0) = \underset{p_{2o}^{RMP}}{max}\ p_{2o}^{RMP} \cdot F\left(\frac{s}{1-\theta}\right) \cdot \left[1-\left(\frac{p_{2o}^{RMP}}{\theta}\right)\right] \\ s.t.\ \theta v - p_{2o}^{RMP} > 0\end{gathered} \tag{29}$$

$$\begin{gathered}\pi_{1o}^{RMP}(0,1) = max\ p_{1o}^{RMP} \cdot F\left(\frac{s}{1-\theta}\right) = p_{2r}^{RMP} \cdot F\left(\frac{s}{1-\theta}\right) \\ s.t.\ \theta v - p_{1o}^{RMP} > 0\end{gathered} \tag{30}$$

$$\pi_{2r}^{RMP}(0,1) = \max_{p_{2r}^{RMP}} p_{2r}^{RMP}\left[1 - F\left(\tfrac{s}{1-\theta}\right)\right]\cdot\left[1 - F\left(p_{2r}^{RMP} + s\right)\right]$$
$$s.t.\ v - p_{2r}^{RMP} - s > 0 \tag{31}$$

Theorem 7. *Under the RMP strategy, when there is only one consumer in the market, the optimal profits of the two sellers are as follows:*

$$\pi_{1r}^{RMP^*}(1,0) = \pi_{1r}^{BMP^*}(1,0) = \frac{\theta(1-\theta-s)}{2(1-\theta)},\ \pi_{1o}^{RMP^*}(1,0) = \pi_{1o}^{OMP^*}(1,0) = \frac{(1-s)s}{2(1-\theta)},$$

$$\pi_{2r}^{RMP^*}(0,1) = \pi_{2r}^{OMP^*}(0,1) = \frac{(1-s)^2(1-\theta-s)}{4(1-\theta)},\ \pi_{2o}^{RMP^*}(0,1) = \pi_{2o}^{BMP^*}(0,1) = \frac{\theta s}{4(1-\theta)}$$

The proof of Theorem 7 is in Appendix A.

Theorem 8. *Under the RMP strategy, when there are N consumers in the market, the expected profits of the two sellers are as follows:*

$$\pi_{r}^{RMP^*}(N_r, N_o) = N_r\cdot\pi_{1r}^{RMP^*}(1,0) + N_o\cdot\pi_{2r}^{RMP^*}(0,1) \tag{32}$$

$$\pi_{o}^{RMP^*}(N_r, N_o) = N_o\cdot\pi_{1o}^{RMP^*}(1,0) + N_r\cdot\pi_{2o}^{RMP^*}(0,1) \tag{33}$$

The proof of Theorem 8 is in Appendix A.

5. Numerical Examples

5.1. The Impact of Product Differences between New Products and Remanufactured Products and Consumers Learning Costs on Profits

As previously defined, as θ becomes smaller, the differences between the two sellers is greater, and with the increase in s, consumers do not receive authoritative information on time, and it is more difficult to learn new product information. In order to analyze the impact of a single parameter on the sellers' pricing decision and highlight the disadvantages of the two sellers as much as possible, we assume that $s = 0$, $N_R = N_O = 50$, and the difference between the two sellers is only reflected in product differences. The impact of product differences between new products and remanufactured products on sellers' profits under the four strategies is shown in Figure 3.

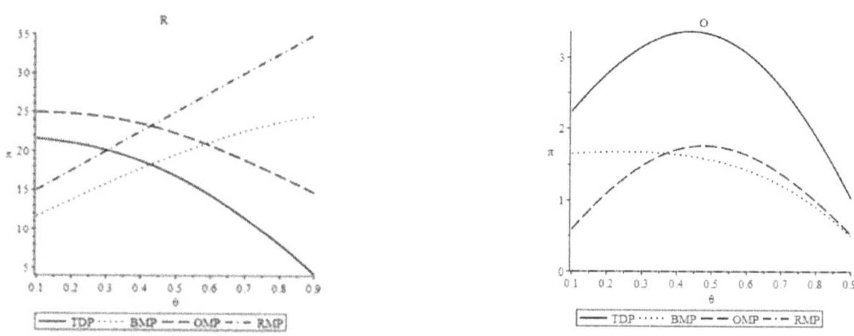

(**a**) (**b**)

Figure 3. (**a**) $s = 0$, changes of the manufacturer's profits with θ; (**b**) $s = 0$, changes of the remanufacturer's profits with θ.

Figure 3a implies that with the increase in θ, the profits of the manufacturer using the TDP as well as OMP strategies decrease, while the profits of the manufacturer using the RMP as well as BMP strategies increase. Figure 3b shows that the increase in θ may be detrimental for the remanufacturer's profit, especially when the profits fall to zero under the RMP strategy. With different strategies, the influence of θ on profits varies widely owing to the consumer's switching behavior. Using the TDP strategy, with θ increases, consumers have an incentive to switch simply because of the low price of the second period, and the two sellers gradually form a symmetrical equilibrium in the competition. However, using the OMP strategy, sellers have the same price ($p_{1o}^{OMP} = p_{2r}^{OMP}$), and consumers visiting the remanufacturer have little incentive to buy the new products; thus, the profits of the manufacturer have declined. Using the BMP strategy, the manufacturer loses some consumers, which leads to a decline in profits.

In addition, s also reflects the consumers' learning ability, which has an impact on the consumers' switching purchase behavior. $s = 0$ means that there is no difference in the ability of consumers to learn about new products and remanufactured products, and $s > 0$ means that it is more difficult for consumers to learn about new products. In general, in order to introduce the impact of s on the seller, under the condition that the product pricing and the sellers' profit are not negative, it can be assumed that s = 0.1. The impact of search cost on sellers' profits is shown in Figure 4. ($N_R = N_O = 50$).

Owing to the presence of s, the two sellers' profit curves have changed compared to Figure 3, which means that the cost of learning for consumers prevents manufacturers from increasing profits, but helps the remanufacturer to gain more profits. Figure 4 shows that with the increase in θ, the manufacturer's profit curves using the TDP and RMP strategies become two parabolas. The remanufacturer uses the price matching strategy to increase their profit. Considering the influence of s and θ on profits, also reflects the strength of the characteristics of the two sellers in the competition. When s is small, the advantage of remanufactured products is weak. Consumers who prefer high-tech or novel products can easily switch to manufacturers. When θ is large, the advantage of the new product is weak. Thus, consumers who seek low prices are more inclined to switch to the remanufacturer; the two sellers may achieve a win–win situation in competition.

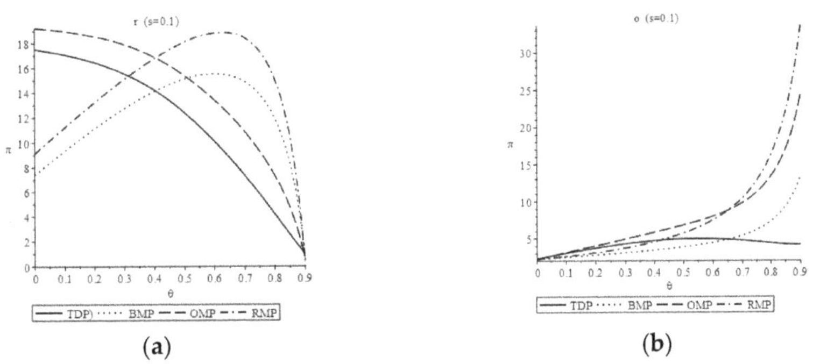

Figure 4. (**a**) $s = 0.1$, changes of the manufacturer's profits with θ; (**b**) $s = 0.1$, changes of the remanufacturer's profits with θ.

5.2. Optimal Strategy for the Manufacturer and the Remanufacturer

The size of the initial consumers may vary due to the difference in sellers. To observe the change trend of the sellers' profits there are three conditions: (1) The manufacturer has more initial consumers; (2) The manufacturers and the remanufacturer have the same number of initial consumers; (3) The remanufacturers have more initial consumers, we list three typical cases: $N_r : N_o = 10 : 90$, $N_r : N_o = 50 : 50$, and $N_r : N_o = 90 : 10$. The profits (π_r, π_o) are shown in Tables 2 and 3.

Table 2. The equilibrium profits of the manufacturer under a different strategy ($s = 0.1$).

θ	N_r:N_o	Strategy			
		TDP	BMP	OMP	RMP
	10:90	15.1	13.3	18.3	16.5
0.05	50:50	17.3	8.4	19.1	13.2
	90:10	19.6	3.5	19.9	10.2
	10:90	12.7	12.4	17.2	16.4
0.35	50:50	14.9	13.5	17.4	16.0
	90:10	17.2	14.5	17.7	15.0
	10:90	5.3	7.0	11.6	13.2
0.75	50:50	5.8	13.9	9.3	17.3
	90:10	6.3	20.8	7.7	21.5

Table 3. The equilibrium profits of the remanufacturer under a different strategy ($s = 0.1$).

θ	N_r:N_o	Strategy			
		TDP	BMP	OMP	RMP
	10:90	4.3	4.3	4.3	4.3
0.05	50:50	2.7	2.4	2.7	2.4
	90:10	1.1	0.6	1.1	0.6
	10:90	5.1	4.8	6.6	6.4
0.35	50:50	4.6	3.3	5.4	4.1
	90:10	4.0	1.7	4.2	1.9
	10:90	4.8	5.1	16.6	17.0
0.75	50:50	4.6	5.2	11.2	12.8
	90:10	4.4	7.2	5.8	8.6

Tables 2 and 3 show that it is not always advantageous for sellers to have a large number of initial consumers. Especially when θ is low, the manufacturer's profits decrease with the increase in the initial consumer scale under the BMP and RMP strategies; when θ is moderate, the manufacturer's profits decrease with the increase in the initial consumer scale under the RMP strategy; when θ is high, the manufacturer's profits decrease with the increase in the initial consumer scales under the OMP strategy, and the remanufacturer's profits decrease with the increase in the initial consumer scale under the BMP strategy. Taking the BMP strategy as an example, the decrease in seller's profits can be explained as the manufacturer implementing the price matching in the first period; when the difference between new products and remanufactured products becomes less obvious, the manufacturer will lose the advantage of charging higher prices. As the initial consumers increase, the consumers' influence on the remanufacturer's purchase is greater, which leads to the decline in the manufacturer's profits because the low price becomes the most important factor in attracting consumers. With the increase in initial consumers, the greater the negative impact of consumers purchasing new products, the greater the decrease in the remanufacturers' profits. The change in sellers' profits in the case of other strategies can also be explained by the above analysis method. Therefore, sellers should maintain a certain range of consumers and adopt reasonable price matching strategies to increase profits.

The change in the initial consumer size will not affect the choice of sellers' equilibrium strategy. When θ is small or moderate, the OMP strategy is the most common choice of the manufacturer and the remanufacturer; when θ is large, the RMP strategy is the most common choice of sellers. This means that, regardless of the product differences between new products and remanufactured products, the remanufacturer should choose price matching, and the manufacturer should choose dynamic pricing or price matching according to the product differences (when θ is small or moderate, the manufacturer should choose dynamic pricing; when θ is large, the manufacturer should choose price matching). Next, we will analyze the reasons why the remanufacturer is unable to flexibly adjust their pricing strategies. When new products and remanufactured products are becoming more and more similar, the remanufacturer will try to meet the consumers' demand for novelty to narrow the service gap. However, in the actual operation process, it is difficult for the remanufacturers to invest a lot of money, time and innovation again to promote products, and consumers are likely to have aesthetic and technical fatigue. Therefore, the remanufacturer, who has lost service advantages, may further exert their low price advantages, turn passivity into initiative, and promote the manufacturer to match their price, which achieves a Pareto improvement.

In general, considering consumers' switching purchase behavior, the only choice for the remanufacturer is price matching, while the manufacturer can choose dynamic pricing or price matching according to product differences. In actual production, the original equipment manufacturer (OEM) usually lacks sufficient funds, specialized equipment and the technical level to make profits from remanufacturing. For example, the Ford Company tried to enter the field of remanufacturing abandoned vehicles, but failed. Unlike OEM, the third-party remanufacturer (TPR) not only has advanced remanufacturing technology, but also tends to form economies of scale for remanufacturing the used products of many brands. "Remanufacturing" recycling can optimize the performance of waste products and minimize the consumption of resources. Meanwhile, remanufactured products have a large space for value preservation and appreciation. Therefore, TPR becomes the main competitor of OEM. Huawei's P40 Pro 5G mobile phone (Huawei, Shenzhen, China) was successfully re-launched at its original price, providing a new way for the competition between OME and TRI.

6. Conclusions

Consumers' switching purchase behavior often occurs in the actual sales activities of new and remanufactured products. Due to the expectation of price reduction in remanufactured products or the uncertainty of the performance of new products, consumers often choose to postpone the purchase of products. In the case that manufactured products and remanufactured products are sold at the same time, some consumers who pursue novelty may delay their purchase and opt for new products from the manufacturer, while some consumers who prefer practicality and low prices, may switch to remanufactured products. To maximize profits, sellers should consider switching purchase behaviors; they should also make use of their own characteristics and advantages to choose appropriate pricing strategies. Given this background, this study analyzes the factors that affect sellers' equilibrium profits, and discusses the optimal pricing strategy for sellers.

The key point of the price matching strategy, compared with traditional dynamic pricing, is that it uses the price of a competitor's product as a reference. Our conclusions are divided into two categories; one is the impact of various parameters on the sellers, and the other is the sellers' optimal pricing strategy. Considering the influence of parameters, we find that consumer learning costs, initial consumers and product differences can affect the sellers' pricing decisions. Specifically, consumer learning costs reduce the profits of the manufacturer but increases that of the remanufacturer. The large number of initial consumers is not always advantageous for sellers to make more profits. Product differences affect the determination of the seller's equilibrium strategy, when the differences between new products and remanufactured products are obvious (θ is small or moderate), OMP is the equilibrium strategy for sellers and when the differences between new products and remanufactured products are not obvious (θ is large), the RMP strategy is the equilibrium strategy for sellers. In the optimal strategy, the remanufacturer should insist on price matching, while the manufacturer should choose dynamic pricing or price matching according to the product differences.

The impact of consumer learning costs on the manufacturer and the remanufacturer are obvious. Learning about new products is not easy for consumers, so the remanufacturer can take advantage of this to gain more profits. At the beginning, sellers face more consumers, and in the future, they may encounter more consumers' switching purchase, and their profits will be damaged. In short, considering consumers' switching purchase behavior, the remanufacturer should choose price matching and the manufacturer should choose dynamic pricing or price matching based on differences between new products and remanufactured products. In fact, whether sellers sell new products or remanufactured products in a competitive environment, the purpose is to eliminate consumers' uncertainty and encourage them to buy products. Therefore, sellers need to choose product pricing according to their own sales characteristics. For example, manufacturers can provide live demonstrations and free trial services of different series of products for consumers who buy TVs, mobile phones and other electronic products, adapt to the functions of new products. Traditionally, compared with remanufactured products, the learning process of new products can make consumers feel the performance of these products more intuitively, and attract many consumers to purchase products. At this time, the market price of remanufactured products is more competitive than that of new products, which can attract price-sensitive consumers to buy and create considerable profit space for enterprises. Some remanufacturers are also using other technologies to improve the consumer perception of remanufactured products. For example, Dell uses the diversified marketing methods of third-party platforms to sell remanufactured products. At this time, the differences between new products and remanufactured products are no longer obvious, and the two sellers return to a new round of price competition, which is the RMP strategy mentioned in this paper. Of course, with the improvement of the accuracy of price forecasting in the future, price matching can not only be limited to this, but also can be set in a certain range in combination with the sensitivity of consumers to prices, which is likely to extend the concept of price matching.

In this research, we only consider the duopoly case. It would be interesting to investigate the pricing of new products and remanufactured products by considering different channel structures. Now, many sellers have begun to implement "online and offline" sales patterns, and the case of a seller considering adding another channel might be particularly interesting. It has to be said that the psychology of consumers is quite complex, and the final behavior is not completely rational; thus, there are some limitations in our interpretation of consumers. In the future, we can combine some irrational factors to further explain the problem.

Author Contributions: All authors conceptualized this study; all authors made contributions throughout all sections, especially H.L. and Q.X. jointly developed the product pricing model and drafted the article, T.P. checked English spelling and grammar. All authors have read and agreed to the published version of the manuscript.

Funding: This work was supported by the National Natural Science Foundation of China (Grant No. 71402012).

Data Availability Statement: The data and cases about new products and remanufactured products in the Introduction are from the Global New Products Database. The profit data of manufacturers and remanufacturers obtained by calculation in this paper are from simulation analysis. The data on Huawei's sales of P40 Pro 5G mobile phone can perfectly support the conclusion of this paper.

Acknowledgments: The authors would like to thank The National Natural Science Foundation of China (Grant No. 71402012)and The National Social Science Fund of China (Grant No. 19XGL016). The Fifth International Conference on Business Intelligence & Financial Engineering, the 12th International Conference on Service Systems and Service Management (ICSSSM), and the 2020 ITEA Webinars were also beneficial.

Conflicts of Interest: The authors declare no conflict of interest.

Appendix A

Proof of Theorem 1. When the consumer first visits the manufacturer, the expected profits of sellers can be obtained from (3) and (4). The two sellers' first-order conditions for profit maximization are $\frac{\partial \pi_{1r}^{TDP}(1,0)}{\partial p_{1r}^{TDP}} = 0$ and $\frac{\partial \pi_{2o}^{TDP}(1,0)}{\partial p_{2o}^{TDP}} = 0$, and we can solve p_{1r}^{TDP} and p_{2o}^{TDP} jointly. Moreover, the second derivative of profit with respect to price is less than zero. Therefore, the profit of the manufacturer in the first period is maximized at $p_{1r}^{TDP^*} = -\frac{3}{8}s - \frac{7}{16}\theta + \frac{5}{8} - \frac{\sqrt{4s^2-12s\theta+17\theta^2+8s-12\theta+4}}{16}$, and the profit of the remanufacturer in the second period is maximized at $p_{2o}^{TDP^*} = \frac{1}{4}s + \frac{1}{8}\theta + \frac{1}{4} - \frac{\sqrt{4s^2-12s\theta+17\theta^2+8s-12\theta+4}}{8}$. Substituting $p_{1r}^{TDP^*}$ and $p_{2o}^{TDP^*}$ into (3) and (4), we obtain $\pi_{1r}^{TDP^*}(1,0)$ and $\pi_{2o}^{TDP^*}(1,0)$.

When the consumer first visits the remanufacturer, the expected profits of sellers can be obtained from (5) and (6). Similar to the above derivation process, we obtain $p_{1o}^{TDP^*} = \frac{3}{16}s - \frac{1}{4}\theta + \frac{7}{16} - \frac{\sqrt{9s^2+8s\theta+16\theta^2-22s-24\theta+17}}{16}$, $p_{2r}^{TDP^*} = -\frac{5}{8}s - \frac{1}{2}\theta + \frac{7}{8} - \frac{\sqrt{9s^2+8s\theta+16\theta^2-22s-24\theta+17}}{8}$. Then, substituting $p_{1o}^{TDP^*}$ and $p_{2r}^{TDP^*}$ into (5) and (6), we obtain $\pi_{1o}^{TDP^*}(1,0)$ and $\pi_{2r}^{TDP^*}(1,0)$.
□

Proof of Theorem 2. According to the profit function of Manufacturer (1), when there are N consumers in the market, the final profits of the manufacturer are equal to the sum of the profits of the two periods. Next, we expand the manufacturer's profit function and then merge again. The details are as follows:

$$\begin{aligned}
\pi_r^{TDP}(N_r,N_o) &= p_{1R}^{TDP}\cdot\sum_{n_{1r}=0}^{N_r}\left(n_{1r}\cdot q_{(n_{1r},N_r)}^{TDP}\right) + \sum_{n_{1r}=0}^{N_r} q_{(n_{1r},N_r)}^{TDP} \sum_{n_{1o}=0}^{N_o} q_{(n_{1o},N_o)}^{TDP}\cdot p_{2r}^{TDP}\cdot\sum_{n_{2r}=0}^{N_o-n_{1o}}\left(n_{2r}\cdot q_{(n_{2r},N_o-n_{1o})}^{TDP}\right)\\
&= p_{1r}^{TDP}\cdot N_r\cdot\left[1-F\left(\tfrac{p_{1r}^{TDP}-p_{2o}^{TDP}+s}{1-\theta}\right)\right] + \sum_{n_{1r}=0}^{N_r} q_{(n_{1r},N_r)}^{TDP}\sum_{n_{1o}=0}^{N_o} q_{(n_{1o},N_o)}^{TDP}\cdot p_{2r}^{TDP}\cdot(N_o-n_{1o})\cdot\left[1-F(p_{2r}^{TDP}+s)\right]\\
&= p_{1r}^{TDP}\cdot N_r\cdot\left[1-F\left(\tfrac{p_{1r}^{TDP}-p_{2o}^{TDP}+s}{1-\theta}\right)\right] + \sum_{n_{1r}=0}^{N_r} q_{(n_{1r},N_r)}^{TDP} N_o\cdot\left[1-F\left(\tfrac{p_{2r}^{TDP}-p_{1o}^{TDP}+s}{1-\theta}\right)\right]\cdot p_{2r}^{TDP}\cdot\left[1-F(p_{2r}^{TDP}+s)\right]\\
&= p_{1r}^{TDP}\cdot N_r\cdot\left[1-F\left(\tfrac{p_{1r}^{TDP}-p_{2o}^{TDP}+s}{1-\theta}\right)\right] + p_{2r}^{TDP}\cdot N_o\cdot\left[\left[1-F\left(\tfrac{p_{2r}^{TDP}-p_{1o}^{TDP}+s}{1-\theta}\right)\right]\right]\cdot\left[1-F(p_{2r}^{TDP}+s)\right]\\
&= N_r\pi_{1r}^{TDP} + N_o\pi_{2r}^{TDP}
\end{aligned}$$

where

$$\begin{aligned}
(N_o-n_{1o})\cdot\sum_{n_{1o}=0}^{N_o} q_{(n_{1o},N_o)}^{TDP} &= N_o\cdot q_{(0,N_o)}^{TDP} + (N_o-1)\cdot q_{(1,N_o)}^{TDP} + (N_o-2)\cdot q_{(2,N_o)}^{TDP} + \ldots + 2\cdot q_{(N_o-2,N_o)}^{TDP} + 1\cdot q_{(N_o-1,N_o)}^{TDP} + 0\cdot q_{(N_o,N_o)}^{TDP}\\
&= \sum_{n_{1o}=0}^{N_o} n_{1o}\cdot q_{(N_o-n_{1o},N_o)}^{TDP}\\
&= N_o\cdot\left[1-F\left(\tfrac{p_{2r}^{TDP}-p_{1o}^{TDP}+s}{1-\theta}\right)\right]
\end{aligned}$$

And

$$\begin{aligned}
\sum_{n_{1r}=0}^{N_r} q_{(n_{1r},N_r)}^{TDP} &= q_{(0,N_r)}^{TDP} + q_{(1,N_r)}^{TDP} + \ldots + q_{(N_r-1,N_r)}^{TDP} + q_{(N_r,N_r)}^{TDP}\\
&= \sum_{n_{1r}=0}^{N_r} C_{N_r}^{n_{1r}}\cdot\left[1-F\left(\tfrac{p_{1r}^{TDP}-p_{2o}^{TDP}+s}{1-\theta}\right)\right]^{n_{1r}}\cdot\left[F\left(\tfrac{p_{1r}^{TDP}-p_{2o}^{TDP}+s}{1-\theta}\right)\right]^{N_r-n_{1r}}\\
&= \left[1-F\left(\tfrac{p_{1r}^{TDP}-p_{2o}^{TDP}+s}{1-\theta}\right)+F\left(\tfrac{p_{1r}^{TDP}-p_{2o}^{TDP}+s}{1-\theta}\right)\right]^{N_r} = 1
\end{aligned}$$

Thus, Equation (8) can be obtained. Equation (9) can also be obtained by the remanufacturer's profit function and then by merging again. The details are as follows:

$$\begin{aligned}\pi_{o}^{TDP}(N_r,N_o) &= p_{1o}^{TDP}\cdot\sum_{n_{1o}=0}^{N_o}\left(n_{1r}\cdot q_{(n_{1o},N_o)}^{TDP}\right)+\sum_{n_{1o}=0}^{N_o} q_{(n_{1o},N_o)}^{TDP}\sum_{n_{1r}=0}^{N_r} q_{(n_{1r},N_r)}^{TDP}\cdot p_{2o}^{TDP}\cdot\sum_{n_{2o}=0}^{N_r-n_{1r}}\left(n_{2o}\cdot q_{(n_{2o},N_r-n_{1r})}^{TDP}\right)\\ &= p_{1o}^{TDP}\cdot N_o\cdot F\left(\frac{p_{2r}^{TDP}-p_{1o}^{TDP}+s}{1-\theta}\right)+\sum_{n_{1o}=0}^{N_o} q_{(n_{1o},N_o)}^{TDP}\sum_{n_{1r}=0}^{N_r} q_{(n_{1r},N_r)}^{TDP}\cdot p_{2o}^{TDP}\cdot(N_r-n_{1r})\cdot\left[1-F\left(\frac{p_{2o}^{TDP}}{\theta}\right)\right]\\ &= p_{1o}^{TDP}\cdot N_o\cdot F\left(\frac{p_{2r}^{TDP}-p_{1o}^{TDP}+s}{1-\theta}\right)+\sum_{n_{1o}=0}^{N_o} q_{(n_{1o},N_o)}^{TDP}\cdot N_r\cdot F\left(\frac{p_{1r}^{TDP}-p_{2o}^{TDP}+s}{1-\theta}\right)\cdot p_{2o}^{TDP}\cdot\left[1-F\left(\frac{p_{2o}^{TDP}}{\theta}\right)\right]\\ &= p_{1o}^{TDP}\cdot N_o\cdot F\left(\frac{p_{2r}^{TDP}-p_{1o}^{TDP}+s}{1-\theta}\right)+p_{2o}^{TDP}\cdot N_r\cdot F\left(\frac{p_{1r}^{TDP}-p_{2o}^{TDP}+s}{1-\theta}\right)\cdot\left[1-F\left(\frac{p_{2o}^{TDP}}{\theta}\right)\right]\\ &= N_o\pi_{1o}^{TDP}+N_r\pi_{2o}^{TDP}\end{aligned}$$

where

$$\begin{aligned}(N_r-n_{1r})\cdot\sum_{n_{1r}=0}^{N_r} q_{(n_{1r},N_r)}^{TDP} &= N_r\cdot q_{(0,N_r)}^{TDP}+(N_r-1)\cdot q_{(1,N_r)}^{TDP}+\ldots+2\cdot q_{(N_r-2,N_r)}^{TDP}+1\cdot q_{(N_r-1,N_r)}^{TDP}+0\cdot q_{(N_r,N_r)}^{TDP}\\ &= \sum_{n_{1r}=0}^{N_r} n_{1r}\cdot q_{(N_r-n_{1r},N_r)}^{TDP}\\ &= N_r\cdot F\left(\frac{p_{1r}^{TDP}-p_{2o}^{TDP}+s}{1-\theta}\right)\end{aligned}$$

And

$$\begin{aligned}\sum_{n_{1o}=0}^{N_o} q_{(n_{1o},N_o)}^{TDP} &= q_{(0,N_o)}^{TDP}+q_{(1,N_o)}^{TDP}+\ldots+q_{(N_o-1,N_o)}^{TDP}+q_{(N_o,N_o)}^{TDP}\\ &= \sum_{n_{1o}=0}^{N_o} C_{N_o}^{n_{1o}}\cdot\left[F\left(\frac{p_{2r}^{TDP}-p_{1o}^{TDP}+s}{1-\theta}\right)\right]^{n_{1o}}\cdot\left[1-F\left(\frac{p_{2r}^{TDP}-p_{1o}^{TDP}+s}{1-\theta}\right)\right]^{N_o-n_{1o}}\\ &= \left[F\left(\frac{p_{2r}^{TDP}-p_{1o}^{TDP}+s}{1-\theta}\right)+1-F\left(\frac{p_{2r}^{TDP}-p_{1o}^{TDP}+s}{1-\theta}\right)\right]^{N_o}=1\end{aligned}$$

□

Proof of Theorem 3. When the consumer first visits the manufacturer, the expected profit of sellers can be obtained from (12) and (13). Since $p_{1r}^{BMP}=p_{2o}^{BMP}$, the pricing of the manufacturer is determined by the remanufacturer; the remanufacturer's first-order condition for profit maximization is $\frac{\partial\pi_{2o}^{BMP}(1,0)}{\partial p_{2o}^{BMP}}=0$. We can solve p_{2o}^{BMP} directly. Moreover, the second derivative of profit with respect to price is less than zero. As a result, the profit of the remanufacturer in the second period is maximized at $p_{2o}^{BMP^*}=\frac{1}{2}\theta$, so $p_{1r}^{BMP^*}=\frac{1}{2}\theta$. Substituting $p_{2o}^{BMP^*}$ into (12) and (13), we can obtain $\pi_{1r}^{BMP^*}(1,0)$ and $\pi_{2o}^{BMP^*}(1,0)$.

When the consumer first visits the remanufacturer, the expected profit of sellers can be obtained from (14) and (15). Similar to the above derivation process, we obtain $p_{1o}^{BMP^*}=\frac{3}{16}s-\frac{1}{4}\theta+\frac{7}{16}-\frac{\sqrt{9s^2+8s\theta+16\theta^2-22s-24\theta+17}}{16}$, $p_{2r}^{BMP^*}=-\frac{5}{8}s-\frac{1}{2}\theta+\frac{7}{8}-\frac{\sqrt{9s^2+8s\theta+16\theta^2-22s-24\theta+17}}{8}$. Then, substituting $p_{1o}^{BMP^*}$ and $p_{2r}^{BMP^*}$ into (14) and (15), we obtain $\pi_{1o}^{BMP^*}(1,0)$ and $\pi_{2r}^{BMP^*}(1,0)$. □

Proof of Theorem 4. According to the profit function of Manufacturer (10), when there are N consumers in the market, the final profit of the manufacturer is equal to the sum of the profit of the two periods. Next, we expand the manufacturer's profit function and then merge again. The details are as follows:

$$
\begin{aligned}
\pi_r^{BMP}(N_r,N_o) &= p_{1R}^{BMP}\cdot\sum_{n_{1r}=0}^{N_r}\left(n_{1r}\cdot q_{(n_{1r},N_r)}^{BMP}\right)+\sum_{n_{1r}=0}^{N_r} q_{(n_{1r},N_r)}^{BMP}\sum_{n_{1o}=0}^{N_o} q_{(n_{1o},N_o)}^{BMP}\cdot p_{2r}^{BMP}\cdot\sum_{n_{2r}=0}^{N_o-n_{1o}}\left(n_{2r}\cdot q_{(n_{2r},N_o-n_{1o})}^{BMP}\right)\\
&= p_{1r}^{BMP}\cdot N_r\cdot\left[1-F(\tfrac{s}{1-\theta})\right]+\sum_{n_{1r}=0}^{N_r} q_{(n_{1r},N_r)}^{BMP}\sum_{n_{1o}=0}^{N_o} q_{(n_{1o},N_o)}^{BMP}\cdot p_{2r}^{BMP}\cdot(N_o-n_{1o})\cdot\left[1-F(p_{2r}^{BMP}+s)\right]\\
&= p_{1r}^{BMP}\cdot N_r\cdot\left[1-F(\tfrac{s}{1-\theta})\right]+\sum_{n_{1r}=0}^{N_r} q_{(n_{1r},N_r)}^{BMP} N_o\cdot\left[1-F(\tfrac{p_{2r}^{BMP}-p_{1o}^{BMP}+s}{1-\theta})\right]\cdot p_{2r}^{BMP}\cdot\left[1-F(p_{2r}^{BMP}+s)\right]\\
&= p_{1r}^{BMP}\cdot N_r\cdot\left[1-F(\tfrac{s}{1-\theta})\right]+p_{2r}^{BMP}\cdot N_o\cdot\left[\left[1-F(\tfrac{p_{2r}^{BMP}-p_{1o}^{BMP}+s}{1-\theta})\right]\right]\cdot\left[1-F(p_{2r}^{BMP}+s)\right]\\
&= N_r\pi_{1r}^{BMP}+N_o\pi_{2r}^{BMP}
\end{aligned}
$$

where

$$
\begin{aligned}
(N_o-n_{1o})\cdot\sum_{n_{1o}=0}^{N_o} q_{(n_{1o},N_o)}^{BMP} &= N_o\cdot q_{(0,N_o)}^{BMP}+(N_o-1)\cdot q_{(1,N_o)}^{BMP}+(N_o-2)\cdot q_{(2,N_o)}^{BMP}+\ldots+2\cdot q_{(N_o-2,N_o)}^{BMP}+1\cdot q_{(N_o-1,N_o)}^{BMP}+0\cdot q_{(N_o,N_o)}^{BMP}\\
&= \sum_{n_{1o}=0}^{N_o} n_{1o}\cdot q_{(N_o-n_{1o},N_o)}^{BMP}\\
&= N_o\cdot\left[1-F(\tfrac{p_{2r}^{BMP}-p_{1o}^{BMP}+s}{1-\theta})\right]
\end{aligned}
$$

And

$$
\begin{aligned}
\sum_{n_{1r}=0}^{N_r} q_{(n_{1r},N_r)}^{BMP} &= q_{(0,N_r)}^{BMP}+q_{(1,N_r)}^{BMP}+\ldots+q_{(N_r-1,N_r)}^{BMP}+q_{(N_r,N_r)}^{BMP}\\
&= \sum_{n_{1r}=0}^{N_r} C_{N_r}^{n_{1r}}\cdot\left[1-F(\tfrac{s}{1-\theta})\right]^{n_{1r}}\cdot\left[F(\tfrac{s}{1-\theta})\right]^{N_r-n_{1r}}\\
&= \left[1-F(\tfrac{s}{1-\theta})+F(\tfrac{s}{1-\theta})\right]^{N_r}=1
\end{aligned}
$$

Thus, Equation (16) can be obtained. Equation (17) can also be obtained by the remanufacturer's profit function and then merge again. The details are as follows:

$$
\begin{aligned}
\pi_o^{BMP}(N_r,N_o) &= p_{1o}^{BMP}\cdot\sum_{n_{1o}=0}^{N_o}\left(n_{1r}\cdot q_{(n_{1o},N_o)}^{BMP}\right)+\sum_{n_{1o}=0}^{N_o} q_{(n_{1o},N_o)}^{BMP}\sum_{n_{1r}=0}^{N_r} q_{(n_{1r},N_r)}^{BMP}\cdot p_{2o}^{BMP}\cdot\sum_{n_{2o}=0}^{N_r-n_{1r}}\left(n_{2o}\cdot q_{(n_{2o},N_r-n_{1r})}^{BMP}\right)\\
&= p_{1o}^{BMP}\cdot N_o\cdot F(\tfrac{p_{2r}^{BMP}-p_{1o}^{BMP}+s}{1-\theta})+\sum_{n_{1o}=0}^{N_o} q_{(n_{1o},N_o)}^{BMP}\sum_{n_{1r}=0}^{N_r} q_{(n_{1r},N_r)}^{BMP}\cdot p_{2o}^{BMP}\cdot(N_r-n_{1r})\cdot\left[1-F(\tfrac{p_{2o}^{BMP}}{\theta})\right]\\
&= p_{1o}^{BMP}\cdot N_o\cdot F(\tfrac{p_{2r}^{BMP}-p_{1o}^{BMP}+s}{1-\theta})+\sum_{n_{1o}=0}^{N_o} q_{(n_{1o},N_o)}^{BMP}\cdot N_r\cdot F(\tfrac{s}{1-\theta})\cdot p_{2o}^{BMP}\cdot\left[1-F(\tfrac{p_{2o}^{BMP}}{\theta})\right]\\
&= p_{1o}^{BMP}\cdot N_o\cdot F(\tfrac{p_{2r}^{BMP}-p_{1o}^{BMP}+s}{1-\theta})+p_{2o}^{BMP}\cdot N_r\cdot F(\tfrac{s}{1-\theta})\cdot\left[1,-,F,(,\tfrac{p_{2o}^{BMP}}{\theta},)\right]\\
&= N_o\pi_{1o}^{BMP}+N_r\pi_{2o}^{BMP}
\end{aligned}
$$

where

$$
\begin{aligned}
(N_r-n_{1r})\cdot\sum_{n_{1r}=0}^{N_r} q_{(n_{1r},N_r)}^{BMP} &= N_r\cdot q_{(0,N_r)}^{BMP}+(N_r-1)\cdot q_{(1,N_r)}^{BMP}+\ldots+2\cdot q_{(N_r-2,N_r)}^{BMP}+1\cdot q_{(N_r-1,N_r)}^{BMP}+0\cdot q_{(N_r,N_r)}^{BMP}\\
&= \sum_{n_{1r}=0}^{N_r} n_{1r}\cdot q_{(N_r-n_{1r},N_r)}^{BMP}\\
&= N_r\cdot F\left(\tfrac{s}{1-\theta}\right)
\end{aligned}
$$

and

$$\begin{aligned}\sum_{n_{1o}=0}^{N_o} q_{(n_{1o},N_o)}^{BMP} &= q_{(0,N_o)}^{BMP} + q_{(1,N_o)}^{BMP} + \ldots + q_{(N_o-1,N_o)}^{BMP} + q_{(N_o,N_o)}^{BMP} \\ &= \sum_{n_{1o}=0}^{N_o} C_{N_o}^{n_{1o}} \cdot \left[F(\tfrac{p_{2r}^{BMP}-p_{1o}^{BMP}+s}{1-\theta})\right]^{n_{1o}} \cdot \left[1-F(\tfrac{p_{2r}^{BMP}-p_{1o}^{BMP}+s}{1-\theta})\right]^{N_o-n_{1o}} \\ &= \left[F(\tfrac{p_{2r}^{BMP}-p_{1o}^{BMP}+s}{1-\theta}) + 1 - F(\tfrac{p_{2r}^{BMP}-p_{1o}^{BMP}+s}{1-\theta})\right]^{N_o} = 1\end{aligned}$$

□

Proof of Theorem 5. When the consumer first visits the manufacturer, the expected profits of sellers can be obtained from (20) and (21). The two sellers' first-order conditions for profit maximization are $\frac{\partial \pi_{1r}^{OMP}(1,0)}{\partial p_{1r}^{OMP}} = 0$ and $\frac{\partial \pi_{2o}^{OMP}(1,0)}{\partial p_{2o}^{OMP}} = 0$. We can solve p_{1r}^{OMP} and p_{2o}^{OMP} jointly. Moreover, the second derivative of profit with respect to price is less than zero. As a result, the profit of the manufacturer in the first period is maximized at $p_{1r}^{OMP^*} = -\frac{3}{8}s - \frac{7}{16}\theta + \frac{5}{8} - \frac{\sqrt{4s^2-12s\theta+17\theta^2+8s-12\theta+4}}{16}$, and the profit of the remanufacturer in the second period is maximized at $p_{2o}^{OMP^*} = \frac{1}{4}s + \frac{1}{8}\theta + \frac{1}{4} - \frac{\sqrt{4s^2-12s\theta+17\theta^2+8s-12\theta+4}}{8}$. Substituting $p_{1r}^{OMP^*}$ and $p_{2o}^{OMP^*}$ into (20) and (21), we obtain $\pi_{1r}^{OMP^*}(1,0)$ and $\pi_{2o}^{TDP^*}(1,0)$.

When the consumer first visits the remanufacturer, the expected profits of sellers can be obtained from (22) and (23). Since $p_{1o}^{OMP} = p_{2r}^{OMP}$, the pricing of the remanufacturer is determined by the remanufacturer, whose first-order condition for profit maximization is $\frac{\partial \pi_{2r}^{OMP}(1,0)}{\partial p_{2r}^{OMP}} = 0$; we can solve p_{2r}^{OMP} directly. Moreover, the second derivative of profit with respect to price is less than zero. As a result, the profit of the remanufacturer in the second period is maximized at $p_{2r}^{OMP^*} = \frac{1}{2} - \frac{1}{2}s$, so $p_{1o}^{OMP^*} = \frac{1}{2} - \frac{1}{2}s$. Substituting $p_{2o}^{OMP^*}$ into (22) and (23), we can obtain $\pi_{1o}^{OMP^*}(1,0)$ and $\pi_{2r}^{OMP^*}(1,0)$. □

Proof of Theorem 6. According to the profit function of the manufacturer, when there are N consumers in the market, the final profits of the manufacturer are equal to the sum of the profits of the two periods. Next, we expand the remanufacturer's profit function and then merge again. The details are as follows:

$$\begin{aligned}\pi_r^{OMP}(N_r,N_o) &= p_{1R}^{OMP} \cdot \sum_{n_{1r}=0}^{N_r} \left(n_{1r} \cdot q_{(n_{1r},N_r)}^{OMP}\right) + \sum_{n_{1r}=0}^{N_r} q_{(n_{1r},N_r)}^{OMP} \sum_{n_{1o}=0}^{N_o} q_{(n_{1o},N_o)}^{OMP} \cdot p_{2r}^{OMP} \cdot \sum_{n_{2r}=0}^{N_o-n_{1o}} \left(n_{2r} \cdot q_{(n_{2r},N_o-n_{1o})}^{OMP}\right) \\ &= p_{1r}^{OMP} \cdot N_r \cdot \left[1-F(\tfrac{p_{1r}^{OMP}-p_{2o}^{OMP}+s}{1-\theta})\right] + \sum_{n_{1r}=0}^{N_r} q_{(n_{1r},N_r)}^{OMP} \sum_{n_{1o}=0}^{N_o} q_{(n_{1o},N_o)}^{OMP} \cdot p_{2r}^{OMP} \cdot (N_o - n_{1o}) \cdot [1-F(p_{2r}^{OMP}+s)] \\ &= p_{1r}^{OMP} \cdot N_r \cdot \left[1-F(\tfrac{p_{1r}^{OMP}-p_{2o}^{OMP}+s}{1-\theta})\right] + \sum_{n_{1r}=0}^{N_r} q_{(n_{1r},N_r)}^{OMP} N_o \cdot \left[1-F(\tfrac{s}{1-\theta})\right] \cdot p_{2r}^{OMP} \cdot [1-F(p_{2r}^{OMP}+s)] \\ &= p_{1r}^{OMP} \cdot N_r \cdot \left[1-F(\tfrac{p_{1r}^{OMP}-p_{2o}^{OMP}+s}{1-\theta})\right] + p_{2r}^{OMP} \cdot N_o \cdot \left[1-F(\tfrac{s}{1-\theta})\right] \cdot [1-F(p_{2r}^{OMP}+s)] \\ &= N_r \pi_{1r}^{OMP} + N_o \pi_{2r}^{OMP}\end{aligned}$$

where

$$\begin{aligned}(N_o,-,n_{1o}) \cdot \sum_{n_{1o}=0}^{N_o} q_{(n_{1o},N_o)}^{OMP} &= N_o \cdot q_{(0,N_o)}^{OMP} + (N_o-1) \cdot q_{(1,N_o)}^{OMP} + (N_o-2) \cdot q_{(2,N_o)}^{OMP} + \ldots + 2 \cdot q_{(N_o-2,N_o)}^{OMP} + 1 \cdot q_{(N_o-1,N_o)}^{OMP} + 0 \cdot q_{(N_o,N_o)}^{OMP} \\ &= \sum_{n_{1o}=0}^{N_o} n_{1o} \cdot q_{(N_o-n_{1o},N_o)}^{OMP} \\ &= N_o \cdot \left[1-F(\tfrac{s}{1-\theta})\right]\end{aligned}$$

and

$$\begin{aligned}\sum_{n_{1r}=0}^{N_r} q^{OMP}_{(n_{1r},N_r)} &= q^{OMP}_{(0,N_r)} + q^{OMP}_{(1,N_r)} + \ldots + q^{OMP}_{(N_r-1,N_r)} + q^{OMP}_{(N_r,N_r)} \\ &= \sum_{n_{1r}=0}^{N_r} C^{n_{1r}}_{N_r} \cdot \left[1 - F\left(\frac{p^{OMP}_{1r} - p^{OMP}_{2o} + s}{1-\theta}\right)\right]^{n_{1r}} \cdot \left[F\left(\frac{p^{OMP}_{1r} - p^{OMP}_{2o} + s}{1-\theta}\right)\right]^{N_r - n_{1r}} \\ &= \left[1 - F\left(\frac{p^{OMP}_{1r} - p^{OMP}_{2o} + s}{1-\theta}\right) + F\frac{p^{OMP}_{1r} - p^{OMP}_{2o} + s}{1-\theta}\right)\right]^{N_r} = 1\end{aligned}$$

Thus, Equation (22) can be obtained. Equation (23) can also be obtained by the remanufacturer's profit function and then merge again. The details are as follows:

$$\begin{aligned}\pi^{OMP}_{o}(N_r, N_o) &= p^{OMP}_{1o} \cdot \sum_{n_{1o}=0}^{N_o} \left(n_{1r} \cdot q^{OMP}_{(n_{1o},N_o)}\right) + \sum_{n_{1o}=0}^{N_o} q^{OMP}_{(n_{1o},N_o)} \sum_{n_{1r}=0}^{N_r} q^{OMP}_{(n_{1r},N_r)} \cdot p^{OMP}_{2o} \cdot \sum_{n_{2o}=0}^{N_r - n_{1r}} \left(n_{2o} \cdot q^{OMP}_{(n_{2o},N_r - n_{1r})}\right) \\ &= p^{OMP}_{1o} \cdot N_o \cdot F\left(\frac{s}{1-\theta}\right) + \sum_{n_{1o}=0}^{N_o} q^{OMP}_{(n_{1o},N_o)} \sum_{n_{1r}=0}^{N_r} q^{OMP}_{(n_{1r},N_r)} \cdot p^{OMP}_{2o} \cdot (N_r - n_{1r}) \cdot \left[1 - F\left(\frac{p^{OMP}_{2o}}{\theta}\right)\right] \\ &= p^{OMP}_{1o} \cdot N_o \cdot F\left(\frac{s}{1-\theta}\right) + \sum_{n_{1o}=0}^{N_o} q^{OMP}_{(n_{1o},N_o)} \cdot N_r \cdot F\left(\frac{s}{1-\theta}\right) \cdot p^{OMP}_{2o} \cdot \left[1 - F\left(\frac{p^{OMP}_{2o}}{\theta}\right)\right] \\ &= p^{OMP}_{1o} \cdot N_o \cdot F\left(\frac{s}{1-\theta}\right) + p^{OMP}_{2o} \cdot N_r \cdot F\left(\frac{s}{1-\theta}\right) \cdot \left[1 - F\left(\frac{p^{OMP}_{2o}}{\theta}\right)\right] \\ &= N_o \pi^{OMP}_{1o} + N_r \pi^{OMP}_{2o}\end{aligned}$$

where

$$\begin{aligned}(N_r - n_{1r}) \cdot \sum_{n_{1r}=0}^{N_r} q^{OMP}_{(n_{1r},N_r)} &= N_r \cdot q^{OMP}_{(0,N_r)} + (N_r - 1) \cdot q^{OMP}_{(1,N_r)} + \ldots + 2 \cdot q^{OMP}_{(N_r-2,N_r)} + 1 \cdot q^{OMP}_{(N_r-1,N_r)} + 0 \cdot q^{OMP}_{(N_r,N_r)} \\ &= \sum_{n_{1r}=0}^{N_r} n_{1r} \cdot q^{OMP}_{(N_r - n_{1r},N_r)} \\ &= N_r \cdot F\left(\frac{s}{1-\theta}\right)\end{aligned}$$

and

$$\begin{aligned}\sum_{n_{1o}=0}^{N_o} q^{OMP}_{(n_{1o},N_o)} &= q^{OMP}_{(0,N_o)} + q^{OMP}_{(1,N_o)} + \ldots + q^{OMP}_{(N_o-1,N_o)} + q^{OMP}_{(N_o,N_o)} \\ &= \sum_{n_{1o}=0}^{N_o} C^{n_{1o}}_{N_o} \cdot \left[F\left(\frac{s}{1-\theta}\right)\right]^{n_{1o}} \cdot \left[1 - F\left(\frac{s}{1-\theta}\right)\right]^{N_o - n_{1o}} \\ &= \left[F\left(\frac{s}{1-\theta}\right) + 1 - F\left(\frac{s}{1-\theta}\right)\right]^{N_o} = 1\end{aligned}$$

□

Proof of Theorem 7. When the consumer first visits the manufacturer, the expected profits of sellers can be obtained from (28) and (29). Since $p^{RMP}_{1r} = p^{RMP}_{2o}$, the pricing of the manufacturer is determined by the remanufacturer, the solution process is similar to the BMP strategy, the profit of the remanufacturer in the second period is maximized at $p^{RMP^*}_{2o} = \frac{1}{2}\theta$, and the manufacturer in the first period is $p^{RMP^*}_{1r} = \frac{1}{2}\theta$. Substituting $p^{RMP^*}_{2o}$ into (12) and (13), we can obtain $\pi^{RMP^*}_{1r}(1,0)$ and $\pi^{RMP^*}_{2o}(1,0)$.

When the consumer first visits the remanufacturer, the expected profits of sellers can be obtained from (30) and (31). Since $p^{RMP}_{1o} = p^{RMP}_{2r}$, the pricing of the remanufacturer is determined by the manufacturer, the solution process is similar to the OMP strategy, the profit of the remanufacturer in the second period is maximized at $p^{RMP^*}_{2r} = \frac{1}{2} - \frac{1}{2}s$, and the profit of the remanufacturer in the first period is $p^{RMP^*}_{1o} = \frac{1}{2} - \frac{1}{2}s$. Substituting $p^{RMP^*}_{2r}$ into (30) and (31), we obtain $\pi^{RMP^*}_{1o}(1,0)$ and $\pi^{RMP^*}_{2r}(1,0)$. □

Proof of Theorem 8. According to the profit function of Manufacturer (26), when there are N consumers in the market, the final profits of the manufacturer are equal to the sum of the profits of the two periods. Next, we expand the manufacturer's profit function and then merge again. The details are as follows:

$$
\begin{aligned}
\pi_r^{RMP}(N_r,N_o) &= p_{1R}^{RMP}\cdot\sum_{n_{1r}=0}^{N_r}\left(n_{1r}\cdot q^{RMP}_{(n_{1r},N_r)}\right)+\sum_{n_{1r}=0}^{N_r} q^{RMP}_{(n_{1r},N_r)}\sum_{n_{1o}=0}^{N_o} q^{RMP}_{(n_{1o},N_o)}\cdot p_{2r}^{RMP}\cdot\sum_{n_{2r}=0}^{N_o-n_{1o}}\left(n_{2r}\cdot q^{RMP}_{(n_{2r},N_o-n_{1o})}\right)\\
&= p_{1r}^{RMP}\cdot N_r\cdot\left[1-F(\tfrac{s}{1-\theta})\right]+\sum_{n_{1r}=0}^{N_r} q^{RMP}_{(n_{1r},N_r)}\sum_{n_{1o}=0}^{N_o} q^{RMP}_{(n_{1o},N_o)}\cdot p_{2r}^{RMP}\cdot(N_o-n_{1o})\cdot\left[1-F(p_{2r}^{RMP}+s)\right]\\
&= p_{1r}^{RMP}\cdot N_r\cdot\left[1-F(\tfrac{s}{1-\theta})\right]+\sum_{n_{1r}=0}^{N_r} q^{RMP}_{(n_{1r},N_r)} N_o\cdot\left[1-F(\tfrac{s}{1-\theta})\right]\cdot p_{2r}^{RMP}\cdot\left[1-F(p_{2r}^{RMP}+s)\right]\\
&= p_{1r}^{RMP}\cdot N_r\cdot\left[1-F(\tfrac{s}{1-\theta})\right]+p_{2r}^{RMP}\cdot N_o\cdot\left[1-F(\tfrac{s}{1-\theta})\right]\cdot\left[1-F(p_{2r}^{RMP}+s)\right]\\
&= N_r\pi_{1r}^{RMP}+N_o\pi_{2r}^{RMP}
\end{aligned}
$$

where

$$
\begin{aligned}
(N_o-n_{1o})\cdot\sum_{n_{1o}=0}^{N_o} q^{RMP}_{(n_{1o},N_o)} &= N_o\cdot q^{RMP}_{(0,N_o)}+(N_o-1)\cdot q^{RMP}_{(1,N_o)}+(N_o-2)\cdot q^{RMP}_{(2,N_o)}+\ldots+2\cdot q^{RMP}_{(N_o-2,N_o)}+1\cdot q^{RMP}_{(N_o-1,N_o)}+0\cdot q^{RMP}_{(N_o,N_o)}\\
&= \sum_{n_{1o}=0}^{N_o} n_{1o}\cdot q^{RMP}_{(N_o-n_{1o},N_o)}\\
&= N_o\cdot\left[1-F(\tfrac{s}{1-\theta})\right]
\end{aligned}
$$

and

$$
\begin{aligned}
\sum_{n_{1r}=0}^{N_r} q^{RMP}_{(n_{1r},N_r)} &= q^{RMP}_{(0,N_r)}+q^{RMP}_{(1,N_r)}+\ldots+q^{RMP}_{(N_r-1,N_r)}+q^{RMP}_{(N_r,N_r)}\\
&= \sum_{n_{1r}=0}^{N_r} C_{N_r}^{n_{1r}}\cdot\left[1-F(\tfrac{s}{1-\theta})\right]^{n_{1r}}\cdot\left[F(\tfrac{s}{1-\theta})\right]^{N_r-n_{1r}}\\
&= \left[1-F(\tfrac{s}{1-\theta})+F(\tfrac{s}{1-\theta})\right]^{N_r}=1
\end{aligned}
$$

Thus, Equation (26) can be obtained. Equation (27) can also be obtained by the remanufacturer's profit function and then merge again. The details are as follows:

$$
\begin{aligned}
\pi_o^{RMP}(N_r,N_o) &= p_{1o}^{RMP}\cdot\sum_{n_{1o}=0}^{N_o}\left(n_{1r}\cdot q^{RMP}_{(n_{1o},N_o)}\right)+\sum_{n_{1o}=0}^{N_o} q^{RMP}_{(n_{1o},N_o)}\sum_{n_{1r}=0}^{N_r} q^{RMP}_{(n_{1r},N_r)}\cdot p_{2o}^{RMP}\cdot\sum_{n_{2o}=0}^{N_r-n_{1r}}\left(n_{2o}\cdot q^{RMP}_{(n_{2o},N_r-n_{1r})}\right)\\
&= p_{1o}^{RMP}\cdot N_o\cdot F(\tfrac{p_{2r}^{TDP}-p_{1o}^{TDP}+s}{1-\theta})+\sum_{n_{1o}=0}^{N_o} q^{RMP}_{(n_{1o},N_o)}\sum_{n_{1r}=0}^{N_r} q^{RMP}_{(n_{1r},N_r)}\cdot p_{2o}^{RMP}\cdot(N_r-n_{1r})\cdot\left[1-F(\tfrac{p_{2o}^{RMP}}{\theta})\right]\\
&= p_{1o}^{RMP}\cdot N_o\cdot F(\tfrac{s}{1-\theta})+\sum_{n_{1o}=0}^{N_o} q^{RMP}_{(n_{1o},N_o)}\cdot N_r\cdot F(\tfrac{s}{1-\theta})\cdot p_{2o}^{RMP}\cdot\left[1-F(\tfrac{p_{2o}^{RMP}}{\theta})\right]\\
&= p_{1o}^{RMP}\cdot N_o\cdot F(\tfrac{s}{1-\theta})+p_{2o}^{RMP}\cdot N_r\cdot F(\tfrac{s}{1-\theta})\cdot\left[1-F(\tfrac{p_{2o}^{RMP}}{\theta})\right]\\
&= N_o\pi_{1o}^{RMP}+N_r\pi_{2o}^{RMP}
\end{aligned}
$$

where

$$
\begin{aligned}
(N_r-n_{1r})\cdot\sum_{n_{1r}=0}^{N_r} q^{RMP}_{(n_{1r},N_r)} &= N_r\cdot q^{RMP}_{(0,N_r)}+(N_r-1)\cdot q^{RMP}_{(1,N_r)}+\ldots+2\cdot q^{RMP}_{(N_r-2,N_r)}+1\cdot q^{RMP}_{(N_r-1,N_r)}+0\cdot q^{RMP}_{(N_r,N_r)}\\
&= \sum_{n_{1r}=0}^{N_r} n_{1r}\cdot q^{RMP}_{(N_r-n_{1r},N_r)}\\
&= N_r\cdot F\left(\tfrac{s}{1-\theta}\right)
\end{aligned}
$$

and

$$
\begin{aligned}
\sum_{n_{1o}=0}^{N_o} q^{RMP}_{(n_{1o},N_o)} &= q^{RMP}_{(0,N_o)}+q^{RMP}_{(1,N_o)}+\ldots+q^{RMP}_{(N_o-1,N_o)}+q^{RMP}_{(N_o,N_o)}\\
&= \sum_{n_{1o}=0}^{N_o} C_{N_o}^{n_{1o}}\cdot\left[F(\tfrac{s}{1-\theta})\right]^{n_{1o}}\cdot\left[1-F(\tfrac{s}{1-\theta})\right]^{N_o-n_{1o}}\\
&= \left[F(\tfrac{s}{1-\theta})+1-F(\tfrac{s}{1-\theta})\right]^{N_o}=1
\end{aligned}
$$

□

References

1. Besanko, D.; Winston, W.L. Optimal Price Skimming by a Monopolist Facing Rational Consumers. *Manag. Sci.* **1990**, *36*, 555–567. [CrossRef]
2. Li, H.; Yang, X.; Tu, Y.; Peng, T. Two-Period Dynamic versus Fixed-Ratio Pricing Policies under Duopoly Competition. *Math. Probl. Eng.* **2019**, *2019*, 6267952. [CrossRef]
3. Anne, T.; Shaffer, G. Price-Matching Guarantees, Retail Competition, and Product-Line Assortment. *Mark. Sci.* **2009**, *28*, 580–588.
4. Chen, Y.; Narasimhan, C.; Zhang, Z. Research Note: Consumer Heterogeneity and Competitive Price-Matching Guarantees. *Mark. Sci.* **2001**, *20*, 300–314. [CrossRef]
5. Sivakumar, K.; Weigand, R.E. Model of retail price match guarantees. *J. Bus. Res.* **1997**, *39*, 241–255. [CrossRef]
6. McWilliams, B.; Gerstner, E. Offering Low Price Guarantees to Improve Customer Retention. *J. Retail.* **2006**, *82*, 105–113. [CrossRef]
7. Jiang, J.; Kumar, N.; Ratchford, B.T. Price-Matching Guarantees with Endogenous Consumer Search. *Soc. Sci. Electron. Publ.* **2017**, *63*, 3147–3529. [CrossRef]
8. Levin, Y.; McGill, J.; Nediak, M. Optimal Dynamic Pricing of Perishable Items by a Monopolist Facing Strategic Consumers. *Prod. Oper. Manag.* **2010**, *19*, 40–60. [CrossRef]
9. Dasu, S.; Tong, C. Dynamic pricing when consumers are strategic: Analysis of posted and contingent pricing schemes. *Eur. J. Oper. Res.* **2010**, *204*, 662–671. [CrossRef]
10. Chew, E.P.; Lee, C.; Liu, R.; Hong, K.S.; Zhang, A. Optimal dynamic pricing and ordering decisions for perishable products. *Int. J. Prod. Econ.* **2014**, *157*, 39–48. [CrossRef]
11. Prasad, A.; Venkatesh, R.; Mahajan, V. Product bundling or reserved product pricing? price discrimination with myopic and strategic consumers. *Int. J. Res. Mark.* **2015**, *32*, 1–8. [CrossRef]
12. Liu, Q.; Zhang, D. Dynamic Pricing Competition with Strategic Consumers under Vertical Product Differentiation. *Manag. Sci.* **2013**, *59*, 84–101. [CrossRef]
13. Moorthy, K.S. Product and price competition in a duopoly. *Mark. Sci.* **1998**, *7*, 141–168. [CrossRef]
14. Li, S.; Liu, Y.; Bandyopadhyay, S. Network effects in online two-sided market platforms: A research note. *Decis. Support Syst.* **2010**, *49*, 245–249. [CrossRef]
15. Wen, D.; Xiao, T.; Dastani, M. Channel choice for an independent remanufacturer considering environmentally responsible consumers. *Int. J. Prod. Econ.* **2021**, *232*, 107941. [CrossRef]
16. Liu, Z.; Chen, J.; Diallo, C.; Venkatadri, U. Pricing and production decisions in a dual-channel closed-loop supply chain with (re) manufacturing. *Int. J. Prod. Econ.* **2021**, *232*, 107935. [CrossRef]
17. Niu, B.; Bao, J.; Cao, B. Retailer's Make-or-Buy Decision for Remanufactured Products under In-house Yield Uncertainty. *Omega* **2022**, *110*, 102627. [CrossRef]
18. Qiao, H.; Su, Q. Distribution channel and licensing strategy choice considering consumer online reviews in a closed-loop supply chain. *Transp. Res. Part E Logist. Transp. Rev.* **2021**, *151*, 102338. [CrossRef]
19. Ma, P.; Gong, Y.; Mirchandani, P. Trade-in for remanufactured products: Pricing with double reference effects. *Int. J. Prod. Econ.* **2020**, *230*, 107800. [CrossRef]
20. Baghdadi, E.; Shafiee, M.; Alkali, B. Upgrading Strategy, Warranty Policy and Pricing Decisions for Remanufactured Products Sold with Two-Dimensional Warranty. *Sustainability* **2022**, *14*, 7232. [CrossRef]
21. Cai, J.H.; Sun, H.N.; Hu, X.Q.; Zhang, W.Q. How reverse information sharing supports pricing and sales effort decisions: Signaling game-based incentive mechanism design. *Comput. Ind. Eng.* **2023**, *177*, 108992. [CrossRef]
22. Han, Y.; Zheng, H.; Huang, Y.; Li, X. Considering consumers' green preferences and government subsidies in the decision making of the construction and demolition waste recycling supply chain: A stackelberg game approach. *Buildings* **2022**, *12*, 832. [CrossRef]
23. Yang, F.; Wang, M.; Sheng, A. Optimal remanufacturing decisions in supply chains considering consumers' anticipated regret and power structures. *Transp. Res. Part E Logist. Transp. Rev.* **2021**, *148*, 102267. [CrossRef]
24. Kwak, M. Monte Carlo Simulation of the Effect of Heterogeneous Too-Cheap Prices on the Average Price Preference for Remanufactured Products. *Sustainability* **2021**, *13*, 9498. [CrossRef]
25. Shin, J. How Does Free Riding on Customer Service Affect Competition? *Mark. Sci.* **2007**, *26*, 488–503. [CrossRef]
26. Liu, X.; Ling, G.U. The study of pricing strategy in online-offline channel based on the consumer's switching behaviour. *J. Manag. Sci.* **2016**, *29*, 27–36.
27. Hviid, M.; Shaffer, G. Hassle Costs: The Achilles' Heel of Price-Matching Guarantees. *J. Econ. Manag. Strategy* **1999**, *8*, 489–521. [CrossRef]
28. Hess, D.; Gerstner, E. Price-Matching Policies: An Empirical Case. *Manag. Decis. Econ.* **1991**, *12*, 305–315. [CrossRef]
29. Janssen, M.; Parakhonyak, A. Price Matching Guarantees and Consumer Search. *Int. J. Ind. Organ.* **2013**, *31*, 1–11. [CrossRef]
30. Xing, D.; Liu, T. Sales effort free riding and coordination with price match and channel rebate. *Eur. J. Oper. Res.* **2012**, *219*, 264–271. [CrossRef]
31. Chen, J.; Chen, B. When Should the Offline Retailer Implement Price Matching? *Eur. J. Oper. Res.* **2019**, *227*, 996–1009. [CrossRef]

32. Mehra, A.; Kumar, S.; Raju, J.S. Competitive Strategies for Brick-and-Mortar Stores to Counter "Showrooming". *Manag. Sci.* **2018**, *64*, 3076–3090. [CrossRef]
33. Zhao, N.; Wang, Q.; Cao, P.; Wu, J. Dynamic pricing with reference price effect and price-matching policy in the presence of strategic consumers. *J. Oper. Res. Soc.* **2019**, *70*, 2069–2083. [CrossRef]

sustainability

Article

Decision-Making of Cross-Border E-Commerce Platform Supply Chains Considering Information Sharing and Free Shipping

Libin Guo * and Yuxiao Shang

School of Management, Qufu Normal University, Rizhao 276826, China
* Correspondence: guolibin@qfnu.edu.cn

Abstract: For cross-border e-commerce companies with high shipping costs, the existing retailer and the new entrant retailer on the platform are usually concerned with information sharing and free shipping due to the uncertainty of market demand. For this, by establishing a Stackelberg game model between two competing retailers, we analyze the strategy of retailers and explore the business strategies of the cross-border e-commerce platform. The study shows that regarding information-sharing strategies, retailer A's willingness to share information is positively related to initial market potential and negatively related to market competition intensity. Moreover, retailer B is willing to spend higher information costs to purchase information when the necessity of the product is more elevated. As for a free shipping strategy, if the existing retailer offers free shipping, the new entrant retailer should also offer free shipping service to consumers when the initial market potential is larger. Conversely, when the initial market potential is smaller, the retailer's willingness to offer free shipping decreases when the intensity of competition in the market increases. When the market tends to be perfectly competitive, the new entrant retailer will not choose a free shipping strategy, and the platform is most profitable when information sharing and free shipping occur simultaneously. However, when the carrier charges a higher shipping fee to customers, the existing retailer is more profitable when the new entrant does not offer free shipping. Therefore, in order to achieve a win-win situation for all parties, the platform needs to develop appropriate operational strategies to influence the decisions of retailers and carriers. Some numerical experiments are made to test the validity of the model and the effect of the parameters involved in the model.

Keywords: cross-border e-commerce platform; information sharing; free shipping; Stackelberg game

Citation: Guo, L.; Shang, Y. Decision-Making of Cross-Border E-Commerce Platform Supply Chains Considering Information Sharing and Free Shipping. *Sustainability* **2023**, *15*, 3350. https://doi.org/10.3390/su15043350

Academic Editors: Shaojian Qu, Qingguo Bai, Ying Ji and Congjun Rao

Received: 14 January 2023
Revised: 2 February 2023
Accepted: 9 February 2023
Published: 11 February 2023

1. Introduction

E-commerce has become a top priority for many businesses around the world [1]. Developing e-commerce can increase sales and profits and offer the opportunity to overcome obstacles to growth [2]. With the development of internet technology, e-commerce has enabled cross-border transactions and become the fastest-growing industry in the global economy [3]. Global cross-border e-commerce (CBEC) was worth over $780 billion in 2019 [4]. The e-commerce business rapidly grew due to COVID-19. Even the most basic survival products, such as food, turned to the Internet because e-commerce was considered more convenient than offline [5]. Significantly affected by COVID-19, Amazon's net sales in 2020 were $386.06 billion, an increase of 37.6% from $280.52 billion in 2019. Meanwhile, data from the General Administration of Customs show that in 2021, the scale of China's cross-border e-commerce exports reached 1.98 trillion yuan, up 15% year-on-year. The boom in CBEC has dramatically enriched the trading market in various countries. However, ICT, the political and regulatory environment, and human resource influence the development of global e-commerce [6], therefore, supply chain shortages and many other global supply chain issues in the post-epidemic era suggest that cross-border e-commerce will slow down in the near future. How to effectively control and manage the platform logistics and supply chain is a critical issue for an e-commerce platform.

To better serve overseas customers, many retailers offer free shipping services. For example, Xiaomi, a famous Chinese smartphone brand, offers free shipping through the cross-border e-commerce platform AliExpress. Temu, the cross-border e-commerce export platform of Pinduoduo, launches with the slogan "free shipping" prominently. Note that shipping and handling costs are causing 52% of people to abandon online shopping. Thus, free shipping has a more significant impact on buyers than price discounts [7]. Shehu et al. [8] believe that free shipping promotions encourage customers to buy riskier items. Thus, offering free shipping services may be an important marketing decision [9].

E-commerce retailers selling products through the platform will accumulate a large amount of data information on the platform. With the growth of e-commerce, information sharing has a great influence on online sales [10]. In the face of uncertain market demand, retailers can process past sales data through machine learning such as the Apriori-based clustering algorithm [11] and the improved genetic algorithm [12] to make more accurate predictions. For cross-border companies with long distances and high costs, information sharing is likely to influence their free shipping and quantitative decisions. Empirical evidence suggests that only 27% of retailers share POS (Point of Sale) data with other members [13]. How to facilitate information sharing among retailers on the platform is an essential issue for the platform to consider.

For competition in the CBEC platform, Lv [14] studies the revenue management of the CBEC platform and the competition and coordination between platform suppliers and e-retailers through supply chain decision-making. Lu et al. [15] explore consumers' willingness to select CBEC platforms and the factors effecting them. Zha et al. [16] research the information sharing strategies of CBEC platforms and the choices of overseas suppliers regarding direct-mail logistics mode and bonded warehouse logistics mode. Qi et al. [17] and Cao et al. [18] study the willingness of traditional retailers to enter e-commerce platforms, and the willingness of retailers and platforms to share their private information if they do. Jiang et al. [19] study the Amazon platform's choice of sales model in the face of uncertain market demand.

For free shipping in e-commerce, Li et al. [20] research free shipping policies for e-commerce under the add-on item recommendation service. Wang and Li [21] study supply chain decisions in e-commerce considering the cost of logistics services and the "free shipping" strategy. Han et al. [22] study the free shipping policy of CBEC based on consumer decision theory. Hua et al. [7] study the optimal order quantity and optimal retail price for retailers, assuming that suppliers offer free shipping, and find that free shipping can benefit suppliers, retailers, and end customers. Etumnu [23] studies the impact of free shipping on grocery sales on Amazon's CBEC platform and found that products with free shipping garnered more sales. Leng and Becerril-Arreola [24] examine the effect of e-commerce retailers' free shipping strategies on consumer purchase behavior. Rui et al. [25] analyze the impact of the free shipping strategy on supply chain members based on the Stackelberg game, and provide theoretical support for e-retailers and consumers to make optimal decisions. Niu et al. [9] examine the free shipping decisions of two competing retailers in CBEC when relying on the same carrier to transport goods.

As for information sharing, the previous literature primarily focuses on the motivations for vertical information sharing upstream and downstream of the supply chain [26–31]. A few studies address horizontal information sharing motives between competitors. Kirby [32] examines the incentives of firms competing in an oligopolistic industry to share information about an unknown demand when such sharing takes place on the basis of exchange. Niu et al. [33] examine whether incumbent carriers with demand information have an incentive to disclose demand information to new carriers in cross-border logistics. Liu et al. [34] study strategies for retail platforms to share demand information to sellers on the platform. Mukhopadhyay et al. [35] study the information sharing strategies of two independent companies with market demand information under the Stackelberg game and devise a "simple to implement" information sharing scheme under which both firms and the total system are better off. To summarize, the relevant studies are shown in Table 1. Our paper integrates the competition among retailers considering information sharing and free shipping on CBEC platforms, and analyzes the business decision-making of the e-commerce platform.

Table 1. Contribution of this paper and related literature.

	E-Commerce	E-Commerce Platform	Free Shipping	Information Sharing	Competition
Li and Zhang [26]	√			√	√
Mukhopadhyay et al. [35]	√			√	√
Liu et al. [34]	√			√	√
Wang and Li [21]	√		√		√
Rui et al. [25]	√		√		√
Niu et al. [9]	√		√		√
Cao et al. [18]	√	√		√	√
Zha et al. [16]	√	√		√	√
Our paper	√	√	√	√	√

For cross-border e-commerce companies with high shipping costs, due to market demand uncertainty, retailers on the platform usually care about information sharing and free shipping. Based on the above analysis, we study the competition between an existing retailer and a new retailer on the CBEC platform under information asymmetry. The retailer needs to decide whether to share information and whether to provide free shipping. We assume the CBEC platform operates using a marketplace model similar to a revenue-sharing scheme, where the platform receives a commission from each online sale [36]. For this model, we consider the following issues:

(1) Considering the impact of free shipping on consumers' willingness to buy, under what circumstances would retailers choose to offer free shipping services?
(2) Is the retailer with information willing to share it? Is the new entrant willing to obtain information? Is the platform willing to enable information sharing?
(3) Which is an equilibrium strategy for competing retailers' decisions on information sharing, free shipping, and quantity?
(4) How should the CBEC platform develop business strategies to achieve a win–win situation in the marketplace mode?

Business strategies represent drivers and constraints of strategic choice [37]. This study constructs four Stackelberg game supply chain models to address these issues, such as the research approach adopted by Qu et al. [38]. We derive analytical expressions for optimal decision-making and profit in different scenarios [39], investigate the strategies of existing and new entrant retailers in terms of information sharing and free shipping, and further analyze the platform's business strategy.

The remainder of this paper is organized as follows. Section 2 describes the model. Section 3 analyzes the decision preferences of retailers by comparing the equilibrium results under the four scenarios. Section 4 presents the numerical studies. The derivation procedure of the equilibrium results, the proof of propositions, and the equilibrium results of the extended model are listed in the Appendix A.

2. Model Description and Notations

In the information problem, Niu et al. [40] and Xing et al. [41] describe information asymmetry with mean and variance, where the lower (higher) information variance implies more (less) accurate demand forecasts. In the problem of free shipping, Niu et al. [9] construct a game model between two competing retailers from the perspective of consumers' utility and profit functions of supply chain parties to solve the problem of free shipping in cross-border e-commerce. Therefore, our paper considers the maximization of profit for each party in the supply chain, constructs the game model using the inverse demand function derived from the consumer utility function, and describes the information asymmetry by mean and variance. Suppose the e-commerce platform with a retailer (denoted A) and a new entrant retailer (denoted B), the two retailers sell the same product to overseas customers on the same CBEC platform (denoted P) with the marketplace mode, and they ship their products through a common carrier (denoted C). See Figure 1 for the running pattern of the model.

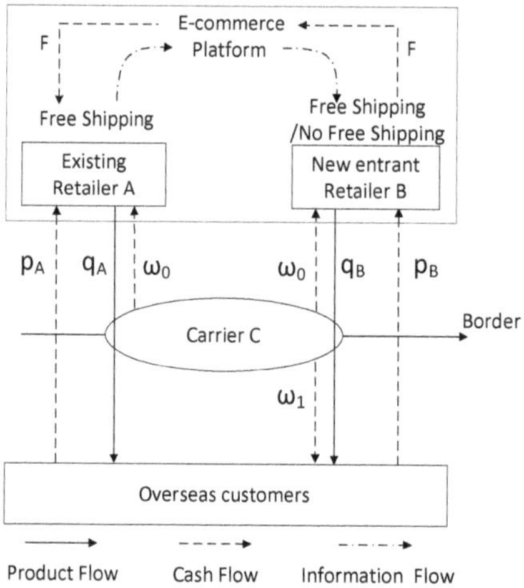

Figure 1. Supply chain structure.

Assume retailer A has been operating on the platform for a long time and has a large amount of sales data, and he can more accurately predict market demand. Retailer B is a newcomer without much information on the market's demand. The platform will create opportunities for information sharing between existing retailers and new entrant retailers to maintain the diversity of the system. If retailer A chooses to share information, retailer B may purchase this information from the platform, and can also decline to buy the information. Certainly, to encourage the existing retailer to share information, the platform will make compensation if the new entrant retailer chooses to purchase sales information. For simplicity, we set the information purchase fee and the amount of compensation to be equal and denote it with F.

Now, we examine the strategic choices of existing retailer A and new entrant retailer B in terms of both information and shipping fees. As for the information interchange, we consider two scenarios: (1) Retailer B obtains information (labeled Scenario Y); (2) Retailer B does not obtain information (labeled Scenario N). As for free shipping, when customers buy products from retailer A, he offers free shipping service. That is, customers need only pay for the products, and the shipping fee is paid by retailer A. For retailer B, she has two options: (1) offering free shipping service like retailer A, denoting this choice as Strategy II; (2) Selling products without free shipping, and the customers need to pay for both product and the shipping service separately. We denote this choice as strategy IN. Combining the information sharing and free shipping strategies, either Strategy II or Strategy IN may occur under both Scenario Y and Scenario N. We further subdivide Scenarios Y and N into four scenarios (Y_II, Y_IN, N_II, N_IN) to describe all possible outcomes.

To proceed, we summarize the notations and descriptions in Table 2 in which subscripts A, B, C, and P are added to refer to solutions for the existing retailer, new entrant retailer, carrier, and platform.

Table 2. Notations and explanations.

Parameters	Meaning
a	Initial market potential
ε	Demand uncertainty
β	Price sensitivity to its own quantity
γ	Price sensitivity to its rivals' quantity
η	Platform's commission rate with marketplace mode
F	Information purchase fee
π_i	The profit of i, where $i = A, B, P$
q_i	Sales quantity of retailers, where $i = A, B$
p_i	Unit product price, where $i = A, B$
ω_0	Unit shipping fee charged to retailers by the carrier
ω_1	Unit shipping fee charged to overseas customers by the carrier

2.1. Demand Function

From the analysis above, we can obtain the following quadratic utility function [42]:

$$U(q_{A,}q_B) = A_A q_A + A_B q_B - \frac{1}{2}\left(a_A q_A^2 + 2bq_A q_B + a_B q_B^2\right) \quad (1)$$

where $A_i(i = A, B)$ denotes the market potential of product i during a period, q_i is the sales amount of product i in the same period, a_i denotes the effect of the quantity of product i on consumers' utility, and parameter b captures the interactions among the products which denotes the measure of product substitutability or complementarity. The consumer utility decreases as products become more substitutable (i.e., when $b > 0$, $U(q_{A,}q_B)$ decreases as b increases). To ensure strict concavity of the utility function and the demand quantities to be positive, we set $A_i > 0$, $a_i > 0$, $a_j A_i - bA_j > 0$, for $i, j = A, B$ and $i \neq j$, $a_A a_B - b^2 > 0$. Assume that products A and B are substitutable, $U(q_A, q_B)$ is symmetric in products A and B. Then we can let $A_A = A_B = a$, $a_A = a_B = \beta$, and the product substitutability $b = r(r > 0)$. As consumers only pay p_A or p_B for the buying product when the retailer offers free shipping, the consumer surplus $R(q_A, q_B)$ under different free shipping strategies can be expressed as consumer utility minus the cost paid [9]:

Strategy II:

$$R(q_{A,}q_B) = aq_A + aq_B - \frac{1}{2}\left(\beta q_A^2 + 2\gamma q_A q_B + \beta q_B^2\right) - p_A q_A - p_B q_B \quad (2)$$

Strategy IN:

$$R(q_{A,}q_B) = aq_A + aq_B - \frac{1}{2}\left(\beta q_A^2 + 2\gamma q_A q_B + \beta q_B^2\right) - p_A q_A - (p_B + \omega_1) q_B \quad (3)$$

where $a > 0, \beta > |\gamma| \geq 0$. As the demand is uncertain, inverse demand functions can be obtained by maximizing the consumer surplus function

Strategy II:

$$p_A^{i\_II} = a + \varepsilon - \beta q_A - \gamma q_B \quad (4)$$

$$p_B^{i\_II} = a + \varepsilon - \beta q_B - \gamma q_A \quad (5)$$

Strategy IN:

$$p_A^{i\_IN} = a + \varepsilon - \beta q_A - \gamma q_B \quad (6)$$

$$p_B^{i\_IN} = a + \varepsilon - \beta q_B - \gamma q_A - \omega_1 \quad (7)$$

for $i \in \{Y, N\}$.

Here, a is the initial market potential, and coefficients β and γ indicate the sensitivity of price to the retailer's own and rival quantities, respectively. We assume that $1 \geq \beta > \gamma > 0$ and denote γ the intensity of competition (a larger γ indicates a more competitive market).

We use the random variable ε to describe the market demand that obeys the normal distribution with $E[\varepsilon] = 0$ and $Var[\varepsilon] = \sigma^2$. Retailer A has the data information, so its demand forecast is $\Gamma = \varepsilon + \varepsilon_1$. If retailer A shares information with retailer B, retailer B's demand forecast is also $\Gamma = \varepsilon + \varepsilon_1$, where ε_1 an independent random variable that follows the normal distribution with $\varepsilon_1 \sim N(0, \sigma_1^2)$, $\sigma^2 > \sigma_1^2$, and $\frac{\sigma^2}{\sigma^2+\sigma_1^2} \in \left(\frac{1}{2}, 1\right)$. According to the discussion by Roy [43] and Niu et al. [40], we can obtain

$$E[\varepsilon|\Gamma] = \frac{\sigma^2 \Gamma}{\sigma^2 + \sigma_1^2} = \frac{\sigma^2(\varepsilon + \varepsilon_1)}{\sigma^2 + \sigma_1^2}, \tag{8}$$

$$V[\varepsilon|\Gamma] = \frac{\sigma^2 \sigma_1^2}{\sigma^2 + \sigma_1^2}. \tag{9}$$

Certainly, when the demand signal is obtained, the variance of the demand forecast becomes smaller, namely, $V[\varepsilon|\Gamma] < \sigma^2$, and the accuracy of the forecast increases.

2.2. Events and Profit Functions

According to Figure 1, we can construct a Stackelberg game to study the supply chain decision as shown in Figure 2 for the sequence of game occurrence. In the first stage, new entrant retailer B chooses whether to purchase demand information from the platform. In the second stage, if retailer B chooses to buy the information, retailer A should decide whether to share the demand information or not. In the third stage, retailer B chooses whether to offer free shipping services. In the fourth stage, retailer A decides the optimal order quantity, and in the fifth stage, retailer B determines the optimal order quantity. Finally, when demand is realized, the market is cleared.

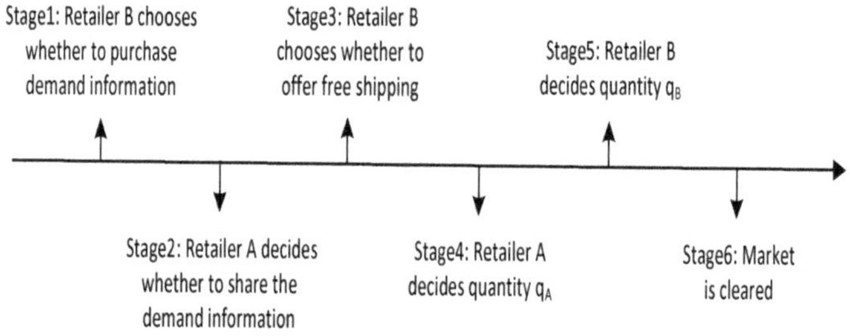

Figure 2. Sequence of events.

With uncertain market demand, the expected profits of retailer A, retailer B, and platform P under different scenarios depend on the demand forecast. Suppose that carrier C transports goods from multiple manufacturers, and its profit is independent of the market demand forecast. We maximize the profits of retailers A and B, respectively, and solve this game using backward induction. The equilibrium results of each situation are summarized in Table A1 in Appendix A (the process of proof is also in Appendix A).

Scenario Y_II: In scenario Y_II, retailer B obtains the information, and at this time, both retailer A and B have the information. Then $\Gamma = \varepsilon + \varepsilon_1$. In the marketplace mode, the platform provides online marketplace services to retailers. Retailers are required to pay the platform for the services in the form of a commission rate η $(1 > \eta > 0)$ [36]. The commission base charged by the platform usually does not include shipping fees. We disregard other fees for now and assume that the platform's profit comes only from the service fees paid by retailers. The expected profits of retailer A, retailer B, and platform P are as follows

$$E\left[\pi_A^{Y\_II}\middle|\Gamma\right] = (1-\eta)E[q_A(p_A-\omega_0)|\Gamma] + F, \Gamma = \varepsilon + \varepsilon_1 \tag{10}$$

$$E\left[\pi_B^{Y\_II}\middle|\Gamma\right] = (1-\eta)E[q_B(p_B-\omega_0)|\Gamma] - F, \Gamma = \varepsilon + \varepsilon_1 \tag{11}$$

$$E\left[\pi_P^{Y\_II}\right] = \eta(E[q_A(p_A-\omega_0)|\Gamma] + E[q_B(p_B-\omega_0)|\Gamma]), \Gamma = \varepsilon + \varepsilon_1 \tag{12}$$

Scenario Y_IN: In this scenario, retailer B obtains the information and chooses the strategy of no free shipping. Retailer A always offers free shipping, and for every purchase from retailer A, the retailer pays a unit shipping fee ω_0 to carrier C. But when a customer purchases from retailer B, retailer B does not have to pay the shipping fee, and the unit shipping fee ω_1 needs to be paid by the customer to the carrier. The expected profits of the supply chain parties are as follows

$$E\left[\pi_A^{Y\_IN}\middle|\Gamma\right] = (1-\eta)E[q_A(p_A-\omega_0)|\Gamma] + F, \Gamma = \varepsilon + \varepsilon_1 \tag{13}$$

$$E\left[\pi_B^{Y\_IN}\middle|\Gamma\right] = (1-\eta)E[q_B p_B|\Gamma] - F, \Gamma = \varepsilon + \varepsilon_1 \tag{14}$$

$$E\left[\pi_P^{Y\_IN}\right] = \eta(E[q_A(p_A-\omega_0)|\Gamma] + E[q_B p_B|\Gamma]), \Gamma = \varepsilon + \varepsilon_1 \tag{15}$$

Scenario N_II: In this scenario, retailer B does not obtain the market demand information, and only retailer A has this information. For demand uncertainty, retailer A's forecast is $\Gamma = \varepsilon + \varepsilon_1$. When retailer B chooses free shipping, the expected profits of the supply chain parties are as follows

$$E\left[\pi_A^{N\_II}\middle|\Gamma\right] = (1-\eta)E[q_A(p_A-\omega_0)|\Gamma], \Gamma = \varepsilon + \varepsilon_1 \tag{16}$$

$$E\left[\pi_B^{N\_II}\right] = (1-\eta)q_B(p_B-\omega_0) \tag{17}$$

$$E\left[\pi_P^{N\_II}\right] = \eta(E[q_A(p_A-\omega_0)|\Gamma] + q_B(p_B-\omega_0)), \Gamma = \varepsilon + \varepsilon_1 \tag{18}$$

Scenario N_IN: For this scenario, retailer B doesn't obtain the information and chooses no free shipping, the expected profits of the supply chain parties are as follows

$$E\left[\pi_A^{N\_IN}\middle|\Gamma\right] = (1-\eta)E[q_A(p_A-\omega_0)|\Gamma], \Gamma = \varepsilon + \varepsilon_1 \tag{19}$$

$$E\left[\pi_B^{N\_IN}\right] = (1-\eta)q_B p_B \tag{20}$$

$$E\left[\pi_P^{N\_IN}\right] = \eta(E[q_A(p_A-\omega_0)|\Gamma] + q_B p_B), \Gamma = \varepsilon + \varepsilon_1 \tag{21}$$

3. Theoretical Analysis of the Model

Now, we analyze the retailers' strategic choices by comparing the performance of two retailers under four scenarios. To guarantee positive equilibrium outcomes, we require the following two feasible regions

(a) $E[\varepsilon|\Gamma] > 0$: $E[\varepsilon|\Gamma] > \frac{4\beta^2\omega_1-2\beta\gamma\omega_0-\gamma^2\omega_1}{4\beta^2-2\beta\gamma-\gamma^2}$, $a > \frac{4\beta^2\omega_1-2\beta\gamma\omega_0-\gamma^2\omega_1}{4\beta^2-2\beta\gamma-\gamma^2}$;

(b) $E[\varepsilon|\Gamma] < 0$: $\frac{8\beta^3\omega_0-8\beta^3\omega_1-4\beta^2\gamma\omega_0+4\beta^2\gamma\omega_1-2\beta\gamma^2\omega_0+2\beta\gamma^2\omega_1+\gamma^3\omega_0-\gamma^3\omega_1}{4\beta^2\gamma-2\beta\gamma^2-\gamma^3} > E[\varepsilon|\Gamma] > \frac{-2\beta\omega_0-\gamma\omega_1}{2\beta-\gamma}$,

$\frac{1+\sqrt{3}}{2}\gamma < \beta \leq 1, 0 < \gamma < \sqrt{3}-1, a > \frac{4\beta^2\omega_1 - 2\beta\gamma\omega_0 - \gamma^2\omega_1}{4\beta^2 - 2\beta\gamma - \gamma^2} - E[\varepsilon|\Gamma]$.

The two feasible regions satisfy the following condition

$F < \frac{(1-\eta)\left(a\left(4\beta^2-2\beta\gamma-\gamma^2\right)+\omega_1\left(\gamma^2-4\beta^2\right)+2\beta\gamma\omega_0\right)^2}{16\beta\left(\gamma^2-2\beta^2\right)^2} + \frac{(1-\eta)\left(-4\beta^2+2\beta\gamma+\gamma^2\right)^2}{16\beta\left(\gamma^2-2\beta^2\right)^2} \times \frac{\sigma^4}{\sigma^2+\sigma_1^2}$.

Proposition 1. *For the optimal quantities and prices for retailers, it holds that*

(a) Quantity: (a_1) $q_A^{Y\_II} > q_B^{Y\_II}$. *(a_2)* $q_A^{Y\_IN} > q_B^{Y\_IN}$. *(a_3)* $q_A^{N\_II} > q_B^{N\_II}$ *if* $\varepsilon+\varepsilon_1 > 0$, *or* $\varepsilon+\varepsilon_1 < 0$, $a > \omega_0 - \frac{4\beta^2}{\gamma^2}(\varepsilon+\varepsilon_1)$; *otherwise* $q_A^{N\_II} < q_B^{N\_II}$. *(a_4)* $q_A^{N\_IN} > q_B^{N\_IN}$ *if* $\varepsilon+\varepsilon_1 > 0$, *or* $\varepsilon+\varepsilon_1 < 0$, $a > \frac{4\beta^2\omega_0 - 4\beta^2\omega_1 + 2\beta\gamma\omega_0 - 2\beta\gamma\omega_1 + \gamma^2\omega_1}{\gamma^2} - \frac{4\beta^2}{\gamma^2}(\varepsilon+\varepsilon_1)$; *otherwise* $q_A^{N\_IN} < q_B^{N\_IN}$.

(b) Price: (b_1) $p_A^{Y\_II} < p_B^{Y\_II}$. *(b_2)* $p_A^{Y\_IN} > p_B^{Y\_IN}$ *if* $0 < \varepsilon+\varepsilon_1 < H$, $a < H + \frac{\gamma^3-\beta\gamma^2}{\beta\gamma^2-\gamma^3}(\varepsilon+\varepsilon_1)$; *otherwise* $p_A^{Y\_IN} < p_B^{Y\_IN}$. *(b_3)* $p_A^{N\_II} > p_B^{N\_II}$ *if* $\varepsilon+\varepsilon_1 > 0$, $a < \omega_0 + \frac{4\beta^2+4\beta\gamma}{\gamma^2}(\varepsilon+\varepsilon_1)$; *otherwise* $p_A^{N\_II} < p_B^{N\_II}$. *(b_4)* $p_A^{N\_IN} > p_B^{N\_IN}$ *if* $a < H + \frac{4\beta^3-4\beta\gamma^2}{\beta\gamma^2-\gamma^3}(\varepsilon+\varepsilon_1)$; *otherwise* $p_A^{N\_IN} < p_B^{N\_IN}$. *Where* $H = \frac{4\beta^3\omega_0 + 4\beta^3\omega_1 - 2\beta^2\gamma\omega_0 + 2\beta^2\gamma\omega_1 - 2\beta\gamma^2\omega_0 - \beta\gamma^2\omega_1 - \gamma^3\omega_1}{\beta\gamma^2-\gamma^3}$.

Proposition 1 summarizes the comparison of equilibrium prices and quantities between two retailers. In particular, Proposition 1 (a) states that the quantitative decision of retailer A is always greater than that of retailer B in the case where retailer B obtains the information, regardless of whether retailer B chooses the free shipping strategy. In the absence of information available to retailer B, retailer A's quantitative decision is greater than retailer B's when retailer A's demand forecast is positive, or the initial market potential is greater than a specific value. Proposition 1 (b_1) says that when retailer B obtains information and offers free shipping, retailer B has a higher retail price than retailer A. This is different from our perception. The possible reason is that retailer B has the same information advantage and free shipping advantage as retailer A after purchasing information and choosing free shipping, but under Stackelberg competition, retailer A has a larger quantity of demand than retailer B, so retailer B will prefer a higher retail price to increase revenue. However, retailer B does not have the advantage of information or free shipping when it does not offer free shipping or does not obtain the information. At an initial market potential less than a particular value, retailer B chooses a lower retail price to increase his competitiveness with a price advantage and obtain a greater demand quantity (Proposition 1 (b_2–b_4)).

Proposition 2. *For the equilibrium market factors for retailers under different scenarios, it holds that*

(a) The equilibrium quantity of retailer A satisfies that

$$\begin{cases} q_A^{Y\_II} < q_A^{N\_II}, & \varepsilon+\varepsilon_1 > 0, \\ q_A^{Y\_II} > q_A^{N\_II}, & \text{otherwise}; \end{cases} \quad \begin{cases} q_A^{Y\_IN} < q_A^{N\_IN}, & \varepsilon+\varepsilon_1 > 0, \\ q_A^{Y\_IN} > q_A^{N\_IN}, & \text{otherwise}; \end{cases} \quad q_A^{Y\_II} < q_A^{Y\_IN}; q_A^{N\_II} < q_A^{N\_IN}.$$

(b) The equilibrium quantity of retailer B satisfies that

$$\begin{cases} q_B^{Y\_II} > q_B^{N\_II}, & \varepsilon+\varepsilon_1 > 0, \\ q_B^{Y\_II} < q_B^{N\_II}, & \text{otherwise}; \end{cases} \quad \begin{cases} q_B^{Y\_IN} > q_B^{N\_IN}, & \varepsilon+\varepsilon_1 > 0; \\ q_B^{Y\_IN} < q_B^{N\_IN}. & \text{otherwise}; \end{cases} \quad q_B^{Y\_II} > q_B^{Y\_IN};\ q_B^{N\_II} > q_B^{N\_IN}.$$

(c) The equilibrium price of retailer A is

$$\begin{cases} p_A^{Y\_II} > p_A^{N\_II}, & \varepsilon+\varepsilon_1 > 0; \sqrt{3}\beta-\beta < \gamma < \beta \text{ or } \varepsilon+\varepsilon_1 < 0, 0 < \gamma < \sqrt{3}\beta-\beta; \\ p_A^{Y\_II} < p_A^{N\_II}, & \text{otherwise}; \end{cases}$$

$$\begin{cases} p_A^{Y\_IN} > p_A^{N\_IN}, & \varepsilon+\varepsilon_1 > 0, \sqrt{3}\beta-\beta < \gamma < \beta \text{ or } \varepsilon+\varepsilon_1 < 0, 0 < \gamma < \sqrt{3}\beta-\beta; \\ p_A^{Y\_IN} < p_A^{N\_IN}, & \text{otherwise}; \end{cases}$$

$$p_A^{Y\_II} < p_A^{Y\_IN};\ p_A^{N\_II} < p_A^{N\_IN}.$$

(d) The equilibrium price of retailer B satisfies that

$$\begin{cases} p_B^{Y\_II} > p_B^{N\_II}, & \varepsilon+\varepsilon_1 > 0; \\ p_B^{Y\_II} < p_B^{N\_II}, & \text{otherwise}; \end{cases} \quad \begin{cases} p_B^{Y\_IN} > p_B^{N\_IN}, & \varepsilon+\varepsilon_1 > 0; \\ p_B^{Y\_IN} < p_B^{N\_IN}, & \text{otherwise}; \end{cases} \quad p_B^{Y\_II} > p_B^{Y\_IN}; p_B^{N\_II} > p_B^{N\_IN}.$$

Regardless of whether retailer B chooses the free shipping strategy, as long as the signal shows positive demand, retailer A always makes more of a minor quantitative decision when retailer B obtains the information than when retailer B does not obtain the information. This is because retailer B makes a greater quantitative decision after receiving the information. Therefore, retailer A's quantitative determination usually decreases to keep the market functioning, but the retail price increases. Retailer A's quantitative and price decisions are greater when retailer B chooses no free shipping. Thus, retailer A expects retailer B to decide not to offer free shipping. However, retailer B has greater demand quantity and retail price under the free shipping strategy than when free shipping is not offered. To find the best solution for the system, we continue the discussion in the later Propositions.

Proposition 3. *For the signal accuracy's influences on equalization, it holds that*

(a) In scenario Y (a₁) $q_i^{Y\_Ij} = M_i^{Y\_Ij} + N_i^{Y\_Ij}(\varepsilon+\varepsilon_1)$, *where* $i \in \{A,B\}, j \in \{I,N\}, N_i^{Y\_Ij}$ *improves the accuracy of demand forecasting. (a₂)* $p_i^{Y\_Ij} = m_i^{Y\_Ij} + n_i^{Y\_Ij}(\varepsilon+\varepsilon_1)$, *where* $i \in \{A,B\}$, $j \in \{I,N\}$, $n_i^{Y\_Ij}$ *improves the accuracy of demand forecasting. (a₃)* $E\left[\pi_i^{Y\_Ij}\middle|\Gamma\right] = M_i^{Y\_Ij} + N_i^{Y\_Ij}(\frac{\sigma^4}{\sigma^2+\sigma_1^2})$, *where* $i \in \{A,B,P\}, j \in \{I,N\}, N_i^{Y\_Ij}$ *is decreasing in the competition degree.*

(b) In scenario N (b₁) $q_A^{N\_Ij} = M_A^{N\_Ij} + N_A^{N\_Ij}(\varepsilon+\varepsilon_1)$, *where* $j \in \{I,N\}, N_A^{N\_Ij}$ *is increasing in the accuracy of the forecast. (b₂)* $p_A^{N\_Ij} = m_A^{N\_Ij} + n_A^{N\_Ij}(\varepsilon+\varepsilon_1)$, *where* $j \in \{I,N\}$, $n_A^{N\_Ij}$ *is increasing in the accuracy of the forecast. (b₃)* $E\left[\pi_i^{N\_Ij}\middle|\Gamma\right] = M_i^{N\_Ij} + N_i^{N\_Ij}(\frac{\sigma^4}{\sigma^2+\sigma_1^2})$, *where* $i \in \{A,P\}, j \in \{I,N\}, N_i^{N\_Ij}$ *is decreasing in the competition degree.*

When there is demand information, we rewrite retailers' demand quantities and retail prices as linear functions of $\varepsilon+\varepsilon_1$ and find that their quantitative and price decisions are positively correlated with the accuracy of the information. And we rewrite retailers' profits as linear functions of $\frac{\sigma^4}{\sigma^2+\sigma_1^2}$. (a₃) and (b₃) of Proposition 3 express the slope of $E\left[\pi_i^{Y\_Ij}\middle|\Gamma\right]$ and $E\left[\pi_i^{N\_Ij}\middle|\Gamma\right]$ are decreasing in the cross-price elasticity (cross-price elasticity means the degree of competition). For example, $E\left[\pi_A^{Y\_II}\middle|\Gamma\right] = M_A^{Y\_II} + N_A^{Y\_II}(\frac{\sigma^4}{\sigma^2+\sigma_1^2})$, where $N_A^{Y\_II} = \frac{(1-\eta)(\gamma-2\beta)^2}{16\beta^3-8\beta\gamma^2}$, $N_A^{Y\_II}$ on monotonic decreasing of γ (see Appendix A for other proofs). Interestingly, the more competitive the demand, the less responsive the retailer's profit is to the accuracy of the demand signal. This is because the more intense the competition, the more sensitive the price is to demand. Retailers will suffer greater losses in fierce competition due to price wars. This finding implies that retailer A's incentive to share information diminishes if demand competition is intense. High competitive pressure and lack of information sharing can also directly affect the profitability of the platform. The platform needs to intervene and adopt specific ways to solve the problem of high competition intensity (such as raising the threshold for platform entry), facilitating the realization of information sharing, and improving the system's overall revenue.

Proposition 4. *Comparison of the supply chain members' profits under different scenarios.*

(a) Retailer A

(a₁) $E\left[\pi_A^{Y\_II}\middle|\Gamma\right] > E\left[\pi_A^{N\_II}\middle|\Gamma\right]$ *if* $F > K_1 \times \frac{\sigma^4}{\sigma^2+\sigma_1^2}$.

$(a_2)\ E\left[\pi_A^{Y\_IN}\middle|\Gamma\right] > E\left[\pi_A^{N\_IN}\middle|\Gamma\right]$ *if* $F > K_1 \times \frac{\sigma^4}{\sigma^2+\sigma_1^2}$.
$(a_3)\ E\left[\pi_A^{Y\_II}\middle|\Gamma\right] < E\left[\pi_A^{Y\_IN}\middle|\Gamma\right]$ *if* $\omega_1 > \frac{4\beta\omega_0-\gamma\omega_0}{\gamma}$.
$(a_4)\ E\left[\pi_A^{N\_II}\middle|\Gamma\right] < E\left[\pi_A^{N\_IN}\middle|\Gamma\right]$ *if* $\omega_1 > \frac{4\beta\omega_0-\gamma\omega_0}{\gamma}$.
(b) Retailer B
$(b_1)E\left[\pi_B^{Y\_II}\middle|\Gamma\right] > E\left[\pi_B^{N\_II}\middle|\Gamma\right]$ *if* $F < K_2 \times \frac{\sigma^4}{\sigma^2+\sigma_1^2}$.
$(b_2)E\left[\pi_B^{Y\_IN}\middle|\Gamma\right] > E\left[\pi_B^{N\_IN}\middle|\Gamma\right]$ *if* $F < K_2 \times \frac{\sigma^4}{\sigma^2+\sigma_1^2}$.
$(b_3)E\left[\pi_B^{Y\_II}\middle|\Gamma\right] > E\left[\pi_B^{Y\_IN}\middle|\Gamma\right]$ *if* $a > K_3$.
$(b_4)E\left[\pi_B^{N\_II}\middle|\Gamma\right] > E\left[\pi_B^{N\_IN}\middle|\Gamma\right]$ *if* $a > K_3$.
where $K_1 = \frac{\gamma(1-\eta)\left(8\beta^3-6\beta^2\gamma-4\beta\gamma^2+\gamma^3\right)}{8\beta(\gamma^2-2\beta^2)^2}$, $K_2 = \frac{(1-\eta)\left(-4\beta^2+2\beta\gamma+\gamma^2\right)^2}{16\beta(\gamma^2-2\beta^2)^2}$, $K_3 = \frac{4\beta^2\omega_0+4\beta^2\omega_1-4\beta\gamma\omega_0-\gamma^2\omega_0-\gamma^2\omega_1}{8\beta^2-4\beta\gamma-2\gamma^2}$.

Regarding the information strategy, when the information compensation fee satisfies that $K_1 < F < K_2$, retailer A is willing to share information, and retailer B is willing to buy information. In terms of the free shipping strategy, when the initial market potential is high, retailer B offers a free shipping strategy. However, when the carrier charges a higher shipping fee to customers, retailer A expects retailer B to refrain from offering a free shipping strategy to maintain its free shipping advantage.

4. Numerical Analysis

To test the reasonableness and validity of the decision in different scenarios, we refer to the research of Bai et al. [44], verify the above propositions through numerical experiments, and analyze the research questions from the perspective of retailers and the platform, respectively.

4.1. Profit Analysis for Retailers

With the guarantee that all variables are positive, the relevant parameters can be set as initial market potential $a = 30$, information purchase fee and compensation fee $F = 20$, unit shipping fee charged by the carrier to retailers $\omega_0 = 5$, unit shipping fee charged by the carrier to overseas customers $\omega_1 = 6$, the commission rate charged by the platform $\eta = 0.02$, forecast of market demand $E[\varepsilon|\Gamma] = 10$ and price sensitivity to its own quantity $\beta = 1$. Since β is set to be greater than γ, we let price sensitivity to its rivals' quantity $\gamma = 0.4$, and the value range of price sensitivity to its own quantity β can be limited to the range $[0.5, 1]$.

Figure 3a shows retailer A can obtain the biggest profit when retailer B obtains the information and chooses the strategy of no free shipping. Moreover, retailer A has a larger profit when retailer B obtains the information regardless of whether retailer B chooses free shipping. However, from Figure 3b, retailer B's profit from offering free shipping is always greater than the profit from not offering, regardless of whether it gets information. Therefore, retailer B prefers to provide free shipping services. Concerning retailer B's willingness to obtain the information, we find that when the information purchase fee $F = 20$, β is located in area a, whether or not free shipping is chosen, the profit when retailer B gets the information is always greater than the profit when it does not. However, when the price has a great influence on its own quantity, that is, β is located in area b, retailer B will choose not to buy the information. We can speculate that retailer B is willing to spend a higher information cost to purchase information when the degree of necessity of the product is more elevated.

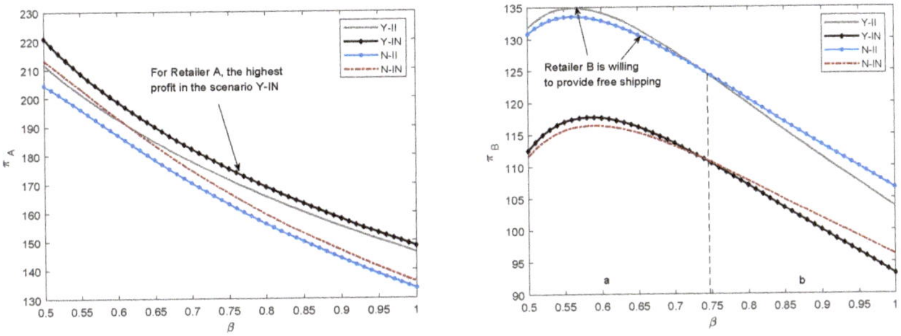

(a) **(b)**

Figure 3. The changes in retailer A and B's profit with respect to β under four scenarios. (**a**) Retailer A. (**b**) Retailer B.

We can see that retailer B is uncertain about whether to purchase information when the information purchase fee is large. Proposition 4 states that the information purchase fee has a greater impact on retailer B's decision to purchase information. We change the fee of purchasing information to $F = 5$, and obtain Figure 4. As shown from Figure 4a,b, regardless of whether retailer B chooses free shipping, the profit of obtaining information is always greater than that of not obtaining information. Therefore, when the information purchase fee is small, retailer B ignores the effect of product necessity and purchases the information. Moreover, reducing the information purchase fee does not affect retailer B's choice of free shipping decision (see Figure 4c,d). From this, we can see that the information purchase fee has an important influence on retailer B's decision on whether to buy the information. That is, when the price is more sensitive to its own quantity (the necessity of the product is low), the platform needs to control the information purchase fee to enable information sharing.

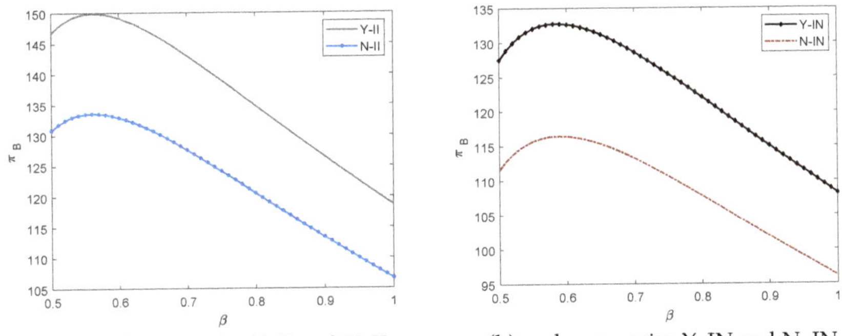

(**a**) under scenarios Y_II and N_II (**b**) under scenarios Y_IN and N_IN

Figure 4. *Cont.*

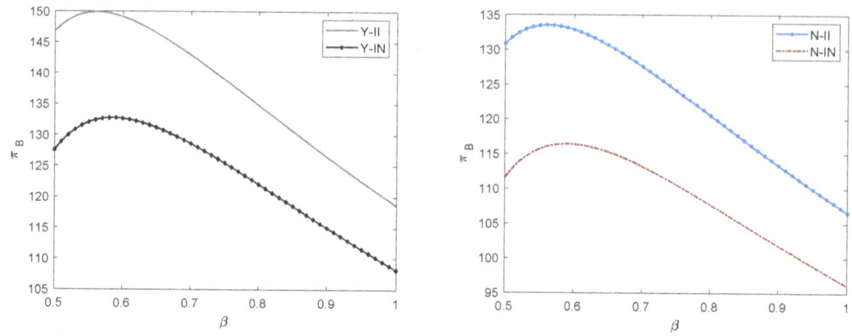

(**c**) under scenarios Y_II and Y_IN (**d**) under scenarios N_II and N_IN

Figure 4. Profits comparison of retailer B under different scenarios ($F = 5$).

4.2. Strategic Choices for Retailers

We observe the effects of initial market potential and competitive intensity on retailer A's willingness to share information, and draw Figure 5. We define the value ranges of $E[\varepsilon|\Gamma]$ and F in the graph to ensure that all variables are positive. Let $F < \frac{(1-\eta)\left(a\left(4\beta^2-2\beta\gamma-\gamma^2\right)+\omega_1\left(\gamma^2-4\beta^2\right)+2\beta\gamma\omega_0\right)^2}{16\beta(\gamma^2-2\beta^2)^2} + \frac{(1-\eta)\left(-4\beta^2+2\beta\gamma+\gamma^2\right)^2}{16\beta(\gamma^2-2\beta^2)^2} \times \frac{\sigma^4}{\sigma^2+\sigma_1^2} = F^+$, and draw the maximum value of F desirable in the graph. Consistent with the above variable settings, let the unit shipping fee charged by the carrier to retailers remain $\omega_0 = 5$, the unit shipping fee charged to overseas customers $\omega_1 = 6$, the commission rate charged by the platform $\eta = 0.02$, and let price sensitivity to its own quantity $\beta = 0.7$.

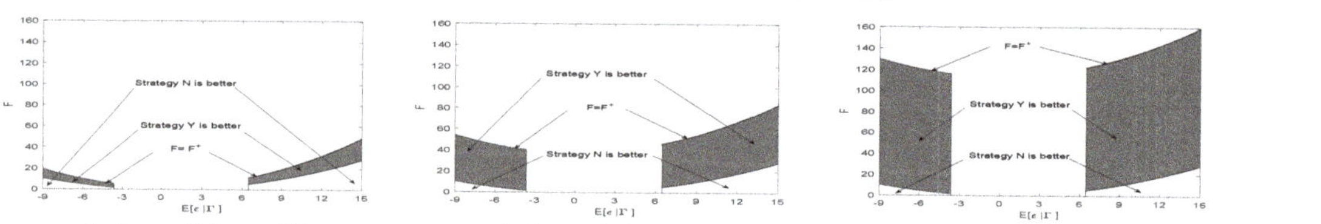

(**a$_1$**) $\gamma = 0.4, a = 10$ (**a$_2$**) $\gamma = 0.4, a = 20$ (**a$_3$**) $\gamma = 0.4, a = 30$

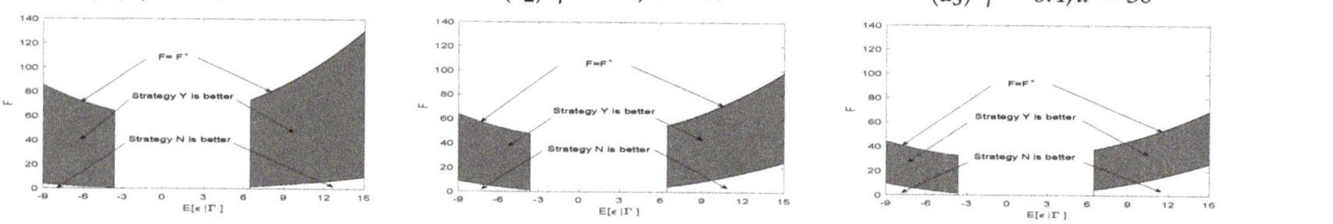

(**b$_1$**) $a = 20, \gamma = 0.1$ (**b$_2$**) $a = 20, \gamma = 0.3$ (**b$_3$**) $a = 20, \gamma = 0.5$

Figure 5. Retailer A's preference over sharing information strategy.

We observe the effect of initial potential changes in the market and changes in competitive intensity on the strategic choices of retailers A (see Figure 5). The greater the initial market potential, the greater the incentive for retailer A to share information. However, the effect of competition intensity on retailer A's desire to share information is negative, the bigger the competition intensity, the less retailer A's incentive to share information desire.

Next, we analyze retailer B's free shipping strategy. With $\omega_0 = 5, \omega_1 = 6, 1 \geq \beta > \gamma > 0$, according to Proposition 4 (b$_3$–b$_4$), we analyze the cases of $a < K_3$ and $a > K_3$ separately. In the range of definition, we let $a = 6$ and $a = 30$, respectively. Under the basic assumption that $\beta > \gamma$ is satisfied, we analyze retailer B's preference by drawing Figure 6. It shows retailer B's choice of free shipping strategy. We find that the probability of retailer B choosing free shipping is greater than the probability of selecting no free shipping when the initial market potential $a < K_3$. And when the competitive intensity of the market γ is located in area m, the probability of retailer B choosing not to have free shipping increases with the competitive intensity. Still, when the competitive intensity is greater (γ is located in area n), retailer B only chooses no free shipping.

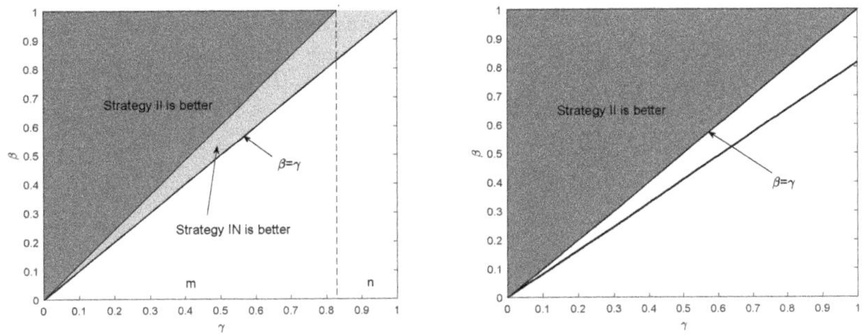

(a) (b)

Figure 6. Retailer B's preference over free shipping strategy. (**a**) $a < K_3$. (**b**) $a > K_3$.

When the initial market potential $a > K_3$, the shape of the graph remains the same regardless of the change in parameters, so retailer B always chooses the free shipping strategy. The possible reason is that when the initial market potential is small, retailer B has just entered the market and is less competitive. Faced with greater competitive pressure, retailer B is unwilling to risk high shipping fees. However, when the initial market potential is larger, the advantage of free shipping will increase the demand quantities, and then retailer B will choose to offer free shipping to be more competitive and maximize its profit.

4.3. Platform Numerical Analysis

The number of sales on the platform is an essential indicator of e-commerce platform operations. To analyze the impact of market competition on the overall number of platform sales under different market forecasts, we set basic market potential $a = 30$, price sensitivity to its own quantity $\beta = 1$, unit shipping fee charged by the carrier to retailers $\omega_0 = 5$, unit shipping fee charged by the carrier to overseas customers $\omega_1 = 6$, and when the market demand forecast is positive, predicted value $E[\varepsilon|\Gamma] = 10$, when it is negative, predicted value $E[\varepsilon|\Gamma] = -5$. The intensity of market competition γ can be limited to a range $(0, 0.7)$. Add the equilibrium quantities of retailers A and B in each case, $Q = q_A^{i\_Ij} + q_B^{i\_Ij}$, where $i \in [Y, N], j \in [I, N]$. Substituting the above parameters into Q. As γ increases gradually in the range of values, the total number of sales will change accordingly. We compare the total number of deals on the platform in four cases. To observe the law of substantial change, the results are mapped to graphs to analyze the influence of competition intensity on the total number of sales under different market forecasts.

From Figure 7, we can see that the overall number of sales on the platform decreases as the intensity of competition increases. Further, the number of sales is greater when the forecast for the market is positive than when it is negative. Figure 7a shows the results for an uncertain market with a positive demand forecast. While the market competition is small (γ is located in area a), the total number of sales on the platform at the time retailer B obtains the information is greater. When the competition intensity is in the middle region (γ is located in area b), the platform has a greater total number of sales when retailer B chooses free shipping. And when the market competition is strong (γ is located in area c), the platform's total number of sales when retailer B chooses free shipping is still greater than when shipping is not free. Still, the platform's total number of sales is larger when retailer B does not obtain the information. This is because, under fierce competition, retailer B will reduce its quantitative decisions based on the principle of risk aversion if market demand information is known. Figure 7b shows the results when the market demand forecast is negative. Interestingly, when there is less competition, the platform sells more when retailer B does not obtain the information. The reason is that here the market demand forecast for uncertainty is negative. After getting information, retailer B naturally makes fewer quantitative decisions. However, when the competitive pressure is particularly high (γ is located in the c area), the total platform sales are greater at the time retailer B gets the information. The possible reason for this is that retailer A will reduce its quantitative decision when faced with greater market pressure and a negative market forecast. Under the Stackelberg game, retailer B's quantitative decision will decrease with retailer A. If retailer B knows the demand information, it may increase part of the quantitative decision when retailer A reduces.

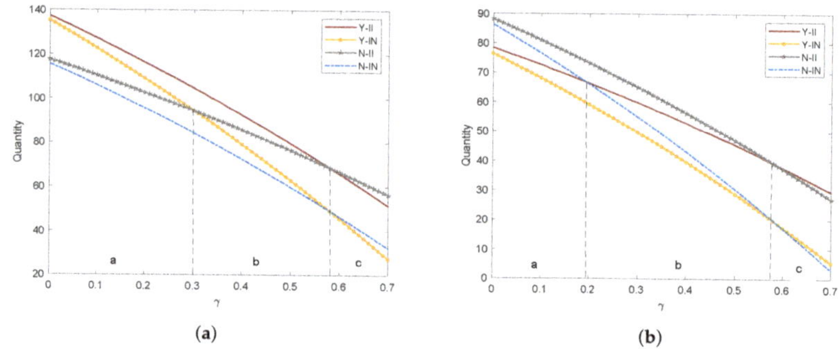

Figure 7. The size of overseas market comparison of four scenarios under different market forecasts. (**a**) $(\varepsilon + \varepsilon_1) > 0$. (**b**) $(\varepsilon + \varepsilon_1) < 0$.

The intensity of market competition has a significant impact on the total number of transactions on the platform, and we further analyze the impact of market competition intensity on platform profits. We plot the profits of the platform in different scenarios under the market mode in Figure 8. Because the profits of the platform are drawn from the profits of retailers A and B, analyzing the profits of retailers A and B in Figure 3 together, we find that although the profits of retailer A are the largest in scenario Y_IN, the profits of the platform are the same as the profit profile of retailer B, which is larger in scenario Y_II. To achieve a win–win situation for all parties, the platform needs to develop appropriate operational strategies to influence factors such as initial market potential and market competition intensity, thus indirectly intervening in the strategic choices of retailers.

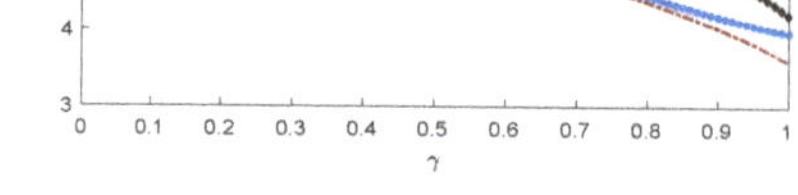

Figure 8. Profit comparison of platform P under four scenarios.

5. Discussion and Conclusions

It is crucial to consider how to operate the CBEC platform supply chain with long distances and high costs. We investigated the decision-making of retailers on CBEC platforms regarding information sharing and free shipping under the marketplace mode, and we examined the willingness of information sharing between the existing retailer and the new entrant retailers. The results obtained using theoretical and numerical analysis show that:

(1) In terms of the information-sharing strategy, we find that the existing retailer's willingness to share information is influenced by initial market potential and competitive intensity. The effect of initial market potential on the willingness to share information is positive. As the initial market potential increases, the willingness of existing retailers to share information also increases. In contrast, the effect of market competition intensity on willingness to share information is negative, i.e., as competition intensity increases, the willingness of existing retailers to share information decreases. The existing retailer will share information when the initial market potential is larger, and the intensity of competition in the market is smaller. In addition, the information fee greatly influences the existing retailer's willingness to share information and the willingness of the new retailer to acquire information. Information sharing can only be realized when the information fee meets certain conditions.

(2) In terms of the free shipping strategy, we find that both initial market potential and competitive intensity affect the choice of the new entrant retailer. When the initial market potential is larger, the new entrant retailer always chooses the free shipping strategy. However, when the initial market potential is smaller, the willingness to choose free shipping decreases as competition intensifies. And when the market tends to be perfectly competitive, the new entrant retailer will decide not to offer free shipping. It is worth noting that if the existing retailer offers free shipping, information sharing will not affect the retailer's free shipping decision in a healthy competitive market environment. The new entrant will always offer free shipping, which is consistent with reality.

From the platform perspective, it is clear that the platform is most profitable when information sharing and free shipping occur simultaneously. For information sharing and free shipping to co-occur, the platform needs to set up reasonable operational strategies to meet the following conditions: the larger initial market potential, the smaller market competition intensity, and the reasonable information fee. Therefore, the platform could seize market share with other platforms by increasing advertising, improving service levels, and enriching products to increase initial market potential and reduce market competition intensity; it can also control the number of e-retailers on the platform and reduce market competition intensity. In addition, while maintaining product diversification, the platform should also consider the attributes of products sold by retailers. When the necessity of the product is higher, the new entrant retailer is willing to spend a higher information cost to purchase information.

The obtained results enrich the theoretical results of platform supply chains by starting from information sharing and free shipping strategies. However, there are still limitations in this paper. First, we set the shipping fee as an exogenous variable that is not generalizable. We consider setting the shipping fee as a decision variable in future studies and considering shipping fee discounts. Second, the service quality of transportation is also a vital issue in CBEC. In the future, we will discuss the competition among retailers by considering the quality of shipping service when free shipping is available. Finally, tax differences are prominent features of cross-border e-commerce. In the future, we will study the competition in the supply chain of B2B e-commerce platforms based on tax planning.

Author Contributions: Conceptualization, L.G. and Y.S.; methodology, Y.S.; software, Y.S.; writing original draft preparation, L.G. and Y.S.; writing review and editing, L.G. and Y.S.; supervision, L.G. All authors have read and agreed to the published version of the manuscript.

Funding: This research was funded by the National Natural Science Foundation of China grant number 12071250, the Natural Science Foundation of Shandong Province grant number ZR2022MG084, the National Statistical Science Research Project grant number 2022LY017, and the Youth Innovation Science and Technology Support Program of Shandong Province Higher Education, grant number 2021RW024.

Institutional Review Board Statement: Not applicable.

Informed Consent Statement: Not applicable.

Data Availability Statement: Not applicable.

Acknowledgments: We greatly appreciate the editor and the anonymous reviewers for their insightful comments and suggestions, which have greatly helped to improve the research.

Conflicts of Interest: The authors declare no conflict of interest.

Abbreviations

The following abbreviations are used in this manuscript:

CBEC Cross-border E-commerce

Appendix A. The Derivation of Equilibriums

We use the solution steps of scenario N_IN as an example to illustrate the process of this game.

In our model, retailers maximize their expected profits through quantity decisions. In the Stackelberg model with retailer A as the leader and retailer B as the follower, we solve this game by reverse induction. The expected profit of members in the supply chain expected profit is given by $E\left[\pi_i^{N\_IN}\right] = \int_{-\infty}^{+\infty}\int_{-\infty}^{+\infty} E\left[\pi_i^{N\_IN}\right] g(\varepsilon,\varepsilon_1)d\varepsilon d\varepsilon_1$.

Since ε and ε_1 are independent of each other, $g(\varepsilon,\varepsilon_1) = g(\varepsilon)g(\varepsilon_1)$. $g(\varepsilon)$ and $g(\varepsilon_1)$ are the normal probability density functions of ε and ε_1, respectively. $g(\varepsilon,\varepsilon_1)$ is a binary normal probability density function.Because of $E[\varepsilon|\Gamma] = \frac{\sigma^2\Gamma}{\sigma^2+\sigma_1^2} = \frac{\sigma^2(\varepsilon+\varepsilon_1)}{\sigma^2+\sigma_1^2}$, we have

$\int_{-\infty}^{+\infty}\int_{-\infty}^{+\infty} E[\varepsilon|\Gamma]g(\varepsilon,\varepsilon_1)d\varepsilon d\varepsilon_1 = 0$. $\int_{-\infty}^{+\infty}\int_{-\infty}^{+\infty} E[\varepsilon|\Gamma]^2 g(\varepsilon,\varepsilon_1)d\varepsilon d\varepsilon_1 = \frac{\sigma^4}{\sigma^2+\sigma_1^2}$.

Step 1: Substituting p_B for $E\left[\pi_B^{N\_IN}\right]$. Since retailer B does not know the market demand information, the profit function for retailer B can be written as: $MaxE\left[\pi_B^{N\_IN}\right] = q_B(1-\eta)(a-\gamma E[q_A]-\beta q_B-\omega_1)$.

Step 2: $E\left[\pi_B^{N\_IN}\right]$ to q_B for which the first-order partial derivative is equal to 0. $\frac{\partial E\left[\pi_B^{N\_IN}\right]}{\partial q_B} = a-\gamma q_A-2\beta q_B-\omega_1 = 0$. $q_B = \frac{a-\gamma E[q_A]-\omega_1}{2\beta}$.

Step 3: Substituting $p_A^{N\_IN}$ and q_B into $E\left[\pi_A^{N\_IN}\middle|\Gamma\right]$. $MaxE\left[\pi_A^{N\_IN}\middle|\Gamma\right] = (1-\eta)E[q_A(p_A-\omega_0)|\Gamma] = q_A(1-\eta)\left(a+E[\varepsilon|\Gamma]-\frac{\gamma(a-\gamma q_A-\omega_1)}{2\beta}-\beta q_A-\omega_0\right), \Gamma = \varepsilon+\varepsilon_1$.

Step 4: $E\left[\pi_A^{N\_IN}\middle|\Gamma\right]$ to q_A for which the first-order partial derivative is equal to 0. $\frac{\partial E\left[\pi_A^{N\_IN}|\Gamma\right]}{\partial q_A} = \frac{2\beta(E[\varepsilon|\Gamma]+a)-a\gamma+2q_A(\gamma^2-2\beta^2)-2\beta\omega_0+\gamma\omega_1}{2\beta} = 0$, $q_A = \frac{2a\beta-a\gamma-2\beta\omega_0+\gamma\omega_1}{4\beta^2-2\gamma^2}+\frac{2\beta E[\varepsilon|\Gamma]}{4\beta^2-2\gamma^2}$.

Step 5: Substituting q_A into the q_B. $q_B = \frac{a(4\beta^2-2\beta\gamma-\gamma^2)+\omega_1(\gamma^2-4\beta^2)+2\beta\gamma\omega_0}{8\beta^3-4\beta\gamma^2}$.

Step 6: Substituting q_A from step 4 and q_B from step 5 into $p_A^{N\_IN}, p_B^{N\_IN}$.

$p_A = \frac{2a\beta-a\gamma+2\beta\omega_0+\gamma\omega_1}{4\beta}+\frac{(4\beta^3-4\beta\gamma^2)E[\varepsilon|\Gamma]}{8\beta^3-4\beta\gamma^2}$, $p_B = \frac{a(4\beta^2-2\beta\gamma-\gamma^2)+\omega_1(\gamma^2-4\beta^2)+2\beta\gamma\omega_0}{8\beta^2-4\gamma^2}$.

Step 7: Substituting q_A, q_B, p_A, p_B into profit function we derive equilibrium profit $E\left[\pi_A^{N\_IN}\middle|\Gamma\right]$, $E\left[\pi_B^{N\_IN}\right]$ and $E\left[\pi_P^{N\_IN}\right]$.The expected profits for members in the supply chain are specified as:

$$E\left[\pi_A^{N\_IN}\middle|\Gamma\right] = \frac{(1-\eta)(2a\beta-a\gamma-2\beta\omega_0+\gamma\omega_1)^2}{16\beta^3-8\beta\gamma^2}+\frac{\beta(1-\eta)(\beta-\gamma)(\beta+\gamma)}{(\gamma^2-2\beta^2)^2}\times\frac{\sigma^4}{\sigma^2+\sigma_1^2},$$

$$E\left[\pi_B^{N\_IN}\right] = \frac{(1-\eta)\left(a(4\beta^2-2\beta\gamma-\gamma^2)+\omega_1(\gamma^2-4\beta^2)+2\beta\gamma\omega_0\right)^2}{(8\beta^2-4\gamma^2)(8\beta^3-4\beta\gamma^2)},$$

$$E\left[\pi_P^{N\_IN}\right] = \frac{\eta\left(a^2(2\beta^2-\gamma^2)(\gamma-2\beta)^2+(2\beta^2-\gamma^2)(2\beta\omega_0-\gamma\omega_1)(2a(\gamma-2\beta)+2\beta\omega_0-\gamma\omega_1)\right)}{8\beta(\gamma^2-2\beta^2)^2}+$$
$$\frac{\eta\left(a(4\beta^2-2\beta\gamma-\gamma^2)+\omega_1(\gamma^2-4\beta^2)+2\beta\gamma\omega_0\right)^2}{(8\beta^2-4\gamma^2)(8\beta^3-4\beta\gamma^2)}+\frac{\eta\beta(\beta-\gamma)(\beta+\gamma)}{(\gamma^2-2\beta^2)^2}\times\frac{\sigma^4}{\sigma^2+\sigma_1^2}.$$

Similarly, other scenario optimal solutions can be derived, and Table A1 summarize the equilibrium results for different scenarios.

Table A1. Equilibrium results under different scenarios.

Scenario	Equilibrium results
Y_II	$q_A^{Y\_II} = \frac{(a-\omega_0)(2\beta-\gamma)}{4\beta^2-2\gamma^2} + \frac{(2\beta-\gamma)E[\varepsilon\mid\Gamma]}{4\beta^2-2\gamma^2}$ $q_B^{Y\_II} = \frac{(a-\omega_0)(4\beta^2-2\beta\gamma-\gamma^2)}{8\beta^3-4\beta\gamma^2} + \frac{(4\beta^2-2\beta\gamma-\gamma^2)E[\varepsilon\mid\Gamma]}{8\beta^3-4\beta\gamma^2}$ $p_A^{Y\_II} = \frac{a(2\beta-\gamma)+\omega_0(2\beta+\gamma)}{4\beta} + \frac{(2\beta-\gamma)E[\varepsilon\mid\Gamma]}{4\beta}$ $p_B^{Y\_II} = \frac{a(4\beta^2-2\beta\gamma-\gamma^2)+\omega_0(4\beta^2+2\beta\gamma-3\gamma^2)}{8\beta^2-4\gamma^2} + \frac{(4\beta^2-2\beta\gamma-\gamma^2)E[\varepsilon\mid\Gamma]}{8\beta^2-4\gamma^2}$ $E\left[\pi_A^{Y\_II}\mid\Gamma\right] = F + \frac{(1-\eta)(\gamma-2\beta)^2(a-\omega_0)^2}{16\beta^3-8\beta\gamma^2} + \frac{(1-\eta)(\gamma-2\beta)^2}{16\beta^3-8\beta\gamma^2} \times \frac{\sigma^4}{\sigma^2+\sigma_1^2}$ $E\left[\pi_B^{Y\_II}\mid\Gamma\right] = -F + \frac{(1-\eta)(-4\beta^2+2\beta\gamma+\gamma^2)^2(a-\omega_0)^2}{16\beta(\gamma^2-2\beta^2)^2} + \frac{(1-\eta)(-4\beta^2+2\beta\gamma+\gamma^2)^2}{16\beta(\gamma^2-2\beta^2)^2} \times \frac{\sigma^4}{\sigma^2+\sigma_1^2}$ $E\left[\pi_P^{Y\_II}\right] = \frac{\eta(a^2-2a\omega_0+\omega_0^2)(32\beta^4-32\beta^3\gamma-8\beta^2\gamma^2+12\beta\gamma^3-\gamma^4)}{16\beta(\gamma^2-2\beta^2)^2} + \frac{\eta(32\beta^4-32\beta^3\gamma-8\beta^2\gamma^2+12\beta\gamma^3-\gamma^4)}{16\beta(\gamma^2-2\beta^2)^2} \times \frac{\sigma^4}{\sigma^2+\sigma_1^2}$
Y_IN	$q_A^{Y\_IN} = \frac{a(2\beta-\gamma)-2\beta\omega_0+\gamma\omega_1}{4\beta^2-2\gamma^2} + \frac{(2\beta-\gamma)E[\varepsilon\mid\Gamma]}{4\beta^2-2\gamma^2}$ $q_B^{Y\_IN} = \frac{a(4\beta^2-2\beta\gamma-\gamma^2)+\omega_1(\gamma^2-4\beta^2)+2\beta\gamma\omega_0}{8\beta^3-4\beta\gamma^2} + \frac{(4\beta^2-2\beta\gamma-\gamma^2)E[\varepsilon\mid\Gamma]}{8\beta^3-4\beta\gamma^2}$ $p_A^{Y\_IN} = \frac{a(2\beta-\gamma)+2\beta\omega_0+\gamma\omega_1}{4\beta} + \frac{(2\beta-\gamma)E[\varepsilon\mid\Gamma]}{4\beta}$ $p_B^{Y\_IN} = \frac{a(4\beta^2-2\beta\gamma-\gamma^2)+\omega_1(\gamma^2-4\beta^2)+2\beta\gamma\omega_0}{8\beta^2-4\gamma^2} + \frac{(4\beta^2-2\beta\gamma-\gamma^2)E[\varepsilon\mid\Gamma]}{8\beta^2-4\gamma^2}$ $E\left[\pi_A^{Y\_IN}\mid\Gamma\right] = F + \frac{(1-\eta)(2a\beta-a\gamma-2\beta\omega_0+\gamma\omega_1)^2}{16\beta^3-8\beta\gamma^2} + \frac{(1-\eta)(\gamma-2\beta)^2}{16\beta^3-8\beta\gamma^2} \times \frac{\sigma^4}{\sigma^2+\sigma_1^2}$ $E\left[\pi_B^{Y\_IN}\mid\Gamma\right] = -F + \frac{(1-\eta)(a(4\beta^2-2\beta\gamma-\gamma^2)+\omega_1(\gamma^2-4\beta^2)+2\beta\gamma\omega_0)^2}{16\beta(\gamma^2-2\beta^2)^2} + \frac{(1-\eta)(-4\beta^2+2\beta\gamma+\gamma^2)^2}{16\beta(\gamma^2-2\beta^2)^2} \times \frac{\sigma^4}{\sigma^2+\sigma_1^2}$ $E\left[\pi_P^{Y\_IN}\right] = \frac{\eta a^2(\gamma-2\beta)^2}{16\beta^3-8\beta\gamma^2} + \frac{\eta a^2(-4\beta^2+2\beta\gamma+\gamma^2)^2}{16\beta(\gamma^2-2\beta^2)^2} + \frac{\eta(2\beta\omega_0-\gamma\omega_1)(2a(\gamma-2\beta)+2\beta\omega_0-\gamma\omega_1)}{16\beta^3-8\beta\gamma^2} +$ $\frac{\eta((\omega_1(4\beta^2-\gamma^2)-2\beta\gamma\omega_0)(2a(-4\beta^2+2\beta\gamma+\gamma^2)+\omega_1(4\beta^2-\gamma^2)-2\beta\gamma\omega_0))}{16\beta(\gamma^2-2\beta^2)^2} - \frac{\eta(-32\beta^4+32\beta^3\gamma+8\beta^2\gamma^2-12\beta\gamma^3+\gamma^4)}{16\beta(\gamma^2-2\beta^2)^2} \times \frac{\sigma^4}{\sigma^2+\sigma_1^2}$
N_II	$q_A^{N\_II} = \frac{(a-\omega_0)(2\beta-\gamma)}{4\beta^2-2\gamma^2} + \frac{2\beta E[\varepsilon\mid\Gamma]}{4\beta^2-2\gamma^2}$ $q_B^{N\_II} = \frac{(a-\omega_0)(4\beta^2-2\beta\gamma-\gamma^2)}{8\beta^3-4\beta\gamma^2}$ $p_A^{N\_II} = \frac{a(2\beta-\gamma)+\omega_0(2\beta+\gamma)}{4\beta} + \frac{(4\beta^3-4\beta\gamma^2)E[\varepsilon\mid\Gamma]}{8\beta^3-4\beta\gamma^2}$ $p_B^{N\_II} = \frac{a(4\beta^2-2\beta\gamma-\gamma^2)+\omega_0(4\beta^2+2\beta\gamma-3\gamma^2)}{8\beta^2-4\gamma^2}$ $E\left[\pi_A^{N\_II}\mid\Gamma\right] = \frac{(1-\eta)(a-\omega_0)^2(\gamma-2\beta)^2}{16\beta^3-8\beta\gamma^2} + \frac{\beta(1-\eta)(\beta-\gamma)(\beta+\gamma)}{(\gamma^2-2\beta^2)^2} \times \frac{\sigma^4}{\sigma^2+\sigma_1^2}$ $E\left[\pi_B^{N\_II}\right] = \frac{(1-\eta)(a-\omega_0)^2(-4\beta^2+2\beta\gamma+\gamma^2)^2}{16\beta(\gamma^2-2\beta^2)^2}$ $E\left[\pi_P^{N\_II}\right] = \frac{\eta(a-\omega_0)^2(32\beta^4-32\beta^3\gamma-8\beta^2\gamma^2+12\beta\gamma^3-\gamma^4)}{16\beta(\gamma^2-2\beta^2)^2} + \frac{\eta\beta(\beta-\gamma)(\beta+\gamma)}{(\gamma^2-2\beta^2)^2} \times \frac{\sigma^4}{\sigma^2+\sigma_1^2}$
N_IN	$q_A^{N\_IN} = \frac{2a\beta-a\gamma-2\beta\omega_0+\gamma\omega_1}{4\beta^2-2\gamma^2} + \frac{2\beta E[\varepsilon\mid\Gamma]}{4\beta^2-2\gamma^2}$ $q_B^{N\_IN} = \frac{a(4\beta^2-2\beta\gamma-\gamma^2)+\omega_1(\gamma^2-4\beta^2)+2\beta\gamma\omega_0}{8\beta^3-4\beta\gamma^2}$ $p_A^{N\_IN} = \frac{2a\beta-a\gamma+2\beta\omega_0+\gamma\omega_1}{4\beta} + \frac{(4\beta^3-4\beta\gamma^2)E[\varepsilon\mid\Gamma]}{8\beta^3-4\beta\gamma^2}$ $p_B^{N\_IN} = \frac{a(4\beta^2-2\beta\gamma-\gamma^2)+\omega_1(\gamma^2-4\beta^2)+2\beta\gamma\omega_0}{8\beta^2-4\gamma^2}$ $E\left[\pi_A^{N\_IN}\mid\Gamma\right] = \frac{(1-\eta)(2a\beta-a\gamma-2\beta\omega_0+\gamma\omega_1)^2}{16\beta^3-8\beta\gamma^2} + \frac{\beta(1-\eta)(\beta-\gamma)(\beta+\gamma)}{(\gamma^2-2\beta^2)^2} \times \frac{\sigma^4}{\sigma^2+\sigma_1^2}$ $E\left[\pi_B^{N\_IN}\right] = \frac{(1-\eta)(a(4\beta^2-2\beta\gamma-\gamma^2)+\omega_1(\gamma^2-4\beta^2)+2\beta\gamma\omega_0)^2}{(8\beta^2-4\gamma^2)(8\beta^3-4\beta\gamma^2)}$ $E\left[\pi_P^{N\_IN}\right] = \frac{\eta(a(4\beta^2-2\beta\gamma-\gamma^2)+\omega_1(\gamma^2-4\beta^2)+2\beta\gamma\omega_0)^2}{(8\beta^2-4\gamma^2)(8\beta^3-4\beta\gamma^2)} +$ $\frac{\eta(a^2(2\beta^2-\gamma^2)(\gamma-2\beta)^2+(2\beta^2-\gamma^2)(2\beta\omega_0-\gamma\omega_1)(2a(\gamma-2\beta)+2\beta\omega_0-\gamma\omega_1))}{8\beta(\gamma^2-2\beta^2)^2} + \frac{\eta\beta(\beta-\gamma)(\beta+\gamma)}{(\gamma^2-2\beta^2)^2} \times \frac{\sigma^4}{\sigma^2+\sigma_1^2}$

Proposition 1

Proof of Proposition 1. Quantity difference:

$$q_A^{Y\_II} - q_B^{Y\_II} = \frac{\gamma^2(a-\omega_0)}{8\beta^3-4\beta\gamma^2} + \frac{\gamma^2 E[\varepsilon|\Gamma]}{8\beta^3-4\beta\gamma^2},$$

$$q_A^{Y\_IN} - q_B^{Y\_IN} = \frac{a\gamma^2+\omega_1(4\beta^2+2\beta\gamma-\gamma^2)-2\beta\omega_0(2\beta+\gamma)}{8\beta^3-4\beta\gamma^2} + \frac{\gamma^2 E[\varepsilon|\Gamma]}{8\beta^3-4\beta\gamma^2},$$
$$q_A^{N\_II} - q_B^{N\_II} = \frac{a\gamma^2-\gamma^2\omega_0}{8\beta^3-4\beta\gamma^2} + \frac{4\beta^2 E[\varepsilon|\Gamma]}{8\beta^3-4\beta\gamma^2},$$
$$q_A^{N\_IN} - q_B^{N\_IN} = \frac{a\gamma^2+\omega_1(4\beta^2+2\beta\gamma-\gamma^2)-2\beta\omega_0(2\beta+\gamma)}{8\beta^3-4\beta\gamma^2} + \frac{4\beta^2 E[\varepsilon|\Gamma]}{8\beta^3-4\beta\gamma^2}.$$

Price difference:

$$p_A^{Y\_II} - p_B^{Y\_II} = -\frac{\gamma^2(a-\omega_0)(\beta-\gamma)}{8\beta^3-4\beta\gamma^2} - \frac{\gamma^2(\beta-\gamma)E[\varepsilon|\Gamma]}{8\beta^3-4\beta\gamma^2},$$
$$p_A^{Y\_IN} - p_B^{Y\_IN} = \frac{-a\gamma^2(\beta-\gamma)+\omega_1(4\beta^3+2\beta^2\gamma-\beta\gamma^2-\gamma^3)+2\beta\omega_0(\beta-\gamma)(2\beta+\gamma)}{8\beta^3-4\beta\gamma^2} + \frac{\gamma^2(\gamma-\beta)E[\varepsilon|\Gamma]}{8\beta^3-4\beta\gamma^2},$$
$$p_A^{N\_II} - p_B^{N\_II} = \frac{(\beta-\gamma)(\gamma^2\omega_0-a\gamma^2)}{8\beta^3-4\beta\gamma^2} + \frac{4\beta(\beta-\gamma)(\beta+\gamma)E[\varepsilon|\Gamma]}{8\beta^3-4\beta\gamma^2},$$
$$p_A^{N\_IN} - p_B^{N\_IN} = \frac{-a\gamma^2(\beta-\gamma)+\omega_1(4\beta^3+2\beta^2\gamma-\beta\gamma^2-\gamma^3)+2\beta\omega_0(\beta-\gamma)(2\beta+\gamma)}{8\beta^3-4\beta\gamma^2} + \frac{4\beta(\beta-\gamma)(\beta+\gamma)E[\varepsilon|\Gamma]}{8\beta^3-4\beta\gamma^2}.$$
$$E[\varepsilon|\Gamma] = \frac{\sigma^2\Gamma}{\sigma^2+\sigma_1^2} = \frac{\sigma^2(\varepsilon+\varepsilon_1)}{\sigma^2+\sigma_1^2},\ \frac{\sigma^2}{\sigma^2+\sigma_1^2} > 0.$$

We analyze the cases where $\varepsilon+\varepsilon_1$ is greater than 0 and less than 0, respectively. Satisfying the condition that all variables are greater than zero, we use the Y_II scenario as an example to illustrate the comparison of the quantities between retailer A and retailer B.

$q_A^{Y\_II} - q_B^{Y\_II} = \frac{\gamma^2(a-\omega_0)}{8\beta^3-4\beta\gamma^2} + \frac{\gamma^2 E[\varepsilon|\Gamma]}{8\beta^3-4\beta\gamma^2}$. If $E[\varepsilon|\Gamma] > 0$, i.e., while $\varepsilon+\varepsilon_1 > 0$, $\beta > \gamma$, and for all variables greater than 0, a must satisfy $a > \frac{4\beta^2\omega_1-2\beta\gamma\omega_0-\gamma^2\omega_1}{4\beta^2-2\beta\gamma-\gamma^2} > \omega_0$, so $q_A^{Y\_II} > q_B^{Y\_II}$. If $E[\varepsilon|\Gamma] < 0$, i.e., while $\varepsilon+\varepsilon_1 < 0$, to satisfy $q_A^{Y\_II} > q_B^{Y\_II}$, it is necessary to satisfy $a > \omega_0 - E[\varepsilon|\Gamma]$. In the range of definition, it is guaranteed that $a > \omega_0 - E[\varepsilon|\Gamma]$, so $q_A^{Y\_II} > q_B^{Y\_II}$.

Therefore, the results of $q_A^{Y\_II} > q_B^{Y\_II}$ can be easily derived. And the process of quantitative comparative analysis regarding the other three scenarios is consistent with the Y_II scenario. Next, we use scenario Y_IN as an example to illustrate the results of comparing the retail prices of retailers A and B.

$$p_A^{Y\_IN} - p_B^{Y\_IN} = \frac{-a\gamma^2(\beta-\gamma)+\omega_1(4\beta^3+2\beta^2\gamma-\beta\gamma^2-\gamma^3)+2\beta\omega_0(\beta-\gamma)(2\beta+\gamma)}{8\beta^3-4\beta\gamma^2} + \frac{\gamma^2(\gamma-\beta)E[\varepsilon|\Gamma]}{8\beta^3-4\beta\gamma^2}.$$

We let $\frac{-a\gamma^2(\beta-\gamma)+\omega_1(4\beta^3+2\beta^2\gamma-\beta\gamma^2-\gamma^3)+2\beta\omega_0(\beta-\gamma)(2\beta+\gamma)}{8\beta^3-4\beta\gamma^2} + \frac{\gamma^2(\gamma-\beta)E[\varepsilon|\Gamma]}{8\beta^3-4\beta\gamma^2} > 0$, and consider both $E[\varepsilon|\Gamma] > 0$ and $E[\varepsilon|\Gamma] < 0$ to get $p_A^{Y\_IN} > p_B^{Y\_IN}$ while $a < \frac{4\beta^3\omega_0+4\beta^3\omega_1-2\beta^2\gamma\omega_0+2\beta^2\gamma\omega_1-2\beta\gamma^2\omega_0-\beta\gamma^2\omega_1-\gamma^3\omega_1}{\beta\gamma^2-\gamma^3} + \frac{\gamma^3-\beta\gamma^2}{\beta\gamma^2-\gamma^3}(\varepsilon+\varepsilon_1)$ and $0 < \varepsilon+\varepsilon_1 < \frac{4\beta^3\omega_0+4\beta^3\omega_1-2\beta^2\gamma\omega_0+2\beta^2\gamma\omega_1-2\beta\gamma^2\omega_0-\beta\gamma^2\omega_1-\gamma^3\omega_1}{\beta\gamma^2-\gamma^3}$. Then we obtain the result of Proposition 1(b_2), the price comparisons in the other scenarios are consistent with Scenario Y_II.

The results in Proposition 1 are proved. □

Proof. Proposition 2

Quantity difference:

$$q_A^{Y\_II} - q_A^{N\_II} = q_A^{Y\_IN} - q_A^{N\_IN} = \frac{\gamma E[\varepsilon|\Gamma]}{2(\gamma^2-2\beta^2)},$$
$$q_A^{Y\_II} - q_A^{Y\_IN} = q_A^{N\_II} - q_A^{N\_IN} = \frac{\gamma(\omega_0-\omega_1)}{4\beta^2-2\gamma^2},$$
$$q_B^{Y\_II} - q_B^{N\_II} = q_B^{Y\_IN} - q_B^{N\_IN} = \frac{(4\beta^2-2\beta\gamma-\gamma^2)E[\varepsilon|\Gamma]}{8\beta^3-4\beta\gamma^2},$$
$$q_B^{Y\_II} - q_B^{Y\_IN} = q_B^{N\_II} - q_B^{N\_IN} = -\frac{(\omega_0-\omega_1)(4\beta^2-\gamma^2)}{8\beta^3-4\beta\gamma^2}.$$

Price difference:

$$p_A^{Y\_II} - p_A^{N\_II} = p_A^{Y\_IN} - p_A^{N\_IN} = \frac{\gamma(-2\beta^2+2\beta\gamma+\gamma^2)E[\varepsilon|\Gamma]}{8\beta^3-4\beta\gamma^2},$$
$$p_A^{Y\_II} - p_A^{Y\_IN} = p_A^{N\_II} - p_A^{N\_IN} = \frac{\gamma(\omega_0-\omega_1)}{4\beta},$$
$$p_B^{Y\_II} - p_B^{N\_II} = p_B^{Y\_IN} - p_B^{N\_IN} = \frac{(4\beta^2-2\beta\gamma-\gamma^2)E[\varepsilon|\Gamma]}{8\beta^2-4\gamma^2},$$
$$p_B^{Y\_II} - p_B^{Y\_IN} = p_B^{N\_II} - p_B^{N\_IN} = \frac{\omega_0(4\beta^2-3\gamma^2)+\omega_1(4\beta^2-\gamma^2)}{8\beta^2-4\gamma^2}.$$

When the assumption $\beta > \gamma$ is satisfied, we can easily obtain Proposition 2. □

Proof. Proposition 3

In scenario Y:

$q_A^{Y\_II} = \frac{(a-\omega_0)(2\beta-\gamma)}{4\beta^2-2\gamma^2} + \frac{(2\beta-\gamma)E[\varepsilon|\Gamma]}{4\beta^2-2\gamma^2}$, $q_A^{Y\_II}$ can be written in terms of $q_A^{Y\_II} = M_A^{Y\_II} + N_A^{Y\_II}(\varepsilon+\varepsilon_1)$, $N_A^{Y\_II} = \frac{(2\beta-\gamma)\bar{\sigma}}{4\beta^2-2\gamma^2}$ while $\beta > \gamma$, $\frac{(2\beta-\gamma)\bar{\sigma}}{4\beta^2-2\gamma^2} > 0$, so $N_A^{Y\_II}$ improves the accuracy of forecasting market demand.

Similarly, the other proofs are as follows:

$q_B^{Y\_II} = \frac{(a-\omega_0)(4\beta^2-2\beta\gamma-\gamma^2)}{8\beta^3-4\beta\gamma^2} + \frac{(4\beta^2-2\beta\gamma-\gamma^2)E[\varepsilon|\Gamma]}{8\beta^3-4\beta\gamma^2} = M_B^{Y\_II} + N_B^{Y\_II}(\varepsilon+\varepsilon_1)$,

$N_B^{Y\_II} = \frac{(4\beta^2-2\beta\gamma-\gamma^2)\bar{\sigma}}{8\beta^3-4\beta\gamma^2} > 0$;

$q_A^{Y\_IN} = \frac{a(2\beta-\gamma)-2\beta\omega_0+\gamma\omega_1}{4\beta^2-2\gamma^2} + \frac{(2\beta-\gamma)E[\varepsilon|\Gamma]}{4\beta^2-2\gamma^2} = M_A^{Y\_IN} + N_A^{Y\_IN}(\varepsilon+\varepsilon_1), N_A^{Y\_IN} = \frac{(2\beta-\gamma)\bar{\sigma}}{4\beta^2-2\gamma^2} > 0$;

$q_B^{Y\_IN} = \frac{a(4\beta^2-2\beta\gamma-\gamma^2)+\omega_1(\gamma^2-4\beta^2)+2\beta\gamma\omega_0}{8\beta^3-4\beta\gamma^2} + \frac{(4\beta^2-2\beta\gamma-\gamma^2)E(\varepsilon|\Gamma)}{8\beta^3-4\beta\gamma^2} = M_B^{Y\_IN} + N_B^{Y\_IN}(\varepsilon+\varepsilon_1)$,

$N_B^{Y\_IN} = \frac{(4\beta^2-2\beta\gamma-\gamma^2)\bar{\sigma}}{8\beta^3-4\beta\gamma^2} > 0$;

$p_A^{Y\_II} = \frac{a(2\beta-\gamma)+\omega_0(2\beta+\gamma)}{4\beta} + \frac{(2\beta-\gamma)E[\varepsilon|\Gamma]}{4\beta} = m_A^{Y\_II} + n_A^{Y\_II}(\varepsilon+\varepsilon_1), n_A^{Y\_II} = \frac{(2\beta-\gamma)\bar{\sigma}}{4\beta} > 0$;

$p_B^{Y\_II} = \frac{a(4\beta^2-2\beta\gamma-\gamma^2)+\omega_0(4\beta^2+2\beta\gamma-3\gamma^2)}{8\beta^2-4\gamma^2} + \frac{(4\beta^2-2\beta\gamma-\gamma^2)E[\varepsilon|\Gamma]}{8\beta^2-4\gamma^2} = m_B^{Y\_II} + n_B^{Y\_II}(\varepsilon+\varepsilon_1)$,

$n_A^{Y\_II} = \frac{(4\beta^2-2\beta\gamma-\gamma^2)\bar{\sigma}}{8\beta^2-4\gamma^2} > 0$;

$p_A^{Y\_IN} = \frac{a(2\beta-\gamma)+2\beta\omega_0+\gamma\omega_1}{4\beta} + \frac{(2\beta-\gamma)E[\varepsilon|\Gamma]}{4\beta} = m_A^{Y\_IN} + n_A^{Y\_IN}(\varepsilon+\varepsilon_1), n_A^{Y\_IN} = \frac{(2\beta-\gamma)\bar{\sigma}}{4\beta} > 0$;

$p_B^{Y\_IN} = \frac{a(4\beta^2-2\beta\gamma-\gamma^2)+\omega_1(\gamma^2-4\beta^2)+2\beta\gamma\omega_0}{8\beta^2-4\gamma^2} + \frac{(4\beta^2-2\beta\gamma-\gamma^2)E(\varepsilon|\Gamma)}{8\beta^2-4\gamma^2} = m_B^{Y\_IN} + n_B^{Y\_IN}(\varepsilon+\varepsilon_1)$,

$n_B^{Y\_IN} = \frac{(4\beta^2-2\beta\gamma-\gamma^2)\bar{\sigma}}{8\beta^2-4\gamma^2} > 0$.

In scenario N:

$q_A^{N\_II} = \frac{(a-\omega_0)(2\beta-\gamma)}{4\beta^2-2\gamma^2} + \frac{2\beta E[\varepsilon|\Gamma]}{4\beta^2-2\gamma^2} = M_A^{N\_II} + N_A^{N\_II}(\varepsilon+\varepsilon_1), N_A^{Y\_II} = \frac{2\beta\bar{\sigma}}{4\beta^2-2\gamma^2} > 0$;

$q_A^{N\_IN} = \frac{2a\beta-a\gamma-2\beta\omega_0+\gamma\omega_1}{4\beta^2-2\gamma^2} + \frac{2\beta E[\varepsilon|\Gamma]}{4\beta^2-2\gamma^2} = M_A^{N\_IN} + N_A^{N\_IN}(\varepsilon+\varepsilon_1), N_A^{N\_IN} = \frac{2\beta\bar{\sigma}}{4\beta^2-2\gamma^2} > 0$;

$p_A^{N\_II} = \frac{2a\beta-a\gamma+2\beta\omega_0+\gamma\omega_0}{4\beta} + \frac{(4\beta^3-4\beta\gamma^2)E[\varepsilon|\Gamma]}{8\beta^3-4\beta\gamma^2} = m_A^{N\_II} + n_A^{N\_II}(\varepsilon+\varepsilon_1)$,

$n_A^{N\_II} = \frac{(4\beta^3-4\beta\gamma^2)\bar{\sigma}}{8\beta^3-4\beta\gamma^2} > 0$;

$p_A^{N\_IN} = \frac{2a\beta-a\gamma+2\beta\omega_0+\gamma\omega_1}{4\beta} + \frac{(4\beta^3-4\beta\gamma^2)E[\varepsilon|\Gamma]}{8\beta^3-4\beta\gamma^2} = m_A^{N\_IN} + n_A^{N\_IN}(\varepsilon+\varepsilon_1)$,

$n_A^{N\_IN} = \frac{(4\beta^3-4\beta\gamma^2)\bar{\sigma}}{8\beta^3-4\beta\gamma^2} > 0$.

The effect of market forecasts on quantities and prices was demonstrated. We next demonstrate how the intensity of competition affects the profits of retailers A and B in the presence of market demand signals. $E\left[\pi_A^{Y\_II}\middle|\Gamma\right] = F + (1-\eta)(\frac{(a-\omega_0)^2(\gamma-2\beta)^2}{16\beta^3-8\beta\gamma^2} + \frac{(\gamma-2\beta)^2}{16\beta^3-8\beta\gamma^2} \times \frac{\sigma^4}{\sigma^2+\sigma_1^2})$. Since $1-\eta > 0$, we calculate the derivative of $\frac{(\gamma-2\beta)^2}{16\beta^3-8\beta\gamma^2}$ with respect to γ and obtain $\frac{16\beta\gamma(\gamma-2\beta)^2}{(16\beta^3-8\beta\gamma^2)^2} + \frac{2(\gamma-2\beta)}{16\beta^3-8\beta\gamma^2}$, while $1 > \beta > \gamma > 0$, $\frac{16\beta\gamma(\gamma-2\beta)^2}{(16\beta^3-8\beta\gamma^2)^2} + \frac{2(\gamma-2\beta)}{16\beta^3-8\beta\gamma^2} < 0$, so $E\left[\pi_A^{Y\_II}\middle|\Gamma\right]$ can be written in the form of $E\left[\pi_A^{Y\_II}\middle|\Gamma\right] = M_A^{Y\_II} + N_A^{Y\_II}(\frac{\sigma^4}{\sigma^2+\sigma_1^2})$, $N_A^{Y\_II} = \frac{(1-\eta)(\gamma-2\beta)^2}{16\beta^3-8\beta\gamma^2}$ and $N_A^{Y\_II}$ is decreasing in the competition degree.

$E\left[\pi_A^{Y\_IN}\middle|\Gamma\right] = F + (1-\eta)(\frac{(2a\beta-a\gamma-2\beta\omega_0+\gamma\omega_1)^2}{16\beta^3-8\beta\gamma^2} + \frac{(\gamma-2\beta)^2}{16\beta^3-8\beta\gamma^2} \times \frac{\sigma^4}{\sigma^2+\sigma_1^2}) = M_A^{Y\_IN} + N_A^{Y\_IN}(\frac{\sigma^4}{\sigma^2+\sigma_1^2})$,

$N_A^{Y\_IN} = \frac{(1-\eta)(\gamma-2\beta)^2}{16\beta^3-8\beta\gamma^2}, \frac{d}{d\gamma}\frac{(\gamma-2\beta)^2}{16\beta^3-8\beta\gamma^2} = \frac{16\beta\gamma(\gamma-2\beta)^2}{(16\beta^3-8\beta\gamma^2)^2} + \frac{2(\gamma-2\beta)}{16\beta^3-8\beta\gamma^2} < 0$.

$E\left[\pi_B^{Y\_II}\middle|\Gamma\right] = -F + (1-\eta)(\frac{(a-\omega_0)^2(-4\beta^2+2\beta\gamma+\gamma^2)^2}{16\beta(\gamma^2-2\beta^2)^2} + \frac{(-4\beta^2+2\beta\gamma+\gamma^2)^2}{16\beta(\gamma^2-2\beta^2)^2} \times \frac{\sigma^4}{\sigma^2+\sigma_1^2}) = M_B^{Y\_II} +$

$N_B^{Y\_II}(\frac{\sigma^4}{\sigma^2+\sigma_1^2}), N_B^{Y\_II} = \frac{(1-\eta)\left(-4\beta^2+2\beta\gamma+\gamma^2\right)^2}{16\beta(\gamma^2-2\beta^2)^2}, \frac{d}{d\gamma}\frac{\left(-4\beta^2+2\beta\gamma+\gamma^2\right)^2}{16\beta(\gamma^2-2\beta^2)^2} = \frac{(2\beta+2\gamma)\left(-4\beta^2+2\beta\gamma+\gamma^2\right)}{8\beta(\gamma^2-2\beta^2)^2} - \frac{\gamma\left(-4\beta^2+2\beta\gamma+\gamma^2\right)^2}{4\beta(\gamma^2-2\beta^2)^3} < 0.$

$E\left[\pi_B^{Y\_IN}\middle|\Gamma\right] = -F + (1-\eta)(\frac{\left(a\left(4\beta^2-2\beta\gamma-\gamma^2\right)+\omega_1\left(\gamma^2-4\beta^2\right)+2\beta\gamma\omega_0\right)^2}{16\beta(\gamma^2-2\beta^2)^2} + \frac{\left(-4\beta^2+2\beta\gamma+\gamma^2\right)^2}{16\beta(\gamma^2-2\beta^2)^2} \times \frac{\sigma^4}{\sigma^2+\sigma_1^2})$
$= M_B^{Y\_IN} + N_B^{Y\_IN}(\frac{\sigma^4}{\sigma^2+\sigma_1^2}), N_B^{Y\_IN} = \frac{(1-\eta)\left(-4\beta^2+2\beta\gamma+\gamma^2\right)^2}{16\beta(\gamma^2-2\beta^2)^2},$
$\frac{d}{d\gamma}\frac{\left(-4\beta^2+2\beta\gamma+\gamma^2\right)^2}{16\beta(\gamma^2-2\beta^2)^2} = \frac{(2\beta+2\gamma)\left(-4\beta^2+2\beta\gamma+\gamma^2\right)}{8\beta(\gamma^2-2\beta^2)^2} - \frac{\gamma\left(-4\beta^2+2\beta\gamma+\gamma^2\right)^2}{4\beta(\gamma^2-2\beta^2)^3} < 0.$

$E\left[\pi_A^{N\_II}\middle|\Gamma\right] = (1-\eta)(\frac{(a-\omega_0)^2(\gamma-2\beta)^2}{16\beta^3-8\beta\gamma^2} + \frac{\beta(\beta-\gamma)(\beta+\gamma)}{(\gamma^2-2\beta^2)^2} \times \frac{\sigma^4}{\sigma^2+\sigma_1^2}) = M_A^{N\_II} + N_A^{N\_II}(\frac{\sigma^4}{\sigma^2+\sigma_1^2}),$
$N_A^{N\_II} = \frac{(1-\eta)\beta(\beta-\gamma)(\beta+\gamma)}{(\gamma^2-2\beta^2)^2}, \frac{d}{d\gamma}\frac{\beta(\beta-\gamma)(\beta+\gamma)}{(\gamma^2-2\beta^2)^2} = \frac{\beta(\beta-\gamma)}{(\gamma^2-2\beta^2)^2} - \frac{4\beta\gamma(\beta+\gamma)(\beta-\gamma)}{(\gamma^2-2\beta^2)^3} - \frac{\beta(\beta+\gamma)}{(\gamma^2-2\beta^2)^2} < 0.$

$E\left[\pi_A^{N\_IN}\middle|\Gamma\right] = (1-\eta)(\frac{(2a\beta-a\gamma-2\beta\omega_0+\gamma\omega_1)^2}{16\beta^3-8\beta\gamma^2} + \frac{\beta(\beta-\gamma)(\beta+\gamma)}{(\gamma^2-2\beta^2)^2} \times \frac{\sigma^4}{\sigma^2+\sigma_1^2}) =$
$M_A^{N\_IN} + N_A^{N\_IN}(\frac{\sigma^4}{\sigma^2+\sigma_1^2}), N_A^{N\_IN} = \frac{(1-\eta)\beta(\beta-\gamma)(\beta+\gamma)}{(\gamma^2-2\beta^2)^2},$
$\frac{d}{d\gamma}\frac{\beta(\beta-\gamma)(\beta+\gamma)}{(\gamma^2-2\beta^2)^2} = \frac{\beta(\beta-\gamma)}{(\gamma^2-2\beta^2)^2} - \frac{4\beta\gamma(\beta+\gamma)(\beta-\gamma)}{(\gamma^2-2\beta^2)^3} - \frac{\beta(\beta+\gamma)}{(\gamma^2-2\beta^2)^2} < 0.$

Then we can get the results of Proposition 3. □

Proof. Proposition 4

We summarize the profit differential between retailer A, retailer B, and Carrier C for different scenarios in Table A2. The Proposition 4 (a_1,a_2,b_1,b_2) can be easily seen in Table A2, and we illustrate the remainder with the profit difference between retailer A in the Y_II and Y_IN scenarios.

Table A2. The profit differences in different scenarios.

Retailer A	$E\left[\pi_A^{Y\_II}\middle|\Gamma\right] - E\left[\pi_A^{N\_II}\middle|\Gamma\right] = F - \frac{\gamma(1-\eta)\left(8\beta^3-6\beta^2\gamma-4\beta\gamma^2+\gamma^3\right)}{8\beta(\gamma^2-2\beta^2)^2} \times \frac{\sigma^4}{\sigma^2+\sigma_1^2}$
	$E\left[\pi_A^{Y\_IN}\middle|\Gamma\right] - E\left[\pi_A^{N\_IN}\middle|\Gamma\right] = F - \frac{\gamma(1-\eta)\left(8\beta^3-6\beta^2\gamma-4\beta\gamma^2+\gamma^3\right)}{\left(8\beta(\gamma^2-2\beta^2)^2\right)} \times \frac{\sigma^4}{\sigma^2+\sigma_1^2}$
	$E\left[\pi_A^{Y\_II}\middle|\Gamma\right] - E\left[\pi_A^{Y\_IN}\middle|\Gamma\right] = \frac{\gamma(1-\eta)(\omega_0-\omega_1)(4a\beta-2a\gamma+\omega_0(\gamma-4\beta)+\gamma\omega_1)}{16\beta^3-8\beta\gamma^2}$
	$E\left[\pi_A^{N\_II}\middle|\Gamma\right] - E\left[\pi_A^{N\_IN}\middle|\Gamma\right] = \frac{\gamma(1-\eta)(\omega_0-\omega_1)(4a\beta-2a\gamma+\omega_0(\gamma-4\beta)+\gamma\omega_1)}{16\beta^3-8\beta\gamma^2}$
Retailer B	$E\left[\pi_B^{Y\_II}\middle|\Gamma\right] - E\left[\pi_B^{N\_II}\right] = -F + \frac{(1-\eta)\left(-4\beta^2+2\beta\gamma+\gamma^2\right)^2}{16\beta(\gamma^2-2\beta^2)^2} \times \frac{\sigma^4}{\sigma^2+\sigma_1^2}$
	$E\left[\pi_B^{Y\_IN}\middle|\Gamma\right] - E\left[\pi_B^{N\_IN}\right] = -F + \frac{(1-\eta)\left(-4\beta^2+2\beta\gamma+\gamma^2\right)^2}{16\beta(\gamma^2-2\beta^2)^2} \frac{\sigma^4}{\sigma^2+\sigma_1^2}$
	$E\left[\pi_B^{Y\_II}\middle|\Gamma\right] - E\left[\pi_B^{Y\_IN}\middle|\Gamma\right] = \frac{(1-\eta)\left((\omega_0-\omega_1)\left(4\beta^2-\gamma^2\right)\left(2a\left(-4\beta^2+2\beta\gamma+\gamma^2\right)+\omega_0\left(4\beta^2-4\beta\gamma-\gamma^2\right)+\omega_1\left(4\beta^2-\gamma^2\right)\right)\right)}{16\beta(\gamma^2-2\beta^2)^2}$
	$E\left[\pi_B^{N\_II}\right] - E\left[\pi_B^{N\_IN}\right] = \frac{(1-\eta)\left((\omega_0-\omega_1)\left(4\beta^2-\gamma^2\right)\left(2a\left(-4\beta^2+2\beta\gamma+\gamma^2\right)+\omega_0\left(4\beta^2-4\beta\gamma-\gamma^2\right)+\omega_1\left(4\beta^2-\gamma^2\right)\right)\right)}{16\beta(\gamma^2-2\beta^2)^2}$

$E\left[\pi_A^{Y\_II}\middle|\Gamma\right] - E\left[\pi_A^{Y\_IN}\middle|\Gamma\right] = \frac{\gamma(1-\eta)(\omega_0-\omega_1)(4a\beta-2a\gamma+\omega_0(\gamma-4\beta)+\gamma\omega_1)}{16\beta^3-8\beta\gamma^2}$. Since $1-\eta > 0$, we solve this inequality on the basis of satisfying $\omega_1 > \omega_0 > 0, 1 > \beta > \gamma > 0, a > 0$ less than 0. And we get that when $\omega_1 > \frac{4\beta\omega_0-\gamma\omega_0}{\gamma}$, $E\left[\pi_A^{Y\_II}\middle|\Gamma\right] < E\left[\pi_A^{Y\_IN}\middle|\Gamma\right]$, otherwise $E\left[\pi_A^{Y\_II}\middle|\Gamma\right] > E\left[\pi_A^{Y\_IN}\middle|\Gamma\right]$. Proposition 4($a_3$) is proven. The analysis process in the other scenarios is the same as the above process, from which we can obtain the content of Proposition 4. □

References

1. Awiagah, R.; Kang, J.; Lim, J.I. Factors affecting e-commerce adoption among SMEs in Ghana. *Inf. Dev.* **2016**, *32*, 815–836. [CrossRef]
2. Alzahrani, J. The impact of e-commerce adoption on business strategy in Saudi Arabian small and medium enterprises (SMEs). *Rev. Econ. Political Sci.* **2019**, *4*, 73–88. [CrossRef]

3. Wang, J. Opportunities and challenges of international e-commerce in the pilot areas of China. *Int. J. Mark. Stud.* **2014**, *6*, 141. [CrossRef]
4. Baek, E.; Lee, H.K.; Choo, H.J. Cross-border online shopping experiences of Chinese shoppers. *Asia Pac. J. Mark. Logist.* **2019**, *32*, 366–385. [CrossRef]
5. Hadasik, B.; Kubiczek, J. E-commerce market environment formed by the COVID-19 pandemic—A strategic analysis. *Forum Sci. Oeconomia* **2022**, *10*, 25–52.
6. Adam, I.O.; Alhassan, M.D.; Afriyie, Y. What drives global B2C E-commerce? An analysis of the effect of ICT access, human resource development and regulatory environment. *Technol. Anal. Strateg. Manag.* **2020**, *32*, 835–850. [CrossRef]
7. Hua, G.; Wang, S.; Cheng, T. Optimal order lot sizing and pricing with free shipping. *Eur. J. Oper. Res.* **2012**, *218*, 435–441. [CrossRef]
8. Shehu, E.; Papies, D.; Neslin, S.A. Free shipping promotions and product returns. *J. Mark. Res.* **2020**, *57*, 640–658. [CrossRef]
9. Niu, B.; Xu, H.; Xie, F. Free shipping in cross-border supply chains considering tax disparity and carrier's pricing decisions. *Transp. Res. Part. Logist. Transp. Rev.* **2021**, *152*, 102369. [CrossRef]
10. Song, Y.; Liu, J.; Zhang, W.; Li, J. Blockchains role in e-commerce sellers decision-making on information disclosure under competition. *Ann. Oper. Res.* **2022**, 1–40 . [CrossRef]
11. Qu, S.; Xu, L.; Mangla, S.K.; Chan, F.T.; Zhu, J.; Arisian, S. Matchmaking in reward-based crowdfunding platforms: A hybrid machine learning approach. *Int. J. Prod. Res.* **2022**, *60*, 7551–7571. [CrossRef]
12. Qu, S.; Li, Y.; Ji, Y. The mixed integer robust maximum expert consensus models for large-scale GDM under uncertainty circumstances. *Appl. Soft Comput.* **2021**, *107*, 107369. [CrossRef]
13. Dominguez, R.; Cannella, S.; Barbosa-Póvoa, A.P.; Framinan, J.M. OVAP: A strategy to implement partial information sharing among supply chain retailers. *Transp. Res. Part. Logist. Transp. Rev.* **2018**, *110*, 122–136. [CrossRef]
14. Lv, Q. Supply chain decision-making of cross-border E-commerce platforms. *Adv. Ind. Eng. Manag.* **2018**, *7*, 1–5.
15. Lu, Y.H.; Yeh, C.C.; Liau, T.W. Exploring the key factors affecting the usage intention for cross-border e-commerce platforms based on DEMATEL and EDAS method. *Electron. Commer. Res.* **2022**, 1–23. [CrossRef]
16. Zha, X.; Zhang, X.; Liu, Y.; Dan, B. Bonded-warehouse or direct-mail? Logistics mode choice in a cross-border e-commerce supply chain with platform information sharing. *Electron. Commer. Res. Appl.* **2022**, *54*, 101181. [CrossRef]
17. Qi, X.; Chan, J.H.; Hu, J.; Li, Y. Motivations for selecting cross-border e-commerce as a foreign market entry mode. *Ind. Mark. Manag.* **2020**, *89*, 50–60. [CrossRef]
18. Cao, K.; Xu, Y.; Cao, J.; Xu, B.; Wang, J. Whether a retailer should enter an e-commerce platform taking into account consumer returns. *Int. Trans. Oper. Res.* **2020**, *27*, 2878–2898. [CrossRef]
19. Jiang, B.; Jerath, K.; Srinivasan, K. Firm strategies in the mid tail of platform-based retailing. *Mark. Sci.* **2011**, *30*, 757–775. [CrossRef]
20. Li, C.; Chu, M.; Zhou, C.; Xie, W. Is it always advantageous to add-on item recommendation service with a contingent free shipping policy in platform retailing? *Electron. Commer. Res. Appl.* **2019**, *37*, 100883. [CrossRef]
21. Wang, Y.Y.; Li, J. Research on pricing, service and logistic decision-making of E-supply chain with Free Shipping strategy. *J. Control Decis.* **2018**, *5*, 319–337. [CrossRef]
22. Han, S.; Chen, S.; Yang, K.; Li, H.; Yang, F.; Luo, Z. Free shipping policy for imported cross-border e-commerce platforms. *Ann. Oper. Res.* **2022** , 1–30. https://doi.10.1007/s10479-022-04774-5. [CrossRef]
23. Etumnu, C.E. Free shipping. *Appl. Econ. Lett.* **2022** , 1–4. [CrossRef]
24. Leng, M.; Becerril-Arreola, R. Joint pricing and contingent free-shipping decisions in B2C transactions. *Prod. Oper. Manag.* **2010**, *19*, 390–405. [CrossRef]
25. Rui, G.; XiaXia, M.; TianXing, S. Research on the Impact of Consumers—Time Sensitivity on E-commerce Free Shipping Strategy. In Proceedings of the 2020 Chinese Control And Decision Conference (CCDC), Hefei, China, 22–24 August 2020; pp. 4204–4209.
26. Li, L.; Zhang, H. Confidentiality and information sharing in supply chain coordination. *Manag. Sci.* **2008**, *54*, 1467–1481. [CrossRef]
27. Jain, A.; Seshadri, S.; Sohoni, M. Differential pricing for information sharing under competition. *Prod. Oper. Manag.* **2011**, *20*, 235–252. [CrossRef]
28. Ha, A.Y.; Tong, S.; Zhang, H. Sharing demand information in competing supply chains with production diseconomies. *Manag. Sci.* **2011**, *57*, 566–581. [CrossRef]
29. Shang, W.; Ha, A.Y.; Tong, S. Information sharing in a supply chain with a common retailer. *Manag. Sci.* **2016**, *62*, 245–263. [CrossRef]
30. Huang, S.; Guan, X.; Chen, Y.J. Retailer information sharing with supplier encroachment. *Prod. Oper. Manag.* **2018**, *27*, 1133–1147. [CrossRef]
31. Min, Z.; Mikai, X.; Debao, D. Research on game strategy of information sharing in cross-border e-commerce supply chain between manufacturers and retailers. In *Proceedings of the Advances in Intelligent Systems and Interactive Applications, Proceedings of the 4th International Conference on Intelligent, Interactive Systems and Applications (IISA2019), Bangkok, Thailand, 28–30 June 2019*; Springer: Berlin/Heidelberg, Germany, 2020; pp. 399–406.
32. Kirby, A.J. Trade associations as information exchange mechanisms. *Rand J. Econ.* **1988**, *19*, 138–146. [CrossRef]

33. Niu, B.; Dai, Z.; Chen, L. Information leakage in a cross-border logistics supply chain considering demand uncertainty and signal inference. *Ann. Oper. Res.* **2022**, *309*, 785–816. [CrossRef]
34. Liu, Z.; Zhang, D.J.; Zhang, F. Information sharing on retail platforms. *Manuf. Serv. Oper. Manag.* **2021**, *23*, 606–619. [CrossRef]
35. Mukhopadhyay, S.K.; Yue, X.; Zhu, X. A Stackelberg model of pricing of complementary goods under information asymmetry. *Int. J. Prod. Econ.* **2011**, *134*, 424–433. [CrossRef]
36. Xu, X.; Chen, Y.; He, P.; Yu, Y.; Bi, G. The selection of marketplace mode and reselling mode with demand disruptions under cap-and-trade regulation. *Int. J. Prod. Res.* **2021**, 1–20. [CrossRef]
37. Simon, D.; Fischbach, K.; Schoder, D. Enterprise architecture management and its role in corporate strategic management. *Inf. Syst. Bus. Manag.* **2014**, *12*, 5–42. [CrossRef]
38. Qu, S.; Shu, L.; Yao, J. Optimal pricing and service level in supply chain considering misreport behavior and fairness concern. *Comput. Ind. Eng.* **2022**, *174*, 108759. [CrossRef]
39. Bai, Q.; Xu, J.; Gong, Y.; Chauhan, S.S. Robust decisions for regulated sustainable manufacturing with partial demand information: Mandatory emission capacity versus emission tax. *Eur. J. Oper. Res.* **2022**, *298*, 874–893. [CrossRef]
40. Niu, B.; Dai, Z.; Zhuo, X. Co-opetition effect of promised-delivery-time sensitive demand on air cargo carriers—Big data investment and demand signal sharing decisions. *Transp. Res. Part. Logist. Transp. Rev.* **2019**, *123*, 29–44. [CrossRef]
41. Xing, W.; Li, Q.; Zhao, X.; Li, J. Information sale and contract selection under downstream competition. *Transp. Res. Part. Logist. Transp. Rev.* **2020**, *136*, 101870. [CrossRef]
42. Singh, N.; Vives, X. Price and quantity competition in a differentiated duopoly. *Rand J. Econ.* **1984**, *15*, 546–554. [CrossRef]
43. Roy, R.A. Market Information and Firm Performance. *Manag. Sci.* **2000**, *46*, 1075–1084.
44. Bai, Q.; Xu, J.; Chauhan, S.S. Effects of sustainability investment and risk aversion on a two-stage supply chain coordination under a carbon tax policy. *Comput. Ind. Eng.* **2020**, *142*, 106324. [CrossRef]

sustainability

Article

Sustainable Trade Promotions in Case of Negative Demand Disruption in E-Commerce

Saeide Bigdellou [1], Shirin Aslani [2,*] and Mohammad Modarres [1]

1 Department of Industrial Engineering, Sharif University of Technology, Tehran 11365/8639, Iran
2 Graduate School of Management and Economics, Sharif University of Technology, Tehran 11365/8639, Iran
* Correspondence: sh.aslani@sharif.edu

Abstract: In this research, we examine the impact of a negative demand disruption on trade promotions strategy, where suppliers offer discounted prices to online supply chain retailers. To analyze the various factors that affect trade promotion strategies, we develop a Stackelberg game model to determine the optimal pricing for both manufacturers and retailers, as well as the optimal order quantity of the retailers. Our findings indicate that through an appropriate sustainable trade promotion policy, the profit of the supply chain's members can be increased in different scenarios, including various product disposal costs and the time of product delivery. In addition to the trade promotion policies, we consider a new strategy where the manufacturer assists the retailer by paying some part of the delivery cost. Then, we compare these strategies to determine which approach leads to the highest profit for the manufacturer, retailer, and integrated supply chain under the different intensities of negative demand disruption.

Keywords: trade promotion; demand disruption; sustainability; lead time-dependent demand

Citation: Bigdellou, S.; Aslani, S.; Modarres, M. Sustainable Trade Promotions in Case of Negative Demand Disruption in E-Commerce. *Sustainability* **2023**, *15*, 2320. https://doi.org/10.3390/su15032320

Academic Editors: Shaojian Qu, Qingguo Bai, Ying Ji and Congjun Rao

Received: 9 January 2023
Revised: 22 January 2023
Accepted: 25 January 2023
Published: 27 January 2023

1. Introduction

In the digital age, the delivery of products to consumers can affect both demand and sales. One effective sales channel in this area is online sales. With the increasing popularity of online shopping among consumers especially in a post-COVID-19 environment, e-commerce business is also becoming more prevalent. According to reports from the French Confederation of Trade Promotions, online sales by its member retailers increased disproportionately by 35% during the 2020 discount season [1]. In addition, the environmental impact of e-commerce is undeniable, some studies that have analyzed this issue. These studies have found that by optimizing the delivery routes of delivery trucks, e-commerce can reduce the negative impact of consumer shopping and is more sustainable than using one's own car for shopping [2,3]. Moreover, some consumers may be unwilling to shop at physical retail stores due to busy schedules or other inconveniences such as mismanagement, long queues, or inappropriate retailer behavior. A survey report indicates that about 42% of top suppliers in the high-level industry sell their products directly to consumers through online channels [4].

Online shopping has changed the strategies of supply chain members. One of the crucial marketing strategies used to boost product sales is promotion. Promotions can be classified into three categories: trade promotions (supplier-to-retailer), retailer promotions (retailer-to-consumer), and consumer promotions (supplier-to-consumer). This article focuses on trade promotions while the consumer demand is under negative disruption. In this marketing strategy, suppliers offer a discounted price to retailers and expect that retailers also offer a discount to end consumers [5]. The cost of this trade promotion is typically the second highest expenditure after the cost of goods. A survey conducted by MEI Computer Technology Group Inc., a leading provider of trade promotion services, found that 42% of the respondents, who were suppliers of consumer goods, indicated that they invested more in trade promotion than ever before in 2010. This significant spending on trade promotion strategies highlights the importance of optimizing these strategies throughout the process [6]. Additionally, trade promotion practices may result in inefficiencies in cost management and high discounts in trade promotion can lead to inefficient practices and decreases manufacturer's profit [7]. Thus, it is crucial to identify and analyze the optimal strategies for promotions in different situations.

Numerous factors can impact the optimal design of trade promotions, such as sales targets, product characteristics, demand uncertainty, and product delivery [8–13]. Demand disruption can alter market uncertainty and, as a result, influence trade promotion strategies. There are various real-world examples of promotion strategies to manage demand disruption adopted by companies. For instance, the sudden outbreak of avian influenza (H7N9) in China in 2013 resulted in a decline in demand for caged eggs, leading farmers to offer large discounts to retailers and consumers on their products, while the opposite occurred for seafood and certain medications. Another example is the COVID-19 pandemic, during which companies with online sales were able to maintain their sales while many physical stores had to close, particularly in the second quarter of 2020 [14]. These and other real-world situations demonstrate the significant economic impact that studying the effects of demand disruption on trade promotion can have. The examination of negative demand disruption within the context of trade promotions is crucial for several reasons. Firstly, trade promotions may be utilized as a means to generate demand in the face of a negative demand disruption. Secondly, firms may have a predetermined budget for marketing and promotions efforts, making it important to consider the impact of demand disruption in the formulation of sustainable trade promotion strategies.

Off-invoice and scan-back are two common types of trade promotion strategies in supply chain management. In the off-invoice, manufacturers offer direct subsidies to retailers. This means that for a limited period of time, a certain discount is given for each product purchased from the manufacturer, with no limit on the quantity of purchased products. The structure of the off-invoice strategy encourages retailers to focus on buying rather than marketing. In the scan-back strategy, manufacturers reimburse retailers a certain amount of money for each unit of goods sold during a promotion. This requires manufacturers to communicate with retailers and verify goods sales based on retailers' scan data [13]. This indicates that markets without a retail scanning system (such as point of sale, POS) cannot use the scan-back method.

This study aims to contribute to the literature on trade promotion strategies by addressing the following four questions:

1. What are the optimal pricing strategies for manufacturers and retailers in a supply chain with uncertain demand when a negative demand disruption occurs?
2. What is the optimal order quantity for the retailer in this scenario?
3. What promotional strategy, either scan-back or off-invoice, or contribution to the retailer's supply costs would be profitable for the manufacturer, retailer, or integrated supply chain in the presence of a negative demand disruption?
4. How does delivery lead time affect the profits of the manufacturer and retailer during negative demand disruption?

To address these questions, we develop a Stackelberg model to study the impact of demand disruption on different trade promotion strategies in a decentralized channel. Delivery lead time is also investigated as an affected factor in analyzing the above mentioned questions. The model is based on an online retail channel during demand disruption and determines optimal pricing strategies for both manufacturers and retailers, as well as the optimal order quantity. The stochastic demand model considered in the model depends on various factors, including retail channel selling prices and online delivery lead time.

The paper is structured as follows: A literature review is presented in Section 2. In Section 3, we explain the problem statement. The model is presented in Section 4, while an analysis of the model is provided in Section 5. Finally, concluding remarks are made in Section 6.

2. Literature Review

Trade promotions are price incentives that manufacturers offer to their retailers. These promotions can be implemented through various methods, such as off-invoice, co-op advertising, and scan-back, and are used in a variety of industries. Previous research in these areas will be reviewed in this section.

Trade promotions in different areas of the supply chain have been explored in several publications. For example, researchers examined a retail incentive in a two-stage supply chain dealing with scan-back trade. Their study found that the manufacturer and supplier in the supply chain can benefit from the scan-back trade mode, but this is not always the case for the retailer. However, they also found that scan-back trade can be profitable for both parties if it is accompanied by a buy-back agreement [15]. In the study [16], a two-stage supply chain model in which a manufacturer supplies a product to a retailer is presented and authors used a trade promotion method to supply chain by determining the optimal prices for multi-period wholesale. Their results demonstrated the benefits of supply chain coordination when the cost of set-up or reordering by the consumer is high and demand is low on average. Authors of [17] designed a new model based on a promotional strategy that considers the benefits of the manufacturer and retailer separately. In their proposed strategy, they aimed to maximize both the manufacturer's and retailer's profits. They found that the retailer's profit in their proposed strategy was better than it would be in an off-invoice manner. The effect of non-monetary product promotion based on the two channels was also studied in the literature. In [18], authors proposed a strategy for a two-level supply chain that includes manufacturers and retailers and studied the buy-one-get-one (BOGO) process.

Although several studies in the literature examined various supply chain structures, including multiple manufacturers, wholesalers, retailers, and online retailers but few of them considered promotion strategies. Most of the studies in the area of trade promotion, studied a supply chain with a single manufacturer and retailer, and few studies examined a supply chain with more than two actors. However, some studies focused on other types of promotions in the supply chain with more than two actors. In [19], a supply chain network with a dual-channel for remanufacturing was developed. They used two separate channels, the direct channel and the retail channel, where the manufacturer in the proposed network delivers and sells the goods to the consumer through both channels at the same price. The effect of price discount contracts and pricing policies in a dual-channel supply chain is examined in [20]. Two Stackelberg and one Nash game were investigated, and authors showed that all scenarios with price discount contracts overcome the other scenarios. In the study [21], authors investigated a two-echelon, multiple-retailer distribution channel while taking into account the promotional strategies used by the retailers and their sales learning curve. They stated that keeping a portion of promotion costs within a reasonable range increased the profits of all the channel coordination participants. The impact of asymmetric demand information on a multi-period price promotion in a supply chain with a single supplier but several retailers was studied in [5]. They developed a stochastic bi-level optimization model while taking into account a Stackelbeg game, and they employed the linearization technique to find the precise solution. A three-echelon supply chain with two retailers in the downstream of the supply chain was studied in [22]. The supplied products of the manufacturer include some imperfect quality items. Their findings indicated that while all unit quality discounts with franchise fees can resolve channel conflict, they are unable to generate profits that benefit all chain members equally.

As mentioned previously, optimal trade promotion is influenced by various factors, such as the uncertainty of demand, marketing budget, and product characteristics. Considering demand uncertainty, a trade promotion technique based on manufacturer-retailer channels is developed in [13] and a technical analysis is conducted for two separate markets using off-invoice and scan-back discounts. The results of this study showed that manufacturers offer better discounts for off-invoice than for scan-back. Authors of [23] also applied two separate models to analyze the promotional trade of retailers and manufacturers using an uncertain supply chain demand approach. The objective of the first model is to maximize the retailer's profit by finding promotional efforts of the retailer based on trade promotion manner. In addition, the second model is an extension of the first model that considers how the manufacturer's trade promotion manner is.

The delivery time of the product can also impact the uncertainty of demand. Several studies have examined the delivery time-dependent stochastic demand in the supply chain. The study [24] examined the benefits of sharing demand forecast information in a two-level dual-channel supply chain. A dual-channel supply chain model for the newsvendor problem also is developed in [25], considering the delivery lead time of the online channel. However, to the best of our knowledge, there are no articles in the literature that consider trade promotion in conjunction with delivery time-dependent stochastic demand.

Regarding demand uncertainty and disruption, some studies have investigated sustainable operations in the supply chain. Sustainable operations concepts investigated in [26] and authors proposed appropriate guidelines for their implementation in Brazilian supply chains. The study [27] analyzed the role of supply chain inventory control systems in a sustainable approach under nonlinear backorder costs. Authors of [28] presented a new game-theoretic model to examine the impact of sustainability aspects on product line length and found that sustainability can reduce the length of the product line and the conflict between society and businesses, as well as the impact of channel decentralization. In Ref. [29], authors studied the price discount of a sustainable supplier in a competitive environment in a two-level supply chain with two suppliers and one retailer, examining the impact of the supplier's price discount on the retailer's profit under uncertain demand. They found that the expected profit of a sustainable supplier increases as the price discount increases, but also found that when both suppliers offer a price discount and the retailer's purchasing cost decreases, the supplier's expected sustainable profit decreases. In Ref. [6], researchers evaluated a trade promotion strategy based on sustainability, demand uncertainty, and capacity constraints in a supply chain network (while focusing on positive demand disruption), which is closely related to the current study. They analyzed three types of trade promotion separately and derived the optimal decision for each strategy. The results of their study confirm that both levels of the supply chain, including manufacturers and retailers, adopt aggressive pricing strategies when demand is disrupted. The results of their study indicate that, for a given level of demand disruption, the off-invoice method is preferred over revenue sharing.

This research makes several contributions to the existing literature on trade promotion. It determines the optimal pricing for the manufacturer and the optimal pricing and quantity for the retailer and compares the profits of different strategies under negative demand disruption. It also examines the coordination of delivery costs between the manufacturer and the retailer and compares trade promotion strategies, which has not been previously explored. Finally, it considers delivery time-dependent stochastic demand in the presence of sudden disruption in demand, which has not been addressed in previous studies. These contributions provide new insights into the impacts of trade promotion strategies on supply chain profitability under different demand and delivery scenarios.

3. Problem Statement

In this study, we examine a supply chain model comprising a manufacturer and a retailer. The model assumes two periods of operation. In the first period, the manufacturer produces a product based on forecast demand, which is a function of the retailer's price and the delivery lead time and follows a uniform random variable. In the second period, the forecasted demand may be disrupted, resulting in a deviation from the original forecast. This disruption can take the form of a positive or negative deviation in demand. If the market experiences a sudden boom, the manufacturer must incur additional costs to increase production in order to meet the increased demand. This additional production is more costly. However, in the event of positive disruption, both the manufacturer and retailer have the option to raise their prices.

In the event of a negative disruption, however, the demand in the market would experience a sudden decrease. As a response, managers may consider implementing incentives for retailers in an effort to stimulate demand. In this study, we examine an appropriate trade promotion strategy for manufacturers to follow in case of negative demand disruption by setting an optimal wholesale price. The retailer, in turn, determines an optimal price and quantity of orders in order to maximize profit.

It is worth noting that the manufacturer may choose to utilize an off-invoice or scan-back promotion, or may opt to provide support to the retailer by covering a portion of delivery costs. The off-invoice policy involves offering discounts on products, while the scan-back policy involves manufacturers reimbursing retailers for each unit of goods sold during a promotion (Figures 1 and 2).

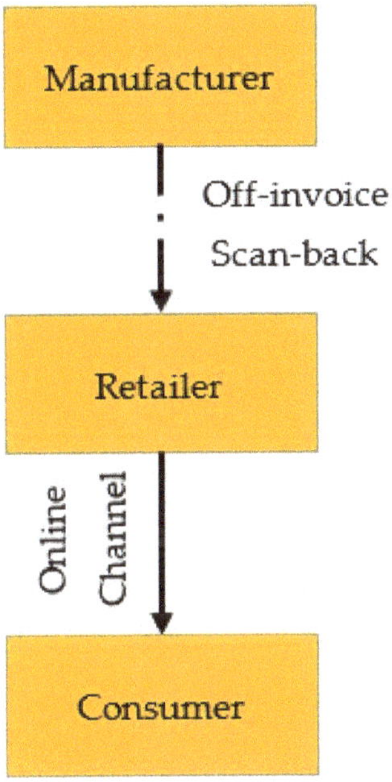

Figure 1. Sales channel structure.

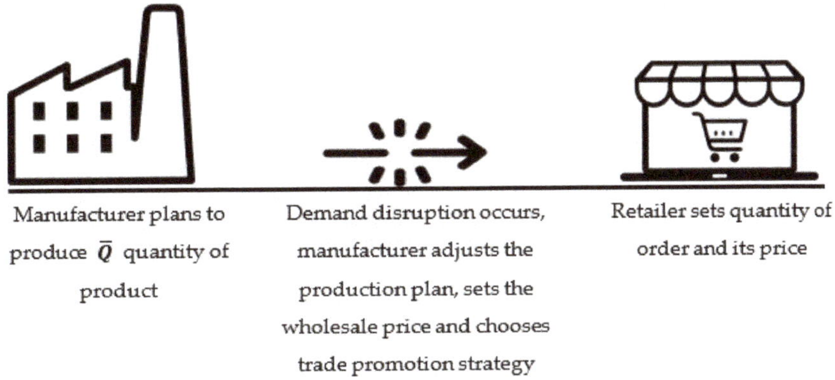

Figure 2. Order of manufacturer and retailer decisions under demand disruption.

Assumptions, Parameters, and Decision Variables

- **Assumptions**

1. Initial delivery lead time is a given parameter that the retailer cannot change due to the related constraint of the system and its budget.
2. A given budget for marketing of manufacturer is considered and the discount for each product is fixed.

- **Parameters and decision variables**

c	Per-unit finished cost of the product for the manufacturer
a_0	Initial demand of the market
a_1	Initial demand of the market after a sudden change in demand
b	Price sensitivity in the online channel
γ	Delivery lead time sensitivity parameter of Demand in the retail channel
α	Discount in the off-invoice policy
β	Discount in the scan-back policy
ϵ	A continuous random variable that follows a uniform distribution
$\overline{Q}$	Previous production plan of the manufacturer
λ_1	Additional unit cost of producing additional products, when $\Delta Q = Q - \overline{Q} < 0$
λ_2	Per-unit disposal cost of products, when $\Delta Q = Q - \overline{Q} > 0$
p_{rf}: Price of the retail channel in the off-invoice policy	p_{rs}: Price of the retail channel in the scan-back policy
Q_{rf}: Order of the retail channel in the off-invoice policy	Q_{rs}: Order of the retail channel in the scan-back policy
W_f: Wholesale price in the off-invoice policy	W_s: Wholesale price in the scan-back policy
Π_{R_f}: Retailer's profit in the off-invoice policy	Π_{R_s}: Retailer's profit in the scan-back policy
Π_{M_f}: Manufacturer's profit in the off-invoice policy	Π_{M_s}: Manufacturer's profit in the scan-back policy

The demand function is given as follows [6,30]:

$$d = a_0 - bp_r - \gamma L + \epsilon \tag{1}$$

where p_r denotes the retailer price and ϵ is a continuous random variable that follows a uniform distribution on $[-(a_0 - bp_r - \gamma L), a_0 - bp_r - \gamma L]$.

We assume that the manufacturer plans production considering demand as $d = a_0 - bp_r - \gamma L + \epsilon$. Then an abrupt disruption occurs and demand changes to $d = a_1 - bp_r - \gamma L + \epsilon_1$, where $a_1 = a_0 + \Delta a$, Δa captures the disruption intensity. In the event of a positive disruption, the market suddenly increases and $\Delta a > 0$, while in the event of a negative disruption, the market suddenly decreases and $\Delta a < 0$. The demand uncertainty in the event of a disruption is assumed to follow a uniform distribution on $[-(a_1 - bp_r - \gamma L), a_1 - bp_r - \gamma L]$. This assumption is based on observations in the internet industry, where high demand leads to high returns and high risk. It is also assumed that the disposal cost is lower than the unit cost of production, such that $\lambda_2 - c \le c$, and $\lambda_2 \le 2c$ [6].

4. The Stackelberg Model for Off-Invoice and Scan-Back

In this section, we explore the optimal decisions for the manufacturer and retailer in the context of a Stackelberg model under two trade promotion strategies: off-invoice and scan-back.

We consider a decentralized setting within a Stackelberg channel (in a decentralized setting, supply chain members are treated as an individual company. They make their decisions based on their information independently and aim to maximize their own profit). We assume the manufacturer to be the leader and first to announce the wholesale price, and the downstream channel member, retailer, to act as the follower and to determine the retail price and ordering decision of the retail channel based on the manufacturer's decision. This assumption is similar to the literature (see [6,13,23,31]).

In other words, the retailer and manufacturer aim to maximize their profit by determining their decision variables, with the retailer determining the optimal retail price and order quantity and the manufacturer determining the optimal wholesale price. We consider the scenario where demand is disrupted and the retailer operates an online sales channel, with the delivery lead time fixed due to budget constraints.

4.1. Off-Invoice

The off-invoice policy refers to a discount applied to the standard price of goods that are sold to a retailer. This discount is applied to all products. The retailer's profit would be as follows:

$$\Pi_{R_f} = p_{rf}\int_0^{Q_{rf}} \frac{t}{2\left(a_1 - bp_{rf} - \gamma L\right)}dt + p_{rf}\int_{Q_{rf}}^{2(a_1 - bp_{rf} - \gamma L)} \frac{Q_{rf}}{2\left(a_1 - bp_{rf} - \gamma L\right)}dt - \left(W_f - \alpha\right)Q_{rf} - (r_0 - r_1 L)^2 \quad (2)$$

The retailer's profit is represented by the first term when the order quantity exceeds the consumer demand, and by the second term when the order quantity is lower than the demand. The third term represents the cost of delivering the products, while the fourth term, which is in quadratic form $(r_0 - r_1 L)^2$, represents the retailer's delivery cost. It is worth noting that $\frac{r_0}{r_1} > L$ [30,32,33].

The manufacturer's profit can be represented by Equation (3).

$$\Pi_{M_f} = \left(W_f - \alpha - c\right)Q_{rf} - \lambda_1\left(Q_{rf} - \overline{Q}_{rf}\right)^+ - \lambda_2\left(\overline{Q}_{rf} - Q_{rf}\right)^+ \quad (3)$$

In this model, the manufacturer is the leader and the retailer is the follower. Equations (4), (5), and (7) depict the optimal retail price, order quantity, and wholesale price, respectively.

$$\frac{\partial \Pi_{R_f}}{\partial Q_{rf}} = 0 \;\; and \;\; \frac{\partial \Pi_{R_f}}{\partial p_{rf}} = 0$$

$$p^*_{rf} = \frac{(a_1 - \gamma L) + \sqrt{(a_1 - \gamma L)^2 + 8b(a_1 - \gamma L)\left(W^*_f - \alpha\right)}}{4b} \quad (4)$$

$$Q^*_{rf} = \frac{5(a_1 - \gamma L) + 4b\left(W^*_f - \alpha\right) - 3\sqrt{(a_1 - \gamma L)^2 + 8b(a_1 - \gamma L)(\beta_2 + \delta)\left(W^*_f - \alpha\right)}}{2} = \frac{\left(3\sqrt{(a_1 - \gamma L)} - \sqrt{(a_1 - \gamma L) + 8b\left(W^*_f - \alpha\right)}\right)^2}{4} \quad (5)$$

In addition, the manufacturer's profit would be:

$$\Pi_{M_f} = \begin{cases} \left(W_f - \alpha - c\right)Q_{rf} - \lambda_1\left(Q_{rf} - \overline{Q}_{rf}\right) & \Delta a > 0 \\ \left(W_f - \alpha - c\right)Q_{rf} - \lambda_2\left(\overline{Q}_{rf} - Q_{rf}\right) & \Delta a < 0 \end{cases} \quad (6)$$

Then we will have:

$$\frac{\partial \Pi_{M_f}}{\partial W_f} = 0$$

$$W_f^* : \begin{cases} \frac{5(a_1-\gamma L)+3\sqrt{(a_1-\gamma L)(17(a_1-\gamma L)+64b(c+\lambda_1))}+32b(c+\lambda_1+2\alpha)}{64b}, \\ \Delta a > 0 \\ \frac{5(a_1-\gamma L)+3\sqrt{(a_1-\gamma L)(17(a_1-\gamma L)+64b(c-\lambda_2))}+32b(c-\lambda_2+2\alpha)}{64b}, \\ \Delta a < 0 \end{cases} \quad (7)$$

4.2. Scan-Back

The scan-back policy involves a discount applied to the standard price of goods that are sold to consumers. The retailer's profit can be expressed as follows:

$$\Pi_{R_s} = (p_{rs}+\beta)\int_0^{Q_s} \frac{t}{2(a_1-bp_{rs}-\gamma L)}dt + (p_{rs}+\beta)\int_{Q_{rs}}^{2(a_1-bp_{rs}-\gamma L)} \frac{Q_{rs}}{2(a_1-bp_{rs}-\gamma L)}dt - W_sQ_{rs} - (r_0-r_1L)^2 \quad (8)$$

The first term represents the retailer's profit when the order quantity exceeds the consumer demand, while the second term represents the retailer's profit when the demand is higher than the retailer's order quantity. The third term represents the cost of supplying the products, and the fourth term, which follows a quadratic form $(r_0 - r_1L)^2$, represents the retailer's delivery cost. It is important to note that $\frac{r_0}{r_1} > L$ [30,32,33].

The manufacturer's profit can be represented by Equation (9).

$$\Pi_{M_s} = (W_s-c)Q_{rs} - \beta\int_0^{Q_s} \frac{t}{2(a_1-bp_{rs}-\gamma L)}dt - \beta\int_{Q_{rs}}^{2(a_1-bp_{rs}-\gamma L)} \frac{Q_{rs}}{2(a_1-bp_{rs}-\gamma L)}dt - \lambda_1(Q_{rs}-\overline{Q}_{rs})^+ - \lambda_2(\overline{Q}_{rs}-Q_{rs})^+ \quad (9)$$

In this model, the manufacturer is the leader and retailer is the follower in this model, the optimal retail price, order quantity, and wholesale price are determined as follows and presented in Equations (10), (11), and (13), respectively.

$$\frac{\partial \Pi_{R_s}}{\partial Q_{rs}} = 0 \;\; and \;\; \frac{\partial \Pi_{R_s}}{\partial p_{rs}} = 0$$

$$p_{rs}^* = \frac{((a_1-\gamma L)+b\beta) - 4b\beta\sqrt{(a_1-\gamma L)+b\beta}\sqrt{(a_1-\gamma L)+b\beta+8bW_s^*}}{4b} \quad (10)$$

$$Q_{rs}^* = \frac{5((a_1-\gamma L)+b\beta)+8bW_s^* - 3\sqrt{((a_1-\gamma L)+b\beta)^2+8((a_1-\gamma L)+b\beta)bW_s^*}}{2} \quad (11)$$

In addition, the manufacturer profit would be:

$$\Pi_{M_s} = \begin{cases} (W_s-\beta-c)Q_{rs} + \frac{\beta Q_{rs}^2}{(a_1-bp_{rs}-\gamma L)} - \lambda_1(Q_{rs}-\overline{Q}_{rs}) & \Delta a > 0 \\ (W_s-\beta-c)Q_{rs} + \frac{\beta Q_{rs}^2}{(a_1-bp_{rs}-\gamma L)} - \lambda_2(\overline{Q}_{rs}-Q_{rs}) & \Delta a < 0 \end{cases} \quad (12)$$

Then we will have:

$$\frac{\partial \Pi_{M_s}}{\partial W_s} = 0$$

$$W_s^* : \begin{cases} \frac{1}{64b((a_1-\gamma L)+b\beta)}((5(a_1-\gamma L)+32b(c+\lambda_1)+19b\beta)((a_1-\gamma L)+b\beta)+9b^2\beta^2 \\ +3((a_1-\gamma L)+2b\beta)\sqrt{((a_1-\gamma L)+b\beta)(17(a_1-\gamma L)+64b(c+\lambda_1)+27b\beta)+9b^2\beta^2}), \\ \Delta a > 0 \\ \\ \frac{1}{64b((a_1-\gamma L)+b\beta)}((5(a_1-\gamma L)+32b(c-\lambda_2)+19b\beta)((a_1-\gamma L)+b\beta)+9b^2\beta^2 \\ +3((a_1-\gamma L)+2b\beta)\sqrt{((a_1-\gamma L)+b\beta)(17(a_1-\gamma L)+64b(c-\lambda_2)+27b\beta)+9b^2\beta^2}), \\ \Delta a < 0 \end{cases} \quad (13)$$

5. Model Implementation and Results

In this section, we investigate the effect of negative demand disruption and delivery lead time on the trade promotion strategies and profits of manufacturers and retailers. We also compare various marketing strategies under negative demand disruption in terms of the manufacturer's profit, retailer's profit, and integrated profit.

To conduct such a test, we compare the result of our model for different intensities of demand disruption, costs on trade promotions, and delivery lead times, using the parameters of [6,30], which are as follows: $a_0 = 200$, $b = 0.75$, $c = 50$, $\gamma = 0.7$, $L = 9.5$, $r_0 = 70$, $r_1 = 7$, α, $\beta = 5$, $\lambda_1 = \lambda_2 = 10$.

For the sake of simplicity we just present the absolute value of negative disruption in this section.

5.1. Impact of Demand Disruption on Profit

This section aims to examine the impact of negative demand disruption on the profits of manufacturers, retailers, and the integrated supply chain.

The integrated supply chain refers to the total profit of the supply chain, including the profits of both the manufacturer and the retailer. In this scenario, there is a single decision maker who possesses all relevant information about the supply chain, manages it, and is able to optimize the performance of the entire system (channel coordination) [34]. The profit of the integrated supply chain is obtained by summing the profits of the manufacturer and the retailer.

As illustrated in Figure 3a–c, a negative demand disruption in the range [0–30] or equally 0% to −26.8% negative disruption in demand leads to a decrease in the profit of the manufacturer, retailer, and integrated supply chain by −34.13%, −41.91%, and −37.81%, respectively. This is due to the fact that higher levels of disruption intensity result in a greater impact on the customer demand and manufacturer's production schedule, leading to an increase in the manufacturer's costs and a corresponding decrease in the profit of the manufacturer, supply chain, and its members. In addition, because disruption affects more factors of the retailer's profit, including the wholesale price, order quantity, and retailer price, causes to decrease in the retailer's profit more than the manufacturer's.

5.2. Impact of Demand Disruption and Manufacturer's Cost on Trade Promotion and Profit

Figure 4 displays the profit of manufacturers under different trade promotion policies in the presence of negative demand disruption. As shown in both Figure 4a,b, the profit of manufacturers decreases as the intensity of negative demand disruption increases. This is due to the fact that demand decreases with negative demand disruption. In the off-invoice policy, higher disposal costs lead to lower profit for manufacturers before $\Delta a = 15$ due to increased costs. However, after this point, disposal costs of $\lambda_2 = 10$ lead to higher profit as the wholesale price decreases and stimulates more consumers' demand that compensates for the associated costs. In contrast, this phenomenon occurs later in the scan-back policy. This may be because, in the scan-back policy, discounts are only paid after the retailer has sold the products. Furthermore, the rate of diminishing returns is higher under the scan-back policy compared to the off-invoice policy, particularly when disposal costs are higher. This suggests that the off-invoice policy may be a more sustainable strategy in such circumstances. This may be because the retailer places larger orders in the off-invoice strategy than in the scan-back back [23], allowing the manufacturer to earn more profit even in the presence of negative demand disruption.

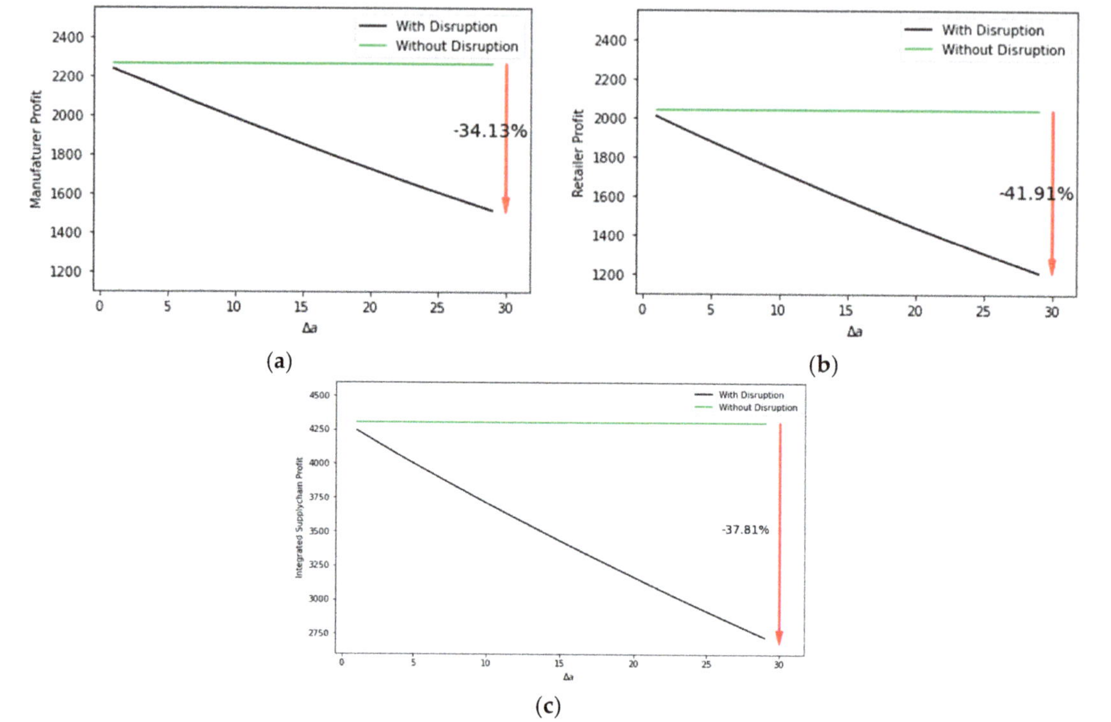

Figure 3. Profits of the (**a**) manufacturer, (**b**) retailer, and (**c**) integrated supply chain, under demand disruption and without disruption ($c = 50$, $a = 200$, $\gamma = 0.7, \lambda_2 = 10$).

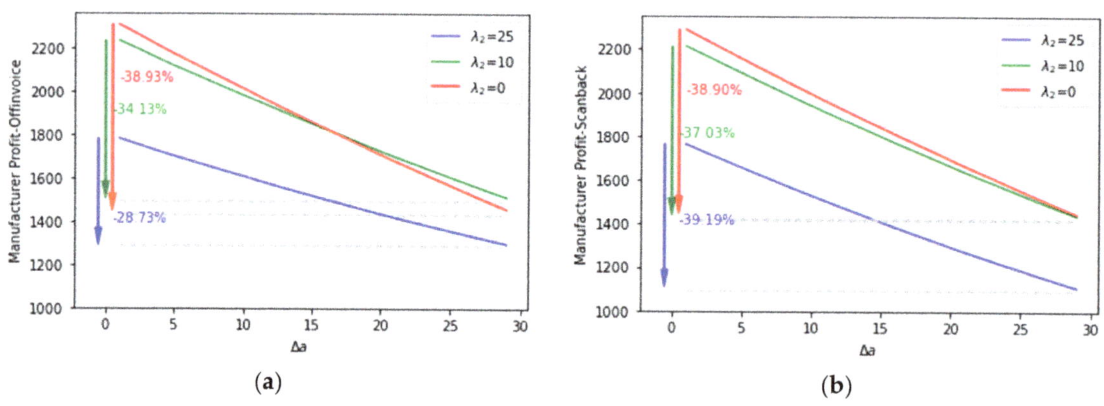

Figure 4. Manufacturer's profit with different disposal costs in (**a**) off-invoice and (**b**) scan-back policies (α, $\beta = 5$, $c = 50$, $a = 200$, $\gamma = 0.7$, $L = 9.5$).

The profits of retailers under negative demand disruption for different trade promotion policies are depicted in Figure 5a,b. As shown in these figures, retailers' profits are higher when disposal costs are higher due to lower wholesale prices and increased order quantities and profits. In both off-invoice and scan-back policies, retailers' profits exhibit similar behavior. It can also be observed that retailers' profits decrease as the intensity of the disruption increases, resulting in a decline in demand.

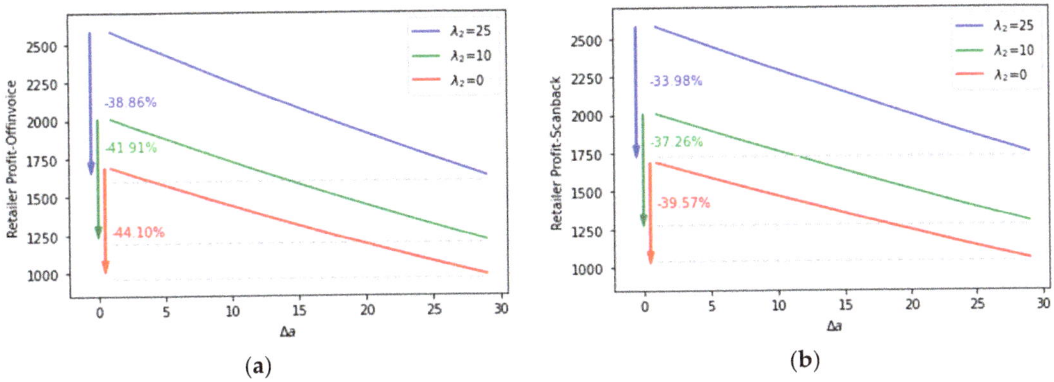

Figure 5. Retailer's profit with different disposal costs in (**a**) off-invoice and (**b**) scan-back ($\alpha, \beta = 5,\ c = 50,\ a = 200,\ \gamma\ = 0.7,\ L = 9.5$).

Additionally, the rate of profit decline is higher under the off-invoice policy in the presence of negative demand disruption. Therefore, the manufacturer's scan-back policy may be a more sustainable strategy in this case, potentially due to the lower order quantity of the retailer under the scan-back policy.

5.3. Impact of Demand Disruption and Online Delivery Lead Time on Trade Promotion and Profit

This section investigates the impact of a ±30% change in delivery lead time L on the profits of retailers and manufacturers. The profits of manufacturers under negative demand disruption for different trade promotion policies are depicted in Figure 6a,b. As shown, negative demand disruption leads to a decrease in manufacturers' profits regardless of the trade promotion strategy employed. Because a reduction in the retailer's delivery lead time leads to an increase in demand and therefore the manufacturer's profit. Additionally, for both policies in the range of negative demand disruption considered, the changes in the manufacturer's profit relative to the demand disruption exhibit approximately linear behavior.

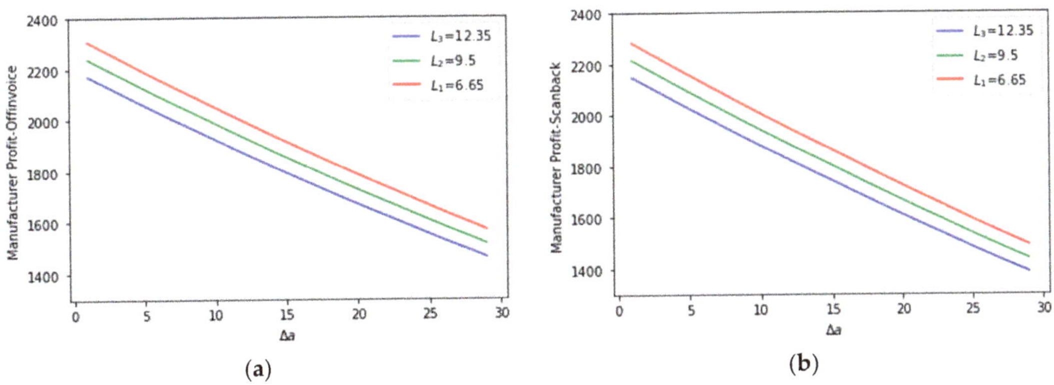

Figure 6. Manufacturer's profit with different delivery lead times (L) in (**a**) off-invoice and (**b**) scan-back ($\alpha,\ \beta = 5,\ c = 50,\ a = 200,\ \gamma\ = 0.7,\ \lambda_2 = 10$).

As demonstrated in Figure 7a,b, negative demand disruption and a decline in demand lead to a reduction in retailers' profits in both off-invoice and scan-back policies. However, the impact of a change in delivery lead time on retailers' profits exhibits a non-linear pattern in the range considered in this study. Specifically, $L_2 = 9.5$ results in higher retailer profit compared to $L_3 = 12.35$ and $L_1 = 6.65$, and $L_3 = 12.35$ leads to higher profit compared to $L_1 = 6.65$. This behavior is expected due to the quadratic form of the delivery lead time in the retailer's profit.

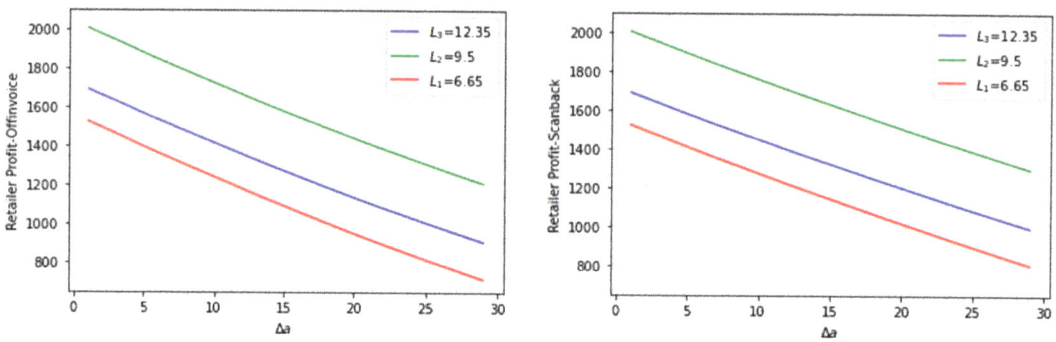

(**a**) (**b**)

Figure 7. Retailer's profits with different delivery lead times (L) in (**a**) off-invoice and (**b**) scan-back policies ($\alpha,\ \beta = 5,\ c = 50,\ a = 200,\ \gamma\ = 0.7, \lambda_2 = 10$).

5.4. Comparing Different Strategies

In this section, we aim to determine which policy results in the highest profit for manufacturers, retailers, or the integrated supply chain under negative demand disruption.

In addition to the off-invoice and scan-back policies, we consider a new strategy called "Coordination in L" where the manufacturer assists the retailer by paying half of the delivery cost. In this proposed policy, the manufacturer is aware of the retailer's delivery costs to consumers and coordinates with the retailer on these costs in order to improve the performance of the supply chain. In this study, we assume that the manufacturer assists the retailer by paying 50% of the delivery cost.

As depicted in Figure 8a, the manufacturer's profit decreases as the intensity of negative demand disruption increases. Figure 8a also demonstrates that in the range of negative demand disruption considered, the scan-back policy leads to lower profit compared to the other policies when the intensity of disruption is low. In addition, when the intensity of disruption is low, the manufacturer's profit is higher when the manufacturer contributes to the retailer's delivery costs (which may be due to the lower cost to the manufacturer compared to offering discounts on products). On the other hand, the off-invoice strategy is more profitable than the other strategies at higher intensities of disruption.

The profit of the retailer for the various strategies is depicted in Figure 8b. This figure illustrates that when negative demand disruption is high, the manufacturer's offer of a scan-back promotion leads to higher profit for the retailer, particularly at high intensities of disruption. In such situations, the wholesale price would be lower with the scan-back promotion compared to the off-invoice policy [23], leading to increased profit for the retailer in the presence of negative demand disruption. Additionally, supply cost support results in higher profit for the retailer compared to the off-invoice strategy, which may be due to the lower cost (as shown in Figure 9).

The profit of the integrated supply chain under negative demand disruption for the various strategies is depicted in Figure 8c. As in the previous results, negative demand disruption leads to lower supply chain profit. The results also show that when the intensity of demand disruption is high, the manufacturer's support of the retailer's supply costs leads to lower profit compared to the other two strategies, while the opposite is true when the intensity of demand disruption is low. Additionally, the results indicate that the scan-back promotion yields more profit when the intensity of disruption is higher (as shown in Figure 10). This may be due to the nature of the scan-back strategy, which offers a discount after the product has been sold, leading to increased benefits for the retailer and impact on supply chain profits.

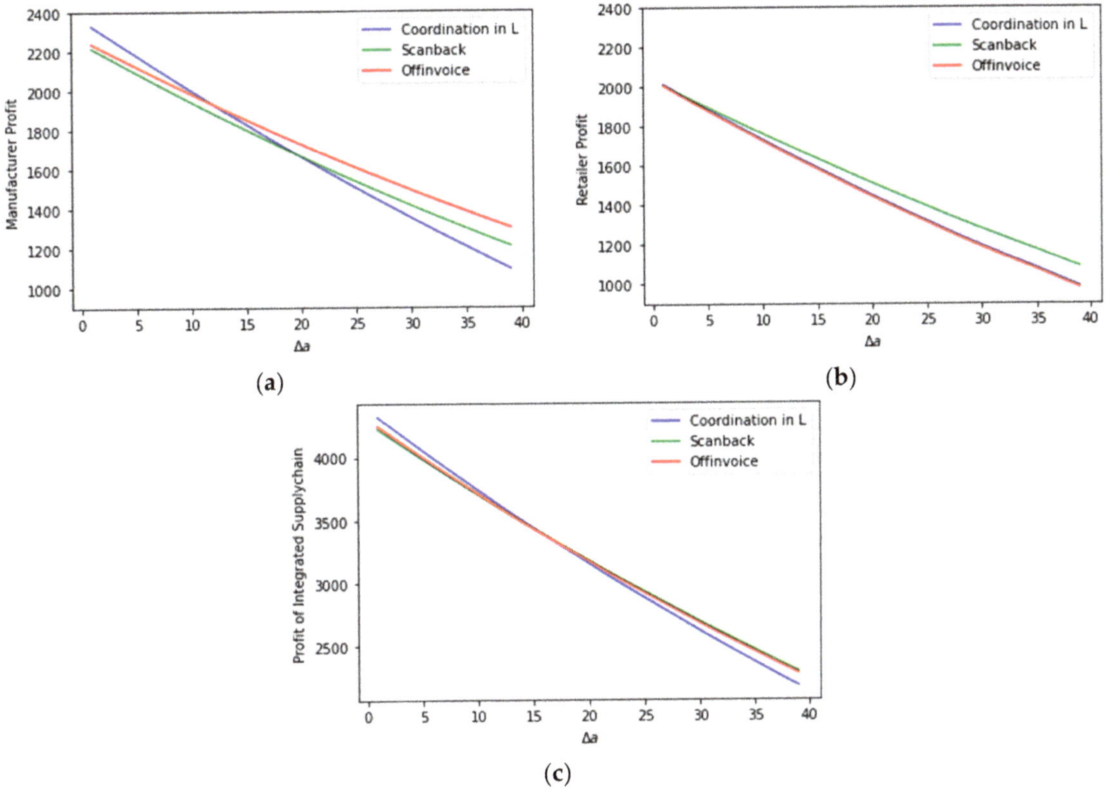

Figure 8. Profits of (**a**) manufacturer, (**b**) retailer, and (**c**) integrated supply chain under demand disruption with different strategies ($\alpha,\ \beta = 5,\ c = 50,\ a = 200,\ \gamma\ = 0.7, \lambda_2 = 10$).

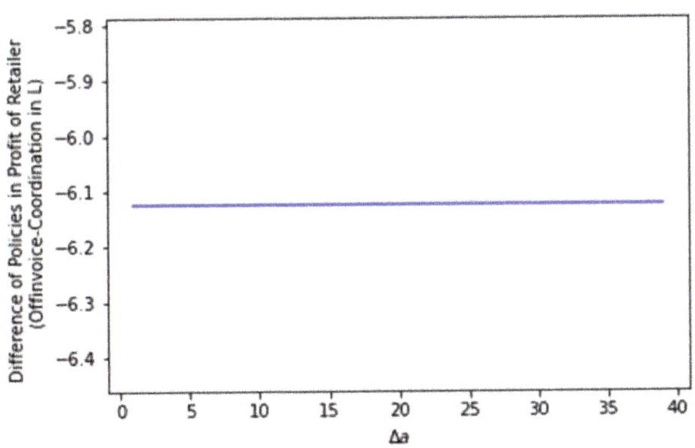

Figure 9. Difference in the retailer's profit in different policies ("off-invoice"–"Coordination in L") under demand disruption ($\alpha,\ \beta = 5,\ c = 50,\ a = 200,\ \gamma\ = 0.7, \lambda_2 = 10$).

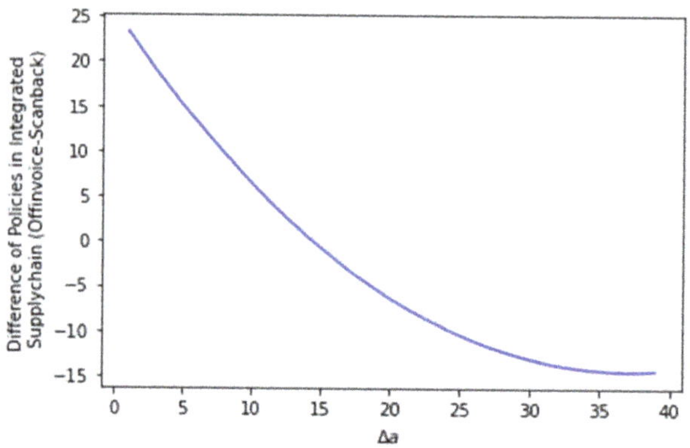

Figure 10. Difference in the profit of integrated supply chain in different policies ("off-invoice"—"scan-back") under demand disruption (α, $\beta = 5$, $c = 50$, $a = 200$, $\gamma = 0.7, \lambda_2 = 10$).

6. Discussion and Conclusions

6.1. Managerial Insights

As previously mentioned, the results of this study indicate that negative demand disruption lead to profit losses for both individual supply chain members and the integrated supply chain as a whole. Based on these results, there are several implications for management to consider.

First, manufacturers should consider taking action to reduce disposal costs when negative demand disruption occurs, as higher disposal costs lead to lower profits. The results of this study suggest that in sudden crises such as a global or national outbreak of diseases (similar to COVID-19), massive earthquakes, or economic crises which causes huge negative disruption in demand [6], the off-invoice policy in higher disposal costs may be a more beneficial and sustainable strategy for manufacturers. Thus, it is recommended that manufacturers use off-invoice policy in countries with high economical fluctuations. On the other hand, the scan-back policy may be more profitable for retailers, particularly when the intensity of negative disruption is high.

Managers in retailers should also consider the optimal delivery lead time, as there is an optimal point at which the retailer's profit is maximized. Reducing delivery lead time can increase demand, but it may not offset the associated cost. Therefore, retailers should carefully manage delivery lead time, particularly during demand disruption. In the range of delivery times studied in this research, a shorter delivery time for the retailer leads to higher profit for the manufacturer. Thus, managers may want to consider actions to reduce the retailer's delivery time in order to increase their profits.

In terms of the overall supply chain, our results indicate that the off-invoice policy is preferable for the manufacturer when the intensity of negative demand disruption is high, while the scan-back policy may be more profitable for the retailer under such conditions. On the other hand, when a sudden rumor or bad news spread among consumers and reduces demand, if the intensity of negative disruption is low, paying the retailer a portion of the delivery lead time cost is more beneficial for the manufacturer. For the integrated supply chain, the scan-back policy is the most profitable option when the intensity of negative disruption is high, when is low paying the retailer a portion of the delivery lead time cost may be beneficial, while the off-invoice policy may be preferred under other circumstances (Figure 11).

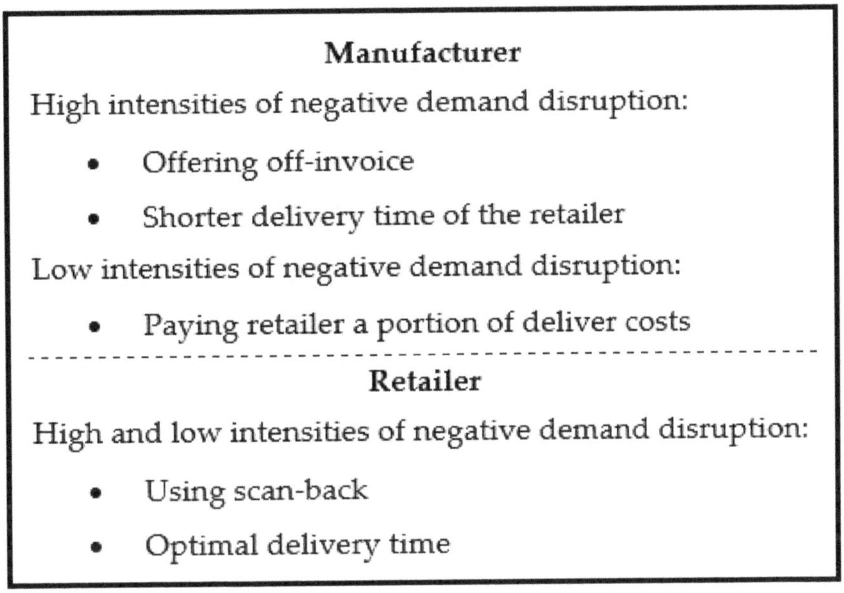

Figure 11. Sustainable promotion strategies of manufacturer and retailer.

6.2. Conclusions and Future Research Directions

In this study, we examined the impact of negative demand disruption on the trade promotion strategies of a manufacturer in a supply chain with a retailer who has an online distribution channel. We constructed a theoretical game model and determined the optimal pricing for both the manufacturer and the retailer, as well as the optimal order quantity for the retailer. Our numerical studies demonstrated that negative demand disruption can alter demand, and the choice of trade promotion policy can have a significant effect on the profits of supply chain members.

It should be noted that this study has certain limitations. We have only considered a single manufacturer and retailer in our model, and the inclusion of market competition could potentially alter the results. Future research could consider the impact of market competition and online sales channels on trade promotion strategies in a supply chain, as well as other forms of consumer demand and the stochastic distributions of these and disruption parameters.

It has been observed that there are various types of consumers in the market, and some of them may be hesitant to make purchases from retail stores due to factors such as time constraints or negative experiences in these settings, such as poor management or rude behavior from retailers. On the other hand, some consumers are opposed to online shopping for various reasons. To accommodate both of these groups, many manufacturers have implemented dual channels for selling their products, comprising both traditional and online channels [35]. As such, further research could focus on examining a dual channel approach or other supply chain structures with multiple manufacturers, wholesalers, distributers, classic retailers, and online retailers that takes into account decision variables such as delivery lead time, in the context of addressing this problem.

Additionally, an examination of trade promotions while considering promotions offered to end consumers could potentially influence the results of this study. Further research could also consider other types of promotions, such as non-monetary promotions offered by manufacturers or retailers.

Author Contributions: Conceptualization, S.B.; methodology, S.B.; supervision, S.A.; supervision, M.M.; visualization, S.B.; writing—original draft, S.B.; review and editing, M.M. and S.A. All authors have read and agreed to the published version of the manuscript.

Funding: This research received no external funding.

Institutional Review Board Statement: Not applicable.

Data Availability Statement: The authors confirm that the data supporting the findings of this study are available upon request.

Conflicts of Interest: The authors declare no conflict of interest.

References

1. Mu, Z.; Zheng, Y.; Sun, H. Cooperative Green Technology Innovation of an E-Commerce Sales Channel in a Two-Stage Supply Chain. *Sustainability* **2021**, *13*, 7499. [CrossRef]
2. Jaller, M.; Pahwa, A. Evaluating the Environmental Impacts of Online Shopping: A Behavioral and Transportation Approach. *Transp. Res. Part D Transp. Environ.* **2020**, *80*, 102223. [CrossRef]
3. Wygonik, E.; Goodchild, A.V. Urban Form and Last-Mile Goods Movement: Factors Affecting Vehicle Miles Travelled and Emissions. *Transp. Res. Part D Transp. Environ.* **2018**, *61*, 217–229. [CrossRef]
4. Dan, B.; Xu, G.; Liu, C. Pricing Policies in a Dual-Channel Supply Chain with Retail Services. *Int. J. Prod. Econ.* **2012**, *139*, 312–320. [CrossRef]
5. Su, Y.; Geunes, J. Multi-Period Price Promotions in a Single-Supplier, Multi-Retailer Supply Chain under Asymmetric Demand Information. *Ann. Oper. Res.* **2013**, *211*, 447–472. [CrossRef]
6. Wu, J.; Chen, Z.; Ji, X. Sustainable Trade Promotion Decisions under Demand Disruption in Manufacturer-Retailer Supply Chains. *Ann. Oper. Res.* **2020**, *290*, 115–143. [CrossRef]
7. Cui, T.H.; Raju, J.S.; Zhang, Z.J. A Price Discrimination Model of Trade Promotions. *Mark. Sci.* **2008**, *27*, 779–795. [CrossRef]
8. Chen, J.; Bell, P.C. The Impact of Customer Returns on Supply Chain Decisions under Various Channel Interactions. *Ann. Oper. Res.* **2013**, *206*, 59–74. [CrossRef]
9. Chen, X.; Wang, A. Trade Credit Contract with Limited Liability in the Supply Chain with Budget Constraints. *Ann. Oper. Res.* **2012**, *196*, 153–165. [CrossRef]
10. Varadarajan, P. Horizontal Cooperative Sales Promotion: A Framework for Classification and Additional Perspectives. *J. Mark.* **1986**, *50*, 61–73. [CrossRef]
11. Wakefield, K.L.; Barnes, J.H. Retailing Hedonic Consumption: A Model of Sales Promotion of a Leisure Service. *J. Retail.* **1996**, *72*, 409–427. [CrossRef]
12. Wu, S.; Liu, Q.; Zhang, R.Q. The Reference Effects on a Retailer's Dynamic Pricing and Inventory Strategies with Strategic Consumers. *Oper. Res.* **2015**, *63*, 1320–1335. [CrossRef]
13. Yuan, H.; Gómez, M.I.; Rao, V.R. Trade Promotion Decisions under Demand Uncertainty: A Market Experiment Approach. *Manage. Sci.* **2013**, *59*, 1709–1724. [CrossRef]
14. Iordache Platis, M.; Olteanu, C.; Hotoi, A.L. Evolution of the Online Sales of Sustainable Products in the COVID-19 Pandemic. *Sustainability* **2022**, *14*, 15291. [CrossRef]
15. Kurata, H.; Yue, X. Trade Promotion Mode Choice and Information Sharing in Fashion Retail Supply Chains. *Int. J. Prod. Econ.* **2008**, *114*, 507–519. [CrossRef]
16. Tantiwattanakul, P.; Dumrongsiri, A. Multi-Period Supply Chain Coordination Using Trade Promotion: Complementary Slackness Approach. *Int. J. Ind. Eng.* **2019**, *26*, 3800.
17. Bambal, A.; Kumar, S.; Abraham, J. Design of Trade Promotion to Maximize Overall Supply Chain Profitability Using Mathematical Optimization. In Proceedings of the 2021 9th International Conference on Reliability, Infocom Technologies and Optimization (Trends and Future Directions) (ICRITO), Noida, India, 3–4 September 2021; pp. 1–6. [CrossRef]
18. Li, Y.; Khouja, M.J.; Pan, J.; Zhou, J. Buy-One-Get-One Promotions in a Two-Echelon Supply Chain. *Manage. Sci.* **2022**. [CrossRef]
19. Yang, L.; Wang, G.; Ke, C. Remanufacturing and Promotion in Dual-Channel Supply Chains under Cap-and-Trade Regulation. *J. Clean. Prod.* **2018**, *204*, 939–957. [CrossRef]
20. Cai, G.G.; Zhang, Z.G.; Zhang, M. Game Theoretical Perspectives on Dual-Channel Supply Chain Competition with Price Discounts and Pricing Schemes. *Int. J. Prod. Econ.* **2009**, *117*, 80–96. [CrossRef]
21. Tsao, Y.-C.; Sheen, G.-J. Effects of Promotion Cost Sharing Policy with the Sales Learning Curve on Supply Chain Coordination. *Comput. Oper. Res.* **2012**, *39*, 1872–1878. [CrossRef]
22. Modak, N.M.; Panda, S.; Sana, S.S. Three-Echelon Supply Chain Coordination Considering Duopolistic Retailers with Perfect Quality Products. *Int. J. Prod. Econ.* **2016**, *182*, 564–578. [CrossRef]
23. Tsao, Y.-C.; Lu, J.-C. Trade Promotion Policies in Manufacturer-Retailer Supply Chains. *Transp. Res. Part E Logist. Transp. Rev.* **2016**, *96*, 20–39. [CrossRef]
24. Yao, D.-Q.; Yue, X.; Wang, X.; Liu, J.J. The Impact of Information Sharing on a Returns Policy with the Addition of a Direct Channel. *Int. J. Prod. Econ.* **2005**, *97*, 196–209. [CrossRef]

25. Yang, J.Q.; Zhang, X.M.; Fu, H.Y.; Liu, C. Inventory Competition in a Dual-Channel Supply Chain with Delivery Lead Time Consideration. *Appl. Math. Model.* **2017**, *42*, 675–692. [CrossRef]
26. Jabbour, C.J.C.; Janeiro, R.C.; de Sousa Jabbour, A.B.L.; Junior, J.A.G.; Salgado, M.H.; Jugend, D. Social Aspects of Sustainable Supply Chains: Unveiling Potential Relationships in the Brazilian Context. *Ann. Oper. Res.* **2020**, *290*, 327–341. [CrossRef]
27. Johansson, L.; Olsson, F. Quantifying Sustainable Control of Inventory Systems with Non-Linear Backorder Costs. *Ann. Oper. Res.* **2017**, *259*, 217–239. [CrossRef]
28. Ji, X.; Wu, J.; Liang, L.; Zhu, Q. The Impacts of Public Sustainability Concerns on Length of Product Line. *Eur. J. Oper. Res.* **2018**, *269*, 16–23. [CrossRef]
29. Li, Y.; Zhou, J. Sustainable Supplier's Equilibrium Discount Strategy under Random Demand. *Sustainability* **2022**, *14*, 4802. [CrossRef]
30. Modak, N.M.; Kelle, P. Managing a Dual-Channel Supply Chain under Price and Delivery-Time Dependent Stochastic Demand. *Eur. J. Oper. Res.* **2019**, *272*, 147–161. [CrossRef]
31. Cachon, G.P.; Netessine, S. Game Theory in Supply Chain Analysis. *Model. Methods Appl. Innov. Decis. Mak.* **2006**, *23*, 200–233. [CrossRef]
32. Desiraju, R.; Moorthy, S. Managing a Distribution Channel under Asymmetric Information with Performance Requirements. *Manage. Sci.* **1997**, *43*, 1628–1644. [CrossRef]
33. Savaskan, R.C.; Van Wassenhove, L.N. Reverse Channel Design: The Case of Competing Retailers. *Manage. Sci.* **2006**, *52*, 1–14. [CrossRef]
34. Pellegrino, R.; Costantino, N.; Tauro, D. Advance Purchase Discounts for Supply Chain Finance System Coordination. *Sustainability* **2020**, *12*, 10156. [CrossRef]
35. Nair, G.; Pleasance, D. Mitigating Channel Conflict. *McKinsey Q.* **2005**, *3*, 16–17.

sustainability

Article

The Robust Emergency Medical Facilities Location-Allocation Models under Uncertain Environment: A Hybrid Approach

Fang Xu [1], Mengfan Yan [2], Lun Wang [2,3,*] and Shaojian Qu [4]

1 Sino-German College, University of Shanghai for Science and Technology, Shanghai 200093, China
2 Business School, University of Shanghai for Science and Technology, Shanghai 200093, China
3 School of Management, Shanghai University, Shanghai 200444, China
4 School of Management Science and Engineering, Nanjing University of Information Science and Technology, Nanjing 210044, China
* Correspondence: 203781404@st.usst.edu.cn

Abstract: In emergency medical facilities location, the hierarchical diagnosis and treatment system plays an obvious role in the rational allocation of medical resources and improving the use efficiency of medical resources. However, few studies have investigated the operational mechanism of hierarchical medical systems in uncertain environments. To address this research gap, this paper proposes a hybrid approach for emergency medical facilities' location-allocation. In the first stage, in order to concentrate on the utilization of medical resources, we choose alternative facility points from the whole facilities through the entropy weight method (EWM). In the second stage, uncertainty sets are used to describe the uncertain number of patients at emergency medical points more accurately. We propose a robust model to configure large base hospitals based on the robust optimization method. Furthermore, the proposed robust models are applied to the emergency management of Huanggang City under COVID-19. The results show that the optimal emergency medical facility location-allocation scheme meets the actual treatment needs. Simultaneously, the disturbance ratio and uncertainty level have a significant impact on the configuration scheme.

Keywords: emergency medical facilities; entropy weight method; robust optimization; location-allocation; hierarchical diagnosis and treatment system

Citation: Xu, F.; Yan, M.; Wang, L.; Qu, S. The Robust Emergency Medical Facilities Location-Allocation Models under Uncertain Environment: A Hybrid Approach. *Sustainability* **2023**, *15*, 624. https://doi.org/10.3390/su15010624

Academic Editors: Mark A. Bonn and Ting Chi

Received: 2 November 2022
Revised: 11 December 2022
Accepted: 26 December 2022
Published: 29 December 2022

1. Introduction

Public health emergencies such as COVID-19 have brought great threats to people and society [1–5]. For timely responses, a hierarchical diagnosis and treatment mode should be established to isolate, control, and treat patients [6–12]. During the epidemic period, the hierarchical diagnosis and treatment mode [13,14] avoided the paralysis of large hospitals caused by the concentration of a large number of patients, and the use efficiency of medical resources was significantly improved. At the same time, medical resources have been reasonably allocated [15]. It is crucial to improve residents' satisfaction and happiness [16].

At present, many scholars have investigated facility location problems [17]. Biswas and Pamucar [18] studied the factors affecting the school location decision from the perspective of students. They developed an integrated group decision-making framework, that is, a pivot pairwise relative criteria importance assessment (PIPRECIA). Pamucar et al. [19] conducted location selection for wind farms using a GIS multi-criteria hybrid model based on fuzzy and rough numbers. Boonmee et al. [20] summarized the humanitarian facility location problem. They divided the location problem into a deterministic facility location problem, dynamic facility location problem, stochastic facility location problem, and robust facility location problem, respectively. Deterministic facility location problems form the basis for dynamic, stochastic, and robust models. However, the medical facility location problem is facing more and more uncertainties (e.g., the uncertain number of patients in facility points, the uncertainty of transportation costs, etc.). The deterministic facility location model cannot describe the impact that uncertain parameter changes have on the facility location problem, which has a certain gap with the actual situation. Therefore, the deterministic facility location model has some disadvantages. Stochastic, dynamic, and robust facility location models can be used to respond to real situations. The dynamic programming model is effective for solving multi-stage decision problems. However, the calculation amount of the dynamic facility location problem increases dramatically when the dimension of decision variables increases. Today's computers still cannot effectively solve large-scale dynamic facility location problems in actual emergency medical responses. For the stochastic facility location, the probability distribution of random parameters needs to be precisely known in advance. In the emergency medical facilities' location, it is difficult to obtain sufficient historical data to estimate the distribution function of random parameters. In order to overcome the shortcomings of the above three methods, a robust facility location model is proposed in this paper. We take into account the uncertain number of patients at the facility points and use the uncertainty sets to describe the uncertain number of patients more accurately. Therefore, this paper focuses on the robust facility location. Extant examples of the literature have studied the emergency medical facilities' location through multi-objective programming [21], the analytic hierarchy process (AHP) and technique for order preference by similarity to an ideal solution (TOPSIS) [22], mixed integer linear programming [23,24], and so on. However, the uncertain number of patients in emergency medical sites during the epidemic situation was not taken into account in the above literature, which increased the risk of decision-making in the emergency medical facilities location and was not good for the life and health of patients. Accordingly, we need to focus on decision-making under an uncertain number of patients to reduce the uncertainty risk. On the other hand, the above-mentioned literature rarely utilized the hierarchical diagnosis and treatment system to locate the emergency medical facilities, so medical resources may not be reasonably allocated, and the use efficiency of medical resources may be reduced. Therefore, in order to deal with the impact of the epidemic more economically and effectively, the improvement of the community medical care level and the completion of the system should be the priority task, which is also beneficial to reduce the burden of large hospitals.

The robust optimization theory is widely used to deal with uncertain optimization problems. The solution of robust optimization is such that all the constraints still hold in the worst case. Unlike stochastic programming [25], robust optimization does not require the probability distribution function of the random parameters. However, it assumes that the uncertain parameters fluctuate in an interval [26–30]. Since its emergence, robust optimization theories have been applied to many fields, such as group decision-making [31–36], portfolios [37–41], efficiency evaluation [42,43], supply chain management [44–49], etc. In emergency medical location decisions, some scholars have adopted the stochastic programming method for modeling [50–54]. However, the stochastic programming needs to know the probability distribution of the patients' number at the facility point. Due to the urgency of the event, it is impossible to accurately obtain the probability function of the patient's number at the facility point. Consequently, the stochastic programming method describing the fluctuation of the patient's number has defects. Hence, in order to overcome the shortcomings of the stochastic programming, we adopted the robust optimization method to handle the uncertainty of the patients' number in the emergency medical facilities' location.

In order to effectively avoid the paralysis of large hospitals caused by the concentration of a large number of patients and to significantly improve the use efficiency of medical resources, this paper proposes a hierarchical diagnosis and treatment system for the emergency medical facilities located under the background of the epidemic. Therefore, a hybrid evaluation method, including EWM and the robust optimization method, is proposed for modeling. We have a two-step plan for post-outbreak isolation and treatment. In the first stage, 10 facilities with the highest scores are selected from 30 facilities by EWM as community emergency medical points. When there are critical patients who cannot be handled by community medical centers, the second stage of decision-making is to send the critical patients to large base hospitals for treatment. We construct a robust location model with capacity and time window constraints with the presence of an uncertain number of patients to configure a large rear hospital.

Different from previous studies, this paper proposes a hybrid approach to cope with the emergency facilities' location problems. This approach contains two-stage decisions under a hierarchical diagnosis system. The first stage decision is to obtain reasonable alternative points from all possible facility points. The second stage decision is to optimally configure the rear hospital under uncertain demand. The contributions of this paper are as follows: Firstly, this paper studies the location-allocation of emergency medical facilities under the hierarchical diagnosis and treatment mode. A hybrid location-allocation decision for emergency medical facilities is also investigated. In the first stage, alternatives are selected from all facility points by EWM. In the second stage, considering the uncertain number of patients at emergency facility points, this paper uses uncertainty sets to describe the number of patients more accurately. On this basis, a robust model with capacity and time window constraints is constructed to allocate large base hospitals. The proposed method fully takes into account the uncertainties when the epidemic occurs. The results of location-allocation significantly reduce the risk of decision-making and provide a strong guarantee for people's health. Therefore, the optimization problem in this paper is in line with the actual situation when the epidemic occurs. Secondly, the robust optimization model is equivalently transformed into a mixed integer linear programming problem by utilizing the duality theory. The robust counterpart model can be solved in polynomial time. Finally, we conduct simulation experiments on the proposed model through the location-allocation scheme of emergency medical facilities in Huanggang City during the COVID-19 epidemic. The results verify the feasibility and robustness of the proposed model. Sensitivity analysis also shows the effectiveness of the proposed method. The proposed method of this paper can provide reference and compliance for health departments to effectively carry out regular epidemic prevention and control.

The remainder of this paper is organized as follows. Section 2 presents the framework and preliminaries; Section 3 derives the emergency facilities location modeling. Section 4 specifies numerical experiments; Section 5 concludes.

2. The Framework and Preliminaries

2.1. The Framewrok of This Paper

The framework of this paper is shown in Figure 1. A hierarchical diagnosis and treatment mode is proposed to cope with the impact of the pandemic. Firstly, it is unrealistic to establish emergency medical facilities at every point, considering the ease of the centralized utilization of medical resources. Hence, EWM was utilized to choose alternative facilities from the whole facilities in the first stage. Secondly, when patients at the facility points are critically ill, the robust optimization approach is used to configure the large rear hospital in the case of the uncertain number of patients in the second stage. The time window constraint is also constructed to ensure the timely treatment of patients. When patients at the facility are mild patients, they are directly isolated at the emergency medical point. Accordingly, a hybrid approach for emergency medical facilities location-allocation is proposed.

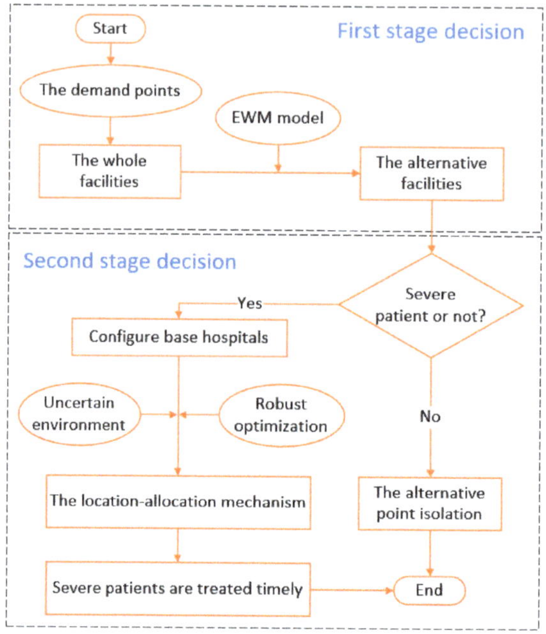

Figure 1. The resolution framework of the proposed hybrid approach.

2.2. Assumptions and Notation

In order to facilitate modeling, the following assumptions were made:

1. The established emergency medical facilities can meet the medical needs of patients in the city, regardless of the situation of transferring patients to other cities.
2. The radiation range of each facility is a small area, and the patients' number, which they receive, is the sum of the patients' number in the area.
3. All critically ill patients are treated by large rear hospitals, which do not occupy the medical resources of the facility point. Additionally, the rear hospitals can meet the treatment needs of the assigned critically ill patients.

The notation in this paper is illustrated in Table 1.

Table 1. The utilized notation in this paper.

Sets	
I	The collection of the whole emergency medical facilities (reconfigurable convention and exhibition centers, sports venues, schools), $i \in I, i = 1, 2, \ldots, n$.
J	The collection of existing large rear hospitals (Grade II and above), $j \in J, j = 1, 2, \ldots, m$.
K	The collection of patient types (mild, moderate, and severe three disease grades, represented by 1, 2, 3), $k \in K, k = 1, 2, 3$.
Decision variables	
$x_i = \begin{cases} 1, & \text{Open emergency medical facility point } i. \\ 0, & \text{Otherwise.} \end{cases}$	
$y_{ij} = \begin{cases} 1, & \text{Patients at facility point } i \text{ are serviced by hospital } j. \\ 0, & \text{Otherwise.} \end{cases}$	
$z_j = \begin{cases} 1, & \text{Select hospital } j \text{ to treat critically ill patients.} \\ 0, & \text{Otherwise.} \end{cases}$	

Table 1. *Cont.*

Parameters	
S	Number of emergency medical facilities opened.
f_i	Operating cost of emergency medical facility point i.
h_{ij}	Distance from facility point i to hospital j.
c_t	Unit driving cost from facility point i to hospital j.
d_{ik}	Patients' number of k type at facility point i.
θ_k	The proportion of k type patients, respectively represents the severity level of patients.
c_j	The maximum service capacity of large rear hospital j.

3. Emergencies Facilities Location Modeling

3.1. Emergency Medical Facilities Alternatives Selection Based on EWM

In this section, we will reveal the first stage of the hybrid approach. Taking into account the ease of the centralized utilization of medical resources, it is impossible for public health departments to establish emergency medical facilities at every point. The scientific and rational decision is to select some facilities from the total of facilities as alternatives. Consequently, this paper utilizes EWM to make location decisions.

As an objective and comprehensive weighting method, EWM determines weight mainly based on the information amount transmitted to decision-makers by each index, which can effectively avoid the influence of subjective factors. Then, we can make weight calculations more scientific and reasonable. The advantages of EWM are as follows: (1) EWM can deeply reflect the ability to distinguish indicators and determine a good weight; (2) Wight assignment is more objective, theoretical, and reliable; (3) The procedure is simple and easy to practice, EWM does not require other software analysis. Therefore, the EWM method was utilized to make the first-stage decision in this paper.

The selection principles of the evaluation indicators are as follows: (1) Objective and true principles. The selected indicators should be objective and true. The data sources should be based on the official data information so as to ensure that the indicators can objectively reflect the real situation of each region and avoid deviations between the data that are caused by personal subjective assumptions and the actual situation. (2) Operability principle. Ensure that the selected index data can be obtained from statistical information released by national government departments or official media. Avoid indicators with vague information or a different statistical caliber. Additionally, we should adopt relatively easy-to-obtain and relatively stable indicator information. (3) Principle of representativeness. The selected indicators should have a certain logical relationship with each other to reflect the overall situation of each region to the greatest extent. The number of indicators should be moderate. Too many indicators will lead to high similarity, which greatly reduces computational efficiency. However, too few indicators will lead to a lack of convincing evaluation results, which is not conducive to reflecting the real situation.

According to the selection principle of evaluation indicators, this paper mainly considers three influential factors of facility location, namely cost factor, capacity factor, and infrastructure factor. Then, we constructed the evaluation index system of emergency medical locations during the epidemic situation, including the construction cost of facilities, transportation convenience, the patients' number that the facility can accommodate, regional population density, accessibility of patients, and the number of hospitals within 10 km. In terms of cost factors, due to the particularity of emergency medical facilities, this paper only considers the construction cost of facilities. Construction costs are determined according to the scale of the facility point. It is the most representative of all the cost-influencing factors. The capacity factor mainly considers the population density in the region and the number of people that can be accommodated at the facility. The capacity limit of the emergency medical point determines its maximum service capacity. The emergency treatment demand needed to be met in the administrative area to the greatest extent. As infrastructure factors, the transportation convenience degree and the accessibility of the patients are sufficient to ensure that the infected are treated and isolated in the first place to prevent more people from becoming infected. The number of hospitals within 10 km can ensure that patients have access to transfer care and medical supplies at the large base hospital when the infected experience deterioration.

The procedure for selecting emergency medical facilities alternatives with the EWM are as follows.

Step 1: Construct the index matrix X. Let the $Y = \{y_1, y_2, \cdots, y_n\}$ indicate the necessary numbers of health care providers (i.e., all the facilities points) and $A = \{a_1, a_2, \cdots, a_m\}$ means the evaluation indicators. Let $I = \{1, 2, \cdots, n\}$ and $J = \{1, 2, \cdots, m\}$ be number sets. x_{ij} is the value of the $j-th$ evaluation index under the facility point i, and the index matrix X is as follows:

$$X = (x_{ij})_{n\times m} = \begin{bmatrix} x_{11} & x_{12} & \cdots & x_{1m} \\ x_{21} & x_{22} & \cdots & x_{2m} \\ \vdots & \vdots & \vdots & \vdots \\ x_{n1} & x_{n2} & \cdots & x_{nm} \end{bmatrix} \tag{1}$$

Step 2: Normalize the index matrix. Since the measurement units of each index are not uniform, it is necessary to standardize them to homogenize the heterogeneous indexes. Positive indicators and negative indicators are utilized for data standardization processing. In addition, the higher the positive indicator value is, the better. Additionally, the lower the negative indicator value is, the better. The specific methods are as follows.

Positive indicators:

$$x'_{ij} = \frac{x_{ij} - \min\{x_{ij}, \ldots, x_{nj}\}}{\max\{x_{1j}, \ldots, x_{nj}\} - \min\{x_{1j}, \ldots, x_{nj}\}} \tag{2}$$

Negative indicators:

$$x'_{ij} = \frac{\max\{x_{1j}, \ldots, x_{nj}\} - x_{ij}}{\max\{x_{1j}, \ldots, x_{nj}\} - \min\{x_{ij}, \ldots, x_{nj}\}} \tag{3}$$

Then, the normalized index matrix is:

$$X' = (x'_{ij})_{n\times m} = \begin{bmatrix} x'_{11} & x'_{12} & \cdots & x'_{1m} \\ x'_{21} & x'_{22} & \cdots & x'_{2m} \\ \vdots & \vdots & \vdots & \vdots \\ x'_{n1} & x'_{n2} & \cdots & x'_{nm} \end{bmatrix} \tag{4}$$

Step 3: Calculate the information entropy value of the $j-th$ index.

$$\varepsilon_j = -\frac{1}{\ln(n)} \left| \sum_{i=1}^{n} p_{ij} \ln(p_{ij}) \right|, j = 1, 2, \ldots, m. \tag{5}$$

Here, $p_{ij} = x'_{ij} / \sum_{i=1}^{n} x'_{ij}$ is the proportion of the $i-th$ emergency medical facility points under the $j-th$ indicator when $p_{ij} = 0$, $\ln(p_{ij})$ is meaningless. In this case, the definition of p_{ij} needs to be amended, that is, $p_{ij} = (1 + x'_{ij}) / \sum_{i=1}^{n} \left(1 + x'_{ij}\right)$.

Step 4: Calculate the entropy weight ω_j of each index.

$$\omega_j = \frac{1 - \varepsilon_j}{\left| m - \sum_{j=1}^{m} \varepsilon_j \right|} \tag{6}$$

Step 5: Calculate the comprehensive score s_i for each emergency medical facility point.

$$s_i = \sum_{j=1}^{m} \omega_j p_{ij}, i = 1, 2, \ldots, n. \tag{7}$$

Therefore, if the information entropy of one index is smaller, it indicates that the variation degree of its index value is greater. The more information it provides, the greater the role it plays in the comprehensive evaluation, and the greater its weight should be. Hence, in the specific analysis process, entropy can be used to calculate the weight of each index according to the variation degree of each index value. Additionally, all the indexes are then weighted to obtain a more objective, comprehensive evaluation result.

3.2. The Deterministic Model

In this section, we allocate the large rear hospital for the alternative facility points by using the robust optimization method to ensure the timely treatment of patients in the second stage.

After a comprehensive evaluation by EWM, the alternative sites were selected. Due to limited medical conditions, emergency medical centers can only be used as a place for treating ordinary patients. When patients are seriously ill, they still need to be sent to a large rear hospital for treatment. Accordingly, we also need to configure the large rear hospitals. We should rationally allocate these emergency medical facility points to rear large hospitals through quantitative analysis. On the basis of considering the capacity limitation of base hospitals and the time window limitation of treating patients, the emergency medical facilities configuration (EMFC) model was constructed with the goal of minimizing the total cost. Thus, a complete emergency medical security system was formed that can respond to public health emergencies.

When the patients' number of k type in the emergency medical facility i is known as d_{ik}, the deterministic model (DM) is as follows:

$$\min \sum_{i \in I} f_i x_i + \sum_{i \in I} \sum_{j \in J} \sum_{k \in K} c_t h_{ij} \theta_k d_{ik} y_{ij} + \sum_{i \in I} \sum_{j \in J} p(t_{ij}) x_i \tag{8}$$

$$s.t. \sum_{i \in I} x_i \leq S \tag{9}$$

$$\sum_{j \in J} y_{ij} = 1, \forall i \in I \tag{10}$$

$$y_{ij} \leq x_i, \forall i \in I, j \in J \tag{11}$$

$$\sum_{i \in I} \sum_{k \in K} \theta_k d_{ik} y_{ij} \leq c_j, \forall j \in J \tag{12}$$

$$p(t_{ij}) = \begin{cases} 0, & 0 \leq t_{ij} < ET \\ c_p(t_{ij} - ET), & ET \leq t_{ij} < LT, \forall i \in I, j \in J \\ +\infty, & t_{ij} \geq LT \end{cases} \tag{13}$$

$$x_i \in \{0, 1\}, \forall i \in I \tag{14}$$

$$y_{ij} \in \{0, 1\}, \forall i \in I, j \in J \tag{15}$$

The objective function minimizes the total cost, which is composed of the operating cost of the emergency medical facility point, the patient transfer cost from the emergency facility point to the large rear hospital, and the penalty cost that fails to meet the optimal treatment time window.

The constraint conditions are represented from Equation (9) to Equation (15). Specifically, Equation (13) is the penalty cost function defined in this paper. The travel time t_{ij} is determined by the ratio between the distance from the facility point to the hospital and the average speed of the transport vehicle, i.e., $(t_{ij} = h_{ij}/\ \overline{v}_j, \forall i \in I, j \in J)$. When the patient's condition becomes worse, the optimal treatment time is ET and the recoverable time window is $[ET, LT]$. When $0 \leq t_{ij} < ET$, the patient can arrive at the base hospital for treatment with no penalty cost, i.e., $(p(t_{ij}) = 0)$. When the patient arrives in the time window $[ET, LT]$, a punishment cost $c_p(t_{ij} - ET)$ is generated. Additionally, $p(t_{ij})$ will increase with the increase in the arrival time. Once the patient's arrival time exceeds the latest recoverable time LT, the patient's life safety is endangered, and the cost increases infinitely.

In addition, Equation (9) represents the maximum number of opened emergency medical facility points. Equation (10) indicates that each emergency medical facility point is serviced by one large rear hospital and can only be served by one rear hospital. Equation (11) means that patients can be sent to the rear hospital for treatment only when the emergency medical facilities have opened. Equation (12) indicates that the number of patients sent from the emergency facility point to the large rear hospital does not exceed the maximum service capacity of the hospital. Equation (14) and Equation (15) are both a 0–1 integer decision variable.

3.3. The Robust Model

When a public health emergency occurs, the number of patients is highly uncertain. Therefore, this paper draws on the robust decision idea of Bertsimas and Sim; we adopted the absolute robust criterion to optimize the target from the worst case [30]. Specifically, we used $\widetilde{d}_{ik}$ to represent the patients' number of k type in the emergency medical facility point i under uncertain circumstances. Then, we had $\widetilde{d}_{ik} \subseteq [d_{ik} - \hat{d}_{ik}\xi_{ik}, d_{ik} + \hat{d}_{ik}\xi_{ik}]$, where d_{ik} is the nominal value and $\hat{d}_{ik}$ is its disturbance value.

Under the disturbance of uncertain parameters, the original deterministic model can be equivalently transformed into the following robust optimization (RM) model:

$$\begin{aligned}
&\min\left\{\sum_{i\in I} f_i x_i + \max_{\xi\in U^p} \sum_{i\in I}\sum_{j\in J}\sum_{k\in K} c_t h_{ij}\theta_k(d_{ik} + \hat{d}_{ik}\xi_{ik})y_{ij} + \sum_{i\in I}\sum_{j\in J} p(t_{ij})x_i\right\} \\
&= \min\left\{\sum_{i\in I} f_i x_i + \sum_{i\in I}\sum_{j\in J}\sum_{k\in K} c_t h_{ij}\theta_k d_{ik} y_{ij} + \sum_{i\in I}\sum_{j\in J} p(t_{ij})x_i \right. \\
&\left. + \max_{\xi\in U^p} \sum_{i\in I}\sum_{j\in J}\sum_{k\in K} c_t h_{ij}\theta_k \hat{d}_{ik}\xi_{ik} y_{ij}\right\}
\end{aligned} \tag{16}$$

$$s.t.(9) \sim (11), (13) \sim (15)$$

$$\sum_{i\in I}\sum_{k\in K} \theta_k d_{ik} y_{ij} + \max_{\xi\in U^p} \sum_{i\in I}\sum_{k\in K} \theta_k \hat{d}_{ik}\xi_{ik} y_{ij} \leq c_j, \forall j \in J \tag{17}$$

Here, Equation (16) minimizes the total cost of the system in the worst case. Equation (17) indicates that the number of patients transported from the emergency medical facility point to the base hospital cannot exceed the maximum service capacity of the hospital in the worst case. In order to further specify the proposed robust model, three models based on different uncertainty sets were introduced as follows.

3.3.1. Budgeted Uncertainty Set

Proposition 1. *If the uncertain patients' number is defined as a budgeted uncertainty set, that is,* $U^p = \left\{\xi : \sum_{i\in I} \xi_{ik} \leq \Gamma_k,\ 0 \leq \xi_{ik} \leq 1,\ \forall k \in K\right\}$, *we can obtain the following robust equivalent model (REM):*

$$\min \sum_{i\in I} f_i x_i + \sum_{i\in I}\sum_{j\in J}\sum_{k\in K} c_t h_{ij}\theta_k d_{ik} y_{ij} + \sum_{i\in I}\sum_{j\in J} p(t_{ij})x_i + \eta \tag{18}$$

$$s.t.\eta \geq \sum_{i \in I}\sum_{k \in K} u_{ik} + \sum_{k \in K} v_k \Gamma_k \tag{19}$$

$$u_{ik} + v_k \geq c_t h_{ij} \theta_k \hat{d}_{ik} y_{ij}, \forall i \in I, j \in J, k \in K \tag{20}$$

$$u_{ik}, v_k \geq 0, \forall i \in I, k \in K \tag{21}$$

$$\sum_{i \in I} x_i \leq S \tag{22}$$

$$\sum_{j \in J} y_{ij} = 1, \forall i \in I \tag{23}$$

$$y_{ij} \leq x_i, \forall i \in I, j \in J \tag{24}$$

$$\sum_{i \in I}\sum_{k \in K} \theta_k d_{ik} y_{ij} + \sum_{i \in I}\sum_{k \in K} u'_{ik} + \sum_{k \in K} v'_k \Gamma_k \leq c_j, \forall j \in J \tag{25}$$

$$u'_{ik} + v'_k \geq \theta_k \hat{d}_{ik} y_{ij}, \forall i \in I, j \in J, k \in K \tag{26}$$

$$u'_{ik}, v'_k \geq 0, \forall i \in I, k \in K \tag{27}$$

$$p(t_{ij}) = \begin{cases} 0\,, & 0 \leq t_{ij} < ET \\ c_p(t_{ij} - ET)\,, & ET \leq t_{ij} < LT\,, \forall i \in I, j \in J \\ +\infty\,, & t_{ij} \geq LT \end{cases} \tag{28}$$

$$x_i \in \{0, 1\}, \forall i \in I \tag{29}$$

$$y_{ij} \in \{0, 1\}, \forall i \in I, j \in J \tag{30}$$

Here, η is the auxiliary variable. u_{ik} and v_k are the dual variable of the problem (16). u'_{ik} and v'_k are the dual variable of the problem (17).

Proof of Proposition 1. Because the definition of the budgeted uncertainty set is $U^p = \left\{\xi : \sum_{i \in I} \xi_{ik} \leq \Gamma_k,\ 0 \leq \xi_{ik} \leq 1,\ \forall k \in K\right\}$, then the maximization problem $\max_{\xi \in U^p}\left\{\sum_{i \in I}\sum_{j \in J}\sum_{k \in K} c_t h_{ij} \theta_k \hat{d}_{ik} \xi_{ik} y_{ij}\right\}$ in Equation (16) is equivalent to Equation (31).

$$\begin{aligned} &\max \sum_{i \in I}\sum_{j \in J}\sum_{k \in K} c_t h_{ij} \theta_k \hat{d}_{ik} \xi_{ik} y_{ij} \\ &s.t. \sum_{i \in I} \xi_{ik} \leq \Gamma_k \\ &\quad 0 \leq \xi_{ik} \leq 1 \qquad \forall i \in I, k \in K \end{aligned} \tag{31}$$

According to the strong duality theory, the dual variables u_{ik} and v_k are introduced, respectively. Additionally, we can further obtain Equation (32).

$$\begin{aligned} &\min \sum_{i \in I}\sum_{k \in K} u_{ik} + \sum_{k \in K} v_k \Gamma_k \\ &s.t. u_{ik} + v_k \geq c_t h_{ij} \theta_k d_{ik} y_{ij} \\ &\quad u_{ik}, v_k \geq 0 \qquad \forall i \in I, j \in J, k \in K \end{aligned} \tag{32}$$

Hence, we can transform the inner layer maximization problem into the minimization problem, and introduce the auxiliary variable η to obtain the robust equivalent model from Equations (18)–(21).

Similarly, according to the strong duality theory, the dual variables u'_{ik} and v'_k are, respectively introduced for Equation (17), and the inner layer maximization problem is transformed into the minimization problem. Thus, Equations (25)–(27) are obtained. □

3.3.2. Box Uncertainty Set

Proposition 2. *If the uncertain patients' number is defined as a box set, that is, $Z_{box} = \{\zeta \in R^M : \|\zeta\|_\infty \leq \psi\}$, ψ is the level of parameter uncertainty and the robust counterpart model in Section 3.3 can be constructed as follows:*

$$\min \sum_{i\in I} f_i x_i + \sum_{i\in I}\sum_{j\in J}\sum_{k\in K} c_t h_{ij}\theta_k d_{ik} y_{ij} + \psi \sum_{i\in I}\sum_{j\in J}\sum_{k\in K} c_t h_{ij}\theta_k \hat{d}_{ik} y_{ij} + \sum_{i\in I}\sum_{j\in J} p(t_{ij}) x_i$$

$$s.t. \sum_{i\in I} x_i \leq S$$

$$\sum_{j\in J} y_{ij} = 1, \forall i \in I$$

$$y_{ij} \leq x_i, \forall i \in I, j \in J$$

$$\sum_{i\in I}\sum_{k\in K} \theta_k d_{ik} y_{ij} + \psi \sum_{i\in I}\sum_{k\in K} \theta_k \hat{d}_{ik} y_{ij} \leq c_j, \forall j \in J$$

$$p(t_{ij}) = \begin{cases} 0, & 0 \leq t_{ij} < ET \\ c_p(t_{ij} - ET), & ET \leq t_{ij} < LT, \forall i \in I, j \in J \\ +\infty, & t_{ij} \geq LT \end{cases}$$

$$x_i \in \{0,1\}, \forall i \in I$$

$$y_{ij} \in \{0,1\}, \forall i \in I, j \in J$$

Proof of Proposition 2. Suppose $\sum_{i\in I} f_i x_i + \sum_{i\in I}\sum_{j\in J} p(t_{ij}) x_i = Q$. According to the definition of the box set, the uncertain patients' number can be written as:

$$\sum_{i\in I}\sum_{j\in J}\sum_{k\in K} c_t h_{ij}\theta_k d_{ik} y_{ij} + \sum_{i\in I}\sum_{j\in J}\sum_{k\in K} \zeta c_t h_{ij}\theta_k \hat{d}_{ik} y_{ij} + Q \leq H, (\zeta \in R^M : \|\zeta\|_\infty \leq \psi)$$

Then, we can obtain:

$$\sum_{i\in I}\sum_{j\in J}\sum_{k\in K} \zeta c_t h_{ij}\theta_k \hat{d}_{ik} y_{ij} \leq H - \sum_{i\in I}\sum_{j\in J}\sum_{k\in K} c_t h_{ij}\theta_k d_{ik} y_{ij} - Q, (\zeta \in R^M : \|\zeta\|_\infty \leq \psi)$$

In the worst case, we have:

$$\max_{\|\zeta\|_\infty \leq \psi} \sum_{i\in I}\sum_{j\in J}\sum_{k\in K} \zeta c_t h_{ij}\theta_k \hat{d}_{ik} y_{ij} \leq H - \sum_{i\in I}\sum_{j\in J}\sum_{k\in K} c_t h_{ij}\theta_k d_{ik} y_{ij} - Q$$

Because the maximum value on the left side of the inequality is $\psi \sum_{i\in I}\sum_{j\in J}\sum_{k\in K} c_t h_{ij}\theta_k \hat{d}_{ik} y_{ij}$, the explicit constraint form can be obtained:

$$\sum_{i\in I}\sum_{j\in J}\sum_{k\in K} c_t h_{ij}\theta_k d_{ik} y_{ij} + \psi \sum_{i\in I}\sum_{j\in J}\sum_{k\in K} c_t h_{ij}\theta_k \hat{d}_{ik} y_{ij} + Q \leq H$$

Similarly, the robust counterpart of constraint 12 can be obtained. Therefore, the model based on the box uncertainty set is proved. □

3.3.3. Ellipsoid Uncertainty Set

Proposition 3. *If the uncertain patients' number is defined as an ellipsoid set, that is, $Z_{ellipsoid} = \{\zeta \in R^M : \|\zeta\|_2 \leq \Omega\}$, Ω is the level of parameter uncertainty and the robust counterpart model in Section 3.3 can be built as follows:*

$$\min \sum_{i\in I} f_i x_i + \sum_{i\in I}\sum_{j\in J}\sum_{k\in K} c_t h_{ij}\theta_k d_{ik} y_{ij} + \Omega \sqrt{\left(\sum_{i\in I}\sum_{j\in J}\sum_{k\in K} c_t h_{ij}\theta_k \hat{d}_{ik} y_{ij}\right)^2} + \sum_{i\in I}\sum_{j\in J} p(t_{ij}) x_i$$

$$s.t. \sum_{i\in I} x_i \leq S$$

$$\sum_{j \in J} y_{ij} = 1, \forall i \in I$$
$$y_{ij} \leq x_i, \forall i \in I, j \in J$$
$$\sum_{i \in I} \sum_{k \in K} \theta_k d_{ik} y_{ij} + \Omega \sqrt{\left(\sum_{i \in I} \sum_{k \in K} \theta_k \hat{d}_{ik} y_{ij} \right)^2} \leq c_j, \forall j \in J$$
$$p(t_{ij}) = \begin{cases} 0, & 0 \leq t_{ij} < ET \\ c_p(t_{ij} - ET), & ET \leq t_{ij} < LT, \forall i \in I, j \in J \\ +\infty, & t_{ij} \geq LT \end{cases}$$
$$x_i \in \{0,1\}, \forall i \in I$$
$$y_{ij} \in \{0,1\}, \forall i \in I, j \in J$$

Proof of Proposition 3. Suppose $\sum_{i \in I} f_i x_i + \sum_{i \in I} \sum_{j \in J} p(t_{ij}) x_i = Q$. According to the definition of the ellipsoid set, the uncertain patients' number can be written as:

$$\sum_{i \in I} \sum_{j \in J} \sum_{k \in K} c_t h_{ij} \theta_k d_{ik} y_{ij} + \sum_{i \in I} \sum_{j \in J} \sum_{k \in K} \zeta c_t h_{ij} \theta_k \hat{d}_{ik} y_{ij} + Q \leq H, (\zeta \in R^M : \|\zeta\|_2 \leq \Omega)$$

Then, we can obtain:

$$\sum_{i \in I} \sum_{j \in J} \sum_{k \in K} \zeta c_t h_{ij} \theta_k \hat{d}_{ik} y_{ij} \leq H - \sum_{i \in I} \sum_{j \in J} \sum_{k \in K} c_t h_{ij} \theta_k d_{ik} y_{ij} - Q, (\zeta \in R^M : \|\zeta\|_2 \leq \Omega)$$

At worst case, we have:

$$\max_{\|\zeta\|_2 \leq \Omega} \sum_{i \in I} \sum_{j \in J} \sum_{k \in K} \zeta c_t h_{ij} \theta_k \hat{d}_{ik} y_{ij} \leq H - \sum_{i \in I} \sum_{j \in J} \sum_{k \in K} c_t h_{ij} \theta_k d_{ik} y_{ij} - Q$$

Consequently, the explicit form of the above formula can be obtained as:

$$\sum_{i \in I} \sum_{j \in J} \sum_{k \in K} c_t h_{ij} \theta_k d_{ik} y_{ij} + \Omega \sqrt{\left(\sum_{i \in I} \sum_{j \in J} \sum_{k \in K} c_t h_{ij} \theta_k \hat{d}_{ik} y_{ij} \right)^2} + Q \leq H$$

Similarly, the robust counterpart of constraint 12 can be obtained. Therefore, the model based on the ellipsoid uncertainty set is proved. □

4. Simulations

In order to verify the proposed method, this section shows an emergency management example under COVID-19.

4.1. Background and Data Sources

This paper chooses Huanggang City to conduct a numerical experiment, which was severely affected by the coronavirus. Huanggang has a total of 10 administrative areas. We took the township as the emergency demand points unit to carry out the detailed division for a total of 127 demand points. The emergency medical facility point is a large open area with flat terrain and convenient transportation. A total of 30 points are selected. Simultaneously, seven large rear hospitals with grades II and above were selected. The number of people per emergency demand point was obtained from the National Bureau of Statistics in 2017, while the number of confirmed COVID-19 patients in Huanggang was obtained from the National Health Commission of the People's Republic of China released on 21 March 2020.

The selection of emergency facilities is based on the service capacity (i.e., Hongshan stadium), that is, $c_e = \frac{venues\ beds}{venues\ area} \times facilities\ point\ area$. The attraction factor of the facility point is calculated by the hospitals' number within 10 km of each facility point. The reference attraction factor was one, and the attraction factor increased by 0.1 for each additional hospital, and so on. The relevant data of the demand points, emergency medical facility points, large rear hospitals, and the number of patients are shown in Tables 2–5, respectively. The detailed distribution of the residents' demand points, candidate facility points, and large rear hospitals is shown in Figure 2. The color distribution of each administrative area is determined according to the number of local patients with COVID-19. The more patients, the darker the color will be.

Table 2. Latitude and longitude coordinates of demand points and population size.

No.	Coordinate	Population	No.	Coordinate	Population	No.	Coordinate	Population
1	114.66064,31.45983	80,798	44	115.65594,30.76541	19,731	87	115.69542,30.30748	50,072
2	114.60284,31.27431	119,425	45	115.63523,30.88881	22,065	88	115.85314,30.51451	30,936
3	114.49922,31.28993	39,658	46	115.75470,30.99033	32,412	89	115.82113,30.31645	36,759
4	114.55444,31.15556	50,883	47	115.90135,31.00211	18,108	90	115.56911,29.85114	161,582
5	114.56646,31.05605	32,145	48	115.75740,30.88672	32,279	91	115.56400,29.85043	34,026
6	114.44895,31.30736	35,447	49	115.76842,30.81582	42,325	92	115.42095,29.91312	18,948
7	114.64585,31.02003	30,639	50	115.61337,30.64183	25,045	93	115.70016,29.88795	4083
8	114.64488,30.96300	23,432	51	115.63833,30.83598	19,318	94	115.61056,30.11381	113,247
9	114.70241,31.14612	56,044	52	115.93407,30.90572	7477	95	115.73690,30.08317	50,297
10	114.53016,31.45716	43,804	53	114.86957,30.63203	71,185	96	115.71488,30.01275	64,690
11	114.64299,31.28841	86,479	54	114.88533,30.74325	57,183	97	115.62547,29.93993	44,857
12	114.66770,31.38754	2049	55	115.07828,30.69643	24,063	98	115.61438,29.99534	35,921
13	114.99998,31.16661	55,601	56	115.19036,30.76037	41,226	99	115.55408,30.02849	32,203
14	115.02587,31.18524	68,485	57	115.09952,30.64720	29,827	100	115.47552,29.94930	47,311
15	115.04128,31.17716	80,722	58	115.02658,30.65196	19,640	101	115.70678,29.86632	37,082
16	114.80704,31.06828	50,109	59	114.98275,30.60512	32,134	102	115.93927,30.07392	141,488
17	115.12917,31.20715	31,297	60	115.08718,30.79427	19,117	103	115.92131,29.88258	98,543
18	115.01422,31.03803	61,839	61	115.05516,30.79163	18,816	104	115.98622,29.75636	97,956
19	114.88605,31.12271	42,918	62	114.93149,30.67896	22,811	105	116.00873,30.05003	21,945
20	114.98878,31.35650	47,632	63	115.26651,30.43888	193,988	106	115.84793,30.08979	59,759
21	115.09380,31.47408	36,169	64	115.02802,30.42591	113,356	107	115.98671,30.21229	33,323
22	115.18852,31.07276	36,567	65	115.34092,30.55166	71,852	108	115.94173,30.17009	13,971
23	115.17771,30.96000	32,079	66	115.12536,30.59354	66,860	109	115.89000,30.00755	65,122
24	115.31886,31.04682	37,066	67	115.17918,30.61747	49,785	110	115.80776,29.87894	71,168
25	115.23286,31.32758	40,083	68	115.23300,30.72709	78,925	111	115.82248,29.81693	55,883
26	115.07628,31.37152	34,513	69	115.44597,30.59109	14,885	112	116.03935,30.07820	27,127
27	114.83977,31.33245	43,233	70	115.41237,30.56442	35,562	113	115.90764,29.78591	47,872
28	114.75598,31.01214	20,520	71	115.47997,30.46645	49,312	114	115.95513,30.27099	12,357
29	115.03054,30.96647	26,470	72	115.27209,30.38250	35,922	115	115.98193,30.10304	34,299
30	115.37487,31.18694	37,992	73	115.11897,30.23280	60,655	116	115.90153,30.13119	27,634
31	114.84998,31.03664	40,457	74	115.14533,30.35013	48,792	117	116.10982,29.83206	15,919
32	115.27363,30.83227	57,980	75	115.53676,30.52150	22,214	118	114.88374,30.44167	156,011
33	115.46262,31.12651	59,219	76	115.44136,30.25094	150,600	119	114.90441,30.47002	21,939
34	115.67112,31.14024	30,041	77	115.33968,30.07489	67,040	120	114.88198,30.47291	22,977
35	115.60163,31.00290	29,204	78	115.38161,30.30637	68,543	121	114.97287,30.45216	16,593
36	115.47994,30.84096	30,825	79	115.50228,30.36850	30,920	122	114.94966,30.48861	21,550
37	115.19510,30.81844	35,918	80	115.61687,30.38513	42,447	123	114.91279,30.54566	25,416
38	115.39093,30.98976	50,823	81	115.79129,30.41924	42,065	124	115.03657,30.59585	25,856
39	115.55835,30.69722	59,270	82	115.80244,30.49354	19,624	125	114.98103,30.53740	19,654
40	115.40694,30.68214	33,132	83	115.42998,30.20307	49,666	126	115.00178,30.58077	4753
41	115.39610,30.78371	114,890	84	115.28774,30.14920	20,547	127	114.91652,30.44885	52,020
42	115.67654,30.74001	121,669	85	115.28122,30.27222	35,569			
43	115.61889,30.59068	16,027	86	115.58902,30.29642	67,756			

Table 3. Coordinates, service capacity, and attraction factors of candidate facility points.

No.	Coordinate	Capacity	Attraction Factor	No.	Coordinate	Capacity	Attraction Factor
1	114.63284,31.31431	27,840	1.3	16	114.89070,30.64937	4478	1.3
2	115.00939,31.16563	17,043	1.2	17	115.70328,30.81313	90,032	1.1
3	114.73256,31.10384	9100	1.1	18	114.95961,30.52801	29,708	1.2
4	115.95857,30.09049	118,450	1	19	115.41705,30.79636	49,879	1.3
5	115.55704,30.00018	6998	1.2	20	115.78056,30.90014	5282	1.2
6	115.10718,31.31427	10,989	1.4	21	115.38103,30.49740	73,502	1
7	114.62315,31.29476	104,917	1.1	22	115.20200,30.48535	8220	1.1
8	114.93146,30.62063	73,426	1.2	23	115.62718,30.00019	64,520	1.1
9	115.08277,30.23464	30,964	1.3	24	114.98103,30.53740	7598	1.2
10	115.02813,31.18019	35,097	1.1	25	115.05760,30.52585	45,008	1.4
11	115.93927,30.11392	9320	1	26	115.02813,31.18019	2890	1
12	115.01814,31.17779	3233	1.2	27	115.09380,30.96408	5354	1.1
13	114.90714,30.43954	52,560	1.1	28	115.41122,30.23686	6210	1.2
14	115.16036,30.64037	39,088	1	29	114.89070,30.64937	10,879	1.2
15	115.43629,30.23262	8352	1.4	30	115.55681,29.85051	4877	1.1

Table 4. Coordinates and number of beds in the large rear hospitals.

No.	Coordinate	Number of Beds	No.	Coordinate	Number of Beds
1	114.62522,31.28687	810	5	114.89880,30.47378	600
2	115.03035,31.18547	560	6	115.59644,29.87249	400
3	115.66802,30.73284	780	7	115.95089,30.08262	350
4	114.88141,30.45194	1050			

Table 5. The number of cases in each region.

Region	Number of Patients	Region	Number of Patients
Huangzhou District	968	Xishui County	303
Tuanfeng County	173	Qichun County	265
Hongan County	316	Huangmei County	284
Luotian County	69	Macheng County	243
Yingshan County	62	Wuxue County	224

The distance between the two points was calculated according to the longitude and latitude coordinates. Equation (33) can be utilized to convert the coordinates of longitude and latitude into the actual traveling distance h_{ij} between the two nodes i and j.

$$h_{ij} = k \cdot \frac{\sqrt{(x_i - x_j)^2 + (y_i - y_j)^2}}{180} \cdot \pi \cdot 6370 \tag{33}$$

Here, $(x_i, y_i), (x_j, y_j)$ is the longitude and latitude coordinates of the two points. The radius of the earth is 6370 (km). The formula $\frac{\sqrt{(x_i - x_j)^2 + (y_i - y_j)^2}}{180} \cdot \pi \cdot 6370$ is used to calculate the linear distance between the two points. The linear distance of the two points for 50 groups was extracted, and we compared this with the actual driving distance obtained from the Baidu map. The error value was obtained.

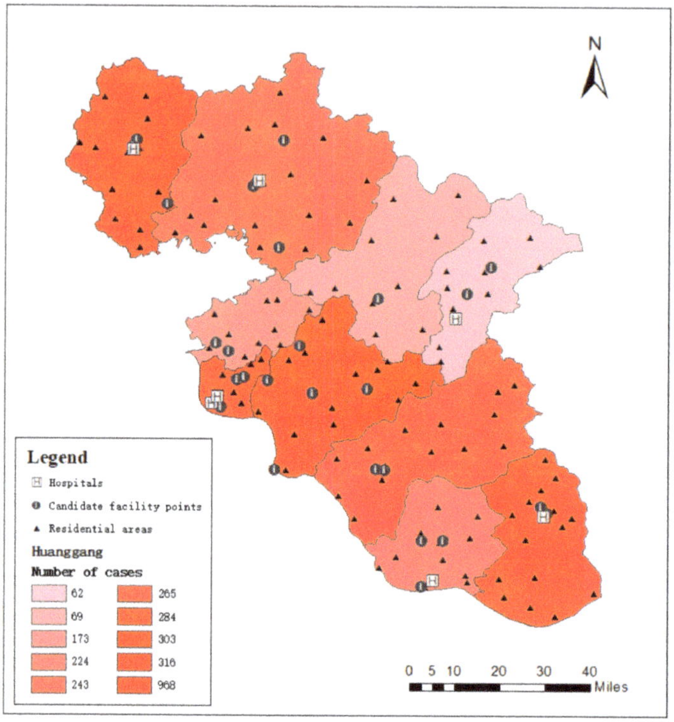

Figure 2. Distribution of demand points, candidate facility points, and base hospitals.

4.2. The Alternative Facilities Selection Based on EWM

The initial index data matrix of the candidate emergency medical facility points is composed of the following factors: the construction cost of facilities, transportation convenience, the patients' number that can be accommodated, regional population density, the accessibility of patients, and the number of hospitals within 10 km. Among them, the construction cost of the facility point is calculated at 1000 yuan per square meter. Transportation convenience is determined by the distance between the facility point and the nearest provincial or national highway. The accessibility of patients is determined based on the maximum time it takes for the demand point to reach the candidate facility point. The regional population density (10,000 people per square kilometer) is obtained according to the area and population of the region.

According to Equations (5) and (6), the information entropy value and weight vector of the six evaluation indexes for the normalized matrix are obtained, as shown in Table 6. Meanwhile, the comprehensive evaluation score of each candidate emergency medical facility is calculated, s_i = (0.3395;0.3860;0.1694;0.8913;0.1325;0.4670;0.7570;0.6649;0.4847;0.6181;0.2177; 0.2641;0.6105;0.6263;0.2178;0.1831;0.6789;0.4127;0.5967;0.1523;0.6507;0.0876;0.6078; 0.1077; 0.3519;0.1659;0.1309;0.2888;0.2024;0.1601).

Table 6. Information entropy and entropy weight.

	1	2	3	4	5	6
Information entropy ε_j	0.86431	0.96559	0.83973	0.95249	0.91932	0.94877
Entropy weight w_j	0.26617	0.06749	0.31439	0.09319	0.15826	0.10049

According to the comprehensive evaluation value S_i, ten emergency medical facilities with high evaluation values were selected: 4, 7, 8, 10, 13, 14, 17, 19, 21, and 23.

4.3. Robust Solution Process

After the alternative emergency medical facilities are selected by EWM, large rear hospitals should be configured rationally to ensure that severe patients can receive timely treatment. According to Equation (33), the distance h_{ij} between each facility point and the base hospital is obtained, as shown in Table 7. Other relevant parameters are set as: $c_t = 10$, $c_p = 6$, $\overline{v}_j = 35$ km/h, $ET = 120$ min, $LT = 480$ min, $\theta_1 = 1$, $\theta_2 = 0.5$, $\theta_3 = 0.1$. When the uncertain level Γ_k is considered, it is assumed that the variation amplitude of the corresponding constraints is equal (i.e., $\Gamma_k = \Gamma$) and Γ is an all integer. In this paper, MATLAB R2016a was utilized for programming, and CPLEX was called to solve the problem under the experimental environment of 8 GB memory and 1.60 GHz CPU with Intel Core i5.

Table 7. The distance between each facility points and the large rear hospital.

Facility Points	Large Base Rear Hospital						
	J_1	J_2	J_3	J_4	J_5	J_6	J_7
I_4	199.1641	319.1826	235.1414	505.2721	626.4589	281.9618	8.558354
I_7	0.906985	93.74792	395.6967	392.0141	481.4094	1149.629	1399.142
I_8	81.52122	127.5054	248.5002	78.25099	46.62485	667.7027	897.0696
I_{10}	46.33759	1.272584	260.4087	330.3708	399.213	951.1418	1115.944
I_{13}	99.2809	168.108	271.98	12.69828	19.58887	595.4059	858.4737
I_{14}	93.30548	124.6055	172.1071	149.7007	72.38121	589.0629	752.9345
I_{17}	130.9173	171.0052	29.24747	399.2292	485.3551	631.5039	600.2825
I_{19}	103.5559	121.9797	86.34575	283.1968	339.3343	627.7912	693.6411
I_{21}	121.5096	171.719	63.80801	223.1036	268.3869	440.9306	548.5269
I_{23}	181.3071	295.079	244.7432	387.752	482.9592	87.61668	259.9675

4.4. Result Analysis

When the disturbance ratio is 2%, and the uncertain level is $\Gamma = 5$, the configuration scheme is (4-7,7-1,8-5,10-2,13-4,14-5,17-3,19-3,21-5,23-6). The specific configuration scenario is shown in Figure 3. The green dot is the demand point of the residents, the blue square is the selected emergency medical facility point, and the red five-pointed star is the large rear hospital. The connecting line indicates the service relationship between the demand point, the facility point, and the base hospital. As can be seen from Figure 3, the needs of residents in each township have been met. The alternative emergency medical facility points (4,7,8,10,13,14,17,19,21,23) have corresponding large base hospitals to provide first-aid support to ensure the further transfer and treatment of critically ill patients. In addition, the optimal facility points are evenly distributed. One emergency medical facility has been established in each of the 10 administrative regions of Huanggang to ensure that the needs of the residents in each administrative region can be effectively covered by the emergency medical facility points. Additionally, the total traveled distance can be reduced. Similarly, we can obtain configuration plans in other scenarios. Due to space limitations, these will not be displayed here

The change in the optimal configuration scheme with a different uncertain level Γ and disturbance proportions is shown in Table 8. The optimal configuration scheme between the large base hospital and emergency medical facilities has also changed with the presence of uncertain patient numbers.

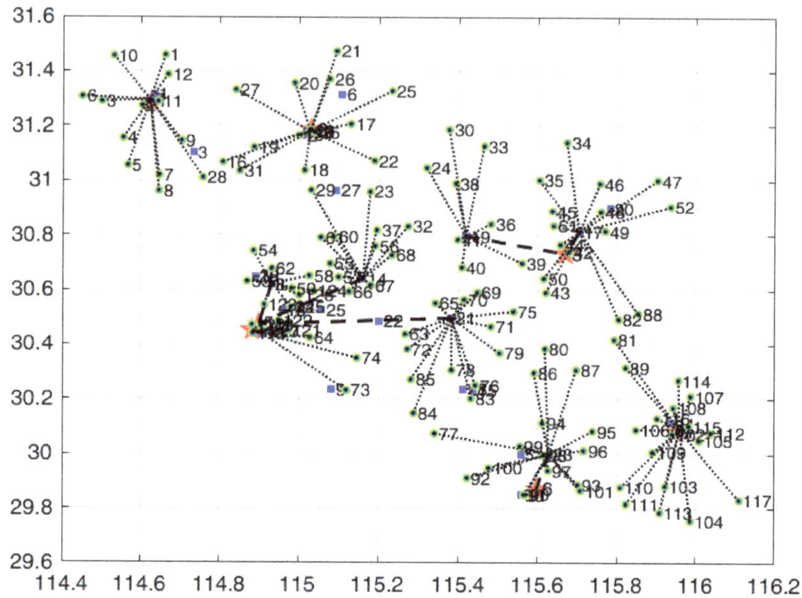

Figure 3. Configuration scenario with 2% disturbance ratio and Γ = 5.

Table 8. Configuration scheme with different disturbance proportions and uncertainty levels.

Γ	Disturbance in Proportion			
	2%	5%	10%	20%
0	4-7,7-1,8-5,10-2,13-4,14-5,17-3,19-3,21-3,23-6	4-7,7-1,8-5,10-2,13-4,14-5,17-3,19-3,21-3,23-6	4-7,7-1,8-5,10-2,13-4,14-5,17-3,19-3,21-3,23-6	4-7,7-1,8-5,10-2,13-4,14-5,17-3,19-3,21-3,23-6
2	4-7,7-1,8-5,10-2,13-4,14-5,17-3,19-3,21-5,23-6	4-7,7-1,8-5,10-2,13-4,14-5,17-3,19-3,21-5,23-6	4-7,7-1,8-4,10-2,13-5,14-5,17-3,19-3,21-5,23-6	4-7,7-1,8-4,10-2,13-5,14-5,17-3,19-3,21-5,23-6
4	4-7,7-1,8-5,10-2,13-4,14-5,17-3,19-3,21-5,23-6	4-7,7-1,8-5,10-2,13-4,14-5,17-3,19-3,21-5,23-6	4-7,7-1,8-4,10-2,13-5,14-5,17-3,19-3,21-5,23-6	4-7,7-1,8-4,10-2,13-5,14-4,17-3,19-2,21-5,23-6
6	4-7,7-1,8-4,10-2,13-5,14-5,17-3,19-3,21-5,23-6	4-7,7-1,8-4,10-2,13-5,14-5,17-3,19-3,21-5,23-6	4-7,7-1,8-4,10-2,13-5,14-4,17-3,19-2,21-5,23-6	4-7,7-1,8-4,10-2,13-5,14-4,17-3,19-2,21-5,23-6
8	4-7,7-1,8-4,10-2,13-5,14-4,17-3,19-2,21-5,23-6	4-7,7-1,8-4,10-2,13-5,14-4,17-3,19-2,21-5,23-6	4-7,7-1,8-4,10-2,13-5,14-4,17-3,19-2,21-5,23-7	4-7,7-1,8-4,10-2,13-5,14-4,17-3,19-2,21-5,23-7
10	4-7,7-1,8-4,10-2,13-5,14-4,17-3,19-2,21-5,23-6	4-7,7-1,8-4,10-2,13-5,14-4,17-3,19-2,21-5,23-7	4-7,7-1,8-4,10-2,13-5,14-4,17-3,19-2,21-5,23-7	4-7,7-1,8-4,10-2,13-5,14-4,17-3,19-2,21-5,23-7

The change in the total cost with a different uncertainty level Γ and disturbance proportions is shown in Figure 4. When Γ = 0, the robust model is equivalent to the deterministic model, and the total cost is 4.47009×10^9. Compared with the robust configuration model, the emergency medical facilities configuration deterministic model (EMFC) is not robust because it does not take into account the uncertain number of patients at the emergency medical points, so it has a certain deviation from the actual situation. As can be seen from Figure 4, the total cost increases with the increase in the uncertainty level Γ when the disturbance proportion remains unchanged. Additionally, the higher the disturbance proportion is, the higher the total cost will be when the uncertainty level remains unchanged. Simultaneously, the uncertainty level Γ can measure the risk preference of decision-makers to some extent. Accordingly, decision-makers can choose the optimal combination of uncertainty levels and the disturbance proportion according to their preference degree to the uncertain risk. If the decision-maker pursues a preference for risk, he can choose a small level of uncertainty and disturbance ratio. However, he must bear the possible losses caused by uncertainty in mind. If the decision-maker has a preference for risk aversion, he can select a large uncertainty level and disturbance proportion to provide a large probability guarantee for the effectiveness and feasibility of the configuration scheme. However, the total cost of the system operation will increase. If the decision-maker is risk neutral, he can choose a compromise.

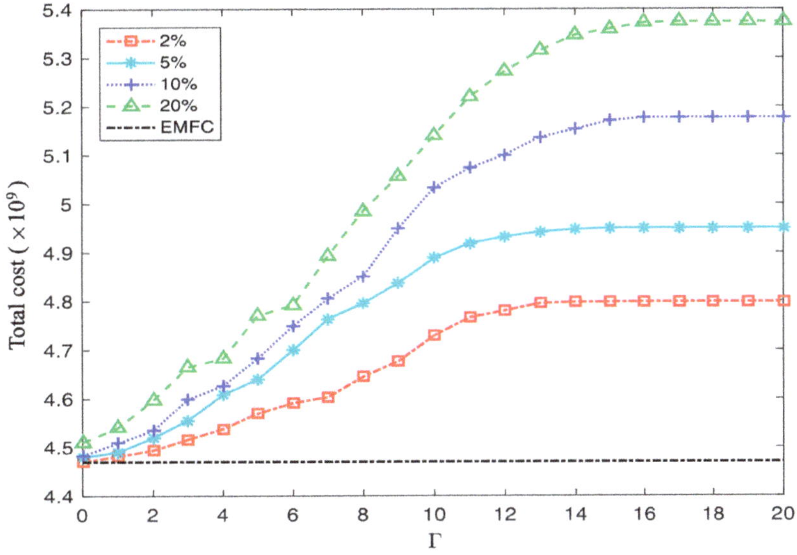

Figure 4. The total cost varies with different disturbance proportions and Γ.

It is worth mentioning that although the total cost varies with different disturbance proportions and uncertainty levels, there are only six configuration schemes. This further indicates that the model has good robustness, and the optimal scheme is not sensitive to parameter perturbation. Among them, the solution of the deterministic model is (4-7,7-1,8-5,10-2,13-4,14-5,17-3,19-3,21-3,23-6), as shown in Figure 5. The former number represents the emergency medical facility point, and the latter number indicates the large rear hospital that serves it when a patient is in an emergency. The blue dot in the figure represents the whole emergency medical facility, the red five-pointed star represents the large rear base hospital, and the black dotted line shows the service relationship between the emergency medical facility and the base hospital. When the disturbance proportion and uncertainty level Γ are small, the configuration scheme is (4-7,7-1,8-5,10-2,13-4,14-5,17-3,19-3,21-5,23-6), as shown in Figure 6. Additionally, the decision-maker with a risk preference can choose this scheme. When the disturbance proportion and uncertainty level Γ are large, the configuration scheme is (4-7,7-1,8-4,10-2,13-5,14-4,17-3,19-2,21-5,23-7), as shown in Figure 7. Additionally, the decision-maker with a risk aversion can choose this scheme. The rest of the configuration schemes are (4-7,7-1,8-5,10-2,13-5,14-5,17-3,19-3,21-5,23-6), (4-7,7-1,8-4,10-2,13-5,14-5,17-3,19-3,21-5,23-6), and (4-7,7-1,8-4,10-2,13-5,14-4,17-3,19-2,21-5,23-6). In this case, the decision-maker with risk neutrality can choose this solution. We will not show the configuration scheme figures here.

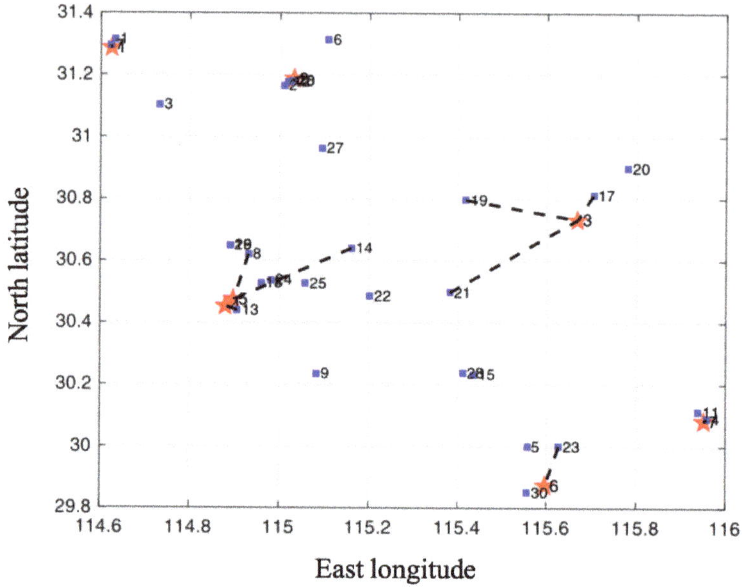

Figure 5. The configuration scheme with $\Gamma = 0$.

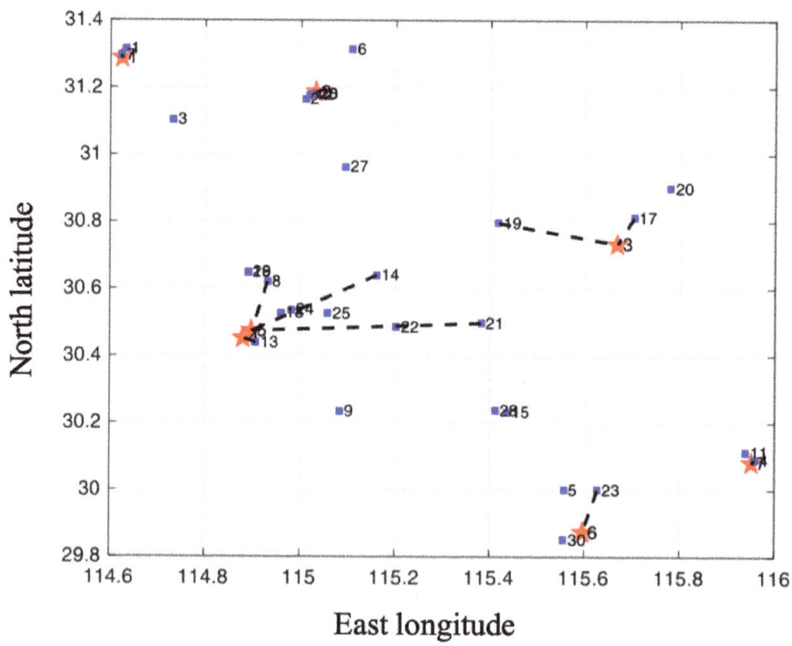

Figure 6. The configuration scheme with $\Gamma = 2$.

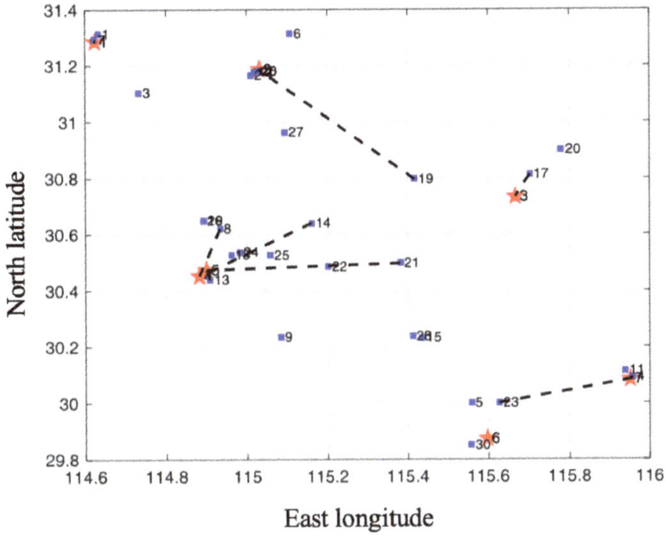

Figure 7. The configuration scheme with $\Gamma = 10$.

The calculation time of each scheme is shown in Figure 8. The shortest time is 1.9127 s, and the longest time is 11.6776 s. Additionally, the average time is 7.54 s, which meets the actual demand. As can be seen from Figure 8, compared with the robust configuration model, the deterministic EMFC model is not robust because it does not take into account the uncertain number of patients at the emergency medical points. Therefore, the solution time of the EMFC model is not sensitive to uncertain level parameters Γ. When the Γ is small, the solution time is relatively short. When the Γ is large, the solution time increases. This is because the increase in the uncertainty level leads to an increase in the search range of the solution, which in turn leads to an increase in the solution time. However, the longest solution time is only about 12 s, which fully meets the actual demand.

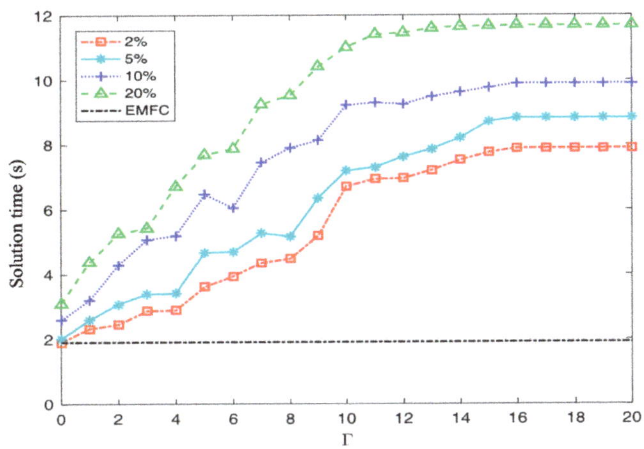

Figure 8. The total solution time varies with different disturbance proportions and Γ.

To sum up, this paper takes Huanggang City as an example to provide the optimal emergency medical facilities location and configuration scheme under COVID-19. Moreover, the impact of uncertain parameters on the total target cost, configuration scheme, and solution time of the model is deeply analyzed. Additionally, the feasibility and robustness of the proposed method are verified.

5. Conclusions

5.1. Discussions

This paper investigates a hierarchical diagnosis and treatment system for emergency medical facilities' location-allocation under uncertain circumstances. Firstly, taking into account the ease of the centralized utilization of medical resources, we adopted EWM to select alternative facilities from the whole of the facilities. Secondly, three uncertainty sets were introduced to describe the uncertainty of the patients' number. A robust optimization model with capacity and time window constraints was constructed to configure the large rear hospital to ensure the timely treatment of patients. The comparison between Figures 4 and 8 shows that although the total cost and solution time of the deterministic location-allocation model is lower, the deterministic model is not robust and cannot effectively describe the uncertain number of patients under the epidemic situation. However, the robust optimization model proposed in this paper not only considers the actual uncertain number of patients but also does not need to know the probability distribution of the number of patients in advance. Additionally, the solution time of the robust model is less than 12 s, which is very consistent with the actual situation. Finally, numerical simulation experiments were conducted to solve the emergency medical facilities' location and configuration in Huanggang City under COVID-19. The results show that the location-allocation decision method proposed in this paper is scientific and effective. The proposed method can meet the treatment needs of patients after public health emergencies and effectively reduce driving time.

During the epidemic period, the hierarchical diagnosis and treatment mode avoids the paralysis of large hospitals caused by the concentration of a large number of patients. It significantly improves the use efficiency of medical resources. This study proposes a hybrid approach of emergency medical facility location-allocation. We have a two-step plan for post-outbreak isolation and treatment. In the first stage, 10 facilities with the highest scores are selected from 30 facilities by EWM, which are regarded as community emergency medical points. When there are critical patients who cannot be handled by community medical centers, the second stage is to send the critical patients to large base hospitals for treatment.

The hierarchical diagnosis and treatment mode plays an obvious role in reversing the unreasonable pattern of medical resource allocation and solving the problem of unbalanced medical resource allocation during the epidemic period. Based on the construction of a coordinated medical and health service network between urban and rural areas, the hierarchical diagnosis and treatment mode has rationally allocated medical resources, effectively revitalized the stock of medical resources, and improved the allocation and use efficiency of medical resources by relying on the majority of hospitals and grassroots medical and health institutions. The most economical and effective measures to deal with the epidemic are to improve the level of community medical care and complete the system. Therefore, this study has a certain practical significance for public health authorities to improve the scientific level of epidemic prevention and control.

5.2. Future Directions

The proposed method in this paper can provide a scientific and reasonable reference for decision-makers to choose the optimal facility layout plan. In order to further improve the practical application value of the proposed model, future research work will refine the factors affecting the location decision. Additionally, we could consider the existence of various factors, such as the traffic time uncertainty under different road congestion conditions and resource constraints, and isolation from the public, so as to further investigate the robust optimization model. In future research directions, we can also consider the impact of facility interruption on the hierarchical diagnosis system, which will make the emergency medical location-allocation model more realistic. Meanwhile, this paper only studies the budgeted uncertainty model. The next work can be compared with the box uncertainty model and ellipsoid uncertainty model, which can further illustrate the effectiveness of the proposed method.

In addition, group consensus plays an important role in decision-making [55–58]. In future studies, we can invite experts from different fields to help emergency management departments make better decisions through the consensus-building process. There are various methods for facility location. This paper only studies the impact of the robust optimization method on facility location. In the future, we can extend the fuzzy rough decision-making approach [59] and multi-criteria decision-making [60] to the emergency medical facilities location. Supply chains have become a hot research field in recent years [61]. In the future, we can study how to improve the fairness and efficiency of supply chains in the transportation of emergency medical supplies. In the future, we can consider adding machine learning [62] methods to the location of emergency medical facilities.

Author Contributions: Conceptualization, L.W. and M.Y.; data curation, F.X. and M.Y.; formal analysis, L.W. and M.Y.; investigation, F.X. and S.Q.; methodology, L.W. and M.Y.; resources, F.X.; validation, F.X. and M.Y.; writing—original draft, L.W. and M.Y.; writing—review and editing, F.X., S.Q. and M.Y. All authors have read and agreed to the published version of the manuscript.

Funding: This research was supported by the National Natural Science Foundation of China (No. 72171149, 72171123).

Institutional Review Board Statement: Not applicable.

Informed Consent Statement: Not applicable.

Data Availability Statement: The data used to support the findings of this study are available from the database with the web address http://www.nhc.gov.cn/ (accessed on 21 March 2020). The data presented in this study are available on request from the corresponding author.

Acknowledgments: The authors would like to thank the anonymous reviewers for their comments and suggestions.

Conflicts of Interest: The authors declare that they have no conflicts of interest.

References

1. Liu, J.; Bai, J.Y.; Wu, D.S. Medical supplies scheduling in major public health emergencies. *Transp. Res. Part E Logist. Transp. Rev.* **2021**, *154*, 102464. [CrossRef]
2. Xu, Y.; Li, J.; Liu, Y.W. Quantitative evaluation of provincial government plans for public health emergencies in China. *Int. J. Disast. Risk Reduct.* **2022**, *81*, 103292. [CrossRef]
3. Alsafi, R.T. Lessons from SARS-CoV, MERS-CoV, and SARS-CoV-2 Infections: What We Know So Far. *Can. J. Infect. Dis. Med.* **2022**, *2022*, 1156273. [CrossRef] [PubMed]
4. Hartt, M. COVID-19: A lonely pandemic. *Cities Health* **2021**, *5*, S80–S82. [CrossRef]
5. Zhang, X.; Ding, Z.J.; Hang, J.Q.; He, Q.Z. How do stock price indices absorb the COVID-19 pandemic shocks? *N. Am. J. Econ. Financ.* **2022**, *60*, 101672. [CrossRef]
6. Zhu, L.; Chen, P.L.; Dong, D.D.; Wang, Z.X. Can artificial intelligence enable the government to respond more effectively to major public health emergencies? Taking the prevention and control of Covid-19 in China as an example. *Socio Econ. Plan. Sci.* **2022**, *80*, 101029. [CrossRef]
7. Liu, M.; Xu, X.F.; Cao, J.; Zhang, D. Integrated planning for public health emergencies: A modified model for controlling H1N1 pandemic. *J. Oper. Res. Soc.* **2020**, *71*, 748–761. [CrossRef]
8. Zhang, L.; Zhao, W.J.; Sun, B.B.; Huang, Y.; Glänzel, W. How scientific research reacts to international public health emergencies: A global analysis of response patterns. *Scientometrics* **2020**, *124*, 747–773. [CrossRef]
9. Jin, R.; Xia, T.; Liu, X.; Murata, T.; Kim, K.S. Predicting Emergency Medical Service Demand With Bipartite Graph Convolutional Networks. *IEEE Access* **2021**, *9*, 9903–9915. [CrossRef]
10. Xu, Z.Q.; Cheng, Y.K.; Yao, S.L. Tripartite Evolutionary Game Model for Public Health Emergencies. *Discrete Dyn. Nat. Soc.* **2021**, *2021*, 6693597. [CrossRef]
11. Tessema, F.S.; Koya, P.R.; Bole, B.K. Optimal Control and Cost-Effectiveness Analysis of Cholera with Vaccination. *J. Math.* **2022**, *2022*, 1705277. [CrossRef]

12. Deng, X.C. Location-Routing Problem of Emergency Facilities under Uncertain Demand by Branch-Price and Cut. *J. Math.* **2021**, *2021*, 9152989. [CrossRef]
13. Yang, N.; Shen, L.Y.; Shu, T.H.; Liao, S.J.; Peng, Y.; Wang, J.H. An integrative method for analyzing spatial accessibility in the hierarchical diagnosis and treatment system in China. *Soc. Sci. Med.* **2021**, *270*, 113656. [CrossRef] [PubMed]
14. Schullerus, G.; Supavatanakul, P.; Krebs, V.; Lunze, J. Modelling and hierarchical diagnosis of timed discrete-event systems. *Math. Comp. Model. Dyn.* **2006**, *12*, 519–542. [CrossRef]
15. Toregas, C.; Swain, R.; ReVelle, C.; Bergman, L. The Location of Emergency Service Facilities. *Oper. Res.* **1971**, *19*, 1363–1373. [CrossRef]
16. He, Q.Z.; Tong, H.; Liu, J.B. How Does Inequality Affect the Residents' Subjective Well-Being: Inequality of Opportunity and Inequality of Effort. *Front. Psychol.* **2022**, *13*, 843854. [CrossRef] [PubMed]
17. Gigovic, L.; Pamucar, D.; Bozanic, D.; Ljubojevic, S. Application of the GIS-DANP-MABAC multi-criteria model for selecting the location of wind farms: A case study of Vojvodina, Serbia. *Renew. Energ.* **2017**, *103*, 501–521. [CrossRef]
18. Biswas, S.; Pamucar, D. Facility Location Selection for B-Schools in Indian Context: A Multi-Criteria Group Decision Based Analysis. *Axioms* **2020**, *9*, 77. [CrossRef]
19. Pamucar, D.; Gigovic, L.; Bajic, Z.; Janosevic, M. Location Selection for Wind Farms Using GIS Multi-Criteria Hybrid Model: An Approach Based on Fuzzy and Rough Numbers. *Sustainability* **2017**, *9*, 1315. [CrossRef]
20. Boonmee, C.; Arimura, M.; Asada, T. Facility location optimization model for emergency humanitarian logistics. *Int. J. Disast. Risk Reduct.* **2017**, *24*, 485–498. [CrossRef]
21. Zhang, H.Z.; Zhang, K.; Chen, Y.T.; Ma, L. Multi-objective two-level medical facility location problem and tabu search algorithm. *Inf. Sci.* **2022**, *608*, 734–756. [CrossRef]
22. Liu, K.N. GIS-based MCDM framework combined with coupled multi-hazard assessment for site selection of post-earthquake emergency medical service facilities in Wenchuan, China. *Int. J. Disast. Risk Reduct.* **2022**, *73*, 102873. [CrossRef]
23. Yücel, E.; Salman, F.S.; Bozkaya, B.; Gökalp, C. A data-driven optimization framework for routing mobile medical facilities. *Ann. Oper. Res.* **2020**, *291*, 1077–1102. [CrossRef]
24. Zhen, L.; Wang, K.; Liu, H.C. Disaster Relief Facility Network Design in Metropolises. *IEEE Trans. Syst. Man Cybern. Syst.* **2015**, *45*, 751–761. [CrossRef]
25. Ji, Y.; Li, H.H.; Zhang, H.J. Risk-Averse Two-Stage Stochastic Minimum Cost Consensus Models with Asymmetric Adjustment Cost. *Group Decis. Negot.* **2022**, *31*, 261–291. [CrossRef]
26. Soyster, A.L. Technical Note—Convex Programming with Set-Inclusive Constraints and Applications to Inexact Linear Programming. *Oper. Res.* **1973**, *21*, 1154–1157. [CrossRef]
27. Ben-Tal, A.; Nemirovski, A. Robust solutions of uncertain linear programs. *Oper. Res. Lett.* **1999**, *88*, 411–424. [CrossRef]
28. Ben-Tal, A.; Nemirovski, A. Robust Convex Optimization. *Math. Oper. Res.* **1998**, *23*, 769–805. [CrossRef]
29. Bertsimas, D.; Sim, M. Robust discrete optimization and network flows. *Math. Program.* **2003**, *98*, 49–71. [CrossRef]
30. Bertsimas, D.; Sim, M. The Price of Robustness. *Oper. Res.* **2004**, *52*, 35–53. [CrossRef]
31. Ji, Y.; Jin, X.W.; Xu, Z.S.; Qu, S.J. A mixed 0-1 programming approach for multiple attribute strategic weight manipulation based on uncertainty theory. *J. Intell. Fuzzy Syst.* **2021**, *41*, 6739–6754. [CrossRef]
32. Qu, S.J.; Li, Y.M.; Ji, Y. The mixed integer robust maximum expert consensus models for large-scale GDM under uncertainty circumstances. *Appl. Soft Comput.* **2021**, *107*, 107369. [CrossRef]
33. Zhang, H.J.; Ji, Y.; Qu, S.J.; Li, H.H.; Huang, R.P. The robust minimum cost consensus model with risk aversion. *Inf. Sci.* **2022**, *587*, 283–299. [CrossRef]
34. Han, Y.; Qu, S.J.; Wu, Z.; Huang, R.P. Robust consensus models based on minimum cost with an application to marketing plan. *J. Intell. Fuzzy Syst.* **2019**, *37*, 5655–5668. [CrossRef]
35. Li, C.N.; Guo, T.Y.; Chen, Y. Robust Emission Reduction Strategies under Cap-and-Trade and Demand Uncertainty. *Sustainability* **2022**, *14*, 13445. [CrossRef]
36. Qu, S.J.; Wei, J.P.; Wang, Q.H.; Li, Y.M.; Jin, X.W.; Chaib, L. Robust minimum cost consensus models with various individual preference scenarios under unit adjustment cost uncertainty. *Inf. Fusion* **2023**, *89*, 510–526. [CrossRef]
37. Huang, R.P.; Qu, S.J.; Yang, X.G.; Xu, F.M.; Xu, Z.S.; Zhou, W. Sparse portfolio selection with uncertain probability distribution. *Appl. Intell.* **2021**, *51*, 6665–6684. [CrossRef]
38. Cacador, S.; Dias, J.M.; Godinho, P. Portfolio selection under uncertainty: A new methodology for computing relative-robust solutions. *Int. T. Oper. Res.* **2021**, *28*, 1296–1329. [CrossRef]
39. Gubu, L.; Rosadi, D.; Abdurakhman. A New Approach for Robust Mean-Variance Portfolio Selection Using Trimmed k-Means Clustering. *Ind. Eng. Manag. Syst.* **2021**, *20*, 782–794. [CrossRef]
40. Luan, D.Q.; Wang, C.M.; Wu, Z.; Xia, Z.J. Two-Stage Robust Optimization Model for Uncertainty Investment Portfolio Problems. *J. Math.* **2021**, *2021*, 3087066. [CrossRef]
41. Xu, L.J.; Zhou, Y.J. New Robust Reward-Risk Ratio Models with CVaR and Standard Deviation. *J. Math.* **2022**, *2022*, 8304411. [CrossRef]
42. Qu, S.J.; Xu, Y.T.; Ji, Y.; Feng, C.; Wei, J.P.; Jiang, S. Data-Driven Robust Data Envelopment Analysis for Evaluating the Carbon Emissions Efficiency of Provinces in China. *Sustainability* **2022**, *14*, 13318. [CrossRef]

43. Qu, S.J.; Feng, C.; Jiang, S.; Wei, J.P.; Xu, Y.T. Data-Driven Robust DEA Models for Measuring Operational Efficiency of Endowment Insurance System of Different Provinces in China. *Sustainability* **2022**, *14*, 9954. [CrossRef]
44. Ji, Y.; Du, J.H.; Han, X.Y.; Wu, X.Q.; Huang, R.P.; Wang, S.L.; Liu, Z.M. A mixed integer robust programming model for two-echelon inventory routing problem of perishable products. *Physica A* **2020**, *548*, 124481. [CrossRef]
45. Liu, Z.M.; Qu, S.J.; Raza, H.; Wu, Z.; Qu, D.Q.; Du, J.H. Two-stage mean-risk stochastic mixed integer optimization model for location-allocation problems under uncertain environment. *J. Ind. Manag. Optim.* **2021**, *17*, 2783–2804. [CrossRef]
46. Liu, H.R.; Qu, S.J.; Li, R.J.; Razaa, H. Bi-objective robust project scheduling with resource constraints and flexible activity execution lists. *Comput. Ind. Eng.* **2021**, *156*, 107288. [CrossRef]
47. Kumar Tarei, P.; Kumar, G.; Ramkumar, M. A Mean-Variance robust model to minimize operational risk and supply chain cost under aleatory uncertainty: A real-life case application in petroleum supply chain. *Comput. Ind. Eng.* **2022**, *166*, 107949. [CrossRef]
48. Savoji, H.; Mousavi, S.M.; Antucheviciene, J.; Pavlovskis, M. A Robust Possibilistic Bi-Objective Mixed Integer Model for Green Biofuel Supply Chain Design under Uncertain Conditions. *Sustainability* **2022**, *14*, 13675. [CrossRef]
49. Kaoud, E.; Abdel-Aal, M.A.M.; Sakaguchi, T.; Uchiyama, N. Robust Optimization for a Bi-Objective Green Closed-Loop Supply Chain with Heterogeneous Transportation System and Presorting Consideration. *Sustainability* **2022**, *14*, 10281. [CrossRef]
50. Sun, H.P.; Li, Y.C.; Zhang, J.H. Collaboration-based reliable optimal casualty evacuation network design for large-scale emergency preparedness. *Socio. Econ. Plan. Sci.* **2021**, *81*, 101192. [CrossRef]
51. Zhang, J.H.; Liu, H.Y.; Yu, G.D.; Ruan, J.H.; Chan, F.T.S. A three-stage and multi-objective stochastic programming model to improve the sustainable rescue ability by considering secondary disasters in emergency logistics. *Comput. Ind. Eng.* **2019**, *135*, 1145–1154. [CrossRef]
52. Zhang, Y.W.; Li, Z.P.; Jiao, P.B.; Zhu, S. Two-stage stochastic programming approach for limited medical reserves allocation under uncertainties. *Complex Intell. Syst.* **2021**, *7*, 3003–3013. [CrossRef]
53. Jamali, A.; Ranjbar, A.; Heydari, J.; Nayeri, S. A multi-objective stochastic programming model to configure a sustainable humanitarian logistics considering deprivation cost and patient severity. *Ann. Oper. Res.* **2021**, *2021*, 1–35. [CrossRef]
54. Yin, X.C.; Büyüktahtakın, İ.E. A multi-stage stochastic programming approach to epidemic resource allocation with equity considerations. *Health Care Manag. Sci.* **2021**, *24*, 597–622. [CrossRef]
55. Zhang, Z.; Li, Z.L. Consensus-based TOPSIS-Sort-B for multi-criteria sorting in the context of group decision-making. *Ann. Oper. Res.* **2022**, *2022*, 1–28. [CrossRef]
56. Gai, T.T.; Cao, M.S.; Chiclana, F.; Zhang, Z.; Dong, Y.C.; Herrera-Viedma, E.; Wu, J. Consensus-trust Driven Bidirectional Feedback Mechanism for Improving Consensus in Social Network Large-group Decision Making. *Group Decis. Negot.* **2022**, *2022*, 1–30. [CrossRef]
57. Zhang, Z.; Li, Z.L. Personalized Individual Semantics-Based Consistency Control and Consensus Reaching in Linguistic Group Decision Making. *IEEE Trans. Syst. Man Cybern. Syst.* **2022**, *52*, 5623–5635. [CrossRef]
58. Gao, Y.; Zhang, Z. Consensus reaching with non-cooperative behavior management for personalized individual semantics-based social network group decision making. *J. Oper. Res. Soc.* **2021**, *2021*, 1–18. [CrossRef]
59. Pamucar, D.; Torkayesh, A.E.; Biswas, S. Supplier selection in healthcare supply chain management during the COVID-19 pandemic: A novel fuzzy rough decision-making approach. *Ann. Oper. Res.* **2022**, *2022*, 1–43. [CrossRef]
60. Biswas, S.; Pamucar, D.; Bozanic, D.; Halder, B. A New Spherical Fuzzy LBWA-MULTIMOOSRAL Framework: Application in Evaluation of Leanness of MSMEs in India. *Math. Probl. Eng.* **2022**, 5480848. [CrossRef]
61. Qu, S.J.; Shu, L.L.; Yao, J.Y. Optimal pricing and service level in supply chain considering misreport behavior and fairness concern. *Comput. Ind. Eng.* **2022**, *174*, 108759. [CrossRef]
62. Qu, S.J.; Xu, L.; Mangla, S.K.; Chan, F.T.S.; Zhu, J.L.; Arisian, S. Matchmaking in reward-based crowdfunding platforms: A hybrid machine learning approach. *Int. J. Prod. Res.* **2022**, *2022*, 1–21. [CrossRef]

sustainability

MDPI

Article

Nonlinear Diffusion Evolution Model of Unethical Behavior among Green Food Enterprise

Qi Yang [1,2], Yuejuan Hou [2], Haoran Wei [2], Tingqiang Chen [2,3,*] and Jining Wang [2,*]

1 School of Safety Science and Engineering, Nanjing Tech University, Nanjing 211816, China
2 School of Economics and Management, Nanjing Tech University, Nanjing 211816, China
3 School of Intellectual Property, Nanjing Tech University, Nanjing 211816, China
* Correspondence: tingqiang88888888@163.com (T.C.); wangjn163@126.com (J.W.)

Abstract: Under the background of low-carbon economy, the unethical behavior of green food enterprises has aggravated the uncertainty and frequency of green food safety problems and even triggered a contagion of unethical behavior among green food enterprises. In view of this, considering the characteristics of organizational behavior, external environmental intervention and social networks, we construct an infectious disease model of the nonlinear spread of unethical behavior in green food enterprises and simulated the mechanism and evolution characteristics of the spread of unethical behavior among them. The main conclusions are as follows. (1) Single adjustment of the level of enterprise moral clarity, damage degree of unethical behavior, and enterprise influence can only reduce the diffusion probability of unethical behavior to a certain extent. (2) Enterprise ethical climate plays a crucial role in the diffusion of unethical behavior among green food enterprises and exerts a "strengthening effect" on other organizational behavior and external environmental intervention factors. (3) The strength of external supervision and strength of punishment exert a "suppression effect" on the diffusion of unethical behavior among green food enterprises.

Keywords: nonlinear evolution; unethical behavior; behavior diffusion; complex networks

Citation: Yang, Q.; Hou, Y.; Wei, H.; Chen, T.; Wang, J. Nonlinear Diffusion Evolution Model of Unethical Behavior among Green Food Enterprise. *Sustainability* **2022**, *14*, 16158. https://doi.org/10.3390/su142316158

Academic Editors: Shaojian Qu, Qingguo Bai, Ying Ji and Congjun Rao

Received: 22 October 2022
Accepted: 29 November 2022
Published: 3 December 2022

1. Introduction

In recent years, with the continual exposure of the Enron financial fraud, the GlaxoSmithKline commercial bribery, and other scandals, ethical crises in enterprises have occurred frequently worldwide. Moral and ethical issues of green food enterprises are transcending the traditional research category of philosophy, and unethical behavior has gradually become a common concern in the field of enterprise management theory and practice [1,2]. With the continuous development of green food, green food safety has gradually become the focus of attention [3–6]. Green food enterprise refers to enterprises in accordance with the scientific method to produce and process pollution-free safe, high-quality and nutritious foods. In product production, transportation, storage and packaging are pollution-free, and the entire production process contributes to pollution prevention [7,8]. Unethical behavior refers to the behavior that violates widely accepted social ethics and is not recognized by most people [9]. Many kinds of unethical behaviors are common, such as corruption, cheating on taxes and academic dishonesty. Unethical behaviors widely occur in various enterprises [10,11]. Moreover, unethical behavior diffuses easily [12], which can cause an adverse impact on the long-term performance of an enterprise and the sustainable development of society. Therefore, studying the influencing factors and evolution mechanisms of unethical behavior diffusion is necessary. Such knowledge can provide theoretical reference for the formulation of strategies to prevent and control the diffusion of unethical behavior.

Various researches have been conducted on the generation and diffusion of unethical behavior [13]. These studies are mainly carried out from three perspectives: organizational behavior, external environmental intervention and social network structure. In terms of organizational behavior, enterprises often fail to be completely rational in decision making due to various factors, which may result in behavioral deviations. Organizational behaviors include the micro-behavior of individuals within an enterprise (such as the level of individual moral clarity) and the overall macro-behavior of an enterprise (such as the enterprise management system and enterprise ethical climate) [14,15]. All of them can affect the diffusion of unethical behavior among enterprises [16].

Wiltermuth and Flynn [17] found that different individuals will make different ethical judgments when faced with the same situation, owing to their different behavioral ethical standards, which can affect the generation and diffusion of unethical behavior. Werbel and Balkin [18] confirmed the importance of management systems to the control of unethical behavior in workplaces. Organizational justice, which is an important part of a management system, can weaken the negative emotions of members, thus effectively controlling unethical behaviors [19]. In addition, the enterprise ethical climate is also an important factor that can affect the generation and diffusion of unethical behavior [2,20]. Gorsira et al. [21] found that employees of private enterprises generally believe that their ethical climate is biased toward egoism and they are likely to implement unethical behavior. According to the research above, the generation of unethical behavior is affected by various organizational behavior factors. Unethical behavior can also diffuse within the enterprise network under the influence of the herd effect, thus intensifying the effects of unethical behavior. Studies that only consider organizational behavioral factors are widely questioned and their scope of interpretation is limited; hence, scholars have begun to turn their research directions to the study of external environmental intervention factors [22].

In the aspect of external environmental interventions, Gino et al. [23] found that external supervision can stimulate team members' sense of guilt and collective honor and reduce the risk of collective immorality. In general, individuals must weigh benefits and penalties before implementing unethical behavior. When penalties are greater than benefits, people do not make unethical decisions [24]. In addition, the external competitive environment of enterprises can affect the generation and diffusion of unethical behavior. Li et al. [25] empirically analyzed the impact of competition on unethical decision-making through the data of 727 employees in Chinese hospitals. The results showed that competition orientation can influence employees' unethical decision making through the adjustment of relation conflict and hostile attribution bias. In a case study of Enron, Kulik et al. [12] found that unethical behavior likely diffuses among enterprises, which are in a fiercely competitive industry. The research perspective of external environmental intervention focuses on a wide range of factors and increases the intensity of theoretical interpretation. However, most of the research above is limited to static analysis, which is inconsistent with the dynamic characteristics of unethical behavior diffusion. Therefore, it is necessary to conduct research further.

In recent years, scholars have gained awareness that unethical behavior is a social phenomenon involving complex interactions of enterprises. This relationship is organically embedded in the social network. Therefore, the researches on the diffusion of unethical behavior from the perspective of social networks have become a hotspot. The main studies are as follows. Brass et al. [26] initially studied unethical behavior from the perspective of social networks and defined them as a set of actors with a certain degree of relevance. On the basis of this definition, they proposed that network relevance characteristics (relationship strength, multivariate relationship, asymmetric relationship, and status), network structural characteristics (structural holes, centrality, and network density), and their interactions have great impact on unethical behavior of individuals within a network. After [26], the study of the relationship between social network characteristics and unethical behavior has been expanded. Bizjak et al. [27] found that social networks have an "infectious effect," which means that the network relationship may lead to a consistency of internal individuals' behavior. Sullivan et al. [28] found that network characteristics of enterprises may change when enterprises implement unethical behavior, which may reduce network cohesion.

Brown et al. [29] studied the impact of social networks on the tax avoidance behavior of enterprises. The results show that tax avoidance behaviors can present a convergence effect among highly connected enterprises. Zuber [30] studied the transmission mechanism of unethical behavior among victims, perpetrators, and observers. The result reveals that social network relationships may change after unethical behaviors occur, thus having an indirect negative impact on enterprises. The research perspective of social networks considers dynamic factors, and its research premise is in line with the internal and external environment of enterprises. This type of research has greatly expanded the research pattern of unethical behavior.

However, most of the current articles on unethical behavior are from the perspective of organizational management, studying unethical behavior between leaders and employees [31–33]. For example, by integrating arguments from social identity and moral disengagement theories, Schuh et al. [31]. developed and tested a model to explain how leaders respond to unethical pro-organizational behavior (UPB) among employees. The results showed that leader perceptions of employee UPB were positively related to leader trust in employees when leaders identified strongly with their organization or when they had a strong propensity to morally disengage. Pablo et al. [32] examined personal growth satisfaction as a mediator and responsibility climate as a moderator of the relationship and found that personal growth satisfaction mediated the negative impact of unethical supervision on intention to stay. In terms of moderation, high responsibility climate weakened the negative relationship between unethical leadership and subordinates' personal growth satisfaction, as well as the indirect negative impact of unethical leadership on subordinates' intention to stay. However, from a corporate perspective, there are very few studies on the impact of unethical behavior by companies or organizations regarding consumers and society, and those are as follows:

Olofsson et al. [34] investigated the time-varying volatility and risk measures of ethical and unethical investments and found that ethical investments are less affected during global financial crises compared to unethical and conventional investments. Moreover, these studies do not delve into the change mechanism of the spread of unethical behavior in the process of elaboration, and at the same time, they do not take into account the evolutionary characteristics of the contagion of unethical behavior in green food enterprises under the current green economy conditions.

The current epidemic model that is based on a complex network is not only limited to the study of virus transmission but also widely used in the field of social science [35–37], such as technology and innovation diffusion [38], financial crisis contagion [39,40], and the spread of rumors [41,42]. The epidemic model provides the necessary technological means for solving social problems, and it also offers theoretical basis for devising coupling strategies. In addition, similar diffusion mechanisms are observed between behaviors and viruses. For example, they both diffuse among individuals through social connection in most cases, in which social connection is their diffusion medium [43]. In the environment of enterprises, a large number of physical contacts transmit various information and affect one another's behaviors. Once an unethical behavior is formed, it would bring additional benefits to the implementer and be imitated and learned by other enterprises under appropriate conditions. Therefore, the diffusion of unethical behavior among enterprises has many mechanisms similar to the spread of infectious diseases.

On the basis of the research above, unethical behaviors have adverse impact on enterprises and society. Without effective controlling, unethical behavior may be imitated by an increasing number of enterprises, and its negative effects will spread rapidly. Eventually, this spread will cause an irrational outbreak of unethical behavior [44]. If the unethical behavior of green food enterprises cannot be effectively controlled, the unethical behavior may be imitated by more and more enterprises, thus forming a contagion in the enterprise network, resulting in the aggravation and spread of food safety problems. Therefore, in order to better formulate a reasonable and effective control of the unethical behavior of green food enterprises and the contagion of their unethical behavior, it is necessary to study the influencing factors and evolutionary mechanism of the spread of unethical behavior in depth. However, current researches have three main shortcomings. First, these studies are only conducted from a single perspective of organizational behavior, external environmental intervention, or social network structure, and they lack comprehensive consideration of the three perspectives above. Second, most of the existing researches are static research. They ignore the dynamic evolution characteristics of unethical behavior diffusion. Third, existing researches focus on the diffusion effect of unethical behavior within an enterprise, but they ignore the diffusion among enterprises. Finally, existing researches only focus on the diffusion effect of unethical behavior within ordinary enterprises, and ignore the diffusion effect of green food enterprises on unethical behaviors such as food safety.

In view of these shortcomings, this study analyzes the influencing factors of unethical behavior diffusion from the cross perspective of organizational behavior and organizational management. Based on the SIR epidemic model, a nonlinear diffusion evolution model of unethical behavior among green food enterprise is constructed. Organizational behavior, external environmental intervention, and social network characteristics are considered. Using MATLAB R2018a software, this study simulates the diffusion mechanisms and evolution characteristics of the diffusion of unethical behavior among green food enterprises. This study makes contributions in three ways. (1) Unlike existing studies that are only based on a single perspective, this study conducts research on the diffusion of unethical behavior among green food enterprise with the comprehensive consideration of organizational behavior, external environmental intervention, and social network characteristics. (2) This study analyzes the dynamic evolution characteristics of the diffusion of unethical behavior among green food enterprise, thus making the research conclusion realistic. (3) This study provides novel conclusions that have theoretical and practical application value: single adjustment of the level of enterprise moral clarity, damage degree of unethical behavior, and enterprise influence can only reduce the diffusion probability of unethical behavior to a certain extent. Even if the formulation of control strategies is based on their interaction, the diffusion remains unable to disappear. However, enterprise ethical climate plays a crucial role in the diffusion of unethical behavior among green food enterprises and exerts a "strengthening effect" on other organizational behavior and external environmental intervention factors.

The rest of this study is organized as follows. The second part analyzes the epidemic mechanisms of unethical behavior diffusion among green food enterprise. The third part constructs a nonlinear diffusion evolution model of unethical behavior among green food enterprise under the interaction of organizational behavior and external environmental intervention. The fourth part simulates the evolution characteristics of the diffusion of unethical behavior among green food enterprise and proposes strategies to prevent and control the diffusion of unethical behavior. The last part puts forward the conclusions.

2. Materials and Methods

In this paper, by analyzing the contagion mechanism of unethical behavior of green food enterprises, under the influence of organizational behavior and external environmental intervention factors, a model of the diffusion of unethical behavior of green food enterprises is constructed based on the epidemic model, and the model construction process is shown in Figure 1.

2.1. Epidemic Mechanisms of Unethical Behavior Diffusion among Green Food Enterprise

2.1.1. Adaptability Analysis of the Epidemic Model

The epidemic model, as a classic model of viral transmission, has been widely used in the study of social behavior diffusion [36,45]. The original meaning of contagion is the diffusion of pathogens from infected organisms to other organisms. Assuming that corporate unethical behavior is the contagion virus in this model, the spread of unethical behavior is assumed to be the spread of unethical behavior of green food enterprises in the network of interrelated green food enterprises, and the contagion of unethical behavior will affect the stakeholders in the network. The principal representations are as follows.

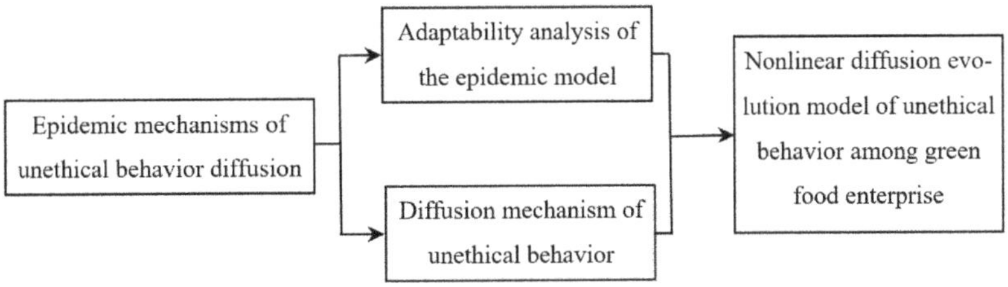

Figure 1. System process flow of the model of unethical behavior diffusion among green food enterprises.

(1) Pathogen–Diffusion Source

In the green food enterprises network, some green food enterprises lack completed management system and internal cohesiveness, which causes some members to be vulnerable to profits. Hence, these members may exhibit behavioral deviations and unethical behaviors. As a source of diffusion, unethical behavior is a "pathogen" with potential transmission ability. It can diffuse among green food enterprises through various diffusion mediums, thus presenting a significant herd effect in the green food enterprises network.

(2) Infectious Medium–Diffusion Medium

A diffusion medium is a carrier of the diffusion source. In the green food enterprises network, each enterprise is not isolated. Direct or indirect associations always exist between them, such as cooperations between green food enterprises and communications between members of different green food enterprises. Unethical behaviors can diffuse rapidly among green food enterprise through the mediums, which has a great impact on society.

(3) Infectious

Green food enterprise with unethical behavior may transmit their status, behaviors, and other information to the external environment during the daily cooperation and communication with other green food enterprises. When associated green food enterprises receive the information, cognitive and behavioral deviations may be generated. Hence, associated green food enterprise may be infected with unethical behavior, indicating that unethical behavior is infectious.

(4) Immunity

Some high-level green food enterprises exist in the green food enterprises network. Such green food enterprises usually have strict and reasonable management systems and first-class leaders. These leaders usually have strong management skills and a strong sense of social responsibility. In addition, they are often good at resolving conflicts of interest in and coordinating team members' thoughts and behaviors. Therefore, unethical behavior is difficult to incite in this type of green food enterprise, because they exhibit immunity to unethical behavior.

In summary, the diffusion of unethical behavior among green food enterprises has a similar infectious mechanism to the contagion of infectious diseases. Therefore, analyzing the diffusion mechanism and evolution characteristics of unethical behavior by using the epidemic model is reasonable and feasible. The analysis provides theoretical reference for preventing the rapid diffusion of unethical behavior among green food enterprises. Table 1 shows that the key concepts in the epidemic model are applied in the network diffusion model of unethical behavior.

Table 1. The corresponding concepts of unethical behavior diffusion.

Diffusion of Unethical Behavior	Meaning
Diffusion source	Unethical behavior
Susceptible green food enterprise	Green food enterprises that have not been infected with unethical behavior
Infected green food enterprise	Green food enterprise that are infected with unethical behavior
Immune green food enterprise	Green food enterprise that are not affected by unethical behavior or had been infected with unethical behavior but eliminated it through adjustment

2.1.2. Diffusion Mechanism of Unethical Behavior

In the green food enterprises network, green food enterprises are divided into three states. S represents a susceptible enterprise, which does not implement unethical behavior but is easily affected by other green food enterprises' unethical behavior. I represents an infected enterprise, which is infected with unethical behavior and can affect other associated green food enterprises. R represents an immune enterprise, which is not affected by unethical behavior or had been infected with unethical behavior and but eliminated it through adjustment. Moreover, S, I, and R are used to express the number of three states of green food enterprises in the network. Figure 2 illustrates the rule of transformation mechanism among the three states of green food enterprises.

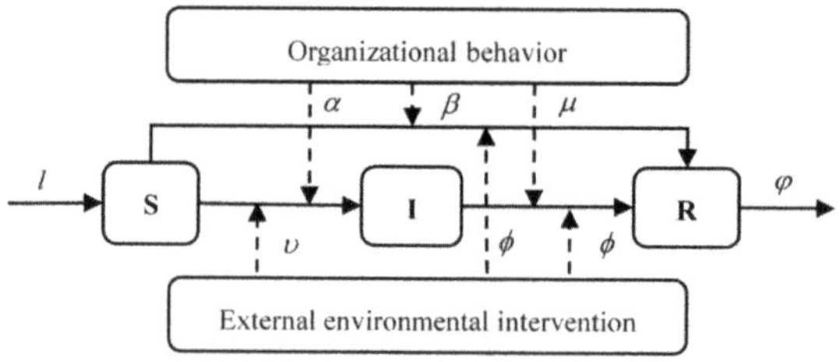

Figure 2. The rule of transformation mechanism among the three states of green food enterprises.

(1) Affected by organizational behavior factors, infected green food enterprises' unethical behavior diffuses to susceptible green food enterprise at the rate of $\alpha(0 \leq \alpha \leq 1)$ during daily cooperating and communicating. Under the influence of external environmental intervention, infected green food enterprise' unethical behavior further diffuses to susceptible green food enterprise at the rate of $\upsilon(0 \leq \upsilon \leq 1)$.

(2) Susceptible green food enterprise may turn into the immune state directly with the probability of $\beta(0 \leq \beta \leq 1)$ by introducing a high-level management team and optimizing the management system. Adjusting the organizational behavior can also make infected green food enterprise immune with the probability of $\mu(0 \leq \mu \leq 1)$.

(3) Under the influence of external environmental intervention factors, susceptible green food enterprise may turn into the immune state directly with the probability of $\phi(0 \leq \phi \leq 1)$. In the process of implementing unethical behavior, infected green food enterprise may eliminate it and become immune with the probability of $\phi(0 \leq \phi \leq 1)$ if they are discovered and punished by external regulators.

(4) In each period, the entry probability of some new green food enterprises is $l(0 \leq l \leq 1)$, and the exit probability of some old green food enterprises is $\varphi(0 \leq \varphi \leq 1)$.

2.2. Nonlinear Diffusion Evolution Model of Unethical Behavior among Green Food Enterprises

To construct the network diffusion model of unethical behavior among green food enterprises, this study assumes that N is the total number of green food enterprise. s, i, and r account for the proportion of susceptible green food enterprise, infected green food enterprise, and immune green food enterprise, namely $s = S/N, i = I/N, r = R/N$, and $s + i + r = 1 (0 \leq s, i, r \leq 1)$. Moreover, the density of infected green food enterprises that have a degree of k is assumed to be $i_k(t)$ at moment t. The probability that susceptible green food enterprise connected to infected green food enterprise is $\Theta(t)$.

In view of the important influence of individual behavior on decision-making or profit, Sundaresan [46] defined a behavior effective function:

$$U(W) = -\frac{1}{\gamma} e^{-\gamma W} \tag{1}$$

where W represents individual patience and γ represents coefficient of individual risk aversion.

Some organizational behavior factors affect the diffusion of unethical behavior. These factors mainly include the level of enterprise moral clarity $\psi(0 < \psi < 1)$ [17,24,47]. The accuracy of determining whether a particular behavior is ethical becomes higher with the increase of ψ. Another factor is damage degree of unethical behavior $\varepsilon(0 < \varepsilon < 1)$ [2,48]. The damage degree of unethical behavior and the impact to the green food enterprises network become greater with the increase of ε. Moreover, the damage degree may have a negative influence on society. Enterprise influence $\theta(0 < \theta < 1)$ [12,28,49] refers to the relationship strength of green food enterprise in the green food enterprises network. The relationship strength of green food enterprise becomes greater with the increase of θ. Furthermore, it will have a more significant influence on other green food enterprises. Strictness of the enterprise management system $\tau(0 < \tau < 1)$ [18,50] becomes stricter with the increase of τ. Enterprise ethical climate $\rho(0 < \rho < 1)$ [2,21] is also included. When ρ is closer to 0, enterprise ethical climate tends to be egocentric. This means that green food enterprise may focus on individual interests without considering the negative social influence. When ρ is closer to 1, the enterprise ethical climate tends to become principled, which represents that green food enterprises have a higher degree of ethical cognitive constraints. Therefore, the function of organizational behavior $g(\psi, \varepsilon, \theta, \tau, \rho)$ is defined as:

$$g(\psi, \varepsilon, \theta, \tau, \rho) = \rho^{1+\tau^{\frac{1}{4}}} e^{-\frac{\varepsilon \psi^{\frac{1}{\varepsilon}} \tau^{\frac{-1}{\rho^2}}}{\theta}} \tag{2}$$

Hence, the infection rate $\alpha(0 \leq \alpha \leq 1)$ under the influence of organizational behavior factors is defined as:

$$\alpha = g(\psi, \varepsilon, \theta, \tau, \rho) = \rho^{1+\tau^{\frac{1}{4}}} e^{-\frac{\varepsilon \psi^{\frac{1}{\varepsilon}} \tau^{\frac{-1}{\rho^2}}}{\theta}} \tag{3}$$

According to the behavioral state transition equation proposed by [51] and external environmental intervention factors that affect the diffusion of unethical behavior among green food enterprises, it mainly includes the strength of external supervision $\delta(0 < \delta < 1)$ [23,52]. The probability of which green food enterprises' unethical behavior is discovered becomes higher with the increase of δ. In addition, strength of punishment $\xi(0 < \xi < 1)$ [24,53] to green food enterprise with unethical behavior is included. Strength of punishment to green food enterprise with unethical behavior becomes greater with the increase of ξ. Moreover, it also includes external competitiveness $\lambda(0 < \lambda < 1)$ [12,25] among green food enterprises. External competitiveness becomes greater with the increase of λ. Therefore, the function of external environmental intervention is defined as:

$$f(\delta, \xi, \lambda) = \frac{1}{1 + e^{\frac{\lambda - \xi}{\delta}}} \tag{4}$$

Therefore, the infection rate $v(0 \leq v \leq 1)$ under the influence of external environmental intervention factors is defined as:

$$v = [1 - f(\delta, \xi, \lambda)] = 1 - \frac{1}{1 + e^{\frac{\lambda - \xi}{\delta}}} \tag{5}$$

Considering that the diffusion of unethical behavior among green food enterprises is influenced by the interaction of organizational behavior and external environmental intervention factors, the total infection rate $\sigma(0 \leq \sigma \leq 1)$ is defined as:

$$\sigma = \alpha \cdot v = g(\psi, \varepsilon, \theta, \tau) \cdot [1 - f(\delta, \xi, \rho)] = \rho^{1+\tau^{\frac{1}{4}}} e^{-\frac{\varepsilon \psi^{\frac{1}{\varepsilon}} \tau \rho^{\frac{-1}{2}}}{\theta}} (1 - \frac{1}{1 + e^{\frac{\lambda - \xi}{\delta}}}) \tag{6}$$

According to the rule of state transition mechanism among green food enterprises discussed in Figure 2 and mean field theory [40,54], the differential equations of network diffusion model of unethical behavior among green food enterprises under the interaction of organizational behavior and external environmental intervention are expressed as follows:

$$\begin{cases} \frac{ds_k(t)}{dt} = l - k\alpha v s_k(t)\Theta(t) - \beta\pi s_k(t) \\ \frac{di_k(t)}{dt} = k\alpha v s_k(t)\Theta(t) - \mu\pi i_k(t) \\ \frac{dr_k(t)}{dt} = \mu\pi i_k(t) + \beta\pi s_k(t) - \varphi r_k(t) \end{cases} \tag{7}$$

According to (7), for the steady-state condition $\frac{di_k(t)}{dt} = 0$, the steady state value becomes $i_k(t)$:

$$i_k(t) = \frac{k\alpha v s_k(t)\Theta(t)}{\mu\phi} = \frac{kl\alpha v\Theta(t)}{\beta\mu\phi^2 + k\alpha v\mu\phi\Theta(t)} \tag{8}$$

The average density of infected green food enterprise becomes $i = \sum_k P(k) i_k(t)$. Based on (8), $\Theta(t)$ becomes:

$$\Theta(t) = \sum_k \frac{kP(k)i_k(t)}{\sum_s sP(s)} = \frac{1}{<k>} \sum_k kP(k)i_k(t) \tag{9}$$

where $<k>$ represents the network average degree of the diffusion of unethical behavior.

Given that $<k> = \sum_k kP(k)$ and $<k^2> = \sum_k k^2 P(k)$, (8) and (9) can be combined as follows:

$$\Theta(t) = \frac{1}{<k>} \sum_k kP(k) \frac{kl\alpha v\Theta(t)}{\beta\mu\phi^2 + k\alpha v\mu\phi\Theta(t)} \tag{10}$$

Given that $\Theta = \Theta(t)$, (10) has a trivial solution: $\Theta = 0$. If (10) has a non-trivial solution, $\Theta \neq 0$, then the necessary condition becomes:

$$\frac{d}{d\Theta} \left(\frac{1}{<k>} \sum_k kP(k) \frac{kl\alpha v\Theta}{\beta\mu\phi^2 + k\alpha v\mu\phi\Theta} \right) |\Theta = 0 \geq 1 \tag{11}$$

Therefore,

$$\frac{1}{<k>} \sum_k kP(k) \frac{kl\alpha v}{\beta\mu\phi^2} \geq 1 \tag{12}$$

Thus, the basic reproduction number of the diffusion of unethical behavior under the influence of organizational behavior and external environmental intervention factors is R_0:

$$R_0 = \frac{l\alpha v \sum_k k^2 P(k)}{\beta\mu\phi^2 \sum_k kP(k)} = \frac{l\rho^{1+\tau^{\frac{1}{4}}} e^{-\frac{\varepsilon \psi^{\frac{1}{\varepsilon}} \tau \rho^{\frac{-1}{2}}}{\theta}} (1 - \frac{1}{1+e^{\frac{\lambda - \xi}{\delta}}}) \sum_k k^2 P(k)}{\beta\mu\phi^2 \sum_k kP(k)} \tag{13}$$

Equation (13) was simulated to study the influence of green food enterprise organizational behavior and external environmental factors on the contagion of the unethical behavior of green food enterprises.

The basic reproduction number $R_0(R_0 > 0)$ indicates the average number of susceptible green food enterprises that are infected by infected green food enterprise [55]. $R_0 = 1$ represents the threshold of the disappearance of diffusion. The diffusion becomes extinct gradually when $R_0 < 1$ and the diffusion occurs with nonzero probability when $R_0 > 1$. Moreover, the greater the value of R_0 is, the greater the diffusion probability becomes.

3. Simulation Analysis and Results Discussion

This study sets the initial parameters as follows: $\rho = 0.1$, $\tau = \theta = \delta = \xi = 0.2$, $l = \psi = 0.3$, $\varepsilon = \mu = \beta = \phi = \lambda = 0.4$. Most nodes in a BA scale-free network have a few connections and only a few nodes have many connections. Moreover, the degree distribution of nodes also follows power law distribution [56]. It is similar to the actual enterprise network in which the number of large-scale and influential green food enterprise is small. However, small-scale green food enterprises are numerous, and yet they only have a few connections with other green food enterprises in the network [57]. Therefore, on the basis of the BA scale-free network ($m = m_0 = 5$, network scale $N = 500$), we use the MATLAB R2018a software to simulate the evolution characteristics of unethical behavior diffusion among green food enterprises under the interaction of organizational behavior and external environmental intervention.

3.1. *Organizational Behavior and Diffusion of Unethical Behavior among Green Food Enterprises*

3.1.1. Single Organizational Behavior Factor and Diffusion of Unethical Behavior among Green Food Enterprises

Figure 3a,b,d demonstrate that the diffusion probability of unethical behavior among green food enterprises shows the decreasing characteristic of increasing margins with the increase of the level of enterprise moral clarity Ψ, damage degree of unethical behavior ε and strictness of enterprise management system τ. Figure 3a indicates that when the level of enterprise moral clarity Ψ is less than 0.4, the change of diffusion probability of unethical behavior is not evident. When it exceeds 0.4, the variation of diffusion probability increases gradually, but its overall change is not significant. Therefore, the method of improving level of enterprise moral clarity Ψ can only reduce the diffusion probability of unethical behavior among green food enterprise to a certain extent. Figure 3b indicates that when damage degree of unethical behavior ε is less than 0.4, the diffusion probability of unethical behavior is less variable. When it exceeds 0.4, the variation of diffusion probability increases gradually, but its overall change is not significant. The reasons of this phenomenon are as follows: unethical behavior with high damage degree may often violate the law and be extremely harmful to the enterprise and society. In addition, its probability of being discovered is also high. Therefore, the implementation of such unethical behavior is often worthless. Figure 3d indicates that the overall change is evident and when strictness of enterprise management system τ is less than 0.4, the diffusion probability of unethical behavior declines. When it exceeds 0.4, the downward trend tends to be flat. This phenomenon reflects that completed enterprise management system can effectively control the diffusion of unethical behavior among green food enterprises. However, when strictness of enterprise management system τ reaches a certain level, the controlling effect of continuously increasing it on the diffusion of unethical behavior is rapidly weakened. Therefore, green food enterprise should follow the principle of appropriateness to avoid invalid allocation of management resources when determining the strictness of enterprise management system.

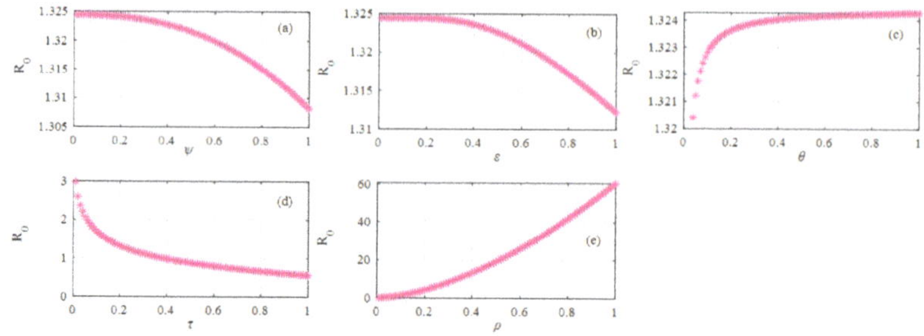

Figure 3. The influence of organizational behavior factors on the diffusion of unethical behavior among green food enterprises. (**a**–**e**) refer to level of enterprise moral clarity ψ, damage degree of unethical behavior ε, enterprise influence θ, strictness of enterprise management system τ and enterprise ethical climate ρ, respectively.

Figure 3c,e demonstrates that the diffusion probability of unethical behavior among green food enterprises shows the increasing characteristic of diminishing margins with the increase of enterprise influence θ and enterprise ethical climate ρ. Figure 3c indicates that when enterprise influence θ is less than 0.2, the diffusion probability of unethical behavior appears jump points, and the upward trend is very obvious. However, when it exceeds 0.2, the upward trend becomes gradually flat and its overall change is not significant. This phenomenon reflects that influential green food enterprise are not only small in number but also stable in status. However, more green food enterprises have small influence in the food green enterprises network. Small enterprise size and lack of a complete management system make them vulnerable to the influence of other green food enterprises, thus causing the diffusion of unethical behavior among green food enterprises. Figure 3e reflects that a large-scale diffusion tendency of unethical behavior in the green food enterprises network as enterprise ethical climate changes from principled to egocentric. Ethical climate of an enterprise is often regarded as the standard of enterprise behavior ethics. A principled enterprise ethical climate is a cognitive constraint on unethical intentions, thus having a controlling effect on the diffusion of unethical behavior. However, under the ethical climate of egoism, green food enterprises tend to focus on individual interests without considering social consequences, which makes them more inclined to implement unethical behavior, thus causing large-scale diffusion of unethical behavior among green food enterprises. This phenomenon also embodies that enterprise ethical climate ρ, is the key factor that affects the diffusion of unethical behavior among green food enterprises. A principled ethical climate has a strong blocking effect on the diffusion of unethical behavior among green food enterprise.

In addition, Figure 3 shows that the single adjustment of the level of enterprise moral clarity ψ, damage degree of unethical behavior ε, and enterprise influence θ affect the diffusion of unethical behavior among green food enterprises, but their blocking effect is weak. The basic reproductive number R_0 remains greater than 1. Unethical behavior can still diffuse with non-zero probability in the enterprise network. If the prevention and control strategies are only based on this, then causing deviation is easy and the effect is limited. However, the strictness of enterprise management system and enterprise ethical climate have a greater impact on the diffusion of unethical behavior among green food enterprise. Adjustment them can make the basic reproductive number R_0 become less than 1. In particular, enterprise ethical climate plays a leading role in the diffusion of unethical behavior among green food enterprises. Therefore, formulating targeted prevention and control strategies from these two aspects is effective.

3.1.2. Multiple Organizational Behavior Factors and Diffusion of Unethical Behavior among Green Food Enterprises

Figure 4 indicates that the single adjustment of level of enterprise moral clarity ψ, damage degree of unethical behavior ε, and enterprise influence θ cannot make unethical behavior disappear gradually in the enterprise network. Figure 4a shows that when level of enterprise moral clarity ψ interacts with damage degree of unethical behavior ε, the diffusion probability of unethical behavior among green food enterprises shows the characteristic of increasing margins with the increase of these two factors. However, the overall change is not significant and the basic reproductive number R_0 is still greater than 1. Figure 4b,e also indicate that when enterprise influence θ interacts level of enterprise moral clarity ψ and damage degree of unethical behavior ε, the diffusion probability of unethical behavior among green food enterprises shows the increasing characteristic with the increase of enterprise influence θ. However, only when level of enterprise moral clarity ψ and damage degree of unethical behavior ε are high, the increasing trend become evident. In other cases, enterprise influence θ has little effect on the diffusion probability of unethical behavior among green food enterprises, and the basic reproductive number R_0 cannot be less than 1. This finding is consistent with the phenomenon reflected in Figure 3, which shows that level of enterprise moral clarity ψ, damage degree of unethical behavior ε, and enterprise influence θ have little influence on the diffusion of unethical behavior among green food enterprises. Even if the formulation of control strategies is based on their interaction, these strategies are unable to eliminate the diffusion of unethical behavior.

Figure 4c,f,h show that when strictness of the enterprise management system τ interacts with level of enterprise moral clarity ψ, damage degree of unethical behavior ε, and enterprise influence θ, the diffusion probability of unethical behavior among green food enterprises shows a significant downward trend with the increase of level of enterprise moral clarity ψ, damage degree of unethical behavior ε, strictness of enterprise management system τ, and the decrease of enterprise influence θ. In addition, the basic reproductive number R_0 can be controlled to become less than 1. This phenomenon reflects the suppression effect of strictness of enterprise management system τ on the diffusion of unethical behavior among green food enterprises. If enterprise management system is established clearly, fairly, and effectively, it will increase the cost of implementing unethical behavior and greatly reduce the possibility of the emergence and diffusion of unethical behavior, thus leading to the gradual elimination of the diffusion of unethical behavior among green food enterprises.

Figure 4d,g,i show that when enterprise ethical climate ρ interacts with level of enterprise moral clarity ψ, damage degree of unethical behavior ε, and enterprise influence θ, the diffusion probability of unethical behavior increases with the change of these organizational behavior factors. Such interaction presents a large-scale diffusion crisis in the green food enterprises network. This phenomenon shows that although under the single influence of the three organizational behavioral factors of level of enterprise moral clarity ψ, damage degree of unethical behavior ε, and enterprise influence θ, the diffusion probability of unethical behavior among green food enterprises is less affected. However, with the transformation of enterprise ethical climate ρ from principled to egocentric, the green food enterprise may pay too much attention to individual interests and neglect social influences. When an egocentric enterprise ethical climate becomes a part of enterprise culture, members will be used to unethical behavior. At this point, ethical climate of the entire green food enterprises network has been almost destroyed. This scenario will result in deviation from enterprise cognition and behavior, which, in turn, leads to the large-scale diffusion of unethical behavior in the network under the influence of the herd effect. This phenomenon also shows that enterprise ethical climate ρ plays a crucial role in the diffusion of unethical behavior among green food enterprises. Therefore, the purpose of eliminating the diffusion of unethical behavior can be achieved by establishing a principled enterprise ethical climate.

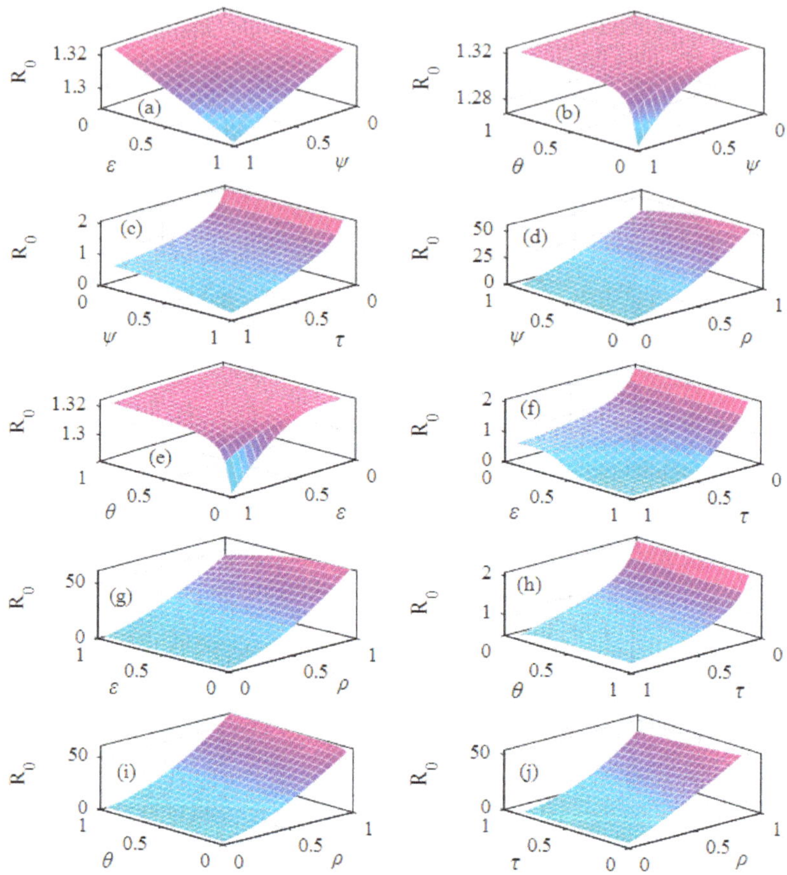

Figure 4. The interactive influence of organizational behavior factors on the diffusion of unethical behavior among green food enterprises. The factors involved in the figure are level of enterprise moral clarity ψ, damage degree of unethical behavior ε, enterprise influence θ, strictness of enterprise management system τ, and enterprise ethical climate ρ.

According to Figure 4, on the one hand, enterprise ethical climate ρ exerts a "strengthening effect" on other organizational behavior factors. Other organizational behavioral factors have an impact on the diffusion probability of unethical behavior among green food enterprises are stronger under an egocentric enterprise ethical climate. Therefore, this scenario may cause irrational outbreaks of unethical behavior among green food enterprises. On the other hand, strictness of the enterprise management system τ exerts a "suppression effect" on the diffusion probability of unethical behavior among green food enterprises. A strict enterprise management system can enhance other organizational behavior factors' controlling effect on the diffusion of unethical behavior, thus gradually eliminating the diffusion of unethical behavior among green food enterprises. In addition, the "strengthening effect" of the enterprise ethical climate ρ is stronger than the "suppression effect" of strictness of the enterprise management system τ. Therefore, when formulating prevention and control strategies, we should take principled enterprise ethical climate as a foundation from a global perspective and increase the level of enterprise moral clarity. Furthermore, we must maintain the strictness of enterprise management system moderately and strengthen the control of the core green food enterprise with greater influence.

3.2. External Environmental Intervention and Diffusion of Unethical Behavior among Green Food Enterprises

3.2.1. Single External Environmental Intervention Factor and Diffusion of Unethical Behavior among Green Food Enterprises

Figure 5a shows that the diffusion probability of unethical behavior among green food enterprises has a decreasing trend of diminishing margins with the increase of strength of external supervision δ and strength of punishment ξ. Figure 5a indicates that when strength of external supervision δ is less than 0.1, the variation of diffusion probability of unethical behavior is almost negligible. When the strength of external supervision δ is greater than 0.1 and less than 0.5, the diffusion probability of unethical behavior diffusion shows a downward trend and the variation is significant. When strength of external supervision δ is greater than 0.5, the downward trend becomes flat gradually. This phenomenon demonstrates that increasing strength of external supervision δ can reduce the diffusion probability of unethical behavior among green food enterprises to a certain extent. However, when strength of external supervision δ has reached a high level, the effect of strengthening supervision becomes limited.

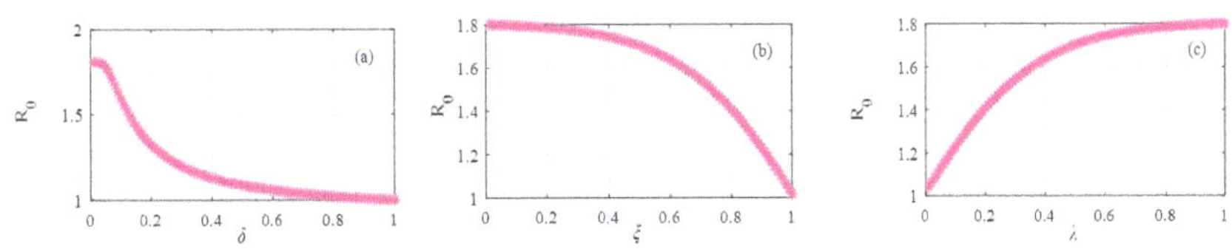

Figure 5. The influence of external environmental intervention factors on the diffusion of unethical behavior among green food enterprises. (**a–c**) refer to strength of external supervision δ, strength of punishment ξ, and external competitiveness λ.

Figure 5b indicates that when strength of punishment ξ is less than 0.5, the diffusion probability of unethical behavior declines slowly. When it is greater than 0.5, the downward trend is evident. This phenomenon demonstrates that the inhibitory effect of slight punishment on the diffusion of unethical behavior is weak. On the contrary, the most severe punishment to unethical behavior can shock and control the diffusion of unethical behavior among green food enterprises effectively.

Figure 5c shows the diminishing trend of the diffusion probability of unethical behavior among green food enterprises with the increase of external competitiveness λ. The analysis of Figure 5c indicates that when external competitiveness λ is less than 0.5, the diffusion probability of unethical behavior shows an upward trend. When it is greater than 0.5, the upward trend becomes flat. This phenomenon demonstrates that unethical behavior is more likely to diffuse among green food enterprises that are in a competitive external environment. Therefore, relevant departments should carry out key supervision on industries with fierce competition to prevent hazards.

In addition, increasing strength of external supervision δ, strength of punishment ξ, and reducing external competitiveness λ can suppress the diffusion of unethical behavior. However, with the single adjustment of these external environmental intervention factors, the basic reproductive number R_0 remains greater than 1. Therefore, the goal of eliminating the diffusion of unethical behavior among green food enterprises cannot be achieved.

3.2.2. Multiple External Environmental Intervention Factors and Diffusion of Unethical Behavior among Green Food Enterprises

Although Figure 5 reflects that the single adjustment of strength of external supervision δ, strength of punishment ξ, and external competitiveness λ can only play a role in reducing the diffusion probability of unethical behavior to a certain extent, and the goal of gradual elimination of diffusion cannot be achieved. Figure 6a shows that when strength of external supervision δ interacts with strength of punishment ξ, the diffusion probability of unethical behavior among green food enterprise shows a significant downward trend with the increase of these two factors, and the basic reproductive number R_0 can be reduced to less than 1. This phenomenon reflects that the supervision and punishment system is an integral organic entirety. If green food enterprises are only supervised without severe punishment, then a shocking effect cannot be achieved. If green food enterprise believe that supervision or punishment is insufficient, and implementing unethical behavior will not involve high risks, then this perception will increase the diffusion possibility of unethical behavior. On the contrary, an effective and enforceable supervision and punishment system will increase the cost of implementing unethical behavior, thereby greatly reducing the diffusion probability of unethical behavior. Therefore, the goal of gradual elimination of unethical behavior can be achieved.

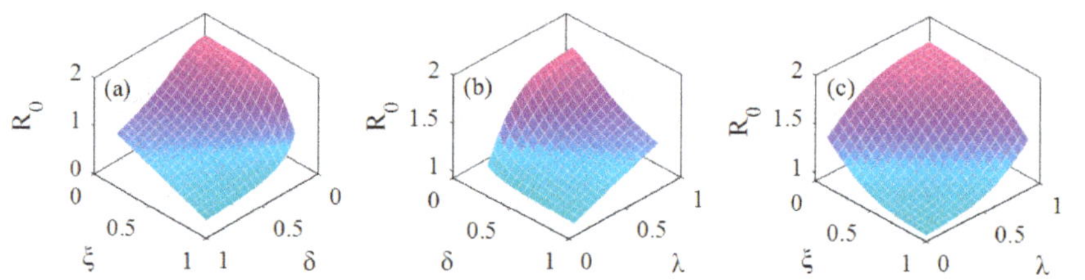

Figure 6. The interactive influence of external environmental intervention factors on the diffusion of unethical behavior among green food enterprises. The factors involved in the figure are strength of external supervision δ, strength of punishment ξ, and external competitiveness λ.

Figure 6b,c show that when external competitiveness λ interacts with strength of external supervision δ and strength of punishment ξ, the diffusion probability of unethical behavior has a significant downward trend with the increase of strength of external supervision δ, strength of punishment ξ, and the decrease of external competitiveness λ. Only when external competitiveness λ is extremely low , adjusting strength of external supervision δ and strength of punishment ξ make the basic reproductive number R_0 less than 1, thus causing the gradual elimination of unethical behavior diffusion. However, when a certain degree of competitiveness exists in the external environment, the single adjustment of strength of external supervision δ and strength of punishment ξ cannot make the basic reproductive number R_0 less than 1. Therefore, adopting a method of adjusting supervision and punishment together is necessary to control the hazard. Relevant authorities need to implement flexible supervision and punishment strategies for green food enterprises in different industries to optimize the allocation of management resources, thereby effectively controlling the social harm of the diffusion of unethical behavior among green food enterprises and maintaining social stability.

On the one hand, Figure 6 shows that external competitiveness λ exerts a "strengthening effect" on other external environmental intervention factors. Other external environmental intervention factors' impacts on the diffusion of unethical behavior among green food enterprises are significant in a competitive external environment. On the other hand, strength of external supervision δ and strength of punishment ξ exert a "suppression effect" on the diffusion of unethical behavior among green food enterprises. Therefore, when formulating prevention and control strategies, combining supervision and punishment organically is necessary to increase the cost of implementing unethical behavior. We need to take the different external competitiveness of different industries into formulating the most reasonable and efficient supervision and punishment strategies to avoid inefficient allocation of management resources, thus effectively controlling the diffusion of unethical behavior among green food enterprises.

3.3. Diffusion of Unethical Behavior among Green Food Enterprises under the Interaction of Organizational Behavior and External Environmental Intervention

Figure 7a,d,g show that when strength of external supervision δ interacts with level of enterprise moral clarity ψ, damage degree of unethical behavior ε, and enterprise influence θ, the diffusion probability of unethical behavior among green food enterprises has a decreasing characteristic with the increase of level of enterprise moral clarity ψ, damage degree of unethical behavior ε, strength of external supervision δ, and the decrease of enterprise influence θ. Moreover, the basic reproductive number R_0 can be reduced to less than 1, which means that gradual elimination of diffusion of unethical behavior can be achieved. This phenomenon reflects that the level of enterprise moral clarity ψ plays a regulatory role in the diffusion of unethical behavior among green food enterprises. On the one hand, the behavioral ethical norms of green food enterprises often have a standard ambiguity, which makes accurate judgments on whether a behavior is ethical or not unethical. On the other hand, given that individuals usually have unique behavioral ethical standards, different individuals may make different moral judgments when faced with the same situation. The ambiguity in ethical judgment affects individuals' moral decision-making, which, in turn, increases the probability of implementing unethical behavior. When enterprise members have a high level of moral clarity, they are more certain on whether they violate ethical norms. Therefore, level of enterprise moral clarity ψ can play a significant role in the prevention and control of the diffusion of unethical behavior.

Figure 7b,e,h show that when strength of punishment ξ interacts with level of enterprise moral clarity ψ, damage degree of unethical behavior ε, and enterprise influence θ, the diffusion probability of unethical behavior among green food enterprises has the decreasing characteristic with the increase of level of enterprise moral clarity ψ, damage degree of unethical behavior ε, strength of punishment ξ, and the decrease of enterprise influence θ. Figure 7c,f,i show that when external competitiveness λ interacts with level of enterprise moral clarity ψ, damage degree of unethical behavior ε and enterprise influence θ, the diffusion probability of unethical behavior among green food enterprises has a decreasing characteristic with the increase of level of enterprise moral clarity ψ, damage degree of unethical behavior ε, decrease of enterprise influence θ, and external competitiveness λ. These results are consistent with the conclusions obtained from Figures 3 and 5. Under the interaction of the organizational behavior and external environmental intervention factors above, the gradual elimination of unethical behavior among green food enterprises can be achieved.

Figure 7j,k,l show that when strictness of enterprise management system τ interacts with strength of external supervision δ, strength of punishment ξ, and external competitiveness λ, the diffusion probability of unethical behavior among green food enterprises has a diminishing characteristic with the increase of strictness of enterprise management system τ, strength of external supervision δ, strength of punishment ξ, and the decrease of external competitiveness λ. Moreover, the basic reproductive number R_0 can be reduced to less than 1. Therefore, strengthening the linkage of internal and external factors is possible by adjusting strictness of enterprise management system τ and external environmental intervention factors simultaneously. This adjustment can achieve the purpose of gradual elimination of unethical behavior.

Figure 7m,n,o show that when enterprise ethical climate ρ interacts with strength of external supervision δ, strength of punishment ξ, and external competitiveness λ, the diffusion probability of unethical behavior among green food enterprises has a large-scale increasing trend with the increase of external competitiveness λ, enterprise ethical climate ρ, and the decrease of strength of external supervision δ and strength of punishment ξ. Thus, enterprise ethical climate ρ is an important regulatory variable that inhibits the generation and diffusion of unethical behavior. Enterprise ethical climate is essentially formed by the superposition of ethical behavioral cognition of general members, managers, and senior leaders of enterprises. Although individual moral values and enterprise ethical climate can affect each other, individual morality must play an active role through the catalysis of enterprise ethical climate. An egocentric enterprise ethical climate will enable green food enterprise to put their own interests above other considerations. Therefore, they may implement unethical behavior to maximize benefits. On the contrary, a principled enterprise ethical climate emphasizes compliance with ethical norms within the enterprise, and the probability of implementing unethical behavior is greatly reduced.

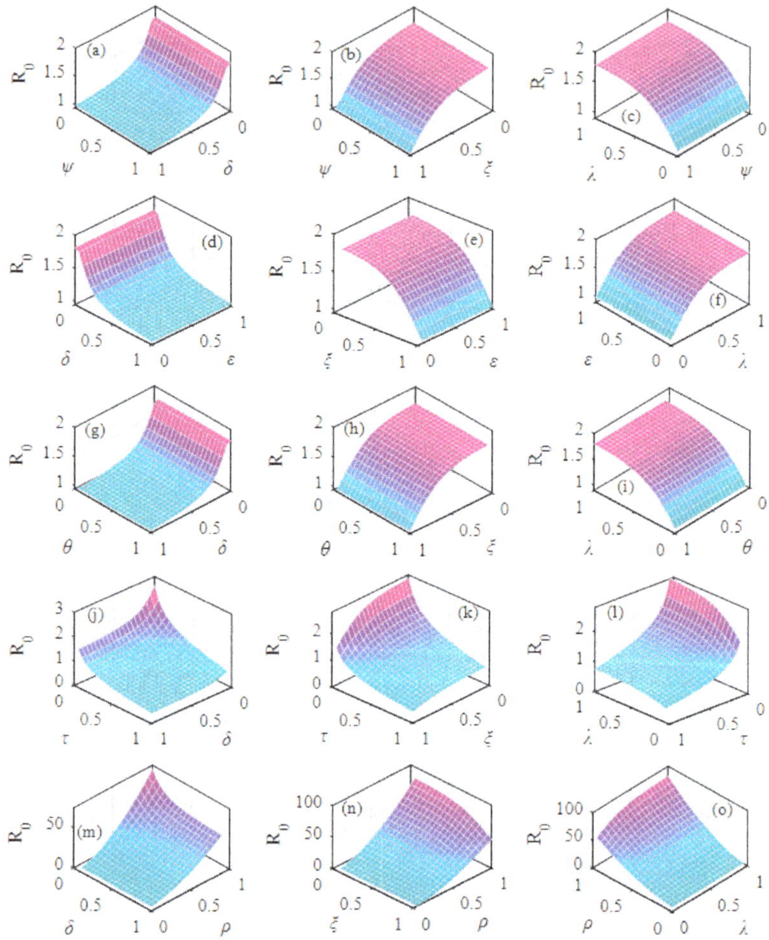

Figure 7. The interactive influence of organizational behavior and external environmental intervention factors on the diffusion of unethical behavior among green food enterprises. The factors involved in the figure are level of enterprise moral clarity ψ, damage degree of unethical behavior ε, enterprise influence θ, strictness of enterprise management system τ, enterprise ethical climate ρ, strength of external supervision δ, strength of punishment ξ, and external competitiveness λ.

In addition, Figure 7 indicates that level of enterprise moral clarity ψ, damage degree of unethical behavior ε, and enterprise influence θ exert a "strengthening effect" on external environmental intervention factors. Thus, external environmental intervention factors' impacts on the diffusion of unethical behavior among green food enterprises can be strengthened, and the diffusion will gradually be eliminated if level of enterprise moral clarity ψ and damage degree of unethical behavior ε are high or enterprise influence θ is low. In addition, enterprise ethical climate ρ exerts a strong "strengthening effect" on external environmental intervention factors. The influence of external environmental intervention factors on the diffusion probability of unethical behavioral among green food enterprises is significant in the egocentric enterprise ethical climate. For such green food enterprises, strengthening external interventions is necessary to prevent large-scale diffusion of unethical behavior. Therefore, when formulating prevention and control strategies, creating a principled enterprise ethical climate from a global perspective is necessary to prevent core green food enterprises from implementing unethical behavior. Moreover, adjusting the strength of external supervision and punishment flexibly according to different competitive environments outside green food enterprise can be significant. We also need to increase the strictness of enterprise management system and the level of enterprise moral clarity. In this scenario, the diffusion of unethical behavior can gradually be eliminated.

3.4. Sensitivity Analysis of Parameters

To describe the evaluation characteristics of unethical behavior diffusion better, we conduct the sensitivity analysis by changing the key parameters of strictness of enterprise management system τ, enterprise ethical climate ρ, strength of external supervision δ, and strength of punishment ξ. Under the circumstance of $m = m_0 = 5$, network scale $N = 500$, $\theta = 0.2$, $l = \psi = 0.3$, $\varepsilon = \mu = \beta = \phi = 0.4$, we change the key parameters as follows: (1) $\rho = 0.1$, $\delta = \xi = \tau = 0.2$, $\lambda = 0.4$; (2) $\rho = 0.15$, $\delta = \xi = \tau = 0.3$, $\lambda = 0.5$; (3) $\rho = 0.2$, $\delta = \xi = \tau = 0.4$, $\lambda = 0.6$; (4) $\rho = 0.25$, $\delta = \xi = \tau = 0.5$, $\lambda = 0.7$.

This section analyzes the sensitivity of these parameters to the nonlinear diffusion of unethical behavior among green food enterprises. Table 2 show the simulation results. The basic reproductive number R_0 maintains the original trend under different key parameters, indicating that the conclusions obtained from the simulation analysis are robust.

Table 2. A sensitivity analysis of the impact of strictness of enterprise management system τ, enterprise ethical climate ρ, strength of external supervision δ, and strength of punishment ξ on the evolution characteristics of unethical behavior diffusion among green food enterprises.

τ	ρ									Expectation	Variance
	0.1	0.2	0.3	0.4	0.5	0.6	0.7	0.8	0.9		
					$\lambda = 0.4$ $\xi = \delta = 0.2$						
0.1	2.98	7.42	12.60	18.40	24.70	31.40	38.40	45.70	53.30	26.10	270.00
0.2	2.20	6.00	10.80	16.30	22.40	29.10	36.30	44.00	52.10	24.40	267.00
0.3	1.75	5.08	9.48	14.70	20.80	27.40	34.70	42.60	51.00	23.10	262.00
0.4	1.43	4.41	8.49	13.50	19.40	26.00	33.40	41.40	50.10	22.00	256.00
0.5	1.20	3.88	7.69	12.50	18.20	24.80	32.20	40.30	49.20	21.10	250.00
0.6	1.02	3.44	7.01	11.60	17.20	23.70	31.10	39.30	48.40	20.30	244.00
0.7	0.87	3.07	6.43	10.90	16.30	22.70	30.10	38.40	47.60	19.60	238.00
0.8	0.75	2.76	5.91	10.20	15.50	21.80	29.10	37.50	46.80	18.90	231.00
0.9	0.65	2.48	5.45	9.53	14.70	20.90	28.30	36.60	46.10	18.30	226.00
					$\lambda = 0.5$ $\xi = \delta = 0.3$						
0.1	2.70	6.71	11.40	16.70	22.30	28.40	34.70	41.30	48.20	23.60	221.00
0.2	1.99	5.42	9.72	14.70	20.30	26.30	32.80	39.80	47.10	22.00	218.00
0.3	1.58	4.59	8.57	13.30	18.80	24.80	31.40	38.50	46.10	20.80	214.00
0.4	1.29	3.98	7.68	12.20	17.50	23.50	30.20	37.40	45.30	19.90	209.00
0.5	1.08	3.50	6.95	11.30	16.50	22.40	29.10	36.40	44.50	19.10	204.00
0.6	0.92	3.11	6.34	10.50	15.60	21.40	28.10	35.50	43.70	18.40	199.00
0.7	0.79	2.78	5.81	9.81	14.70	20.50	27.20	34.70	43.00	17.70	194.00
0.8	0.68	2.49	5.34	9.18	14.00	19.70	26.30	33.90	42.30	17.10	189.00
0.9	0.59	2.24	4.93	8.61	13.30	18.90	25.50	33.10	41.60	16.50	184.00

Table 2. *Cont.*

τ	ρ									Expectation	Variance
	0.1	0.2	0.3	0.4	0.5	0.6	0.7	0.8	0.9		
					$\lambda = 0.6\ \xi = \delta = 0.4$						
0.1	2.54	6.32	10.80	15.70	21.00	26.70	32.70	38.90	45.40	22.20	196.00
0.2	1.88	5.11	9.16	13.90	19.10	24.80	30.90	37.50	44.30	20.70	193.00
0.3	1.49	4.33	8.07	12.60	17.70	23.40	29.60	36.30	43.50	19.70	190.00
0.4	1.22	3.75	7.23	11.50	16.50	22.20	28.40	35.30	42.70	18.80	186.00
0.5	1.02	3.30	6.55	10.60	15.50	21.10	27.40	34.30	41.90	18.00	181.00
0.6	0.87	2.93	5.97	9.90	14.70	20.20	26.50	33.50	41.20	17.30	177.00
0.7	0.74	2.62	5.47	9.24	13.90	19.30	25.60	32.70	40.50	16.70	172.00
0.8	0.64	2.35	5.03	8.65	13.20	18.60	24.80	31.90	39.80	16.10	168.00
0.9	0.55	2.11	4.64	8.11	12.50	17.80	24.10	31.20	39.20	15.60	164.00
					$\lambda = 0.7\ \xi = \delta = 0.5$						
0.1	2.44	6.08	10.40	15.10	20.20	25.70	31.40	37.40	43.70	21.40	181.00
0.2	1.81	4.91	8.81	13.30	18.40	23.90	29.80	36.00	42.70	20.00	179.00
0.3	1.43	4.16	7.76	12.10	17.00	22.50	28.50	34.90	41.80	18.90	176.00
0.4	1.17	3.61	6.96	11.10	15.90	21.30	27.30	33.90	41.00	18.00	171.00
0.5	0.98	3.17	6.30	10.20	14.90	20.30	26.40	33.00	40.30	17.30	168.00
0.6	0.83	2.82	5.74	9.52	14.10	19.40	25.50	32.20	39.60	16.60	164.00
0.7	0.71	2.52	5.26	8.89	13.30	18.60	24.60	31.40	39.00	16.00	159.00
0.8	0.61	2.26	4.84	8.32	12.70	17.90	23.90	30.70	38.30	15.50	155.00
0.9	0.53	2.03	4.46	7.80	12.00	17.20	23.10	30.00	37.70	15.00	151.00

The results in Table 2 further verified the conclusions obtained in Figures 4 and 7. Enterprise ethical climate ρ exerts a "strengthening effect" on other factors. Moreover, Table 2 and Figure 4 indicate that enterprise ethical climate ρ plays a leading role in the diffusion of unethical behavior among green food enterprises. The effect of enterprise ethical climate ρ to inhibit the diffusion of unethical behavior among green food enterprises is much stronger than other factors. Therefore, formulating targeted prevention and control strategies from this aspect is effective.

4. Conclusions

This study comprehensively considers organizational behavior and external environmental intervention factors, and builds a nonlinear diffusion evolution model of unethical behavior among green food enterprises. Then, we conduct a simulation analysis of the diffusion mechanisms and evolution characteristics of the diffusion of unethical behavior among green food enterprises. The main conclusions are as follows:

(1) In the green food enterprises network, the diffusion probability of unethical behavior has a decreasing trend with the increase of the level of enterprise moral clarity, damage degree of unethical behavior, strictness of enterprise management system, and the decrease of enterprise influence and enterprise ethical climate. The single adjustment of the level of enterprise moral clarity, damage degree of unethical behavior, and enterprise influence has little influence on the diffusion of unethical behavior. Even if the formulation of control strategies is based on their interaction, the strategies can only reduce the diffusion probability to a certain extent and is unable to eliminate the diffusion. When the level of enterprise moral clarity, damage degree of unethical behavior, and enterprise influence interact with enterprise ethical climate, the diffusion probability shows an evident variation. The variation reflects that enterprise ethical climate exerts a "strengthening effect" on other organizational behavior factors. Moreover, strictness of the enterprise management system exerts a "suppression effect" on the diffusion of unethical behavior, and the "strengthening effect" of the enterprise ethical climate is stronger than the "suppression effect" of the strictness of the enterprise management system. Therefore, when formulating prevention and control strategies, we should take a principled enterprise ethical climate

as the foundation and increase level of enterprise moral clarity. We must maintain the strictness of the enterprise management system moderately and strengthen the control of the core enterprise with greater influence, reduce the emergence and contagion of unethical behaviors of enterprises, effectively prevent the spread of immoral contagion of green food enterprises in the green food market, and even cause major food safety problems.

(2) In the green food enterprises network, the diffusion probability of unethical behavior shows a decreasing trend with the increase of strength of external supervision, strength of punishment, and the decrease of external competitiveness. However, the single adjustment of strength of external supervision, strength of punishment, and external competitiveness can only play a role in reducing the diffusion probability to a certain extent, and it cannot gradually eliminate unethical behavior. However, when strength of external supervision interacts with strength of punishment, the diffusion probability of unethical behavior shows a significant downward trend with the increase of these two factors. As a result, diffusion can be eliminated gradually. When external competitiveness interacts with strength of external supervision and strength of punishment, the diffusion probability of unethical behavior shows a downward trend. However, when external competitiveness is extremely low, the single adjustment of strength of external supervision and strength of punishment will eliminate the diffusion gradually. In addition, external competitiveness exerts a "strengthening effect" on other external environmental intervention factors. Furthermore, strength of external supervision and strength of punishment exert a "suppression effect" on the diffusion of unethical behavior. Therefore, when formulating prevention and control strategies, relevant departments such as enterprises or governments should combine supervision and punishment organically is necessary to increase the cost of implementing unethical behavior and through the interaction of supervision and punishment mechanisms, the impact of unethical behavior of green enterprises is minimized. On the basis of considering the external competitiveness of green food enterprises, we need to increase the supervision of green food by relevant departments and the punishment of green food enterprises on food safety issues. By doing so, inefficient allocation of management resources and the fluke psychology of green food enterprises can be avoided, and the diffusion of unethical behavior among food enterprise can be effectively controlled.

(3) In the green food enterprises network, when the level of enterprise moral clarity, damage degree of unethical behavior, and enterprise influence interact with external environmental intervention factors, the diffusion probability of unethical behavior shows an evident decreasing characteristic. In addition, such adjustments can gradually eliminate the diffusion. Moreover, when strictness of an enterprise management system and enterprise ethical climate interact with external environmental intervention factors, the diffusion probability of unethical behavior can be reduced to a level that is close to 0, thus controlling the diffusion of unethical behavior effectively. Moreover, level of enterprise moral clarity, damage degree of unethical behavior, enterprise influence, and enterprise ethical climate exert a "strengthening effect" on the external environmental intervention factors. The "strengthening effect" of enterprise ethical climate is much stronger than the other three factors. Therefore, when formulating the prevention and control strategies, constructing a principled enterprise ethical climate is necessary to prevent core green food enterprise from implementing unethical behavior. Adjusting the strength of external supervision and punishment flexibly according to different competitive environments outside green food enterprise is also significant. Finally, we need to increase the strictness of the enterprise management system and the level of enterprise moral clarity. By doing so, the diffusion of unethical behavior can be eliminated gradually.

Under the interaction of organizational behavior and external environmental intervention, this study analyzes the diffusion mechanisms and evolution characteristics of the diffusion of unethical behavior among green food enterprises. The analysis enriches the theoretical results and helps us understand the evolution process of the diffusion of unethical behavior among green food enterprises. The conclusions of this study can provide theoretical references for relevant functional departments to prevent and control the large-scale diffusion of unethical behavior among green food enterprise, so as to provide a theoretical reference for food safety and other food problems in society. At the same time, the infection model constructed in this paper can also be applied to the development of control strategies for problems caused by the contagion of unethical behaviors; for example, the problem of medical data leakage is related to unethical behaviors such as the purchase and sale of private information by related enterprises. However, in order to deeply consider the differences between various corporate ethical standards and their impact on the diffusion of unethical behaviors, and also to consider the influence of consumers, governments, and corporate emotions on moral judgment, this paper can enrich and improve the theoretical analysis framework of the diffusion of unethical behaviors from the perspectives of differences in corporate ethical standards and moral disgust of market entities in the future.

Author Contributions: Q.Y. and Y.H. wrote the manuscript and simulated simulations. H.W. and T.C. designed the model. J.W. provided funding. All authors have read and agreed to the published version of the manuscript.

Funding: This research was funded by the National Natural Science Foundation of China (grant number 71971111), the Humanities and Social Science Planning Foundation of the Ministry of Education of China (grant number 19YJAZH086), the Key Project of Philosophy and Social Science Research in Colleges and Universities of Jiangsu Province (grant number 2018SJZDI063) and the Outstanding Innovation Team of Philosophy and Social Science Research in Colleges and Universities of Jiangsu Province (grant number 2017ZSTD005).

Informed Consent Statement: Informed consent was obtained from all subjects involved in thestudy.

Data Availability Statement: The method in this article is computer mathematical simulation. Numerical simulation analysis is the most effective way to test real-time dynamic data without a large number of empirical validations. The authors simulate the diffusion mechanisms and evolution characteristics of the diffusion of unethical behavior among green food enterprise by using MATLAB R2018a software. This paper does not have data that can be obtained because we directly used the plot function of MATLAB R2018a software to make the images.

Conflicts of Interest: The authors declare that they have no conflict of interest.

References

1. Treviño, L.K.; Weaver, G.R.; Reynolds, S.J. Behavioral ethics in organizations: A review. *J. Manag.* **2006**, *32*, 951–990. [CrossRef]
2. Kish-Gephart, J.J.; Harrison, D.A.; Treviño, L.K. Bad apples, bad cases, and bad barrels: Meta-analytic evidence about sources of unethical decisions at work. *J. Appl. Psychol.* **2010**, *95*, 1–31. [CrossRef] [PubMed]
3. Chang, M.; Huang, L.; Chen, H. Towards More Sustainable Diets: Investigating Consumer Motivations towards the Purchase of Green Food. *Sustainability* **2021**, *13*, 12288. [CrossRef]
4. Kowalska, A.; Ratajczyk, M.; Manning, L.; Bieniek, M.; Macik, R. "Young and Green" a Study of Consumers' Perceptions and Reported Purchasing Behaviour towards Organic Food in Poland and the United Kingdom. *Sustainability* **2021**, *13*, 13022. [CrossRef]
5. Paller, V.; Macalinao-Ramirez, C.; Bandal, M. Environmental contamination with parasites in selected rural farms in the Philippines: Impacts of farming practices on leafy greens food safety. *Parasitology* **2021**, *149*, 482–489. [CrossRef] [PubMed]
6. Tan, B.C.; Lau, T.C.; Sarwar, A.; Khan, N. The effects of consumer consciousness, food safety concern and healthy lifestyle on attitudes toward eating "green". *Br. Food J.* **2022**, *124*, 1187–1203. [CrossRef]
7. Sun, C.; Huang, D.; Li, H.; Chen, C.; Wang, C.; Li, M.; Wang, Z. Green Food Industry in China: Spatial Pattern and Production Concentration Drivers. *Front. Environ. Sci.* **2021**, *9*, 665990. [CrossRef]
8. Wongsaichia, S.; Naruetharadhol, P.; Schrank, J.; Phoomsom, P.; Sirisoonthonkul, K.; Paiyasen, V.; Srichaingwang, S.; Ketkaew, C. Influences of Green Eating Behaviors Underlying the Extended Theory of Planned Behavior: A Study of Market Segmentation and Purchase Intention. *Sustainability* **2022**, *14*, 8050. [CrossRef]
9. Jones, T.M. Ethical decision making by individuals in organizations: An issue-contingent model. *Acad. Manag. Rev.* **1991**, *16*, 366–395. [CrossRef]
10. Lu, J.G.; Lee, J.J.; Gino, F.; Galinsky, A.D. Polluted morality: Air pollution predicts criminal activity and unethical behavior. *Psychol. Sci.* **2018**, *29*, 340–355. [CrossRef]
11. Ripoll, G.; Ballart, X. Judging Unethical Behavior: The Different Effects of External and Public Service Motivation. *Int. Public Manag. J.* **2020**, *23*, 1–23. [CrossRef]

12. Kulik, B.W.; O'Fallon, M.J.; Salimath, M.S. Do competitive environments lead to the rise and spread of unethical behavior? Parallels from Enron. *J. Bus. Ethics* **2008**, *83*, 703–723. [CrossRef]
13. Miao, Q.; Newman, A.; Yu, J.; Xu, L. The relationship between ethical leadership and unethical pro-organizational behavior: Linear or curvilinear effects? *J. Bus. Ethics* **2013**, *116*, 641–653. [CrossRef]
14. Hersey, P.; Blanchard, K.H.; Johnson, D.E. *Management of Organizational Behavior*; Prentice Hall: Upper Saddle River, NJ, USA, 2007.
15. Thanwadee, C. Organizational factors affecting safety implementation in food companies in Thailand. *Int. J. Occup. Saf. Ergon. JOSE* **2014**, *20*, 213–225.
16. Pinto, J.; Leana, C.R.; Pil, F.K. Corrupt organizations or organizations of corrupt individuals? Two types of organization-level corruption. *Acad. Manag. Rev.* **2008**, *33*, 685–709. [CrossRef]
17. Wiltermuth, S.S.; Flynn, F.J. Power, moral clarity, and punishment in the workplace. *Acad. Manag. J.* **2013**, *56*, 1002–1023. [CrossRef]
18. Werbel, J.; Balkin, D.B. Are human resource practices linked to employee misconduct? A rational choice perspective. *Hum. Resour. Manag. Rev.* **2010**, *20*, 317–326.
19. Jacobs, G.; Belschak, F.D.; Den Hartog, D.N. (Un) ethical behavior and performance appraisal: The role of affect, support, and organizational justice. *J. Bus. Ethics* **2014**, *121*, 63–67. [CrossRef]
20. Peterson, D.K. Deviant workplace behavior and the organization's ethical climate. *J. Bus. Psychol.* **2002**, *17*, 47–61. [CrossRef]
21. Gorsira, M.; Steg, L.; Denkers, A.; Huisman, W. Corruption in organizations: Ethical climate and individual motives. *Adm. Sci.* **2018**, *8*, 4. [CrossRef]
22. Birtch, T.A.; Chiang, F.F. The influence of business school's ethical climate on students' unethical behavior. *J. Bus. Ethics* **2014**, *123*, 283–294. [CrossRef]
23. Gino, F.; Ayal, S.; Ariely, D. Contagion and differentiation in unethical behavior: The effect of one bad apple on the barrel. *Psychol. Sci.* **2009**, *20*, 393–398. [CrossRef] [PubMed]
24. Asadullah, M.A.; Siddiquei, A.N.; Hussain, A.; Arain, G.A. Power, moral clarity and punishment severity: A moderated-mediation model. *South Asian J. Bus. Stud.* **2017**, *6*, 38–52. [CrossRef]
25. Li, Y.; Feng, T.; Jiang, W. How Competitive Orientation Influences Unethical Decision-making in Clinical Practices? *Asian Nurs. Res.* **2018**, *12*, 182–189. [CrossRef]
26. Brass, D.J.; Butterfield, K.D.; Skaggs, B.C. Relationships and unethical behavior: A social network perspective. *Acad. Manag. Rev.* **1998**, *23*, 14–31. [CrossRef]
27. Bizjak, J.; Lemmon, M.; Whitby, R. Option backdating and board interlocks. *Rev. Financ. Stud.* **2009**, *22*, 4821–4847. [CrossRef]
28. Sullivan, B.N.; Haunschild, P.; Page, K. Organizations non gratae? The impact of unethical corporate acts on interorganizational networks. *Organ. Sci.* **2007**, *18*, 55–70. [CrossRef]
29. Brown, J.L.; Drake, K.D. Network ties among low-tax firms. *Account. Rev.* **2013**, *89*, 483–510. [CrossRef]
30. Zuber, F. Spread of unethical behavior in organizations: A dynamic social network perspective. *J. Bus. Ethics* **2015**, *131*, 151–172. [CrossRef]
31. Schuh, S.C.; Cai, Y.; Kaluza, A.J.; Steffens, N.K.; David, E.M.; Haslam, S.A. Do leaders condone unethical pro-organizational employee behaviors? The complex interplay between leader organizational identification and moral disengagement. *Hum. Resour. Manag.* **2021**, *60*, 969–989. [CrossRef]
32. Pablo, R.P.; Ricardo, M.C.; Alexis, B.G. Is unethical leadership a negative for Employees' personal growth and intention to stay? The buffering role of responsibility climate. *Eur. Manag. Rev.* **2021**, *18*, 535–549.
33. Khan, S.; Khan, A.K.; Shah, A.M.; Ali, L.; Ullah, R. Impact of employees' perceived threat of market competition on unethical marketing and selling practices: Moral disengagement and ethical leadership. *Bus. Ethics Environ. Responsib.* **2021**, *30*, 758–771. [CrossRef]
34. Olofsson, P.; Råholm, A.; Uddin, G.S.; Troster, V.; Kang, S.H. Ethical and unethical investments under extreme market conditions. *Int. Rev. Financ. Anal.* **2021**, *78*, 101952. [CrossRef]
35. Liu, Y.; Wang, B.; Wu, B.; Shang, S.; Zhang, Y.; Shi, C. Characterizing super-spreading in microblog: An epidemic-based information propagation model. *Phys. A Stat. Mech. Its Appl.* **2016**, *463*, 202–218. [CrossRef]
36. Chen, T.; Wang, L.; Wang, J.; Yang, Q. A Network Diffusion Model of Food Safety Scare Behavior considering Information Transparency. *Complexity* **2017**, *2017*, 5724925. [CrossRef]
37. Wang, L.; Li, S.; Wu, C. Risk contagion in inter-firm credit guarantee network. *Phys. A Stat. Mech. Its Appl.* **2019**, *526*, 120842. [CrossRef]
38. Rode, J.; Weber, A. Does localized imitation drive technology adoption? A case study on rooftop photovoltaic systems in Germany. *J. Environ. Econ. Manag.* **2016**, *78*, 34–48. [CrossRef]
39. Derbali, A.; Hallara, S. Measuring systemic risk of Greek banks: New approach by using the epidemic model "SEIR". *Cogent Bus. Manag.* **2016**, *3*, 1153864. [CrossRef]
40. Wang, L.; Li, S.; Chen, T. Investor behavior, information disclosure strategy and counterparty credit risk contagion. *Chaos Solitons Fractals* **2019**, *119*, 37–49. [CrossRef]
41. Giorno, V.; Spina, S. Rumor spreading models with random denials. *Phys. A Stat. Mech. Its Appl.* **2016**, *461*, 569–576. [CrossRef]
42. Hosseini, S.; Azgomi, M.A. A model for malware propagation in scale-free networks based on rumor spreading process. *Comput. Netw.* **2016**, *108*, 97–107. [CrossRef]

43. Centola, D. The spread of behavior in an online social network experiment. *Science* **2010**, *329*, 1194–1197. [CrossRef] [PubMed]
44. Moore, C.; Gino, F. Approach, ability, aftermath: A psychological process framework of unethical behavior at work. *Acad. Manag. Ann.* **2015**, *9*, 235–289. [CrossRef]
45. Skaza, J.; Blais, B. Modeling the infectiousness of Twitter hashtags. *Phys. A Stat. Mech. Its Appl.* **2017**, *465*, 289–296. [CrossRef]
46. Sundaresan, M. Constant absolute risk aversion preferences and constant equilibrium interest rates. *J. Financ.* **1983**, *38*, 205–212. [CrossRef]
47. Joosten, A.; Van Dijke, M.; Van Hiel, A.; De Cremer, D. Feel good, do-good!? On consistency and compensation in moral self-regulation. *J. Bus. Ethics* **2014**, *123*, 71–84. [CrossRef]
48. Valle, M.; Kacmar, K.M.; Zivnuska, S. Understanding the Effects of Political Environments on Unethical Behavior in Organizations. *J. Bus. Ethics* **2019**, *156*, 173–188. [CrossRef]
49. Allen, R.W.; Porter, L.W.; Angle, H.L. *Organizational Influence Processes*; Routledge: New York, NY, USA, 2016.
50. Berman, E.M.; Bowman, J.S.; West, J.P.; Van Wart, M.R. *Human Resource Management in Public Service: Paradoxes, Processes, and Problems*; CQ Press: London, UK, 2019.
51. Klein, M.C.; Mogles, N.; Treur, J.; van Wissen, A. Contagion of habitual behavior in social networks: An agent-based model. In Proceedings of the 2012 International Conference on Privacy, Security, Risk and Trust and 2012 International Conference on Social Computing, IEEE, Washington, DC, USA, 3–5 September 2012; pp. 538–545.
52. Scheiner, C.W.; Baccarella, C.; Bessant, J.; Voigt, K.I. Participation motives, moral disengagement, and unethical behaviour in idea competitions. *Int. J. Innov. Manag.* **2018**, *22*, 1850043. [CrossRef]
53. Bauman, C.W.; Tost, L.P.; Ong, M. Blame the shepherd not the sheep: Imitating higher-ranking transgressors mitigates punishment for unethical behavior. *Organ. Behav. Hum. Decis. Process.* **2016**, *137*, 123–141. [CrossRef]
54. Vikram, S.V.; Sinha, S. Emergence of universal scaling in financial markets from mean-field dynamics. *Phys. Rev. E* **2011**, *83*, 016101. [CrossRef]
55. Anderson, R.M.; May, R.M. *Infectious Diseases of Humans: Dynamics and Control*; Oxford University Press: Northants, UK, 1992.
56. Barabás, A.L.; Albert, R. Emergence of scaling in random networks. *Science* **1999**, *286*, 509–512. [CrossRef]
57. Kilduff, M.; Brass, D.J. Organizational social network research: Core ideas and key debates. *Acad. Manag. Ann.* **2010**, *4*, 317–357. [CrossRef]

sustainability

MDPI

Article

Robust Counterpart Models for Fresh Agricultural Product Routing Planning Considering Carbon Emissions and Uncertainty

Feng Yang [1], Zhong Wu [2,*], Xiaoyan Teng [1] and Shaojian Qu [3,4,*]

1 Business School, University of Shanghai for Science and Technology, Shanghai 200093, China
2 School of Management, Shanghai University of International Business and Economics, Shanghai 201620, China
3 School of Management Science and Technology, Nanjing University of Information Science and Technology, Nanjing 210044, China
4 School of Management, Shanghai University, Shanghai 200444, China
* Correspondence: wuzhong_1968@163.com (Z.W.); qushaojian@usst.edu.cn (S.Q.)

Abstract: Cold chain transportation guarantees the quality of fresh agricultural products in people's lives, but it comes with huge environmental costs. In order to improve transportation efficiency and reduce environmental impact, it is crucial to quantify the routing planning problem under the impact of carbon emissions. Considering fixed costs, transportation costs, and carbon emission costs, we propose a mixed integer linear programming model with the aim of minimizing costs. However, in real conditions, uncertainty poses a great challenge to the rationality of routing planning. The uncertainty is described through robust optimization theory and several robust counterpart models are proposed. We take the actual transportation enterprises as the research object and verify the validity of the model by constructing a Benders decomposition algorithm. The results reveal that the increase in uncertainty parameter volatility forces enterprises to increase uncontrollable transportation costs and reduce logistics service levels. An increase in the level of security parameters could undermine the downward trend and reduce 1.4% of service level losses.

Keywords: agricultural products; routing planning; robust model

Citation: Yang, F.; Wu, Z.; Teng, X.; Qu, S. Robust Counterpart Models for Fresh Agricultural Product Routing Planning Considering Carbon Emissions and Uncertainty. *Sustainability* **2022**, *14*, 14992. https://doi.org/10.3390/su142214992

Academic Editors: Manuel Pedro Rodríguez Bolívar and Maxim A. Dulebenets

Received: 24 August 2022
Accepted: 7 November 2022
Published: 13 November 2022

1. Introduction

Fresh agricultural products have become a necessity in daily life with huge market demand [1,2]. With the gradual improvement of living conditions, people's demand for the quality of agricultural products has also been soaring. Freshness, greenness, and health have gradually become the focus of people, especially agricultural products such as vegetables, fruits, meat, and seafood. The invention and use of modern transportation equipment, means of transportation, and logistics modes have met people's high quality demand for fresh agricultural products. The effective connection of the distribution link from the production base to the point of sale plays a key role in ensuring the quantity and quality of agricultural products transported [3,4]. The transportation of fresh agricultural products is different from the counterpart of other products. In the process of transportation, it is strictly dependent on cold chain transportation, and transportation conditions are strictly controlled, including measures to strictly ensure the quality of agricultural products, such as constant temperature and humidity [5]. In recent years, the transaction scale of fresh food e-commerce in China has reached 364.13 billion yuan (2020) and 465.81 billion yuan (2021). At the same time, the loss of agricultural products in transit is staggering. Up to 40% of fresh produce losses in North America cost USD 218 billion. Food waste in the EU reaches 89 million tons per year. Fresh agricultural products are perishable, which is the main reason for their loss, and how to reduce the loss of agricultural products has become the focus of research in the industry and academia. The circulation mode of fresh agricultural products in China is still based on the traditional mode of urban farmers' markets. Reasonable planning of the transportation path of agricultural products can improve transportation efficiency and reduce logistics costs.

During the transportation of fresh products, the measures to reduce and delay the decay of products are as follows. First of all, reduce the decay rate of the storage link, and use refrigeration equipment to keep the product fresh [6,7]. Second, in the distribution process, adopt a precise distribution mode, avoid multiple transfers and repeated loading and unloading, and try to deliver the exact number at the exact time and place according to the order quantity. Third, in the sales process, retail terminal sellers and stores take appropriate strategies or measures, including promotional measures such as secondary packaging, or bundled sales, to sell products in time. Fourth, in the process of processing unsold spoiled products, improve the recovery rate of fresh products, effectively classify and harmlessly treat products, and improve resource utilization or secondary use, including establishing biogas digesters and animal feed processors. Among the above-mentioned main links, the most important methods currently adopted are the first two, which are also the most commonly used measures and strategies at present. How to optimize the management strategy, reduce the loss of fresh agricultural products, and improve transportation efficiency has significant value for the current actual demand.

In addition, the environmental impact during the transportation of fresh agricultural products is also a topic worth noting. Fresh agricultural products not only require the use of a large number of vehicles to meet the demand; for example, ordinary product shipping, but also have a greater environmental impact during transportation due to their special requirements for the refrigeration equipment of vehicles. Coupled with its huge market size, if the issue of excessive energy consumption cannot be effectively addressed, it will further aggravate its negative impact on the environment and further aggravate the deterioration of the climate and environment. In response to global climate change, many countries and regions have put forward corresponding policies and measures, including energy structure adjustment, industrial structure adjustment, and operation mode transformation, among others [8]. In response to global climate change, China has solemnly proposed the great goal and vision of carbon neutrality [9–12]. The core of energy transformation under the carbon-neutral vision is the gradual replacement of high-carbon energy with zero-carbon and low-carbon energy, and the optimization and innovation of traditional technology with low-carbon and high-efficiency development technology.

The main contributions of this paper are as follows:

- We include constraints such as carbon emission limits into the research on fresh agricultural product routing planning, and comprehensively consider multiple costs to study the timeliness of transportation, operation economy, and environmental sustainability.
- We extend deterministic parameters to uncertain scenarios and propose to establish two robust corresponding models to solve the thorny problem of parameter data scarcity in uncertain scenarios.
- In order to improve the compatibility of the algorithm, we introduce a Benders decomposition algorithm to verify the model and compare the feasibility of the routing planning scheme through real cases.

The rest of this paper is organized as follows. Section 2 briefly introduces the literature review. Section 3 presents the problem and builds a basic linear programming model for deterministic scenarios. Section 4 extends the model to robust optimization models in uncertain scenarios. In Section 5, we collect data and build the algorithm framework. Section 6 conducts a sensitivity analysis. Section 7 concludes this paper and outlines possible future research.

2. Literature Review

Greenhouse gas emissions in the fresh agricultural products industry has become an important research field. Many scholars have shown that fresh agricultural products in the cold chain transport process will produce a lot of greenhouse gases, which has a great negative impact on the environment. The research on the transportation of agricultural products has attracted a large number of scholars, and the research scope covers cold chain transportation, perishable product transportation, uncertainty, vehicle scheduling, carbon emission issues, and so on [13–16]. Along with the national government and the people's attention to environmental issues, the concepts of "low-carbon life", "low-carbon logistics" and "low-carbon economy" have been proposed especially for carbon emissions. The transportation of fresh agricultural products is still dominated by fossil fuel consumption due to its large transportation volume and long transportation distance. The proportion of clean energy use is still low, which is difficult to meet the needs for environmentally sustainable development. Under the background of a carbon neutral policy, how to effectively measure the carbon emissions of the fresh agricultural product transportation industry and optimize the transportation network has important research significance for protecting the ecological environment and improving resource utilization.

Research related to agricultural products has attracted many scholars, as shown in Table 1. The supply chain of fresh agricultural products involves many complex links. Due to the uncertainty and instability of the environment, the uncertainty of demand or supply is often very difficult. In other words, the market demand or production area supply is often inestimable and is disturbed and affected by the real environment, including many uncertain factors such as weather, climate, emergencies, and [17] uncertainty. At the supply level of fresh agricultural products, the output of agricultural products depends on the stability of climatic conditions. The application of modern advanced equipment has played a good role in promoting the stability of agricultural product supply. The existing research on supply stability focuses on technology, such as equipment and cultivation technology, and pays less attention to management and operation optimization. At the level of demand for fresh agricultural products, the improvement of living standards makes people have higher demand. It is not only the increase in demand, but also the improvement in quality. The demand for fresh products is extremely unstable and is directly affected by consumer preference behavior, which is difficult to directly measure and quantify [16,18]. The heterogeneous preferences of consumers directly lead to the uncertainty of demand [19].

Table 1. Related literature.

Literature	Agricultural Product Transportation	Quantitative Analysis	Carbon Emissions	Uncertainty	Real Cases
[20]	*	*			
[21]	*	*			*
[22]		*			*
[23]	*	*			*
[24]	*	*	*		
[25]	*	*	*		
[26]	*		*		*
[27]	*	*			
[28]	*	*	*	*	
[29]	*	*			*
This paper	*	*	*	*	*

Note: * represent that the content listed in the title of column is studied in the corresponding reference.

Existing literature focuses on the research of consumer behavior analysis in deterministic scenarios, and few literature focuses on uncertain scenarios. Based on the above analysis, it is of practical value to carry out research on the transportation planning of fresh agricultural products under uncertain scenarios. There are still the following research gaps in the research on fresh agricultural product transportation. First, the traditional linear programming model does not take into account the carbon emission of fresh agricultural products transportation. Second, the model in deterministic scenarios is too ideal and it is necessary to research and expand in uncertain scenarios. Third, previous studies mostly use hypothetical numerical cases, which are difficult to guide corporate decision-making.

3. Problem Description

3.1. Problem Description

In the context of carbon neutrality goals, higher requirements have been posed for the innovation of agricultural transportation scheduling. We have considered the production, transportation, and distribution systems of fresh agricultural products, including multiple origin centers (pastures or farms), multiple distribution centers, and multiple demand sites. Fresh agricultural product transportation companies transfer fresh agricultural products through transportation (trucks) and transport them to demand sites (see Figure 1). Affected by the particularity of agricultural products, during the transportation of fresh agricultural products the number of agricultural products arriving is reduced due to the problem of decay. In the process of transportation, the potential decay of fresh food should be considered so that the initial transportation quantity should be increased. Fresh agricultural products finally arrive at the demand site for sale. In terms of traffic scheduling planning, how to effectively avoid risks and improve benefits is a hot research topic at present. At the same time, our research has also promoted the implementation of carbon neutrality goals.

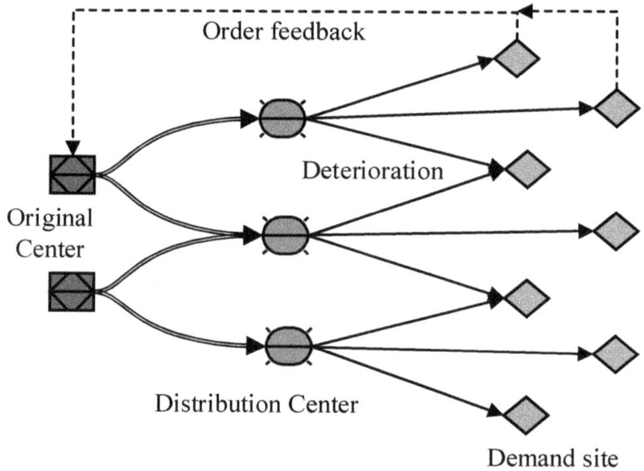

Figure 1. Transportation planning for fresh agricultural products.

3.2. Assumptions

According to the feasibility and scientific nature of mathematical modeling, this section proposes the following basic assumptions:

- It is assumed that logistics transportation is one-way transportation [30], that is, the direction is from the origin to the demand place;
- All agricultural products supplied are from the origin sites, and there is no input from outside the industrial chain [31];
- Transport vehicles of the same model (fuel consumption and capacity) are used in the same phase;
- Know the location of candidate demand sites in advance [32];
- The speed of the vehicle varies depending on the actual situation [33].

4. Model

4.1. Mixed Integer Linear Programming Model

In this section, the routing planning of the fresh agricultural products transportation network is modeled and analyzed. Considering the various costs in the research problem comprehensively, namely the fixed cost of distribution centers, configuration cost of vehicles, transportation cost, and time windows cost, a mixed integer linear programming model (MILP) model is constructed. The objective is to minimize the total cost, as shown in Formula (1).

$$\underset{x_i \in X, y_{ij} \in Y}{\text{minimize}} \left\{ \begin{array}{l} \sum_{i \in I} c_i^f x_i + \sum_{i \in I} c_i^n n_i + \sum_{i \in I} \sum_{j \in J} c_{ij}^v d_{ij} \left\lceil \frac{y_{ij} q_j}{h_v (1-\tau)} \right\rceil \\ + \sum_{i \in I} \sum_{j \in J} y_{ij} c_{ij}^t \left\{ \frac{d_{ij}}{v_{ij}} + w_{ij} - t_0, 0 \right\}^+ + \sum_{i \in I} \sum_{j \in J} c_{ij}^e E_{ij}^e d_{ij} n_i \end{array} \right\} \tag{1}$$

Subject to,

$$s.t. \sum_{i \in I} x_i \leq |I| \tag{2}$$

$$y_{ij} \leq M x_i, \forall i \in I \tag{3}$$

$$t_j^l \leq t_i + y_{ij} d_{ij} {v_{ij}}^{-1} + w_{ij} \leq t_{ij}^u, \forall i \in I, \forall j \in J \tag{4}$$

$$\sum_{i \in I} y_{ij} = 1, \forall j \in J \tag{5}$$

$$\sum_{j \in J} \frac{y_{ij} q_j}{1 - v} \leq n_i h_v, \forall i \in I \tag{6}$$

$$n_i h_v \leq H_i^{\max}, \forall i \in I \tag{7}$$

$$\sum_{i \in I} E_e d_{ij} n_i \leq e^{\max}, \forall j \in J \tag{8}$$

$$\lceil y_{ij} \rceil d_{ij} \leq d^{\max}, \forall i \in I, \forall j \in J \tag{9}$$

$$x_i \in \{0,1\}, n_i \in \{0,1,\ldots,N\}, y_{ij} \in [0,1], \forall i \in I, \forall j \in J \tag{10}$$

The relevant constraints are explained as follows. Constraint (2) is the quantity constraint of available distribution centers. Constraint (3) means that the premise of the routing planning decision is that the distribution center is selected [34,35]. Constraint (4) is a time window constraint, and t_j^l, t_j^u represents the minimum arrival time and the maximum arrival time, respectively [36,37]. Constraint (5) means that the demand of any demand station must be met. Constraint (6) indicates that the actual transportation volume is lower than the rated transportation volume. Constraint (7) is the inventory constraint of the distribution center. Constraint (8) refers to carbon emission constraints, and the cumulative carbon emissions are strictly limited by carbon-neutral policies. Constraint (9) is the maximum mileage constraint for the truck. Constraint (10) represents constraints on related decision variables.

4.2. Interval Robust Counterpart Model

The complete information scenario is the most ideal scenario in theoretical research. On the contrary, the external environment of the market is full of uncertainty, so there is no ideal scenario in real life [38]. There are great uncertainties in the transportation of green agricultural products, especially the uncertainty of demand and the uncertainty of supply. Many scholars use robust optimization theories and methods to resist the influence of uncertain factors [39–41] and verify the feasibility of the robust model through multiple dimensions [42–44], citing the research ideas of relevant scholars [45–47] to expand the demand uncertainty in the transportation of green agricultural products.

Proposition 1. *Introduce a robust optimization theory to define uncertain parameter sets* U_I, *the transformation from deterministic to uncertain parameters can be realized, then the objective function* $\underset{x_i \in X, y_{ij} \in Y}{\text{minimize}} \{ F(x_i, y_{ij}) | q_j \}$ *in the deterministic model can be transformed into*

$\underset{x_i \in X, y_{ij} \in Y, \varepsilon_j \in U_I}{\text{minimize}} \left\{ F\left(x_i, y_{ij} \middle| q_j^0, \varepsilon_j \Delta\right) \right\}$ *in Equation(11) of the interval robust counterpart (IRC) model, with the constraints (6) in the MILP model relaxed to* $\sum_{j \in J} \frac{y_{ij} q_j^0}{1-\tau} + \Psi \sup_{\varepsilon_j \in U_I} \sum_{j \in J} c_v d_{ij} \left\lceil \frac{y_{ij} \varepsilon_j \Delta}{1-\tau} \right\rceil \leq n_i h_v, \forall i \in I.$

Proof of Proposition 1. The complete information situation is further transformed into an incomplete information situation, and the deterministic MILP model is transformed into a robust model, the goal of which is to pursue total cost minimization under the condition of incomplete information. Based on the basic MILP model, which q_j under the complete information scenario defines the interval value uncertainty set U_I, which is $\left\{ \widetilde{q}_j \in \mathbb{R}^n, \widetilde{q}_j = q_j^0 + \varepsilon_j \Delta, \varepsilon_j \in [a, b], \sum_{j \in J} \varepsilon_j \leq \Psi, \forall j \right\}$, where q_j^0 is the nominal value, Δ is the benchmark floating quantity, and ε_j is the uncertain parameter. Only the floating interval $[a, b]$ is known but the specific probability distribution is not known and Ψ is a safety parameter indicating the amount of uncertainty, so we made $q_j \to \widetilde{q}_j$. Then, the related constraints are further relaxed to meet the constraints of real resource constraints. Among them, the constraint $\sum_{j \in J} \frac{y_{ij} q_j}{1-v} \leq n_i h_v, \forall i \in I$ in the MILP model, the relaxation is as Equation (16). Similarly, $q_j \to \widetilde{q}_j$ is also replaced in the objective function, and the objective function is also transformed into a robust problem in an uncertain parameter environment. To simplify the expression, let $F_0\left(y_{ij} \middle| q_j^0\right)$ and $F'[y_{ij}, \varepsilon_j, \Delta]$ represent the expressions of objective functions for deterministic and uncertain scenarios, respectively. Among them, $\sum_{i \in I} \sum_{j \in J} c_{ij}^v d_{ij} \left\lceil \frac{y_{ij} q_j}{h_v(1-\tau)} \right\rceil$ is transferred to $\Psi \sup_{\varepsilon_j \in U_I} \sum_{i \in I} \sum_{j \in J} c_v d_{ij} \left\lceil \frac{y_{ij} \widetilde{q}_j}{h_v(1-\tau)} \right\rceil, \widetilde{q}_j := q_j^0 + \varepsilon_j \Delta$. The objective function is disassembled $\widetilde{q}_j \to q_j^0 + \varepsilon_j \Delta, \varepsilon_j \in U_I$ to obtain the objective under the uncertain situation with safety parameters, as shown in (11), which is $\underset{x_i \in X, y_{ij} \in Y, \varepsilon_j \in U_I}{\text{minimize}} \left\{ F\left(x_i, y_{ij} \middle| q_j^0, \varepsilon_j \Delta\right) \right\}$. □

The objective function (11) of the IRC model is to minimize the total cost.

$$\underset{x_i \in X, y_{ij} \in Y}{\text{minimize}} \left\{ \begin{array}{l} \sum_{i \in I} c_i^f x_i + \sum_{i \in I} c_i^n n_i + \sum_{i \in I} \sum_{j \in J} y_{ij} c_{ij}^t \left\{ \frac{d_{ij}}{v_{ij}} + w_{ij} - t_0, 0 \right\}^+ \\ + \sum_{i \in I} \sum_{j \in J} c_{ij}^e E_{ij}^e d_{ij} n_i + \sum_{i \in I} \sum_{j \in J} c_{ij}^v d_{ij} \left\lceil \frac{y_{ij} q_j^0}{h_v(1-\tau)} \right\rceil \\ + \Psi \sup_{\varepsilon_j \in U_I} \sum_{i \in I} \sum_{j \in J} c_{ij}^v d_{ij} \left\lceil \frac{y_{ij} \varepsilon_j \Delta}{h_v(1-\tau)} \right\rceil \end{array} \right\} \tag{11}$$

Subject to,
Constraints (2), (3), (4), (5)

$$\sum_{j \in J} \frac{y_{ij} q_j^0}{1-\tau} + \Psi \sup_{\varepsilon_j \in U_I} \sum_{j \in J} c_v d_{ij} \left\lceil \frac{y_{ij} \varepsilon_j \Delta}{1-\tau} \right\rceil \leq n_i h_v, \forall i \in I \tag{12}$$

Constraints (7), (8), (9), (10)

Interpretations of the relevant constraints of the IRC model are as follows. Constraints (2)–(5) have the same meaning in the MILP model. They are quantity constraints of distribution centers, association constraints of routes and distribution centers, time windows constraints, and demand satisfaction constraints. Constraint (16) indicates that the actual traffic volume under uncertain conditions is lower than the rated traffic volume constraint. The meaning of constraint (7)–(10) is the same as in the MILP model.

4.3. Ellipsoid Robust Counterpart Model

In addition to interval uncertainty sets, ellipsoidal uncertainty sets are also commonly used tools to characterize uncertain parameters. In this section, we build the ellipsoidal robust counterpart (ERC) model.

Proposition 2. *Introduce a robust optimization theory to define an ellipsoid uncertainty set* U_E*, which can realize the transformation of deterministic parameters to uncertain parameters, then the objective function* $\underset{x_i \in X, y_{ij} \in Y}{\text{minimize}}\{x_i, y_{ij}|q_j\}$ *in the MILP model can be converted to* $\underset{x_i \in X, y_{ij} \in Y, \xi_j \in U_E}{\text{minimize}}\left\{F\left(x_i, y_{ij}\middle|q_j^0, \xi_j\right)\right\} \underset{x_i \in X, y_{ij} \in Y}{\text{minimize}}\left\{G_0(xi, yij) + G_0\left(xi, yij\middle|q_j^0\right) + \Omega \sup_{\xi j \in UE} G'(xi, yij|\xi j)\right\}$ *in the ERC model, with the constraints (6) in the MILP model will be relaxed to* $\sum_{i \in I}\sum_{j \in J} c_{ij}^v d_{ij}\left\lceil \frac{y_{ij}q_j^0}{h_v(1-\tau)} \right\rceil + \Omega \sup_{\xi_j \in U_E} \sqrt{\sum_{i \in I}\sum_{j \in J}\left(c_{ij}^v d_{ij}\left\lceil \frac{y_{ij}\varepsilon_j \Delta}{h_v(1-\tau)} \right\rceil\right)^2}$.

Proof of Proposition 2. The complete information scenario is extended to the incomplete information scenario, and the demand parameters are expanded, represented by q_j [48,49]. The basic MILP model is further transformed into the ERC model, where the ellipsoid set is defined as the $U_E = \left\{\tilde{q}_j \in \mathbb{R}^n, q_j := q_j^0 + \xi\Delta, \|\xi\| \le \Omega\right\}$ uncertainty parameter covariance satisfying $\left\{q_j \in \mathbb{R}^n, \left(q_j - q_j^0\right)^T \Sigma^{-1}\left(q_j - q_j^0\right) \le \Omega^2\right\}$, where q_j^0 is a nominal value, $\Delta = \Sigma^{\frac{1}{2}}$, Σ is a positive definite matrix, and Ω is a safety parameter indicating the amount of uncertainty. By using an affine transformation, it can also be expressed as a ball of radius Ω. Considering $q_j \to \tilde{q}_j$, the constraints associated with it are further relaxed to satisfy the constraint constraints, where $\sum_{j \in J} \frac{y_{ij}q_j}{1-v} \le n_i h_v$ is constrained, which is relaxing as $\sum_{j \in J} \frac{y_{ij}q_j^0}{1-\tau} + \Omega \sup_{\xi_j \in U_E} \sqrt{\sum_{j \in J}\left(c_v d_{ij}\left\lceil \frac{y_{ij}\xi_j \Delta}{1-\tau} \right\rceil\right)^2} \le n_i h_v$. Similarly, we let $q_j \to \tilde{q}_j$ be replaced in the objective function, and the objective function is also transformed accordingly. To simplify the expression, set $G_0(x_i, y_{ij}) = \sum_{i \in I} c_i^f x_i + \sum_{i \in I} c_i^n n_i + \sum_{i \in I}\sum_{j \in J} y_{ij} c_{ij}^t \left\{\frac{d_{ij}}{v_{ij}} + w_{ij} - t_0, 0\right\}^+$ and $G\left(y_{ij}, q_j^0\right) = \sum_{i \in I}\sum_{j \in J} c_{ij}^e E_{ij}^e d_{ij} n_i + \sum_{i \in I}\sum_{j \in J} c_{ij}^v d_{ij}\left\lceil \frac{y_{ij}q_j^0}{h_v(1-\tau)} \right\rceil$ and consider the robustness of the objective function $\sqrt{\sum_{i \in I}\sum_{j \in J}\left(c_{ij}^v d_{ij}\left\lceil \frac{y_{ij}\xi_j \Delta}{h_v(1-\tau)} \right\rceil\right)^2} \le \Omega \sup_{\xi_j \in U_E} G'(x_i, y_{ij}|\xi_j \Delta)$, to represent the optimal strategy under the worst case, $\underset{x_i \in X, y_{ij} \in Y}{\text{minimize}}\{F(x_i, y_{ij})\}$ that is changed to $\underset{x_i \in X, y_{ij} \in Y, \xi_j \in U_E}{\text{minimize}}\left\{F\left(x_i, y_{ij}\middle|q_j^0, \xi_j\right)\right\} \underset{x_i \in X, y_{ij} \in Y}{\text{minimize}}\left\{G_0(xi, yij) + G_0\left(xi, yij\middle|q_j^0\right) + \Omega \sup_{\xi j \in UE} G'(xi, yij|\xi j)\right\}$.
□

The objective function of the ERC model is to minimize the total cost (13).

$$\underset{x_i \in X, y_{ij} \in Y}{\text{minimize}}\left\{\begin{array}{l} \sum_{i \in I} c_i^f x_i + \sum_{i \in I} c_i^n n_i + \sum_{i \in I}\sum_{j \in J} y_{ij} c_{ij}^t \left\{\frac{d_{ij}}{v_{ij}} + w_{ij} - t_0, 0\right\}^+ \\ +\sum_{i \in I}\sum_{j \in J} c_{ij}^e E_{ij}^e d_{ij} n_i + \sum_{i \in I}\sum_{j \in J} c_{ij}^v d_{ij}\left\lceil \frac{y_{ij}q_j^0}{h_v(1-\tau)} \right\rceil \\ +\Omega \sup_{\xi_j \in U_E} \sqrt{\sum_{i \in I}\sum_{j \in J}\left(c_{ij}^v d_{ij}\left\lceil \frac{y_{ij}\varepsilon_j \Delta}{h_v(1-\tau)} \right\rceil\right)^2} \end{array}\right\} \tag{13}$$

Subject to,

Constraints (2), (3), (4), (5)

$$\sum_{j\in J}\frac{y_{ij}q_j^0}{1-\tau}+\Omega\sup_{\xi_j\in U_E}\sqrt{\sum_{j\in J}\left(c_v d_{ij}\left\lceil\frac{y_{ij}\xi_j\Delta}{1-\tau}\right\rceil\right)^2}\le n_i h_v,\forall i\in I \tag{14}$$

Constraints (7), (8), (9), (10)

Interpretations of the relevant constraints are as follows. Constraints (2)–(5) in the ERC model mean the same as those in the MILP model, which are, respectively, the quantity constraints of distribution centers, routing and distribution center association constraints, time windows, and demand satisfaction constraints. The constraint (14) indicates that the actual traffic volume under uncertain parameters is lower than the rated traffic volume constraint. The meaning of constraint (7)–(10) is the same as in the MILP model.

5. Data and Algorithm

In this section, we will verify the effectiveness of the proposed model and theory in real scenes. A transportation enterprise located in the famous vegetable production base in Shouguang, Shandong Province, was selected as the case study object. Vegetables are famous for their high level of industrialization, large area, high yield, high quality, and complete varieties [50]. This paper studies an agricultural production and marketing enterprise, which is engaged in the production and marketing operation services of agricultural products. The specific process is to purchase agricultural products from the original center (OC), transfer them through the transit distribution center (DC), and transport them to the target demand site (DS). The specific form is shown in Figure 2. The production base uses new production technology and advanced production equipment, which can not only produce high-quality organic products but also realize the off-season production of some vegetables, which can be continuously distributed to the distribution. The department supplies fresh produce. The selection of candidate transit warehouses refers to a variety of criteria, including the accessibility of the traffic location, whether the equipment meets the standard, whether the warehouse capacity meets the demand, and the economy. Through comprehensive screening, five candidate warehouses were finally determined, which were marked with symbol circles. The demand site is the supermarket closest to the user. These supermarkets are densely distributed on the main roads and traffic-intensive areas of the city, respectively.

The contents of this section are outlined below. Section 5.1 introduces the source and data structure of the data used in this case. Section 5.2 designs a solver-based solution algorithm framework.

Figure 2. Geographical location.

5.1. Data

The basic data information involves refrigerated transfer stations, refrigerated trucks, and demand sites. The data comes from field research [51,52], as shown in Tables 2 and 3. The actual distance between two sites can be directly obtained from Google Maps with web crawling technology, as shown in Table 4.

Table 2. Refrigerated truck parameters.

Item	Data
The load of transportation tools	1000–1500 kg
The unit energy consumption	25 L
The maximum speed limit	60 km/h
The average speed	30–45 km/h

Table 3. Data of the distribution center.

The Distribution Center	DC_1	DC_2	DC_3	DC_4	DC_5
The fixed cost	15,000 CNY	20,000 CNY	10,000 CNY	20,000 CNY	20,000 CNY
The maximum stock	9000 kg	7000 kg	8000 kg	6000 kg	6000 kg
Longitude (E)	117.019473	117.531636	117.5187	117.388586	117.391322
Latitude (N)	36.719945	36.678218	36.882706	36.690375	36.740998
The demand site	Ds_1	Ds_2	Ds_3	Ds_4	Ds_5
The nominal demand	2000 kg	3500 kg	4000 kg	2500 kg	3000 kg
The demand site	Ds_6	Ds_7	Ds_8	Ds_9	Ds_{10}
The nominal demand	2500 kg	3000 kg	4500 kg	3000 kg	3500 kg

Table 4. Distance.

	DC_1	DC_2	DC_3	DC_4	DC_5
Ds_1	16.4 km	11.2 km	10.8 km	7.2 km	13.9 km
Ds_2	39.9 km	62.3 km	60.9 km	57.1 km	58.2 km
Ds_3	52.2 km	54.4 km	53.3 km	59.7 km	55.2 km
Ds_4	66.8 km	60.4 km	57.9 km	56.6 km	56.1 km
Ds_5	36.1 km	50.1 km	36.1 km	32.2 km	45.3 km
Ds_6	7.8 km	11.7 km	12.3 km	10.4 km	8.5 km
Ds_7	58.7 km	41.9 km	62.4 km	60.5 km	58.5 km
Ds_8	64.2 km	51.1 km	53.2 km	53.1 km	56.5 km
Ds_9	50.5 km	62.2 km	59.2 km	60.8 km	56.4 km
Ds_{10}	41.2 km	36.9 km	37.5 km	35.6 km	33.5 km

5.2. Algorithm

Based on the above basic data values, this section uses Python as the programming platform to design the solution framework, and calls the solver Gurobi (9.5) to solve the MILP, ERC, and IRC models. To ensure the scientificity of the case, the control variable method is used for verification, and the same environment (Windows 10, Intel (R) Core (TM) i5-8300H, HP, USA) is used on the same computer CPU@2.3 GHz, RAM8 GB, 512 G SSD).

The Benders decomposition algorithm has been widely used in various optimization problems [53–55]. In order to solve the problem of routing planning for large-scale fresh agricultural products, we will build a solution framework based on Gurobi in this section. However, algorithms based on solvers often require high-standard forms of models and are difficult to solve large-scale computing problems. To effectively deal with non-standard model optimization problems, we develop a customized Benders decomposition algorithm to solve them. Firstly, we separate the original problem into a subproblem. Then, we solve the subproblem to obtain the initial feasible solution. Secondly, we trace the initial feasible solution of the subproblem back to the main problem and further construct the overall optimization model considering the relevant constraints of the main problem. Thirdly, in order to improve the efficiency of the model, we adjust the constraints in the model that are difficult to solve directly by adding optimal cuts or feasible cuts to Farkas' lemma [56]. Finally, we derive the optimal or feasible solution of the model. The overall framework of our improved algorithm is shown in Algorithm 1.

Algorithm 1. Benders decomposition algorithm

Input nominal parameters : $c_i^f, c_{ij}^v, c_{ij}^t, c_{ij}^e, d_{ij}, q_j, h_v, \tau, v_{ij}, w_{ij}, t_0, E_{ij}^e$
Initialization boundary : $(LB := -\infty, UB := +\infty)$, Set (I, J)
Identify Master problem $\{\text{MP}, \ F(x_i, n_i | y_{ij})\}$, Sub-problem $\{\text{SP}, \ F(y_{ij})\}$
Repeat $n \leftarrow 1$
  for MILP model, $q_j \leftarrow q_j$
  for IRC model, $q_j \leftarrow q_j^0 + \varepsilon_j \Delta, \varepsilon_j \in U_I$
  for ERC model, $q_j \leftarrow q_j^0 + \xi_j \Delta, \xi_j \in U_E$
  SP $\leftarrow$ Dual Sub-problem (DSP)
    solving DSP by Gurobi get μ_{ij}^*(dual variables of $\overline{y}_{ij}$) with constraints
    if DSP is unbounded, add benders feasibility cut in MP
      if $obj^n{}_{DSP} < UB^{n+1}, UB^{n+1} = obj^n$
      else, $UB^{n+1} = UB^n$
    if DSP is bounded, add benders optimality cut in MP
      if $obj^n{}_{DSP} < UB^{n+1}, UB^{n+1} = obj^n$
      else, $UB^{n+1} = +UB^n$
      if no feasible solution for DSP
      end
  Obtain x_i, n_i by solving MP
    if $obj^n{}_{MP} > LB^{n+1}, LB^{n+1} = obj^n$
    else, $LB^{n+1} = LB^n$
$n \leftarrow n + 1$
until (Iterations = n or $UB - LB < \varepsilon$)
Return value $\{x_i^*, n_i^*, y_{ij}^*\}$

6. Simulation

In Sections 6.1 and 6.2, we analyze the impact of safety parameters and volatility. We present the routing planning scheme in Section 6.3. We compare the impact of carbon tax changes in Section 6.4. In Section 6.5, we compare the impact of changes in storage ratios across warehouses.

6.1. *Influence of Safety Parameters*

We compare the advantages of each model through cost response to safety parameters, as shown in Figure 3. The horizontal axis in the figure is a safety parameter, and the vertical axis is the total transportation cost. It can be seen that the MILP model has strong stability, and its total cost will not change with the change in security parameters. However, at the same time, there are also defects, that is, it is impossible to deal with planning problems in an uncertain environment. In the case of uncertain market parameters, both the IRC and ERC models can be obtained by feasible solutions but at the same time, the price of robust rots is required to resist the interference of uncertain parameters [57,58]. As the level of security parameters increases (that is, the uncertainty expansion), the total transportation costs show an upward trend. When the market is uncertain (SP = 10, all sites are disturbed by uncertain parameters), the maximum increase in total transportation costs is 20.1%, that is, the maximum cost increase in the worst scenario is 20.1%. In terms of the scope of the uncertain set, the larger the range of uncertain parameter fluctuations, ($[-0.10, 0.10]$ belongs to $[-0.20, 0.20]$), the greater its transportation cost, whether it is an ERC or IRC model.

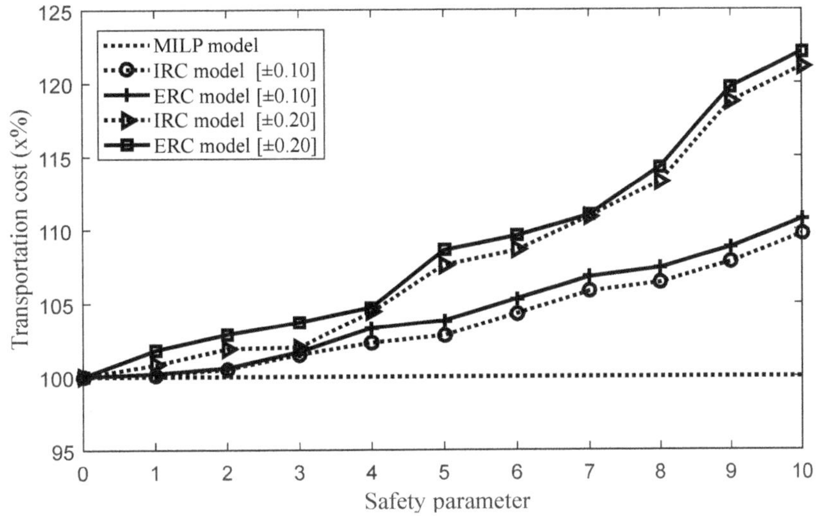

Figure 3. Influence of safety parameters.

6.2. Influence of Volatility

We obtain the average value of the parameters by repeating the ERC and IRC models multiple times, and analyze the impact of uncertain parameters on the level of logistics services, as shown in Figure 4. The relevant parameter setting is as follows, and the horizontal axis represents the percentage of the amplitude of the uncertain parameter fluctuation. The vertical axis indicates the level of logistics service. The safety parameter level is 3, 5, 8, and 10, that is, the parameters of the maximum of 3, 5, 8, and 10 nodes may fluctuate. Compare the logistics service level of the MILP model as the reference (100%). Observe the impact of uncertain parameters on the level of logistics services.

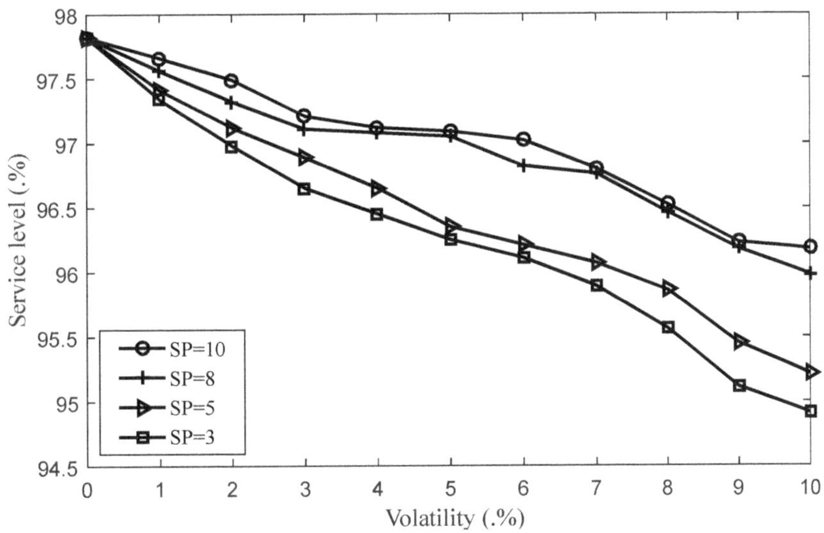

Figure 4. The effect of volatility.

It can be observed that the logistics service level of the robust model (the ERC and IRC models) is affected by the volatility of uncertain parameters. Overall, as the volatility of uncertain parameters increases, the level of the logistics service has shown a downward trend (the level of the service has decreased by 2.8%), which aligns with reality, and it is also confirmed in the research of scholars [59]. Another noteworthy phenomenon is that with the impact of the level of security parameters, the decline in the level of the logistics service levels has slowed down, and about 1.4% of the service level has declined. This discovery can provide guidance for corporate managers in their decision-making. In the case of the same parameter fluctuations, the use of higher security parameters will help improve the level of logistics services. In the same security parameter level, the greater the volatility of the parameter, the lower the level of logistics services. The accurate estimation of enterprises for demand parameters can better serve their retail companies and improve the comprehensive management level.

6.3. Routing Planning Design

Figure 5 depicts the planning and design scheme for the shipping of agricultural products from the origin center (OC) to the distribution center (DC). At this stage, agricultural products are transported in large quantities and with a single route. For direct transportation, large vehicles are generally used for transportation. Nodes with large storage levels at transit sites need to dispatch multiple vehicles to and from to complete transportation tasks. The relevant sites are described based on real maps.

The routing scheme of the MILP model, IRC model, and ERC model agricultural product distribution phase is shown in Figure 6. At this stage, agricultural products were transported from the distribution center (DC) to the demand site (Ds). The location of the node is fixed and definite. It can be seen from the figure that in the MILP model, there is a phenomenon of crossline and long-range transportation in the route planning scheme. The distribution routes of different distribution centers are marked with different colors. In other words, the overlapping transportation route means that extra transportation costs may be generated, causing a waste of transportation resources. This phenomenon also exists in the IRC model. In the ERC model's solution, there are fewer route overlaps, which will make the transportation plan more reasonable and convenient, and it will help improve the effective allocation of transportation resources.

Figure 5. Route planning and design in the original stage.

6.4. Impact of Carbon Tax Changes

Figure 7 is the responsiveness of each of the three models to changes in carbon taxes. Referring to real-world scenarios, we compare the MILP model, the IRC model, and the ERC model with carbon taxes varying from an interval of 0 to 50. It can be seen from the comparison that each model shows an upward trend with the increase of carbon tax. There is a slight difference in the comparison of details, among which, the cost increase in the IRC and ERC models is higher than the counterpart in the MILP model. Compared with the IRC model, the cost of the ERC model shows slighter increases, demonstrating the robustness of the ERC model. The increase in carbon tax costs will compel companies to adjust their transportation strategies to reduce costs by optimizing routes while meeting the requirements of environmental benefits.

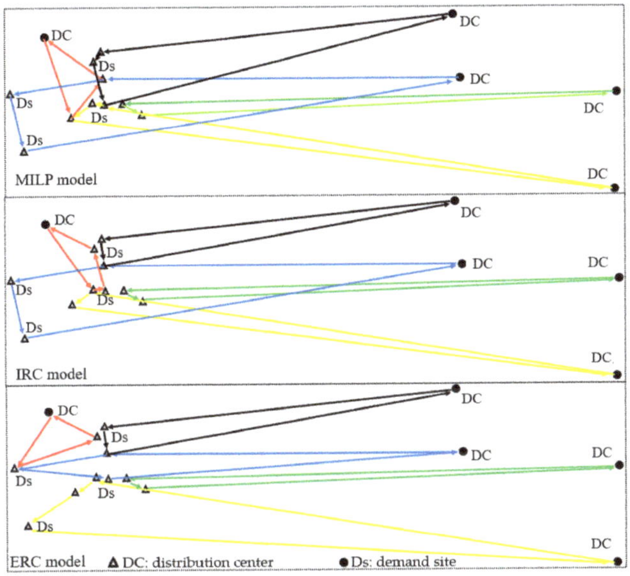

Figure 6. The routing planning.

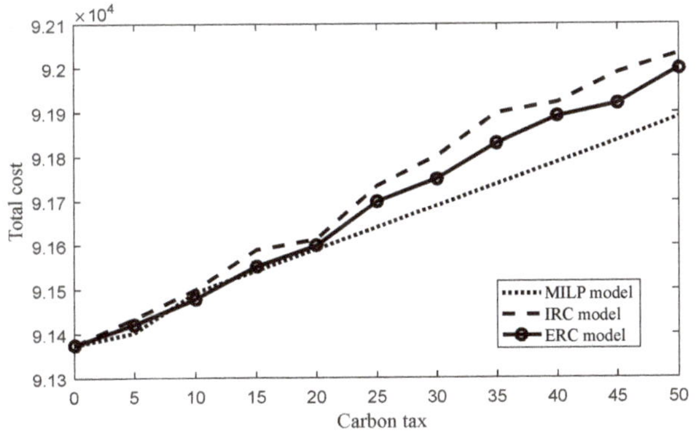

Figure 7. Impact of carbon tax changes.

6.5. Storage Proportion of Distribution Center

Figure 8 depicts the storage proportion of distribution center. In order to compare the changes in the storage proportion of each model, the MILP model and the robust models are compared in this section. It can be seen that there are differences in each model. Among them, in the IRC and ERC models, the storage proportion at distribution centers DC1, DC4, and DC5 all present an upward trend, while distribution center DC2 and DC3 show a downward trend. Changes in the proportion of distribution center directly affect transportation costs and the quality of transportation services. The transportation plan can be effectively optimized by increasing the proportion of the distribution center method service close to the service demand point. In the actual enterprise management process, it provides guidance for the formulation of enterprise operation decision-making plans to some extent.

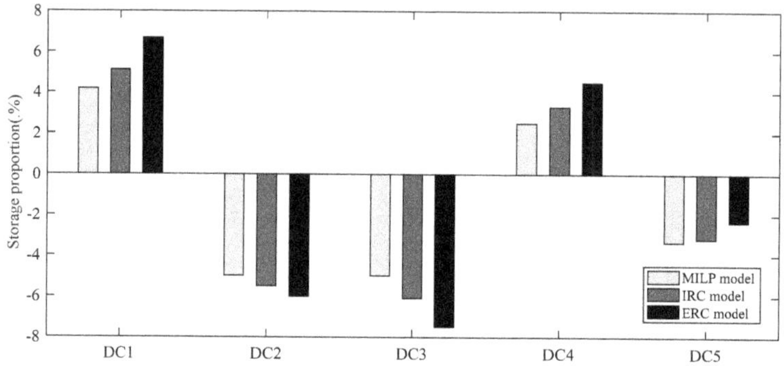

Figure 8. Storage proportion at the distribution center.

7. Conclusions

In this paper, a mixed integer linear programming model for the cold chain transportation of agricultural products that comprehensively considers various costs is proposed. The purpose is to study how to effectively reduce transportation costs and improve transportation efficiency in the parameter determination scenario. Uncertainty in the market environment is difficult to measure directly, which will lead to a rapid increase in transportation costs and unnecessary losses of agricultural products. Additionally, the parameters are extended to uncertain scenarios based on the mixed integer linear programming model, and a robust counterpart model based on uncertain parameter scenarios is proposed. A specific uncertainty robust set is constructed to characterize the fluctuation range of the uncertain parameters. Through the optimization model, the optimal strategy can be made in the worst case, so as to avoid the cost or loss due to increasing too fast. The guarantee of transportation is improved from the perspective of uncertainty optimization. Finally, the proposed model and traffic strategy are analyzed through a design case, the validity of the proposed model is verified, and management suggestions and traffic planning schemes are provided. The research reveals that the increase in carbon tax costs will compel companies to optimize their transportation route planning, improving their transportation efficiency. In addition, uncertain parameters will lead to a significant increase in cost, and accurate evaluation of demand parameters is essential in improving enterprise competitiveness.

However, there are some limitations to this study. First, the agricultural vehicle scheduling problem is analyzed only from the perspective of operations research optimization. In practical conditions, there are still many management problems that need to be further considered. Second, in the real world, there is inventory management under supply uncertainty scenarios due to climatic factors. Third, the multi-objective planning of the fresh agricultural products needs further research. Fourth, in different sales fields, there is the problem of precise rationing of heterogeneous consumer demand. Lastly, in the multimodal transport mode, there are complex scheduling and coordination problems, which have been rarely discussed in previous research and thus can be further researched in the future.

Author Contributions: Funding acquisition, F.Y. and Z.W.; Methodology, F.Y.; Project administration, S.Q.; Software, X.T. All authors have read and agreed to the published version of the manuscript.

Funding: This research received no external funding.

Institutional Review Board Statement: Not applicable.

Informed Consent Statement: Not applicable.

Data Availability Statement: All data generated or analyzed during this study are included in this paper.

Conflicts of Interest: The authors declare no conflict of interest.

Nomenclature

The following abbreviations are used in this manuscript:

Notion	Explanation
I	The set of distribution centers $\forall i \in I$
J	The set of demand sites $\forall j \in J$
c_i^f	The fixed cost of distribution centers i, $\forall i \in I$
c_{ij}^v	The unit transportation cos t of trucks in routing ij
d_{ij}	The distance between distribution centers i to demand sites j
q_j	The demand for the demand site j
h_v	The rated cargo capacity of trucks
τ	The corruption rate of the fresh agricultural product
c_{ij}^t	The penalty cos t per unit time delay in routing ij
v_{ij}	The average travel speed of trucks in routing ij
w_{ij}	The waiting time in the routing ij
t_0	The scheduled arrival time
c_{ij}^e	The unit carbon cos t in routing ij
E_{ij}^e	The unit carbon emissions in routing ij
$d^{\max}$	The accumulated mileage limit
t_j^l, t_j^u	The time windows for delivery
$H_i^{\max}$	The maximum load of the distribution center
$e^{\max}$	The maximum carbon emission limit
x_i	$x_i \in \{0,1\}$, distribution centers i, $\forall i \in I$, selection, or not
n_i	$n_i \in \{0,1,2,\ldots,N\}$, the number of trucks at the distribution center
y_{ij}	$y_{ij} \in [0,1]$, the proportion of the transportation quantity between the distribution center i and the demand site j. When y_{ij} is not equal to 0, it means that therouting ij is selected; otherwise, not.
Ψ	The safety parameter in the IRC model
ε_j	The uncertain parameter in the IRC model
Δ	The float value
q_j^0	The base value of demand
$\tilde{q}_j$	The floating value of demand
F_0	Nominal parameter expression in the IRC model
F'	The volatility parameter expression in the IRC model
Ω	The safety parameter in the ERC model
ξ	The uncertain parameter in the ERC model
G_0	Nominal parameter expression in the ERC model
G'	The volatility parameter expression in the ERC model
U_I	The set of ε_j in the IRC model
U_E	The set of ξ in the ERC model
$\mathbb{R}^n$	N-dimensional complete sets

References

1. Chen, J.; Fan, T.; Pan, F. Urban delivery of fresh products with total deterioration value. *Int. J. Prod. Res.* **2021**, *59*, 2218–2228. [CrossRef]
2. De Oliveira, J.G.; Silva, G.D.; Cipriano, L.; Gomes, M.; Egea, M.B. Control of postharvest fungal diseases in fruits using external application of RNAi. *J. Food Sci.* **2021**, *86*, 3341–3348. [CrossRef] [PubMed]
3. Saif, A.; Elhedhli, S. Cold supply chain design with environmental considerations: A simulation-optimization approach. *Eur. J. Oper. Res.* **2016**, *251*, 274–287. [CrossRef]
4. Teng, S. Route planning method for cross-border e-commerce logistics of agricultural products based on recurrent neural network. *Soft Comput.* **2021**, *25*, 12107–12116. [CrossRef]
5. Hariga, M.; As'Ad, R.; Shamayleh, A. Integrated Economic and Environmental Models for a Multi Stage Cold Supply Chain under Carbon Tax Regulation. *J. Clean. Prod.* **2017**, *166*, 1357–1371. [CrossRef]
6. Evans, J.A.; Russell, S.L.; James, C.; Corry, J.E.L. Microbial contamination of food refrigeration equipment. *J. Food Eng.* **2004**, *62*, 225–232. [CrossRef]
7. Shishov, V.V.; Talyzin, M.S. Efficiency of Refrigeration Equipment on Natural Refrigerants. *Chem. Pet. Eng.* **2020**, *56*, 385–392. [CrossRef]

8. Melillo, J.M.; McGuire, A.D.; Kicklighter, D.W.; Moore, B.; Vorosmarty, C.J.; Schloss, A.L. Global climate change and terrestrial net primary production. *Nature* **1993**, *363*, 234–240. [CrossRef]
9. Johansson, B.; Hman, M. A comparison of technologies for carbon-neutral passenger transport. *Transp. Res. Part D Transp. Environ.* **2002**, *7*, 175–196. [CrossRef]
10. Krammer, P.; Dray, L.; Kohler, M.O. Climate-neutrality versus carbon-neutrality for aviation biofuel policy. *Transp. Res. Part D-Transp. Environ.* **2013**, *23*, 64–72. [CrossRef]
11. Liu, Z.; Deng, Z.; He, G.; Wang, H.L.; Zhang, X.; Lin, J.; Qi, Y.; Liang, X. Challenges and opportunities for carbon neutrality in China. *Nat. Rev. Earth Environ.* **2022**, *3*, 141–155. [CrossRef]
12. Ebrahimi, S.; Mac Kinnon, M.; Brouwer, J. California end-use electrification impacts on carbon neutrality and clean air. *Appl. Energy* **2018**, *213*, 435–449. [CrossRef]
13. Dulebenets, M.A.; Ozguven, E.E.; Moses, R.; Ulak, M.B. Intermodal freight network design for transport of perishable products. *Open J. Optim.* **2016**, *5*, 120. [CrossRef]
14. Theophilus, O.; Dulebenets, M.A.; Pasha, J.; Lau, Y.-Y.; Fathollahi-Fard, A.M.; Mazaheri, A. Truck scheduling optimization at a cold-chain cross-docking terminal with product perishability considerations. *Comput. Ind. Eng.* **2021**, *156*, 107240. [CrossRef]
15. Dulebenets, M.A.; Ozguven, E.E. Vessel scheduling in liner shipping: Modeling transport of perishable assets. *Int. J. Prod. Econ.* **2017**, *184*, 141–156. [CrossRef]
16. Zheng, F.; Pang, Y.; Xu, Y.; Liu, M. Heuristic algorithms for truck scheduling of cross-docking operations in cold-chain logistics. *Int. J. Prod. Res.* **2021**, *59*, 6579–6600. [CrossRef]
17. Qu, S.J.; Xu, L.; Mangla, S.K.; Chan, F.T.S.; Zhu, J.L.; Arisian, S. Matchmaking in reward-based crowdfunding platforms: A hybrid machine learning approach. *Int. J. Prod. Res.* **2022**, 1–22. [CrossRef]
18. Feldmann, C.; Hamm, U. Consumers' perceptions and preferences for local food: A review. *Food Qual. Prefer.* **2015**, *40*, 152–164. [CrossRef]
19. He, Q.Z.; Tong, H.; Liu, J.B. How Does Inequality Affect the Residents' Subjective Well-Being: Inequality of Opportunity and Inequality of Effort. *Front. Psychol.* **2022**, *13*, 843854. [CrossRef]
20. Li, B.X.; Hu, H. Logistics Model Based on Agricultural Product Transportation. *Agro Food Ind. Hi-Tech* **2017**, *28*, 3370–3373.
21. Li, P.J. Logistics transportation route for agricultural products based on an improved ant colony algorithm. *Agro Food Ind. Hi-Tech* **2017**, *28*, 1876–1880.
22. Qi, C.; Hu, L. Optimization of vehicle routing problem for emergency cold chain logistics based on minimum loss. *Phys. Commun.* **2020**, *40*, 101085. [CrossRef]
23. Wang, M.; Cheng, Q.; Huang, J.C.; Cheng, G.Q. Research on optimal hub location of agricultural product transportation network based on hierarchical hub-and-spoke network model. *Phys. A-Stat. Mech. Its Appl.* **2021**, *566*, 125412. [CrossRef]
24. Khan, Z.A.; Koondhar, M.A.; Khan, I.; Ali, U.; Liu, T.J. Dynamic linkage between industrialization, energy consumption, carbon emission, and agricultural products export of Pakistan: An ARDL approach. *Environ. Sci. Pollut. Res.* **2021**, *28*, 43698–43710. [CrossRef] [PubMed]
25. Al-Mansour, F.; Jejcic, V. A model calculation of the carbon footprint of agricultural products: The case of Slovenia. *Energy* **2017**, *136*, 7–15. [CrossRef]
26. Sanye-Mengual, E.; Ceron-Palma, I.; Oliver-Sola, J.; Montero, J.I.; Rieradevall, J. Environmental analysis of the logistics of agricultural products from roof top greenhouses in Mediterranean urban areas. *J. Sci. Food Agric.* **2013**, *93*, 100–109. [CrossRef]
27. Zhu, S.J.; Fu, H.M.; Li, Y.H. Optimization Research on Vehicle Routing for Fresh Agricultural Products Based on the Investment of Freshness-Keeping Cost in the Distribution Process. *Sustainability* **2021**, *13*, 8110. [CrossRef]
28. Ahmed, W.; Sarkar, B. Management of next-generation energy using a triple bottom line approach under a supply chain framework. *Resour. Conserv. Recycl.* **2019**, *150*, 104431. [CrossRef]
29. Pan, F.; Fan, T.; Qi, X.; Chen, J.; Zhang, C. Truck scheduling for cross-docking of fresh produce with repeated loading. *Math. Probl. Eng.* **2021**, *2021*, 5592122. [CrossRef]
30. Drezner, Z.; Wesolowsky, G.O. Selecting an optimum configuration of one-way and two-way routes. *Transp. Sci.* **1997**, *31*, 386–394. [CrossRef]
31. Chen, J.; Gui, P.; Ding, T.; Na, S.; Zhou, Y. Optimization of transportation routing problem for fresh food by improved ant colony algorithm based on tabu search. *Sustainability* **2019**, *11*, 6584. [CrossRef]
32. Toole, J.L.; Colak, S.; Sturt, B.; Alexander, L.P.; Evsukoff, A.; González, M.C. The path most traveled: Travel demand estimation using big data resources. *Transp. Res. Part C Emerg. Technol.* **2015**, *58*, 162–177. [CrossRef]
33. Taghvaeeyan, S.; Rajamani, R. Portable roadside sensors for vehicle counting, classification, and speed measurement. *IEEE Trans. Intell. Transp. Syst.* **2013**, *15*, 73–83. [CrossRef]
34. Purba, D.S.D.; Kontou, E.; Vogiatzis, C. Evacuation route planning for alternative fuel vehicles. *Transp. Res. Part C-Emerg. Technol.* **2022**, *143*, 103837. [CrossRef]
35. Karak, A.; Abdelghany, K. The hybrid vehicle-drone routing problem for pick-up and delivery services. *Transp. Res. Part C-Emerg. Technol.* **2019**, *102*, 427–449. [CrossRef]
36. Schneider, M.; Stenger, A.; Goeke, D. The electric vehicle-routing problem with time windows and recharging stations. *Transp. Sci.* **2014**, *48*, 500–520. [CrossRef]

37. Jiang, L.; Dhiaf, M.; Dong, J.F.; Liang, C.Y.; Zhao, S.P. A traveling salesman problem with time windows for the last mile delivery in online shopping. *Int. J. Prod. Res.* **2020**, *58*, 5077–5088. [CrossRef]
38. Soyster, A.L.; Murphy, F.H. Data driven matrix uncertainty for robust linear programming. *Omega-Int. J. Manag. Sci.* **2017**, *70*, 43–57. [CrossRef]
39. Wiesemann, W.; Kuhn, D.; Sim, M. Distributionally Robust Convex Optimization. *Oper. Res.* **2014**, *62*, 1358–1376. [CrossRef]
40. Ji, Y.; Li, H.H.; Zhang, H.J. Risk-Averse Two-Stage Stochastic Minimum Cost Consensus Models with Asymmetric Adjustment Cost. *Group Decis. Negot.* **2022**, *31*, 261–291. [CrossRef]
41. Qu, S.J.; Wei, J.P.; Wang, Q.H.; Li, Y.M.; Jin, X.W.; Chaib, L. Robust minimum cost consensus models with various individual preference scenarios under unit adjustment cost uncertainty. *Inf. Fusion* **2023**, *89*, 510–526. [CrossRef]
42. Shi, Y.; Boudouh, T.; Grunder, O. A robust optimization for a home health care routing and scheduling problem with consideration of uncertain travel and service times. *Transp. Res. Part E-Logist. Transp. Rev.* **2019**, *128*, 52–95. [CrossRef]
43. Gorissen, B.L.; Yanıkoğlu, İ.; den Hertog, D. A practical guide to robust optimization. *Omega* **2015**, *53*, 124–137. [CrossRef]
44. Bai, D.; Carpenter, T.; Mulvey, J. Making a case for robust optimization models. *Manag. Sci.* **1997**, *43*, 895–907. [CrossRef]
45. Yu, C.-S.; Li, H.-L. A robust optimization model for stochastic logistic problems. *Int. J. Prod. Econ.* **2000**, *64*, 385–397. [CrossRef]
46. Baron, O.; Milner, J.; Naseraldin, H. Facility location: A robust optimization approach. *Prod. Oper. Manag.* **2011**, *20*, 772–785. [CrossRef]
47. Qu, S.; Li, Y.; Ji, Y. The mixed integer robust maximum expert consensus models for large-scale GDM under uncertainty circumstances. *Appl. Soft Comput.* **2021**, *107*, 107369. [CrossRef]
48. Matthews, L.R.; Guzman, Y.A.; Floudas, C.A. Generalized robust counterparts for constraints with bounded and unbounded uncertain parameters. *Comput. Chem. Eng.* **2018**, *116*, 451–467. [CrossRef]
49. Abdel-Aal, M.A.M.; Selim, S.Z. Robust optimization for selective newsvendor problem with uncertain demand. *Comput. Ind. Eng.* **2019**, *135*, 838–854. [CrossRef]
50. Liang, Z.; Wu, J.; Li, J.; Yang, Z.; Li, J.; Li, J. Application and Popularization of New Green Prevention and Control Technology for Greenhouse Vegetables in Shouguang City, Shandong Province. *Asian Agric. Res.* **2019**, *11*, 78–81.
51. Gao, J.; Wang, L.W.; Gao, P.; An, G.L.; Wang, Z.X.; Xu, S.Z.; Wang, R.Z. Performance investigation of a freezing system with novel multi-salt sorbent for refrigerated truck. *Int. J. Refrig.-Rev. Int. Du Froid* **2019**, *98*, 129–138. [CrossRef]
52. Romijn, T.C.J.; Donkers, M.C.F.; Kessels, J.; Weiland, S. A Distributed Optimization Approach for Complete Vehicle Energy Management. *Ieee Trans. Control Syst. Technol.* **2019**, *27*, 964–980. [CrossRef]
53. Rahmaniani, R.; Crainic, T.G.; Gendreau, M.; Rei, W. The Benders decomposition algorithm: A literature review. *Eur. J. Oper. Res.* **2017**, *259*, 801–817. [CrossRef]
54. Hooker, J.N.; Ottosson, G. Logic-based Benders decomposition. *Math. Program.* **2003**, *96*, 33–60. [CrossRef]
55. Hooker, J.N. Planning and scheduling by logic-based Benders decomposition. *Oper. Res.* **2007**, *55*, 588–602. [CrossRef]
56. Jeyakumar, V.; Lee, G.M. Complete characterizations of stable Farkas' lemma and cone-convex programming duality. *Math. Program.* **2008**, *114*, 335–347. [CrossRef]
57. Zhang, X.; Ding, Z.J.; Hang, J.Q.; He, Q.Z. How do stock price indices absorb the COVID-19 pandemic shocks? *N. Am. J. Econ. Financ.* **2022**, *60*, 101672. [CrossRef]
58. Chen, Z.; Wu, L.; Fu, Y. Real-Time Price-Based Demand Response Management for Residential Appliances via Stochastic Optimization and Robust Optimization. *Ieee Trans. Smart Grid* **2012**, *3*, 1822–1831. [CrossRef]
59. Zhang, Z.-H.; Jiang, H. A robust counterpart approach to the bi-objective emergency medical service design problem. *Appl. Math. Model.* **2014**, *38*, 1033–1040. [CrossRef]

sustainability

MDPI

Article

A Credit Risk Contagion Intensity Model of Supply Chain Enterprises under Different Credit Modes

Yuhao Wang [1], Jiaxian Shen [1], Jinnan Pan [1] and Tingqiang Chen [1,2,*]

1 School of Economics and Management, Nanjing Tech University, Nanjing 211816, China
2 Academy of Mathematics and Systems Science, Chinese Academy of Sciences, Beijing 100190, China
* Correspondence: tingqiang88888888@163.com

Abstract: The rapid development of theoretical and practical innovations in corporate finance driven by supply chain finance has exacerbated the complexity of credit default risk contagion among supply chain enterprises. Financial risks in the supply chain greatly hinder its sustainable development; thus, strengthening financial risk management is necessary to ensure the sustainability of the supply chain. Based on the single-channel and dual-channel credit financing models of retailers in the supply chain, the purpose of this paper was to construct a model of the intensity of credit default risk contagion among supply chain enterprises under different credit financing models, and investigate the influencing factors of credit risk contagion among supply chain enterprises and its mechanism of action through a computational simulation system. The results were as follows: (1) there was a positive relationship between the production cost of suppliers and the contagion intensity of the supply chain credit default risk, and the contagion effect of the supply chain credit default risk increased significantly when both retailers defaulted on trade credit to suppliers; (2) the market retail price of the product was negatively related to the contagion intensity of the supply chain credit default risk, and the contagion intensity of the supply chain credit default risk increased significantly when both retailers defaulted on trade credit to the supplier; (3) the intensity of credit default risk contagion in the supply chain was positively correlated with both the commercial bank risk-free rate and the trade credit rate, and retailers' repayment priority on trade credit debt was negatively correlated with suppliers' wholesale prices and positively correlated with retailers' order volumes, with retailers' repayment priority positively affecting retailers' bank credit rates and negatively affecting suppliers' bank credit rates; and (4) retailers' repayment priority on trade credit debt was negatively correlated with the intensity of supply chain credit default risk contagion, and the lower the retailer's bank credit limit, the higher the trade credit limit, and the stronger the credit default contagion effect in the supply chain.

Keywords: supply chain finance; trade credit financing; bank credit financing; credit default; contagion intensity

Citation: Wang, Y.; Shen, J.; Pan, J.; Chen, T. A Credit Risk Contagion Intensity Model of Supply Chain Enterprises under Different Credit Modes. *Sustainability* **2022**, *14*, 13518. https://doi.org/10.3390/su142013518

Academic Editors: Shaojian Qu, Qingguo Bai, Ying Ji and Congjun Rao

Received: 12 September 2022
Accepted: 17 October 2022
Published: 19 October 2022

1. Introduction

Supply chain finance is a product of the integration of supply chain operation and financial business due to their respective development needs, and is a financial service derived from supply chain management [1] (Lamoureux, 2007). With the rapid development of supply chain finance and its innovation, numerous complex financing models have been formed, among which bank credit and trade credit are the most frequently used and the most theoretically researched ways of financing supply chain enterprises. In reality, SMEs in the supply chain face strong financing constraints due to their low credit ratings, which make it difficult for them to obtain credit funds directly from banks [2]. To a certain extent, trade credit financing has solved the problem of financing difficulties for SMEs and has gradually replaced bank credit financing as the only financing channel for SMEs downstream of the supply chain [3,4]. Due to the special industrial transaction mode of supply chain enterprises, the enterprise association is more complex and close, and the relationship between upstream and downstream enterprises presents the characteristics of behavioral dependence, synergy and mutual benefit [5,6]. Upstream core enterprises will take the initiative to reduce production costs to ease the financing needs of downstream enterprises, in order to improve their own profitability and that of the supply chain as a whole [7–9]. However, the special interconnectedness between upstream and downstream enterprises in the supply chain often directly leads to credit risk contagion among supply chain enterprises. Once a credit crisis is triggered by certain enterprises, it can quickly spread to other related enterprises in the supply chain and even affect the stability of the entire supply chain and the economic system, thus triggering a crisis in the industry [10]. What are the credit financing models for upstream and downstream companies in the supply chain driven by supply chain finance? How does supply chain credit default risk develop and become contagious? What are the evolutionary characteristics of supply chain credit risk contagion? In light of the above considerations, this paper attempted to analyze the micromechanism of supply chain credit default risk based on the credit model of SMEs in the supply chain, and to portray the supply chain credit risk contagion mechanism triggered by retailers under different models of bank credit and trade credit. Commercial bank credit is the main and most common form of external financing in supply chain finance. SMEs in the supply chain can be guaranteed by core enterprises with sufficient capital and good credit and business conditions, thus reducing the credit risk assessment of commercial banks on SMEs and ultimately realizing the demand for bank credit financing [11]. The financing decisions of upstream and downstream supply chain enterprises for commercial bank credit are mainly determined by the operational characteristics of the supply chain [5,12,13]. For example, the financing of supply chain enterprises subject to financial constraints and applied to commercial banks for financial support in the form of inventory pledges and accounts receivable financing [14]. And, based on a game between commercial banks and supply chain operators, that in order to improve the overall operational efficiency of the supply chain, firms with capital problems must consider both the operational and financing decisions of supply chain firms [15]. Furthermore, suitable financing solutions can effectively stimulate retailers' order volumes and achieve supply chain coordination [16]. In addition, retailers' decisions are also influenced by the amount of their own capital in both bank credit and equity financing models, with retailers more likely to choose bank credit financing when the level of their own capital is high and to prefer equity financing when the level of capital is low [17,18].

The core of trade credit financing is designed to help SMEs in the supply chain to use their own commercial credit, to reach agreements with core companies. Short-term financing for SMEs is achieved through deferred payment, thereby alleviating the liquidity problems of SMEs [19]. The most typical type of supply chain trade credit finance is deferred payment between suppliers and downstream firms. Using the deferred payment covenant between supplier and retailer as a condition to construct a decision model under the trade credit model, changing the traditional financing model for the first time [20]. Comparing deferred payment financing with bank financing, we can discover that deferred payment financing is better than bank financing and that retailers are more likely to opt for multi-stage trade credit, ordering goods in small quantities and batches, in order to obtain higher trade credit facilities [21]. And, the duration of trade credit covenants between suppliers and retailers is positively related to order demand and increases the risk of retailer default, and there are optimal solutions for both suppliers' and retailers' financing decisions when the product is perishable [22]. Once trade credit is incorporated into an economic ordering model with a full backlog shortage (EOQ model), we can see that retailers' order volumes increase when the allowable delay increases when they are unable to repay trade credit and have to pay additional interest when the credit term is exceeded [23]. This demonstrates how trade credit financing may help capital-constrained merchants solve their financing problems while also boosting the growth of supply chain businesses. With the rapid development and continuous innovation of trade credit financing, the advance financing strategy has become popular and is a common financing strategy adopted by capital-constrained suppliers, which eases the financial pressure of upstream companies through advance payment by downstream companies [24]. For example, prepayment financing plays a large role in alleviating suppliers' financial constraints, and retailers' optimal decisions vary widely with and without prepayment [25]. Advance payment mechanisms with risk compensation can alleviate suppliers' financial difficulties in the face of productivity uncertainty, while revenue-sharing contracts can reduce double marginal effects to achieve supply chain coordination [26]. Existing research has found that APRCs are an effective way to address suppliers' financial constraints. In addition, when deficits are large, an advance payment facility with risk compensation can be more profitable for both parties. When capital deficits are small, suppliers can also better manage bank loan financing.

Supply chain financing has largely solved the financial constraints and financing dilemmas of supply chain enterprises, but it has also deepened the degree of credit linkage among supply chain enterprises. Supply chain financing becomes a channel for credit risk contagion, which can be exacerbated when supply chain companies face unexpected circumstances or an unstable macroenvironment. Financial risks in the supply chain greatly hinder its sustainable development; thus, strengthening financial risk management is a necessary task to ensure the sustainability of the supply chain. The special interconnectedness of supply chain firms is a major factor in the creation of supply chain financial risk and an important channel for its transmission [27]. Credit risk arises from uncertainty changes in the transaction process due to external uncertainties, political policy changes and competitors. This requires supply chain companies to monitor and control the transaction process in advance to prevent the creation and contagion of supply chain finance credit risk. Besides, the trading partners of firms in the supply chain are a potential medium for credit risk contagion in supply chain finance. Trading partners with a poor credit standing tend to be more susceptible to supply chain financial credit risk contagion, which in severe cases may even cause a cluster default effect among supply chain firms [28]. In this regard, existing research has demonstrated that CVaR can be used to control default risk arising from trade credit [29]. In addition, the generation and contagion of credit risk in supply chain finance is affected by the characteristics of supply chain finance networks, and portrayed and measured the impact of supply chain finance network characteristics on the contagion effect of supply chain credit risk based on a Bayesian network model [30]. Furthermore, cascading failure models can accurately measure and validate the contagion of credit risk in supply chain finance networks [31]. And, trade credit insurance can help manufacturers expand sales and significantly reduce the risk of default for supply chain firms [32].

To sum up, the existing theoretical studies on supply chain finance mainly focus on the analysis of financing strategies and their influencing factors on various actors in the supply chain finance model, i.e., banks, suppliers, retailers and other participating actors. Few scholars have systematically studied supply chain credit risk contagion and its evolution mechanism from the perspective of the supply chain. According to the analysis above, this paper draws on the differences in supply chain credit risk contagion between single-channel and dual-channel credit financing models for retailers in the supply chain, and a model of the credit default risk contagion intensity of supply chain enterprises under different credit financing models was constructed. Moreover, with the help of MATLAB R2016b software, the evolutionary characteristics of retailer repayment priority on the decisions of various participants in supply chain credit financing, and on the intensity of supply chain credit default risk contagion, were analyzed. The contributions of this paper are as follows: (1) the association between the contagion effect of credit default risk in the supply chain and the influencing factors under the single-channel financing model of retailers was studied; (2) based on the retailer's dual-channel financing model, a supply chain credit default risk contagion intensity model that considered the retailer's repayment priorities was constructed; and (3) the impact of retailers' repayment priority on supply chain credit financing decisions and the contagion effect of supply chain credit default risk was analyzed.

The second part of this paper portrays the contagion mechanism of the credit risk of supply chain enterprises under the single-channel credit model, proposes a theoretical model for the contagion of credit risk of supply chain enterprises under the single-channel credit model, and finally, simulates the contagion mechanism of supply chain credit risk and its evolution characteristics through computational simulation. In the third part, the contagion mechanism of credit risk of supply chain enterprises under the dual-channel credit model is reviewed, a theoretical model of the credit risk contagion of supply chain enterprises under the dual-channel credit model is proposed, and finally, the contagion mechanism of supply chain credit risk and its evolution characteristics are analyzed through computational simulation. The last part outlines the conclusions of this paper.

2. A Model of Supply Chain Credit Risk Contagion under a Single-Channel Credit Model

2.1. Model Assumptions and Notation

Assume a simple supply chain consisting of a supplier and multiple retailers, where the supplier is a core firm, the retailers are all SMEs, and multiple retailers order goods from the core firm at the same time. Assume that both the supplier and the retailer are financially constrained in their trading process, and are unable to produce and sell independently [33]. For suppliers located in the core enterprises with high profitability and corporate reputation, commercial banks can provide loans directly to them, while retailers who are SMEs cannot obtain financing directly from banks due to their own limitations, so they can consider applying for trade credit financing from the core enterprises in the supply chain to obtain financial support. At the beginning of the sales period, subject to funding constraints, retailer i applies for trade credit financing from a supplier at an interest rate of r_S. After the supplier has determined the wholesale price w, retailer i determines the order quantity Q_i and the financing amount $T_i = wQ_i$. Similarly, a financially constrained supplier receives an order from a retailer and signs a loan agreement to obtain financing through a commercial bank in the amount of $M = \sum_{i=1}^{n} cQ_i$ where C represents the supplier's marginal cost of production and r represents the bank loan rate. It is assumed that the supplier is sufficiently rational such that $M(1 + r_B) \leq \sum_{i=1}^{n} T_i(1 + r_S)$. The supplier receives the funding, organizes production and dispatches the product, which the retailer sells in the retail market at market price M. At the end of the sales period, all of the funds owned by the retailers flow back and repay the trade credit facility debt to the suppliers, who repay the bank credit facility debt. If market conditions are good and all retailers have sufficient funds to repay the principal and interest on the trade credit facility provided by the supplier, then the supplier will also be able to repay the bank credit facility as promised and neither the retailer nor the supplier will be in default at that point. If market conditions are poor and a proportion of retailers is unable to repay their supplier debts on time, this will, to some extent, result in the supplier being unable to repay the commercial bank facility, at which point the credit risk of that proportion of retailers is transmitted to the supplier. If market conditions are adverse and the closing capital flows of all retailers are not sufficient to repay supplier debts, i.e., all retailers default on trade credit, the supplier suffers severe losses and its recoveries are highly unlikely to be sufficient to cover the bank credit, at which point the supplier defaults and there is credit default risk contagion in the supply chain. Therefore, the supply chain decision-making process under the single-channel credit model is shown in Figure 1.

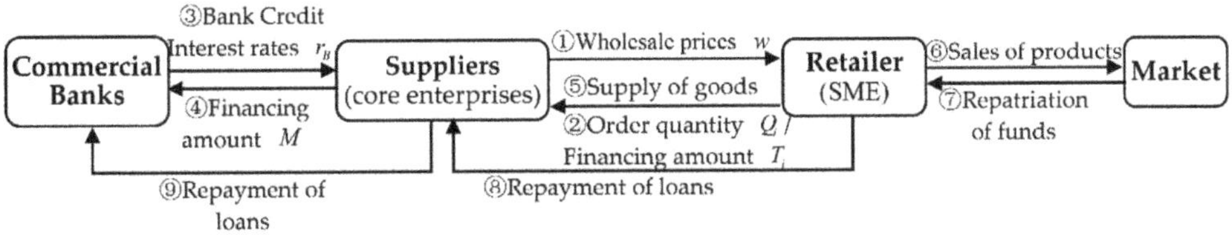

Figure 1. A decision model for supply chain credit financing under a single-channel credit model.

Based on the analysis above, the following basic assumptions were made for the purpose of the subsequent discussion:

Assumption 1. *The product is perishable and the residual value of unsold units of product at the end of the sales period is 0.*

Assumption 2. *The supplier and the retailer have limited liability, and both the retailer and the supplier will only repay their full revenue in the event of a default.*

Assumption 3. *The marginal cost per unit of the product produced by the supplier's organization is c. The wholesale price per unit of the product at the time the retailer makes an order for the product is w, the retail price of the product p remains constant throughout the sale of the product, and $0 < c < w < p$. To avoid price discrimination by the supplier against different retailers, all retailers in this paper face the same wholesale and retail prices.*

Assumption 4. *The market demand faced by the retailer i is $x_i, i = 1, 2, \ldots, n$ and the market demand of individual retailers x_i are independent of each other, and the sum of the market demand of all retailers is $X(\ X = \sum x_i)$.*

Assumption 5. *The market demand for the product is random and the retail price of the product remains constant during the sales process. According to the newsboy model, $f_i(x_i)$ and $F_i(x_i)$ are the probability density function and distribution function of market demand x_i, respectively, and $F_i(x_i)$ has the characteristics of continuity, differentiability and incrementality, satisfying the incremental lapse rate (IFR) property $\overline{F}_i(x_i) = 1 - F_i(x_i)$. In addition, the sum of market demand X of all retailers obeys a distribution $G(X)$ and has a probability density function $g(X)$.*

The above variable symbols are defined in Table 1.

Table 1. A description of symbols.

Symbol	Description
c	Marginal cost of production per unit of product
w	Wholesale price per unit of product
p	Product retail price
Q_i	Order quantity for retailer i
x_i	Market demand for retailer i
T_i	Trade credit facility amount for retailer i
r_S	Trade credit rate for retailer i
M	Amount of bank credit facilities for suppliers
r_B	Bank credit rates for suppliers
r_f	Bank risk-free rate
k_s	Supplier default thresholds
k_r	Retailer default thresholds
β	The intensity of contagion of credit default risk

2.2. Mechanisms of Supply Chain Credit Risk Contagion in a Single-Channel Credit Model

The return of capital to all retailers at the end of the sales period, when there is an oversupply in the market, results in a loss of sales profit for the retailer and thus makes them vulnerable to trade credit default risk. At the same time, suppliers with whom trade credit facilities are contracted can also be exposed to risk contagion and default on credit, resulting in capital losses for commercial banks. The following definitions outline this:

Definition 1. *When a retailer is unable to repay its trade credit facility debt, as evidenced by the fact that the sales revenue of retailer i is less than its trade credit facility debt, retailer i is said to have defaulted on its trade credit facility.*

Definition 2. *When the supplier is unable to repay the bank's credit facility debt, which is expressed as the supplier's capital receipts being less than the bank's credit facility debt, the supplier is said to be in credit default with the bank.*

Definition 3. *As a result of a trade credit default by retailer i, which leads to a credit default by the supplier to the bank, the conditional probability in this scenario is defined as the intensity of credit default risk contagion in the supply chain, denoted as* β.

Individual retailer defaults and supplier defaults are closely connected, and there are three possible cases for each retailer's default status along the supply chain.

Case 1. *No retailer in the supply chain defaults on a trade credit.*

When all retailers in the supply chain do not default on their trade credit, the supplier will be able to successfully recover all the retailers' payment funds to repay the bank credit facility debt, and there is no credit risk contagion in the supply chain, so at this point, $\beta_1 = 0$.

Case 2. *There are trade credit defaults by all retailers in the supply chain.*

It is highly likely that suppliers will default on their trade credit facility debts to the banks when all of the retailer sales revenues are unable to repay them. At the end of the sales period, if retailer i defaults on its trade credit, sales revenue for retailer i is $M(x_i) = p\min(x_i, Q)$, which is less than its trade credit principal and interest $(1 + r_S)T_i$. Therefore, $x_i < \frac{(1+r_S)T_i}{p} = k_{ri}$. If all retailers in the supply chain default on their trade credit, the necessary condition becomes $\sum x_i < \frac{\sum(1+r_S)T_i}{p} = k_r^1$, which means $X < k_r^1$. In this case, the full revenue recovered by the supplier is $N(x) = \sum p\min(x_i, Q)$. If a supplier defaults on a credit facility to a bank, $N(x) < (1 + r_B)M$, the necessary condition becomes $\sum x_i < \frac{(1+r_B)M}{p} = k_s^1$, which means $X < k_s^1$. At this point, credit risk contagion occurs in the supply chain, the intensity of which can be defined as $\beta_2 = p(X < k_s^1 | X < k_r^1)$.

Case 3. *Some retailers in the supply chain default on trade credit while some other retailers do not.*

For the sake of discussion, assume that there are two retailers in the supply chain, with Retailer 1 in default and Retailer 2 not in default. At the end of the sales period, Retailer 1's inflow of funds $M(x_1) = p\min(x_1, Q_1)$ is less than the principal and interest on the debt $(1 + r_S)T_1$. At the end of the sales period, when the inflow of funds to Retailer 1 is less than the principal and interest on the debt, given that $x_1 < \frac{(1+r_S)T_1}{p} = k_{r1}$, then Retailer 1 has defaulted on its trade credit. For Retailer 2, its inflows $M(x_2) = p\min(x_2, Q_2)$ are more than the principal and interest on the debt $(1 + r_S)T_2$. Given that $x_2 \geq \frac{(1+r_S)T_2}{p} = k_{r2}$, Retailer 2 will not default. In this case, the funds that the supplier is able to recover at the end of the sales period are $N(x) = M(x_1) + (1 + r_S)T_2$. Given that $N(x) < (1 + r_B)M$, which means $x_1 < \frac{(1+r_B)M-(1+r_S)T_2}{p} = k_s^1 - k_{r2}$, the supplier defaults on bank credit. At this point, there is a contagion of credit default risk in the supply chain, the intensity of which can be defined as $\beta_3 = p(x_1 < k_s^1 - k_{r2} | x_1 < k_{r1}, x_2 \geq k_{r2})$.

According to the aforementioned investigation, we discovered that there are market demand levels for suppliers and retailers when they default on credit, which are k_r^1, k_{r1}, k_{r2} and k_s^1. When market demand surpasses specific thresholds, supply chain credit risk develops. Hence, these market demand thresholds are also known as default thresholds.

Among them, are $k_r^1 = \frac{\sum(1+r_S)T_i}{p}$, $k_{r1} = \frac{(1+r_S)T_1}{p}$, $k_{r2} = \frac{(1+r_S)T_2}{p}$ and $k_s^1 = \frac{(1+r_B)M}{p}$. In the case of trade credit defaults by all retailers, given that $X < k_r^1$ and $X < k_s^1$, these defaults can lead to credit defaults by suppliers to banks. For situations where some retailers are in default on trade credit, i.e., where only Retailer 1 is in default and Retailer 2 is not, given that $x_1 < k_{r1}$, $x_2 \geq k_{r2}$ and $x_1 < k_s^1 - k_{r2}$, trade credit defaults by retailers can still lead to credit defaults by suppliers to banks. Therefore, this paper focused on the generation of trade credit default risk and bank credit default risk when market demand was uncertain, and on their contagion effects.

2.3. A Model of Supply Chain Credit Risk Contagion under a Single-Channel Credit Model

Based on the Stackelberg game theory, the supply chain financing problem is solved by the inverse solution method according to the decision order of the suppliers and each retailer. Firstly, we solved for the optimal purchase quantity of each retailer given the wholesale price of the supplier, then we solved for the optimal wholesale price of the supplier of the core enterprise, and finally we determined the intensity of supply chain credit risk contagion.

2.3.1. Optimal Decision-Making for Retailers

According to the analysis of the supply chain credit risk contagion mechanism under the single-channel credit financing model, when the inflow of funds from retailer i is insufficient to repay the trade credit, then $x_i < \frac{(1+r_S)T_i}{p} = k_{ri}$. Retailer i defaults on its trade credit when the market demand x_i faced by retailer i is less than the critical demand k_{ri}. Assuming that the retailers are rational, then $w(1+r_S) < p$ and $k_{ri} < Q_i$. Thus, the return to the retailer i can be expressed as follows:

$$\pi_{R_i} = \begin{cases} px_i - T_i(1+r_S) & k_{ri} \leq x_i \leq Q_i \\ pQ_i - T_i(1+r_S) & Q_i < x_i \end{cases} \tag{1}$$

Therefore, the expected profit for Retailer i can be expressed as:

$$E(\pi_{R_i}) = \int_{k_{r_i}}^{Q_i} (px_i - T_i(1+r_S))f_i(x_i)dx_i + \int_{Q_i}^{\infty} (pQ_i - T_i(1+r_S))f_i(x_i)dx_i \tag{2}$$

The optimization problem of retailer i, which maximizes the desired profit through the decision to order quantity Q_i, can be expressed as:

$$\begin{aligned} \max E(\pi_{R_i}) &= \int_{k_{ri}}^{Q_i} (px_i - T_i(1+r_S))f_i(x_i)dx_i + \int_{Q_i}^{\infty} (pQ_i - T_i(1+r_S))f_i(x_i)dx_i \\ & s.t. \;\; k_{ri} < Q_i \end{aligned} \tag{3}$$

Solving the above optimization problem yields the equilibrium result when the retailer i is financed through trade credit. Solving the first-order partial derivative of Equation (3) with respect to Q_i yields: $\frac{\partial E(\pi_{R_i})}{\partial Q_i} = p\bar{F}_i(Q_i^*) - (1+r_S)w\bar{F}_i(k_{ri})$. Since $\frac{\partial^2 E(\pi_{R_i})}{\partial {Q_i}^2} < 0$, the optimal purchase quantity ${Q_i}^*$ for retailer i exists, is unique and satisfies the following equation:

$$\frac{\bar{F}_i(Q_i^*)}{\bar{F}_i(k_{ri})} = \frac{(1+r_S)w}{p} \tag{4}$$

Among them, $k_{ri} = \frac{(1+r_S)T_i}{p}$.

2.3.2. Optimal Decision-Making for Suppliers

Based on the market sales and actual earnings of retailer i, the supplier's earnings are:

$$\pi_{S_i} = \begin{cases} px_i & x_i < k_{ri} \\ (1+r_S)T_i & k_{ri} \le x_i \end{cases} \tag{5}$$

At the end of the sales period, the bank's return needs to be determined based on whether the supplier is insolvent, and the supplier's repayment status is directly related to whether the individual retailers are in default. Based on the analysis above, we know that there are two main scenarios for the default status of all retailers in the supply chain: the first, in which all retailers are in default on their trade credit, and the second scenario, in which some of the retailers in the supply chain default on their trade credit and the remaining retailers repay as promised.

① All retailers are in trade credit default.

In this case, the revenue generated by retailer i can be expressed as:

$$\pi_{S_i} = px_i \tag{6}$$

According to the analysis of the supply chain credit risk contagion mechanism under the single-channel credit financing model, it can be seen that the supplier will experience credit default to the bank when the total market demand of all retailers i meets $X < k_s^1 < k_r^1, x_i < k_{ri}$, and when all retailers i default on trade credit. Therefore, the supplier's revenue at the end of the sales period can be expressed as:

$$\pi_S^1 = \left[\sum \pi_{S_i} - (1+r_B)M\right]^+ = \begin{cases} 0 & x_i < k_{ri}, X < k_s^1 < k_r^1 \\ pX - (1+r_B)M & x_i < k_{ri}, k_s^1 \le X < k_r^1 \end{cases} \tag{7}$$

Among them, $[a]^+ = \max(a, 0)$. Therefore, the expected profit of the supplier at the end of the sales period can be expressed as:

$$E\left(\pi_S^1\right) = \int_{k_s^1}^{k_r^1} [pX - (1+r_B)M] dG(X) \tag{8}$$

At this point, the bank's income at the end of the sales period can be expressed as:

$$\pi_B^1 = \begin{cases} pX & x_i < k_{ri}, X < k_s^1 < k_r^1 \\ (1+r_B)M & x_i < k_{ri}, k_s^1 \le X < k_r^1 \end{cases} \tag{9}$$

The expected profit of the bank at the end of the sales period can be expressed as:

$$E(\pi_B^1) = \int_0^{k_s^1} pX\, dG(X) + \int_{k_s^1}^{k_r^1} (1+r_B)M\, dG(X) \tag{10}$$

The loan interest rate r_B meets the concept of risk compensation $(1+r_f)M = E(\pi_B^1)$, where r_f is the risk-free interest rate, because the bank operates in a competitive equilibrium market.

Combined with Formulas (8) and (10), the optimization problem of the supplier decision can be expressed as:

$$\max E(\pi_S^1) = \int_0^{k_r^1} pX\, dG(X) - (1+r_f)M = pk_r^1 G(k_r^1) - p\int_0^{k_r^1} G(X)\, dX - (1+r_f)M \\ s.t.\ \ 0 < k_s^1 < k_r^1, x_i < k_{ri} \tag{11}$$

② In numerous supply chains, there exist retailers with trade credit defaults and non-defaults.

For the convenience of discussion, assume that there is only one supplier and two retailers in the supply chain, namely Retailer 1 and Retailer 2, in which Retailer 1 defaults and Retailer 2 does not default.

In this case, the revenues of Retailer 1 and Retailer 2 are expressed as follows:

$$\begin{cases} \pi_{S_1} = px_1 & x_1 < k_{r1} \\ \pi_{S_2} = (1+r_S)wQ_2 & x_2 \geq k_{r2} \end{cases} \quad (12)$$

Analysis of the supply chain credit risk contagion mechanism under the single-channel credit financing model shows that when the market demand for Retailer 1 and Retailer 2 satisfies $x_1 < k_s^1 - k_{r2} < k_{r1}$ and $k_{r2} \leq x_2$, the supplier will default on the commercial bank if only Retailer 1 defaults on trade credit and Retailer 2 does not default on trade credit. Therefore, the supplier's revenue at the end of the sales period can be expressed as:

$$\pi_S^2 = [\pi_{S_1} + \pi_{S_2} - (1+r_B)M]^+ = \begin{cases} 0 & x_1 < k_s^1 - k_{r2} < k_{r1}, x_2 \geq k_{r2} \\ px_1 + T_2(1+r_S) - M(1+r_B) & k_s^1 - k_{r2} \leq x_1 < k_{r1}, x_2 \geq k_{r2} \end{cases} \quad (13)$$

Therefore, the expected profit of the supplier at the end of the sales period can be expressed as:

$$E\left(\pi_S^2\right) = \int_{k_{r2}}^{\infty} \int_{k_s^1 - k_{r2}}^{k_{r1}} [px_1 + T_2(1+r_S) - M(1+r_B)]dF(x_1)dF(x_2) \quad (14)$$

At this point, the bank's earnings at the end of the sales period can be expressed as:

$$\pi_B^2 = \begin{cases} px_1 + T_2(1+r_S) & x_1 < k_s^1 - k_{r2} < k_{r1}, x_2 \geq k_{r2} \\ M(1+r_B) & k_s^1 - k_{r2} \leq x_1 < k_{r1}, x_2 \geq k_{r2} \end{cases} \quad (15)$$

Hence, the bank's expected profit at the end of the sales period can be expressed as:

$$E(\pi_B^2) = \int_{k_{r2}}^{\infty} \int_0^{k_s^1 - k_{r2}} px_1 + T_2(1+r_S)dF(x_1)dF(x_2) + \int_{k_{r2}}^{\infty} \int_{k_s^1 - k_{r2}}^{k_{r1}} M(1+r_B)dF(x_1)dF(x_2) \quad (16)$$

Similarly, the loan rate r_B satisfies the risk compensation principle B $(1+r_f)M = E(\pi_B^2)$. Combining Equations (14) and (16), when only some retailers in the supply chain default on trade credit, the optimization problem for the supplier decision can be expressed as:

$$\begin{aligned} \max E(\pi_S^2) &= \int_{k_{r2}}^{\infty} \int_0^{k_{r1}} [px_1 + T_2(1+r_S)]dF(x_1)dF(x_2) - M(1+r_f) \\ &= \bar{F}(k_{r2})\left[(T_1+T_2)(1+r_S)F(k_{r1}) - p\int_0^{k_{r1}} F(x_1)dx_1\right] - M(1+r_f) \\ &s.t.\ \ 0 < k_s^1 - k_{r2} < k_{r1}, k_{r2} \leq x_2 \end{aligned} \quad (17)$$

The analysis above shows that the wholesale price w takes on a range of values $\left[c(1+r_B), \frac{p}{1+r_S}\right]$. Combining Equations (11) and (17), it is found that the supplier's expected profit function in both cases is a continuous function, and there must be a maximum value of the continuous function on the closed interval. Therefore, the optimal wholesale price for the retailer in both cases exists and $w^* \in \left[c(1+r_B), \frac{p}{1+r_S}\right]$.

2.3.3. Contagion Intensity of Supply Chain Credit Default Risk

Based on the analysis above of the mechanism of supply chain credit default risk contagion under the single-channel credit model and the definition of supply chain credit default risk contagion intensity, the supply chain credit default risk contagion intensity model is constructed as:

$$\beta = \begin{cases} 0 & \text{Case 1} \\ p(X < k_s^1 | X < k_r^1) & \text{Case 2} \\ p(x_1 < k_s^1 - k_{r2} | x_1 < k_{r1}, x_2 \geq k_{r2}) & \text{Case 3} \end{cases} \quad (18)$$

Based on the calculated Q^* and w^*, the default thresholds for retailers and suppliers can be obtained as $k_r^1 = \frac{\sum(1+r_S)T_i}{p} = \frac{\sum(1+r_S)wQ_i}{p}$, $k_{r1} = \frac{(1+r_S)T_1}{p} = \frac{(1+r_S)wQ_1}{p}$, $k_{r2} = \frac{(1+r_S)T_2}{p} = \frac{(1+r_S)wQ_2}{p}$ and $k_s^1 = \frac{(1+r_B)M}{p} = \frac{\sum(1+r_B)cQ_i}{p}$. Substituting k_r^1, k_{r1}, k_{r2} and k_s^1 into Equation (18), the contagion intensity of credit default risk in the supply chain can be modeled as:

$$\beta = \begin{cases} 0 & \text{Case 1} \\ \frac{p(X<k_s^1)}{p(X<k_r^1)} = \frac{G(k_s^1)}{G(k_r^1)} & \text{Case 2} \\ \frac{p(x_1<k_s^1-k_{r2},x_2\geq k_{r2})}{p(x_1<k_{r1},x_2\geq k_{r2})} = \frac{F(k_s^1-k_{r2})}{F(k_{r1})} & \text{Case 3} \end{cases} \quad (19)$$

Denoting the intensity of infection under scenarios I, II and III as β_1, β_2 and β_3, respectively, we obtain $\beta_1 = 0$, $\beta_2 = \frac{G(k_s^1)}{G(k_r^1)}$ and $\beta_3 = \frac{F(k_s^1-k_{r2})}{F(k_{r1})}$.

Hence, there is a strong correlation between the intensity of supply chain credit default risk contagion in the single-channel financing model β and parameters, such as supplier production costs c, market retail prices p, trade credit rates r_S and risk-free rates r_f.

2.4. Simulation Analysis

In the previous paper, the influencing factors of supply chain credit risk contagion and its mechanism were analyzed through theoretical derivation. In order to visually portray the supply chain credit risk contagion mechanism and its evolutionary characteristics, based on the assumptions in the previous paper, a computational simulation analysis was conducted by considering that the supply chain system is composed of a core enterprise supplier and two SME retailers. To facilitate the analysis of the influence mechanism of supplier production cost on the supply chain credit default risk contagion, it was assumed that the market demand of both Retailer 1 and Retailer 2 obeyed an exponential distribution with the parameter of 1/700.

2.4.1. Characteristics of the Impact of Suppliers' Marginal Cost of Production Per Unit of Product on the Contagion Intensity of Supply Chain Credit Default Risk

Assuming that the retail price of the product $p = 8$, the retailer trade credit rate $r_S = 0.07$ and the risk-free rate $r_f = 0.04$, the mechanism of the impact of the marginal cost of production of the supplier's product on the contagion intensity of supply chain credit default risk is shown in Figure 2.

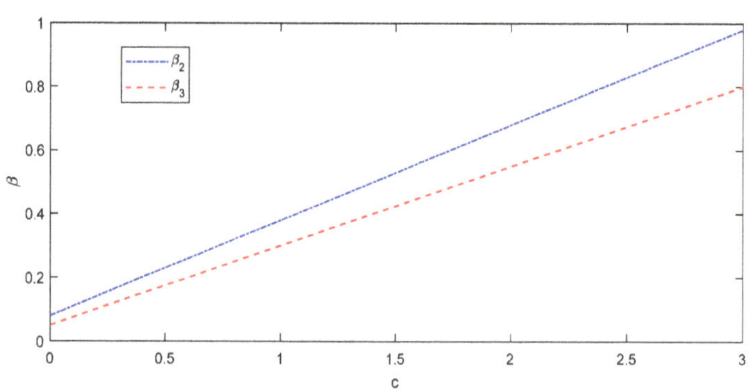

Figure 2. The effect of the marginal cost of production per unit of product for suppliers c on the contagion intensity of credit default risk β in the supply chain.

According to Figure 2, there is a positive relationship between the intensity of supply chain credit default risk contagion and supplier production cost. The supply chain credit default risk contagion effect is more significant when both retailers default on trade credit to their suppliers. Moreover, the increase in supplier production cost increases the supply chain credit default risk contagion intensity. This is because as the supplier production cost increases, the supplier will correspondingly increase the wholesale price of the product. In the case that the product market retail price remains unchanged, the profit of the retailer at the end of the sales period decreases, and the risk of trade credit default of the retailer increases. This leads to a greater possibility of credit default by the suppliers to banks, and once a credit default event occurs, it will inevitably cause an increase in the contagion intensity of supply chain credit default risk, which is not conducive to the sustainable development of the supply chain.

2.4.2. Characteristics of the Impact of the Retail Product Price on the Contagion Intensity of Credit Default Risk in the Supply Chain

Assume that the supplier's marginal cost of production per unit of product $c = 3$, the retailer's trade credit rate $r_S = 0.07$ and the risk-free rate $r_f = 0.04$. Therefore, the mechanism of the effect of the retail price of the product on the contagion intensity of the supply chain credit default risk is shown in Figure 3.

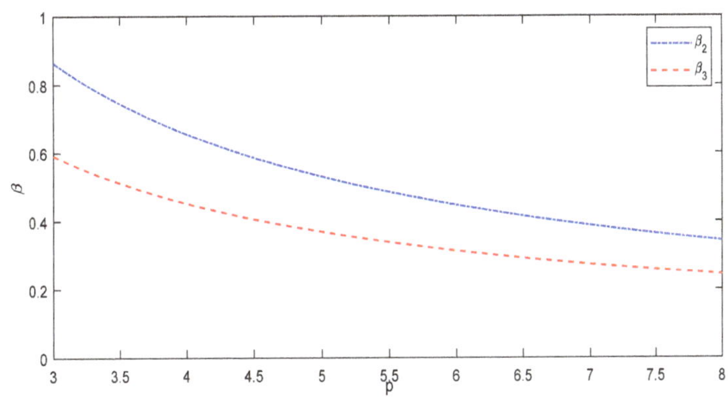

Figure 3. The effect of the product retail price p on the contagion intensity β of the supply chain credit default risk.

From Figure 3, it can be seen that there is a negative relationship between the intensity of supply chain credit default risk contagion and the retail price of products, and the supply chain credit default contagion effect is more significant when both retailers default on trade credit to their suppliers. This is because when the retailer faces a certain market demand for the product, the profit margin of the retailer at the end of the sales period will become larger with the positive fluctuation of the retail price of the product. At this point, retailers are less likely to default on trade credit, suppliers become less exposed to bank credit defaults, and the contagion effect of supply chain credit defaults diminishes and sustainability is enhanced.

2.4.3. Characteristics of the Impact of the Bank's Risk-Free Rates on the Intensity of Credit Default Risk Contagion in the Supply Chain

Assume that the marginal cost of production per unit of the supplier's product $c = 3$, the retail price of the product $p = 8$ and the retailer's trade credit rate $r_S = 0.07$. Therefore, the mechanism of the bank's risk-free rate on the contagion intensity of the supply chain credit default risk is shown in Figure 4.

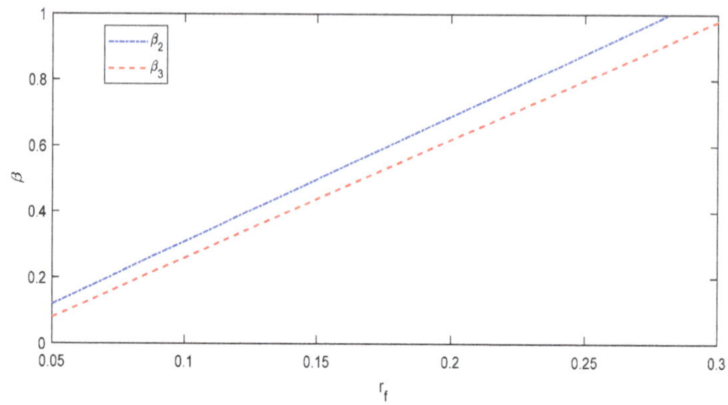

Figure 4. The relationship between the bank's risk-free rate r_f and the intensity β of credit default risk contagion in the supply chain.

According to Figure 4, it can be seen that there is a positive relationship between supply chain credit default risk contagion intensity and the commercial bank's risk-free rate, and the supply chain credit default contagion effect is more significant when both retailers default on trade credit to suppliers. This is because as the risk-free rate of commercial banks increases, the cost of the bank credit for suppliers also increases. As a result, suppliers will increase the wholesale price of their products to ensure their own revenue, which will lead to an increase in the cost of trade credit for retailers and a greater likelihood of trade credit default for them. As a result, suppliers are more likely to default on their credit to banks, which increases the contagion effect of supply chain credit default risk and ultimately affects the sustainability of the supply chain.

2.4.4. Characteristics of the Impact of Retailer Trade Credit Rates on the Intensity of Credit Default Risk Contagion along the Supply Chain

Assume that the supplier's production cost $c = 3$, the retail price of the product $p = 8$ and the bank's risk-free rate $r_f = 0.04$. Therefore, the mechanism of the retailer's trade credit interest rate on the contagion intensity of the supply chain credit default risk is shown in Figure 5.

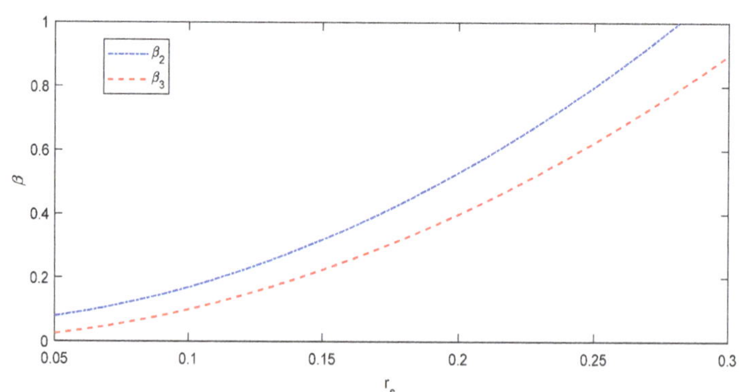

Figure 5. The relationship between the retailer trade credit rate r_S and the intensity of credit default risk contagion in the supply chain β.

From Figure 5, it can be seen that the intensity of supply chain credit default risk contagion is positively related to the retailer's trade credit interest rate, and the supply chain credit default contagion effect is more significant when both retailers default on trade credit to their suppliers. The reason for this is that when the retail price of a product is certain, the cost of trade credit for retailers will increase with the increase in trade credit interest rates, resulting in a reduction in retailers' profit margins. At this time, the possibility of trade credit default by retailers increases, and the probability of credit default risk contagion in the supply chain likewise increases, which is not conducive to the sustainability of the supply chain.

3. A Model of Supply Chain Credit Default Risk Contagion under a Dual-Channel Credit Model

3.1. Model Assumptions and Notation

Assume a simple supply chain consisting of a core firm supplier and an SME retailer, where both the supplier and the retailer have a shortage of funds and are unable to carry out their production and sales activities independently. The supplier, as a core enterprise, can provide loans directly from commercial banks due to its high profitability and corporate reputation. Meanwhile, the retailer, as an SME, cannot obtain financing directly from banks due to its limitations, but can obtain partial loans through the repurchase guarantee of the core enterprise. The retailer can also apply for trade credit directly from the core enterprise upstream of the supply chain to obtain financial support. Therefore, retailers are considered to be financed under the dual-channel credit model, adopting a combination of trade credit and bank credit based on the repurchase mechanism of core enterprises. Thus, the decision process of supply chain financing is as follows:

① At the beginning of the sales period, the supplier makes a guarantee to the commercial bank through a repurchase deed, and at the end of the sales season, will acquire the retailer's surplus products at a repurchase price of m.

② The retailer determines the order quantity as Q, obtains financing from a commercial bank as $M = \alpha wQ$ at an interest rate of r_B, for the retailer's bank credit ratio α, then pays the supplier an advance payment as M, and the remaining purchase price as $N = (1-\alpha)wQ$ is financed by the supplier's trade credit at an interest rate of r_S, $1-\alpha$ for the trade credit ratio.

③ The supplier receives an advance that does not cover the production activity and is financed through bank credit with $R = cQ - \alpha wQ$ where c is the unit cost of the product produced by the supplier at an interest rate r_b. There is an infinitive $R < N$.

④ After the supplier organizes production for shipment, the retailer sells the product at a fixed market retail price p. If there is a surplus of product at the retailer, the supplier buys it back at a buyback price m. The residual value of the surplus product is 0.

⑤ At the end of the sales period, the retailer realizes all the funds back and repays the bank credit and trade credit to the commercial bank and the supplier, respectively, and the supplier repays the commercial bank debt after funds are returned. The whole supply chain credit financing decision process outlined above is shown in Figure 6.

In the production and sales activities of supply chain enterprises, if the market sales are good, the capital flow of the retailer is sufficient to repay all the principal and interest of the bank credit and trade credit. The supplier repays the bank credit debt as promised and neither the retailer nor the supplier will default. If the market sales are bad, the retailer cannot repay all debts and there is a risk of default, which will lead to the supplier defaulting to a certain extent as well. In this case, there is a contagion of credit default risk in the supply chain. Since the retailer applies for credit from both commercial banks and suppliers, when the retailer's sales revenue cannot meet the repayment of both debts, the retailer needs to decide whether to prioritize the repayment of bank credit debts or trade credit debts. Based on this, the probability of a retailer's priority in repaying trade credit debt is defined as the retailer's repayment priority, denoted as θ,$\theta \in [0,1]$, where θ is a continuous variable. Based on the analysis above, the following basic assumptions were made in this section:

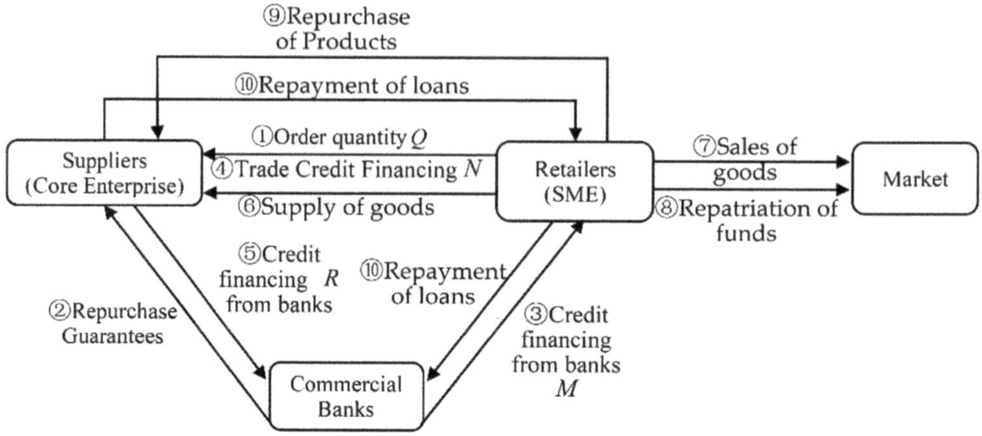

Figure 6. A decision model for supply chain credit financing under a dual-channel credit model.

Assumption 6. *Suppliers and retailers are perfectly rational and only repay their full revenues in the event of a credit default that prevents them from settling their debts.*

Assumption 7. *Both the supplier and the retailer are financially constrained and do not have sufficient initial capital, the supplier's marginal cost of production per unit of product is c, the wholesale price per unit of product is w, the retail price of product is p and the buyback price per unit of product at the end of the sales period is m. The relationship between the magnitude of these price parameters satisfies* $m < c < w < p$.

Assumption 8. *The market demand for the product during the sales process is random and the retail price of the product remains constant during the sales process. According to the characteristics of the newsboy model,* $f(x)$ *and* $F(x)$ *are the probability density function and distribution function of market demand x, respectively, and* $F(x)$ *has the characteristics of continuity, differentiability and incrementality, satisfying the incremental failure rate (IFR), where* $\overline{F}(x) = 1 - F(x)$.

Based on the assumptions above, the symbols and definitions of the variables involved in this section are shown in Table 2.

Table 2. A description of symbols.

Symbol	Implication
m	Repurchase price per unit of product
c	Marginal cost of production per unit of product
w	Wholesale price per unit of product
p	Product retail price
Q	Retailer order quantity
x	Demand in the market
M	Retail merchant bank credit facilities
r_B	Retail merchant bank credit rates
N	Retailers' trade credit facility quantity
r_S	Retailers' trade credit rates
R	Amount of bank credit facilities for suppliers
r_b	Supplier bank credit rates
r_f	Bank risk-free rates
α	Retail merchant bank credit ratio
θ	Retailer repayment priorities
k_s	Supplier default thresholds
k_r	Retailer default thresholds
β	The intensity of contagion of credit default risk in the supply chain

3.2. Mechanisms of Supply Chain Credit Default Risk Contagion under a Dual-Channel Credit Model

At the end of the sales period, the retailer realizes a return of funds and then settles the debt. When actual sales market conditions are poor and market demand is significantly lower than the retailer's order quantity, the retailer is at risk of credit default. At this point, the credit risk is to some extent transmitted to the supplier through the trade credit financing contract between the supplier and the retailer, resulting in the supplier also defaulting on its obligations. In a dual-channel credit model for retailers, the priority of the retailer's repayment of bank credit debt and trade credit debt will directly affect the supplier's repayment to the commercial bank. Therefore, the repayment priority of the retailer under a dual-channel credit model will be a key factor in the contagion of supply chain credit default risk. To provide a clearer picture of supply chain credit default risk contagion, the following definitions were made:

Definition 4. *At the end of the sales period, if a retailer's sales revenue is less than its bank credit obligations or trade credit obligations, the retailer is in credit default. If a supplier's capital receipts are less than its bank credit obligations, the supplier is also in credit default.*

Definition 5. *Due to the unpredictability of the market, default risk for retailers is mostly caused by bank credit failures and trade credit defaults to suppliers. In contrast, the risk arising from the contagion of risk to suppliers is the risk of default on suppliers' credit to banks.*

Definition 6. *As a result of a trade credit default by a retailer, a bank credit default by a supplier also occurs. The conditional probability in this case is defined as the intensity of contagion of credit default risk in the supply chain and is denoted as β.*

In light of the analysis above, in the event of a credit default by a retailer, the repayment of its credit debt can be divided into two scenarios: priority repayment of trade credit debt and priority repayment of bank credit debt.

Case 4. *Retailers prioritize repayment of supplier trade credit debt.*

① If the retailer does not default on the trade credit, then there is no default to the supplier. If the supplier is able to recover all of the remaining balance N from the retailer, the remaining balance N is greater than the supplier's financing R and the supplier is able to repay its bank loan, then the supplier is not in default. Therefore, there is no contagion of credit default risk in the supply chain and the intensity of contagion is $\beta_{\mathrm{I}_1} = 0$.

② If a retailer defaults on a trade credit facility to a supplier, then the retailer is unable to repay the trade credit facility at the end of the sales period, and its inflow of funds $px + m(Q - x)$ is less than the principal and interest sum $N(1 + r_S)$ of the trade credit facility. When $x < \frac{N(1+r_S)-mQ}{p-m} = k_{r1}$ and the total funds $px + m(Q - x)$ recovered by the supplier are less than the principal and interest sum of its bank credit $R(1 + r_b)$, that is, in the case of $x < \frac{R(1+r_b)-mQ}{p-m} = k_{s1}$, the supplier defaults. Therefore, the contagion intensity of credit default risk in the supply chain is $\beta_{\mathrm{I}_2} = P(x < k_{s1} | x < k_{r1})$.

Based on the analysis above and the previous assumptions, the probability that a retailer will repay its suppliers first is θ, in which case the contagion intensity of credit default risk in the supply chain is $\beta_{\mathrm{I}} = \theta P(x < k_{s1} | x < k_{r1})$.

Case 5. *Retailers prioritize repayment of bank credit debt.*

① If the retailer defaults on its credit to the bank at the end of the sales period, in a situation where repayment of bank credit is a priority, the retailer will also be unable to repay the supplier's trade credit facility, resulting in the supplier defaulting on the bank. Therefore, the contagion intensity of credit default risk in the supply chain is $\beta_{\mathrm{II}_1} = 1$.

② If the retailer does not default on its credit facility to the bank at the end of the sales period, the retailer repays the bank credit facility as promised and its funds flow

$px + m(Q - x)$ is greater than the sum of the principal and interest of the bank credit facility $M(1 + r_B)$; that is, $x > \frac{M(1+r_B)-mQ}{p-m} = k_{r2}$. If the retailer's remaining funds $px + m(Q - x) - M(1 + r_B)$ are less than the sum of the principal and interest on the trade credit facility $N(1 + r_S)$, that is, $x < \frac{N(1+r_S)+M(1+r_B)-mQ}{p-m} = k_{r3}$, then the retailer is in default with the supplier. At this point, if the supplier's recovery of funds $px + m(Q - x) - M(1 + r_B)$ is less than the sum of the principal and interest of its bank credit $R(1 + r_b)$, that is, $x < \frac{R(1+r_b)+M(1+r_B)-mQ}{p-m} = k_{s2}$, the supplier defaults on the bank. Therefore, the contagion intensity of the credit default risk in the supply chain is $\beta_{II_2} = P(x < k_{s2}|k_{r2} < x < k_{r3})$.

Based on the analysis above and the previous assumptions, the probability that the retailer will repay the bank first is $1 - \theta$. In this case, the contagion intensity of credit default risk in the supply chain is $\beta_{II} = (1 - \theta)P(x < k_{s2}|k_{r2} < x < k_{r3}) + 1 - \theta$. The analysis shows that there are market demand thresholds for both retailers and suppliers when they default on credit, which are k_{r1}, k_{r2}, k_{r3}, k_{s1} and k_{s2}. Among them, are $k_{r1} = \frac{N(1+r_S)-mQ}{p-m}$, $k_{r2} = \frac{M(1+r_B)-mQ}{p-m}$, $k_{r3} = \frac{N(1+r_S)+M(1+r_B)-mQ}{p-m}$, $k_{s1} = \frac{R(1+r_b)-mQ}{p-m}$ and $k_{s2} = \frac{R(1+r_b)+M(1+r_B)-mQ}{p-m}$. Supply chain credit risk arises when market demand exceeds the default threshold, and the supply chain may be subject to credit default risk contagion. When a retailer gives priority to repaying its suppliers, if $x < k_{r1}$ and $x < k_{s1}$, then a retailer's trade credit default will lead to a supplier's default on bank credit. When the retailer has priority in repaying the bank's debt, if $k_{r2} < x < k_{r3}$ and $x < k_{s2}$, then the retailer's default risk is contagious to the supplier.

3.3. *A Model of Supply Chain Credit Default Risk Contagion under a Dual-Channel Credit Model*

Throughout the supply chain operation and financing process, the commercial bank plays a Stackelberg game with the suppliers and retailers in the supply chain, respectively, as shown in Figure 7. In the whole credit financing process, firstly, the commercial bank decides the bank credit rate r_B for the retailer and the bank credit rate r_b for the supplier. Secondly, the supplier determines the wholesale price of the product w. Finally, the retailer decides the order quantity of the product Q. According to the decision-making sequence of each subject in the game process, the optimal solution of each subject is obtained by the reverse solution method, and the contagion intensity of the credit default risk in the supply chain under a dual-channel credit model is determined.

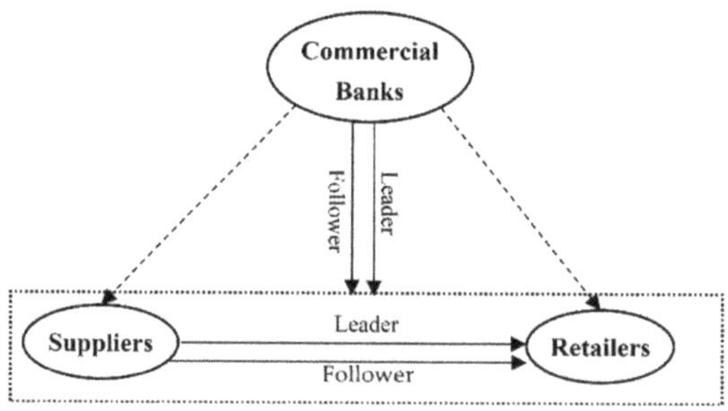

Figure 7. A double-game scheme between supply chain enterprises and banks.

3.3.1. Optimal Decision-Making for Retailers

The retailer obtains its financial support through a dual-channel credit model, where the retailer must repay its bank credit debt and trade credit debt at the end of the sales period. The retailer is at risk of default when the inflow of funds to the retailer, $px + m(Q - x)$, is less than the principal and interest on all debt, $M(1 + r_B) + N(1 + r_S)$, when market demand x is less than the critical demand $k_{r3} = \frac{N(1+r_S)+M(1+r_B)-mQ}{p-m}$. Since the retailer is rational, for the supply chain to operate properly and for the retailer to make a normal profit, there is a condition $N(1+r_S)+M(1+r_B)<px$, which means that $k_{r3}<Q$ holds. Therefore, at the end of the sales period, the retailer's earnings can be expressed as:

$$\Pi_R = \begin{cases} 0 & x < k_{r3} \\ px + m(Q-x) - M(1+r_B) - N(1+r_S) & k_{r3} < x < Q \\ pQ - M(1+r_B) - N(1+r_S) & Q < x \end{cases} \tag{20}$$

Therefore, the retailer's expected profit is:

$$\mathrm{E}(\Pi_R) = \int_{k_{r3}}^{Q} [px + m(Q-x) - M(1+r_B) - N(1+r_S)] f(x)dx + \int_{Q}^{\infty} [pQ - M(1+r_B) - N(1+r_S)] f(x)dx \tag{21}$$

The retailer maximizes its expected profit by deciding on the optimal order quantity Q:

$$\begin{aligned} \max_Q \mathrm{E}(\Pi_R) &= \int_{k_{r3}}^{Q} [px + m(Q-x) - M(1+r_B) - N(1+r_S)] f(x)dx + \int_{Q}^{\infty} [pQ - M(1+r_B) - N(1+r_S)] f(x)dx \\ \text{s.t. } & k_{r3} < Q \end{aligned} \tag{22}$$

The simplification of Equation (22) yields:

$$\begin{aligned} \max_Q \mathrm{E}(\Pi_R) &= (p-m)(Q-k_{r3}) - (p-m)\int_{k_{r3}}^{Q} F(x)dx \\ & \text{s.t. } k_{r3} < Q \end{aligned} \tag{23}$$

The first-order derivative of Equation (23) with respect to the order quantity Q is:

$$\frac{\partial \mathrm{E}(\Pi_R)}{\partial Q} = (p-m)\bar{F}(Q) - (p-m)k_{r3}'\bar{F}(k_{r3}) \tag{24}$$

Since $\frac{\partial^2 \mathrm{E}(\Pi_R)}{\partial Q^2} < 0$, the optimal solution for the retailer's order quantity is calculated to satisfy:

$$\frac{\bar{F}(Q^*)}{\bar{F}(k_{r3})} = \frac{\alpha w(1+r_B) + (1-\alpha)w(1+r_S) - m}{p-m} \tag{25}$$

Among them, is $k_{r3} = \frac{N(1+r_S)+M(1+r_B)-mQ}{p-m}$.

3.3.2. Optimal Decision-Making for Suppliers

The retailer takes dual-channel credit financing, so the retailer's repayment priority directly affects the supplier's revenue profile. When the retailer is not at risk of credit default, $x>k_{r3}$, the supplier recovers all of its funds $N(1+r_S)$. When the retailer is at risk of credit default, $x<k_{r3}$, if the retailer chooses to consider the trade credit debt first, the supplier's revenue is $\theta \min[px, N(1+r_S)-m(Q-x)]$. If the retailer chooses to consider the bank credit debt first, the supplier's revenue is $(1-\theta)[px-M(1+r_B)]^+$. Therefore, based on the retailer's market sales and earnings, the revenue generated by the supplier's transaction with the retailer can be expressed as:

$$\Pi_S = \begin{cases} \theta \min[px, N(1+r_S) - m(Q-x)] + (1-\theta)[px - M(1+r_B)]^+ & 0 < x < k_{r3} \\ N(1+r_S) - m(Q-x) & k_{r3} < x < Q \\ N(1+r_S) & Q < x \end{cases} \tag{26}$$

Therefore, the supplier's expected profit can be expressed as:

$$\begin{aligned} \mathrm{E}(\Pi_S) = & \int_0^{k_{r3}} \left(\theta \min[px, N(1+r_S) - m(Q-x)] + (1-\theta)[px - M(1+r_B)]^+\right) f(x)dx \\ & + \int_{k_{r3}}^{Q} [N(1+r_S) - m(Q-x)] f(x)dx + \int_Q^{\infty} N(1+r_S) f(x)dx - R(1+r_b) \end{aligned} \tag{27}$$

The supplier maximizes its own expected revenue by choosing the optimal wholesale price, and solving for the first-order partial derivative of Equation (27) with respect to w yields:

$$\begin{aligned}\frac{\partial E(\Pi_S)}{\partial w} = & [(p-m)(1-\theta F(k_{r1}))+(1-\theta)pF(k_{r1})]k_{r1}'-(1-\theta)(p-m)F(k_{r3})k_{r3}' \\ & +(1-\theta)m[(Q'-k_{r2}')F(k_{r2})+(Q-k_{r2})f(k_{r2})k_{r2}'] \\ & +mQ'\overline{F}(Q)+\alpha\theta(1+r_b)-(c-\alpha w)(1+r_b)Q'\end{aligned} \tag{28}$$

The combined calculation gives $\frac{\partial^2 E(\Pi_S)}{\partial w^2} < 0$, such that the supplier's optimal wholesale price w^* exists and satisfies:

$$\begin{aligned}& [(p-m)(1-\theta F(k_{r1}))+(1-\theta)pF(k_{r1})]k_{r1}'-(1-\theta)(p-m)F(k_{r3})k_{r3}' \\ & +(1-\theta)m[(Q'-k_{r2}')F(k_{r2})+(Q-k_{r2})f(k_{r2})k_{r2}']+mQ'\overline{F}(Q)+\alpha\theta(1+r_b)-(c-\alpha w)(1+r_b)Q'=0\end{aligned} \tag{29}$$

Therefore, the supplier's maximized expected profit is:

$$\begin{aligned}\max_w E(\Pi_S) = & \int_0^{k_{r3}}\left(\theta\min[px,N(1+r_S)-m(Q-x)]+(1-\theta)[px-M(1+r_B)]^+\right)f(x)dx \\ & +\int_{k_{r3}}^{Q}[N(1+r_S)-m(Q-x)]f(x)dx+\int_Q^{\infty}N(1+r_S)f(x)dx \quad -R(1+r_b)\end{aligned} \tag{30}$$

3.3.3. Optimal Decision-Making for Commercial Banks

① Credit financing decisions between banks and retailers

The retailer applies to a commercial bank for a bank credit facility with funding M and a loan rate of r_B. Because of the retailer's dual-channel credit model, the retailer's repayment priority directly affects the bank's return at the end of the sales period. When the retailer is not at risk of default, $x > k_{r3}$, the bank recovers the sum of the retailer's bank credit principal and interest $M(1+r_B)$. When there is a risk of default by the retailer, $x < k_{r3}$, the bank's return at this point is $\theta[px+m(Q-x)-N(1+r_S)]^+$ if the retailer has priority in repaying the trade credit debt. If the retailer has priority in repaying the bank credit debt, the bank's return at this point is $(1-\theta)\min[px+m(Q-x),M(1+r_B)]$. Based on the retailer's market sales and actual profitability, the return arising from the bank's transaction with the retailer can be expressed as:

$$\Pi_{B1} = \begin{cases}\theta[px+m(Q-x)-N(1+r_S)]^+ + (1-\theta)\min[px+m(Q-x),M(1+r_B)] & 0<x<k_{r3} \\ M(1+r_B) & k_{r3}<x\end{cases} \tag{31}$$

As banks are in a competitive equilibrium market, the retailer bank credit rate r_B satisfies the risk compensation principle $(1+r_f)M = E(\Pi_{B1})$, where r_f is the risk-free rate.

$$(1+r_f)M = \int_0^{k_{r3}}\left(\theta[px+m(Q-x)-N(1+r_S)]^+ + (1-\theta)\min[px+m(Q-x),M(1+r_B)]\right)f(x)dx+\int_{k_{r3}}^{\infty}M(1+r_B)f(x)dx \tag{32}$$

Equation (32) indicates that the principle of competitive equilibrium in the credit market makes the bank's demand for credit principal and interest income equal to the bank's expected return in the retailer's bank credit facility. Simplifying Equation (32) yields:

$$M(r_B-r_f) = (p-m)\left[\theta\int_{k_{r1}}^{k_{r3}}F(x)dx+(1-\theta)\int_0^{k_{r2}}F(x)dx\right]+(1-\theta)mQF(k_{r2}) \tag{33}$$

② Credit financing decisions between banks and suppliers

The supplier applies to the commercial bank for a bank credit facility with funding R at an interest rate of r_b. In a bank credit transaction between a commercial bank and a supplier, the commercial bank's return is directly related to the supplier's default, while the supplier's return is influenced by the retailer's profitability at the end of the sales period and the retailer's repayment priority. When there is no risk of default by the retailer, $x > k_{r3}$, at which point there is no default by the retailer to either the supplier or the bank, the supplier recovers the principal and interest on the trade credit and repays the bank credit debt as promised. Therefore, the bank recovers the principal and interest on the supplier's bank credit $R(1+r_b)$. When there is a risk of default by the retailer, $x < k_{r3}$, if the retailer repays the trade credit debt first, the bank's proceeds are then $\theta\min[px+m(Q-x),R(1+r_b)]$. If the retailer has priority in repaying the bank credit debt, the bank's return at this point is $(1-\theta)\min[px+m(Q-x)-M(1+r_B),R(1+r_b)]$. At the end of the sales period, over the course of a bank credit transaction between a commercial bank and a supplier, the specific gain to the commercial bank can be expressed as:

$$\Pi_{B2} = \begin{cases} \theta\min[px+m(Q-x),R(1+r_b)] & \\ +(1-\theta)\min[px+m(Q-x)-M(1+r_B),R(1+r_b)] & 0 < x < k_{r3} \\ R(1+r_b) & k_{r3} < x \end{cases} \tag{34}$$

As mentioned above, the supplier bank credit rate r_b satisfies the risk compensation principle: $(1+r_f)R = \mathrm{E}(\Pi_{B2})$.

$$(1+r_f)R = \int_0^{k_{r3}} (\theta\min[px+m(Q-x),R(1+r_b)] + (1-\theta)\min[px+m(Q-x)-M(1+r_B),R(1+r_b)])f(x)dx + \int_{k_{r3}}^{\infty} R(1+r_b)f(x)dx \tag{35}$$

Equation (35) indicates that the principle of competitive equilibrium in the credit market makes the bank's demand for credit principal and interest income equal to the bank's expected return in the supplier's bank credit facility.

Simplifying Equation (35) yields:

$$R(r_b - r_f) = (p-m)\left[\theta\int_0^{k_{s1}} F(x)dx + (1-\theta)\int_{k_{r2}}^{k_{s2}} F(x)dx\right] \tag{36}$$

Therefore, the credit rates r_B and r_b offered by commercial banks to retailers and suppliers should satisfy Equations (33) and (36), respectively.

3.3.4. The Intensity of Contagion of Credit Default Risk in the Supply Chain

Based on the analysis of the contagion mechanism of credit default risk in the supply chain under the dual-channel credit model in Section 3.2, and the definition of contagion intensity, the contagion model of credit default risk in the supply chain under the dual-channel credit model is:

$$\beta = \begin{cases} \theta P(x < k_{s1} | x < k_{r1}) & \text{Case } I \\ (1-\theta)P(x < k_{s2} | k_{r2} < x < k_{r3}) + 1 - \theta & \text{Case } II \end{cases} \tag{37}$$

From the above derived Q^* and w^*, the lending rates r_B and r_b for commercial banks and the default thresholds k_{r1}, k_{r2}, k_{r3}, k_{s1} and k_{s2} for retailers, the suppliers can be derived by bringing their values into Equation (37) to obtain the intensity of contagion as:

$$\beta = \theta\frac{F(k_{s1})}{F(k_{r1})} + (1-\theta)\frac{F(k_{s2}) - F(k_{r2})}{F(k_{r3}) - F(k_{r2})} + (1-\theta) \tag{38}$$

Therefore, the first-order partial derivative of contagion intensity β with respect to retailer repayment priority θ is:

$$\frac{\partial\beta}{\partial\theta} = -\frac{F(k_{s2}) - F(k_{r2})}{F(k_{r3}) - F(k_{r2})} + \frac{F(k_{s1})}{F(k_{r1})} - 1 < 0 \tag{39}$$

Therefore, contagion intensity β and retailer repayment priority θ are negatively correlated. When the repayment priority is higher, the intensity of credit risk contagion in the supply chain is lower.

3.4. *Simulation Analysis*

Based on the setup by existing research for the newsboy model, where market demand obeys a distribution, this section assumes that the market demand x faced by the retailer obeys a uniform distribution [21,29]. The retailer's repayment priority θ is set as a variable and all other parameters are fixed. The remaining parameters are set as follows: product market retail price $p = 8$, production cost $c = 3$, repurchase price $m = 1$, retailer trade credit ratio $r_S = 0.07$, commercial bank risk-free rate $r_f = 0.04$, market demand distribution $x \sim U(0, 3000)$, and retailer bank credit ratio $\alpha = 0.3$, $\alpha = 0.5$ and $\alpha = 0.7$. Among them, $\alpha = 0.3$ denotes a low level of bank credit for retailers, $\alpha = 0.5$ denotes a medium level and $\alpha = 0.7$ denotes a high level. Based on the parameters above, numerical simulations were conducted to explore the impact of retailer repayment priority on the decision of each participant in supply chain credit financing and the contagion intensity of supply chain credit default risk.

3.4.1. The Impact of Retailer Repayment Priorities on Supplier Decisions

The relationship between retailer repayment priority and supplier wholesale prices is shown in Figure 8.

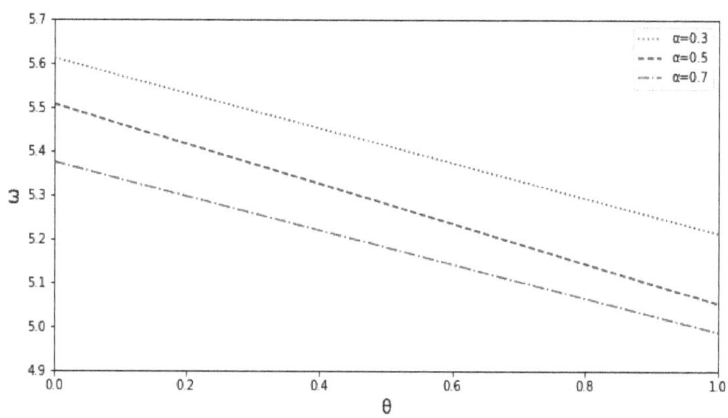

Figure 8. The relationship between retailer repayment priority and supplier wholesale prices.

Figure 8 shows that there is a negative relationship between the retailer's repayment priority θ to the supplier and the supplier's wholesale price w; i.e., as the retailer's repayment priority θ to the supplier increases, the supplier's wholesale price w decreases. In addition, when the retailer's repayment priority θ to the supplier is certain, the supplier's wholesale price w decreases as the retailer's bank credit ratio α increases. This suggests that when α is larger, that is, the retailer applies for a greater proportion of bank credit, the retailer applies for relatively less trade credit from the supplier and the supplier's expected loss is correspondingly reduced, when the supplier will be more willing to offer wholesale price concessions to the retailer. In addition, the supplier's expected loss continues to decrease to some extent when the retailer has a higher probability of repaying the supplier's trade credit debt on a priority basis. In this case, suppliers will choose greater wholesale price incentives to promote the volume of transactions with retailers, ultimately achieving their own profits and supply chain sustainability.

3.4.2. The Impact of Retailer Repayment Priorities on Retailer Decisions

As shown in Figure 9, the retailer's repayment priority θ to suppliers positively affects the retailer's order quantity Q. As the retailer's repayment priority θ to suppliers gradually becomes larger, the retailer's order quantity Q also becomes larger. In addition, when a retailer has a certain probability of repaying its supplier's debt first, the greater the retailer's bank credit ratio α, the greater the retailer's order quantity Q. When the retailer's bank credit ratio is smaller, the retailer's order quantity Q is also relatively smaller. This suggests that when retailers apply for more bank credit, they will apply for less trade credit from their suppliers, when the risk of loss to the supplier will be correspondingly reduced, facilitating the implementation of wholesale price preferences between suppliers and retailers even more. In addition, as retailers become more likely to repay their suppliers' trade credit facilities on a priority basis, suppliers and retailers gradually form a good partnership, with retailers increasing their order volumes on the basis of price concessions offered by suppliers, thereby increasing the sales profitability of both retailers and suppliers, and ensuring the sustainability of the supply chain.

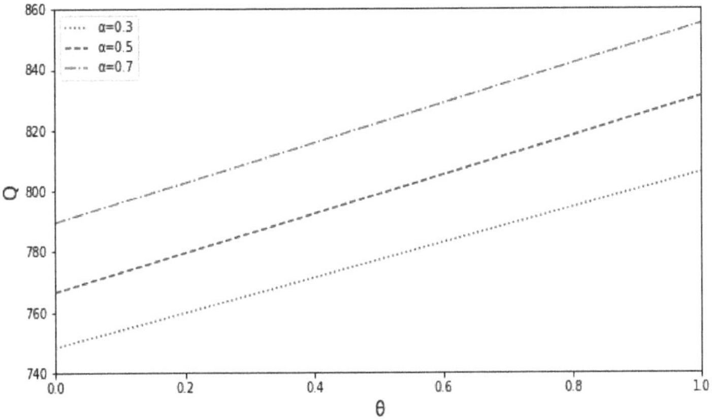

Figure 9. The relationship between retailer repayment priority and retailer order quantity.

3.4.3. The Impact of Retailer Repayment Priorities on Bank Lending Rates

The relationships between the retailer's repayment priority and the retailer's bank credit rate, and the retailer's repayment priority and the supplier's bank credit rate, are shown in Figures 10 and 11, respectively.

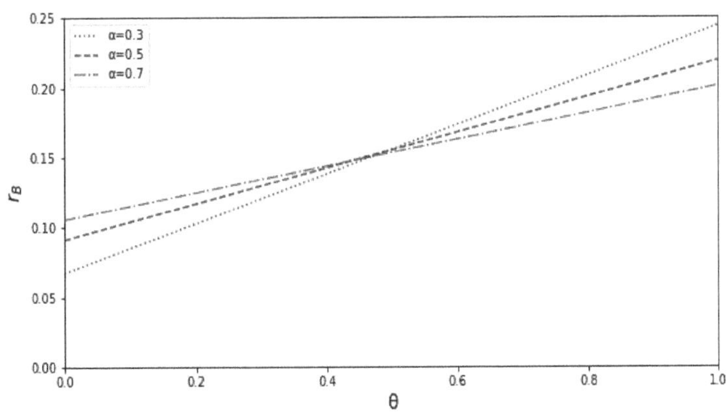

Figure 10. The relationship between retailer repayment priority and retailer bank credit rates.

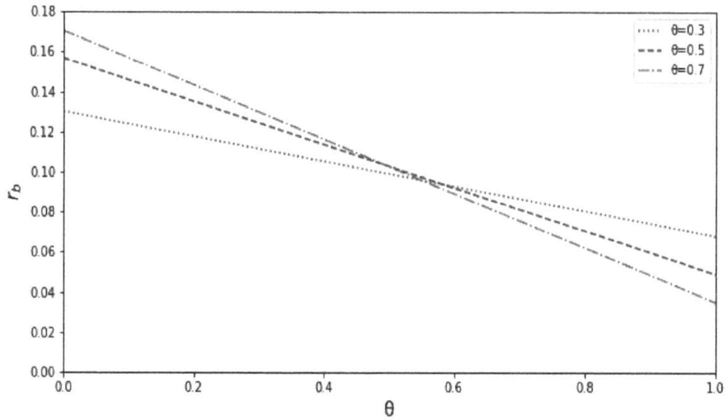

Figure 11. The relationship between retailer repayment priority and supplier bank credit rates.

Figure 10 shows that there is a positive relationship between a retailer's repayment priority θ on its trade credit debt to suppliers and the retailer's bank credit rate r_B. As the retailer's repayment priority θ gradually increases, the retailer's bank credit rate increases. This indicates that when retailers have a lower probability of repaying their suppliers' trade credit debts and a higher probability of repaying their bank credit, retailers are less likely to default on their credit to commercial banks and commercial banks will then consider offering lower bank credit rates to retailers. Conversely, when a retailer is more likely to repay its supplier debt and less likely to repay its commercial bank debt, the commercial bank is more likely to suffer a default loss at the end of the selling season, leading it to increase its bank credit rate to the retailer. The change in the interest rate of retailer bank credit was analyzed for the retailer bank credit ratios of $\alpha = 0.3$, $\alpha = 0.5$ and $\alpha = 0.7$. We found that as the retailer's repayment priority θ increases, the bank credit rate rises more slowly when the retailer's bank credit ratio α is higher, when the retailer's bank credit limit is larger, and when the retailer's bank credit limit is relatively small. When the retailer's repayment priority θ is low, the retailer is less likely to repay its supplier debt and more likely to repay its commercial bank debt. The higher the retailer's bank credit ratio, the higher the retailer's bank credit limit, and the higher its bank loan interest rate. In contrast, when the retailer's repayment priority θ is higher, the retailer has a lower probability of repaying its commercial bank debt on a priority basis, and the higher the retailer's bank credit, the lower its bank credit rate. This suggests that although retailers consider repaying commercial bank debt on a priority basis, due to the lower credit rating and smaller asset size of SMEs such as retailers, banks are exposed to a greater risk of default by retailers when they are granted larger bank credit lines. It was difficult to determine the supply chain's sustainability for this study. As a result, commercial banks will increase their interest rates on bank credit to retailers to control the risk of default by retailers during the transaction. In addition, as the retailer's bank credit line becomes larger, the bank credit rate becomes larger, and the retailer's risk of default becomes greater. Therefore, commercial banks will strictly control the increase in their bank credit rate in order to balance the retailer's risk of default, mainly by controlling the slow increase in the bank credit rate when the retailer applies for a higher bank credit line.

Figure 11 shows that there is a negative correlation between the retailer's repayment priority θ on the supplier's trade credit debt and the supplier's bank credit rate r_b. As the retailer's repayment priority θ on trade credit debt gradually increases, the supplier's bank credit rate decreases. This suggests that the greater the probability that a retailer will prioritize repayment of the supplier's trade credit debt, the lower the probability that the supplier will be exposed to the retailer's trade credit default risk at the end of the sales period. Commercial banks are less likely to face a loss of funds and in turn consider reducing the cost of supplier bank credit by lowering the supplier's credit rate. Conversely, where the probability of the retailer repaying the supplier's trade credit debt on a priority basis is low, the retailer considers repaying the commercial bank's debt on a priority basis, when the supplier faces a higher risk of the retailer defaulting on its trade credit. In this case, the commercial bank will consider increasing the supplier's bank credit rate to reduce its own potential risk of loss. A comparative analysis was conducted of the change in the supplier's bank credit rate when the retailer's bank credit ratio was $\alpha = 0.3$, $\alpha = 0.5$ and $\alpha = 0.7$, respectively. According to Figure 11, as the retailer's repayment priority θ gradually increases, the retailer's bank credit ratio decreases. When the retailer's bank credit limit is low, the supplier's bank credit rate decreases more slowly. When the retailer's bank credit is larger, the supplier's bank credit rate decreases more quickly. At a lower repayment priority θ of the retailer, the higher the retailer's bank credit ratio α, the higher the bank credit rate r_b faced by the supplier, and the lower the retailer's bank credit ratio α, the lower the bank credit rate r_b faced by the supplier. When the retailer's repayment priority θ is higher, the higher the retailer's bank credit ratio α and the lower the supplier's bank credit rate r_b. This phenomenon is mainly due to the fact that commercial banks derive their revenue from the retailer's bank credit and the supplier's bank credit, respectively. When the retailer's repayment priority θ is lower, the retailer has a higher probability of repaying the bank credit debt first, the commercial bank faces less risk of default on the retailer's credit, and the commercial bank gains relatively less benefit from the retailer's bank credit being smaller. Thus, the commercial bank offers a lower bank credit rate to the supplier to enhance its own benefit. When a retailer has a higher probability of repaying the supplier's debt first, the commercial bank is exposed to the risk of default by the retailer, and when the retailer's bank credit line is larger, the commercial bank is exposed to a higher risk of default. Therefore, in order to ensure the sustainability of the supply chain, the commercial bank mitigates its risk by offering a lower interest rate on the supplier's loan.

3.4.4. Characteristics of the Impact of Retailer Repayment Priority on the Intensity of Credit Default Risk Contagion in the Supply Chain

The relationship between retailer repayment priority and the intensity of credit default risk contagion in the supply chain is shown in Figure 12.

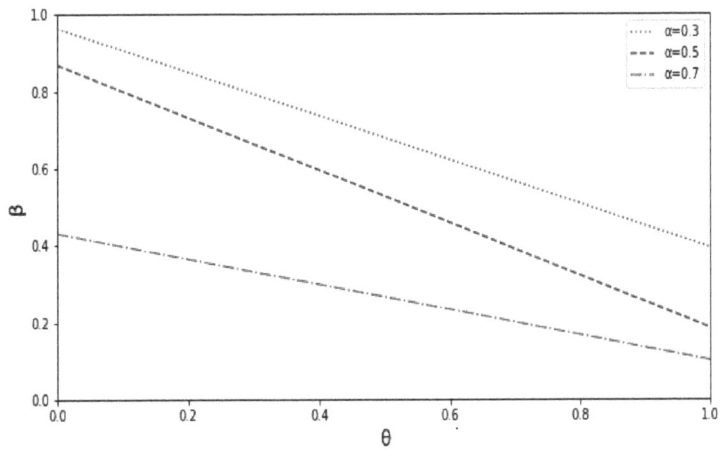

Figure 12. The relationship between retailer repayment priority and the intensity of credit default risk contagion in the supply chain.

Figure 12 shows that there is a negative relationship between the retailer's repayment priority θ to suppliers and the contagion intensity of credit default risk β in the supply chain. As the retailer's repayment priority θ to suppliers increases, the intensity of credit default risk contagion in the supply chain decreases. This indicates that the more likely a retailer is to prioritize trade credit debt for repayment, the lower the retailer's risk of defaulting on trade credit to its suppliers and the supplier's ability to minimize its own losses, which in turn reduces the likelihood of default to the commercial bank. As a result, the contagion effect of credit default risk along the supply chain is reduced. In addition, the relatively higher intensity of credit default risk contagion in the supply chain is not conducive to the sustainability of the supply chain, when the retailer's bank credit ratio α is lower and the retailer's repayment priority θ is certain. This is due to the fact that a lower retailer bank credit ratio means that the retailer applies for a higher amount of trade credit from its suppliers. If the retailer defaults on its trade credit to the supplier, the supplier will suffer greater losses, which in turn leads to a greater likelihood of the supplier defaulting on its credit to the bank, thus increasing the contagion effect of credit default risk in the supply chain.

4. Conclusions

This paper focused on the contagion of supply chain credit default risk under the single-channel credit model of retailers and the dual-channel credit model from the perspective of credit financing for supply chain enterprises, in order to provide a reference to strengthen the risk management of supply chain finance and ensure the sustainable operation of the supply chain. Firstly, under the single-channel financing model of retailers, retailers only obtained trade credit financing from suppliers and suppliers only obtained bank credit financing from banks. After constructing a model of the contagion intensity of credit default risk between trade credit and bank credit in the supply chain, the analysis revealed the correlation characteristics between the contagion effect of credit default risk and the influencing factors in the supply chain. Then, based on the dual-channel financing model of retailers, i.e., retailers obtaining trade credit financing and bank credit financing from suppliers and banks, respectively, and suppliers obtaining bank credit financing from banks, a model of the contagion intensity of credit default risk in the supply chain was constructed by considering the repayment priority of retailers. The impact of the retailer's repayment priority on the supply chain credit financing decision and the contagion effect of supply chain credit default risk were analyzed and obtained. The following research findings were obtained through theoretical deduction and numerical simulation:

(1) The intensity of credit default risk contagion in the supply chain under the retailer's single-channel credit model is positively related to the supplier's production cost, the commercial bank's risk-free interest rate and the trade credit interest rate. When both retailers default on trade credit to their suppliers, the contagion effect of credit default risk in the supply chain becomes more pronounced, which is detrimental to the sustainability of the supply chain. As the risk-free interest rate of commercial banks increases, the cost of bank credit to suppliers becomes larger, which will lead to an increase in the cost of trade credit to retailers. The likelihood of trade credit default by retailers then becomes larger, and the contagion effect of credit default risk in the supply chain increases. When the market retail price of a product is fixed, the higher the interest rate on trade credit, the higher the cost of trade credit to the retailer. Therefore, the retailer's profit will be reduced, which will lead to a greater likelihood of trade credit default by the retailer to the supplier, increasing the likelihood of credit default risk contagion along the supply chain. In addition, the intensity of supply chain credit default risk contagion is negatively correlated with the market retail price of the product, and it is clear that the intensity of credit default risk contagion along the supply chain is greater when both retailers default on trade credit to their suppliers.

(2) Under the retailer's dual-channel credit model, there is a conflict between the retailer's priority for paying back the supplier and the supplier's wholesale price. The greater the probability that the retailer will prioritize the repayment of the supplier's trade credit debt, the lower the supplier's wholesale price will be. There is a positive relationship between the retailer's repayment priority to suppliers and the retailer's

order quantity; i.e., the higher the probability that the retailer will repay the supplier's trade credit debt on a priority basis, the higher the retailer's order quantity will be. The retailer's repayment priority positively affects the retailer's bank credit rate, but inversely affects the supplier's bank credit rate. A retailer's repayment priority is negatively related to the intensity of credit default risk contagion in the supply chain. The lower the probability that a retailer has priority in repaying its suppliers' trade credit debts, the stronger the contagion effect of credit default risk in the supply chain. In addition, the lower the retailer's bank credit limit and the higher the trade credit limit, the stronger the contagion effect of credit default in the supply chain.

The findings of this paper can provide a theoretical reference for reducing the risk of credit default in supply chains under different credit models and utilizing the benign credit financing function of commercial banks. Moreover, the analysis of the correlation characteristics among the influencing factors in the process of credit default risk contagion in the supply chain can provide policymakers with ideas for policy formulation. Finally, as globalization continues to accelerate, the risk of supply chain credit default under different credit models may spread to many countries, and the findings of this paper have important implications for commercial banks and the state to strengthen supply chain financial risk management and maintain national financial security.

Author Contributions: Writing—original draft preparation, Y.W.; writing—review and editing, Y.W.; Model construction and mathematical simulation, J.S. and J.P.; model building design and guidance, T.C.; funding acquisition, T.C. All authors have read and agreed to the published version of the manuscript.

Funding: This work was supported by the National Natural Science Foundation of China (Nos. 71871115), the Key project of Jiangsu Social Science Foundation (No. 22WTA-019), the Major Project of Philosophy and Social Science Research in Colleges and Universities in Jiangsu Province (No. 2019SJZDA035), Young and Middle-aged Academic Leaders of Qinglan Project in Jiangsu Province.

Conflicts of Interest: The authors declare that they have no competing interests.

References

1. Lamoureux, M. A supply chain finance prime. *Supply Chain Financ.* **2007**, *4*, 34–48.
2. Xiao, S.; Sethi, S.P.; Liu, M.; Ma, S. Coordinating contracts for a financially constrained supply chain. *Omega* **2017**, *72*, 71–86. [CrossRef]
3. Ouyang, L.Y.; Chang, C.T.; Teng, J.T. An EOQ model for deteriorating items under trade credits. *J. Oper. Res. Soc.* **2005**, *56*, 719–726. [CrossRef]
4. Ge, Y.; Qiu, J. Financial development, bank discrimination and trade credit. *J. Bank. Financ.* **2007**, *31*, 513–530. [CrossRef]
5. Yang, S.A.; Birge, J.R. Trade credit, risk sharing, and inventory financing portfolios. *Manag. Sci.* **2018**, *64*, 3667–3689. [CrossRef]
6. Qu, S.; Xu, L.; Mangla, S.K.; Chan, F.T.; Zhu, J.; Arisian, S. Matchmaking in reward-based crowdfunding platforms: A hybrid machine learning approach. *Int. J. Prod. Res.* **2022**, 1–21. [CrossRef]
7. Hofmann, E. Supply chain finance: Some conceptual insights. In *Beiträge zu Beschaff. Und Logist*; Springer: Wiesbaden, Germany, 2005; Volume 16, pp. 203–214.
8. Evans, J.; Koch, T. Surviving chapter 11: Why small firms prefer supplier financing. *J. Econ. Financ.* **2007**, *31*, 186–206. [CrossRef]
9. Bates, T.W.; Kahle, K.M.; Stulz, R.M. Why Do U.S. Firms Hold so Much More Cash than They Used to? *J. Financ.* **2009**, *64*, 1985–2021. [CrossRef]
10. Gao, G.X.; Fan, Z.P.; Fang, X.; Lim, Y.F. Optimal Stackelberg strategies for financing a supply chain through online peer-to-peer lending. *Eur. J. Oper. Res.* **2018**, *267*, 585–597. [CrossRef]
11. Chen, P.; Wang, C.; Liu, Y. Real estate prices and firm borrowings: Micro evidence from China. *China Econ. Rev.* **2015**, *36*, 296–308. [CrossRef]
12. Data, M.; Hu, Q. Financing newsvendor inventory. *Oper. Res. Lett.* **2008**, *36*, 569–573. [CrossRef]
13. Lai, G.; Debo, L.G.; Sycara, K. Sharing inventory risk in supply chain: The implication of financial constraint. *Omega* **2009**, *37*, 811–825. [CrossRef]
14. Albert, R.K. Economic aspects of inventory and financing. *Law Contemorary Probl.* **1948**, *13*, 556–578.
15. Buzacott, J.A.; Zhang, R.Q. Inventory management with asset-based financing. *Manag.Sci.* **2004**, *50*, 1274–1292. [CrossRef]
16. Yan, N.; Sun, B. Coordinating loan strategies for supply chain financing with limited credit. *OR Spectr.* **2013**, *35*, 1039–1058. [CrossRef]

17. Jin, Y.; Wang, S.; Hu, Q. Contract type and decision right of sales promotion in supply chain management with a capital constrained retailer. *Eur. J. Oper. Res.* **2015**, *240*, 415–424. [CrossRef]
18. Yan, N.; Sun, B.W.; Zhang, H.; Liu, C.Q. A partial credit guarantee contract in a capital-constrained supply chain: Financing equilibrium and coordinating strategy. *Int. J. Prod. Econ.* **2016**, *173*, 122–133. [CrossRef]
19. Tang, C.S.; Yang, S.A.; Wu, J. Sourcing from suppliers with financial constraints and performance risk. *Manuf. Serv. Oper. Manag.* **2018**, *20*, 70–84. [CrossRef]
20. Goyal, S.K. Economic order quantity under conditions of permissible delay in payments. *J. Oper. Res. Soc.* **1985**, *36*, 335–338. [CrossRef]
21. Kouvelis, P.; Zhao, W. The Newsvendor Problem and Price-Only Contract When Bankruptcy Costs Exist. *Prod. Oper. Manag.* **2011**, *20*, 921–936. [CrossRef]
22. Wang, W.C.; Teng, J.T.; Lou, K.R. Seller's optimal credit period and cycle time in a supply chain for deteriorating items with maximum lifetime. *Eur. J. Oper. Res.* **2014**, *232*, 315–321. [CrossRef]
23. Jaggi, C.K.; Yadavalli VS, S.; Verma, M.; Sharma, A. An eoq model with allowable shortage under trade credit in different scenario. *Appl. Math. Comput.* **2015**, *252*, 541–551. [CrossRef]
24. Maiti, A.K.; Maiti, M.K.; Maiti, M. Inventory model with stochastic lead-time and price dependent demand incorporating advance payment. *Appl. Math. Model.* **2009**, *33*, 2433–2443. [CrossRef]
25. Jiang, L.; Hao, Z. Alleviating supplier's capital restriction by two-order arrangement. *Oper. Res. Lett.* **2014**, *42*, 444–449. [CrossRef]
26. Peng, H.; Pang, T. Financing Strategies for a Capital-Constrained Supplier under Yield Uncertainty. *J. Ind. Manag. Optim.* **2020**, *16*, 887–909. [CrossRef]
27. Martin, J. Suppliers' participation in supply chain finance practices: Predictors and outcomes. *Int. J. Integr. Supply Manag.* **2017**, *11*, 193–216. [CrossRef]
28. Jorion, P.; Zhang, G. Credit contagion from counterparty risk. *J. Financ.* **2009**, *64*, 2053–2087. [CrossRef]
29. Chen, X.; Wang, A. Trade credit contract with limited liability in the supply chain with budge constraints. *Ann. Oper. Res.* **2012**, *196*, 153–165. [CrossRef]
30. Garvey, M.D.; Carnovale, S.; Yeniyurt, S. An Analytical Framework for Supply Network Risk Propagation: A Bayesian Network Approach. *Eur. J. Oper. Res.* **2015**, *243*, 618–627. [CrossRef]
31. Tang, L.; Jing, K.; He, J.; Stanley, H.E. Complex Interdependent Supply Chain Networks: Cascading Failure and Robustness. *Phys. A Stat. Mech. Appl.* **2016**, *443*, 58–69. [CrossRef]
32. Li, Y.; Zhen, X.; Cai, X. Trade credit insurance, capital constraint, and the behavior of manufacturers and banks. *Ann. Oper.Res.* **2016**, *240*, 395–414. [CrossRef]
33. Jing, B.; Seidmann, A. Finance sourcing in a supply chain. *Decis. Support Syst.* **2014**, *58*, 15–20. [CrossRef]

sustainability

MDPI

Article

Data-Driven Robust Data Envelopment Analysis for Evaluating the Carbon Emissions Efficiency of Provinces in China

Shaojian Qu [1], Yuting Xu [1], Ying Ji [2,*], Can Feng [1], Jinpeng Wei [1] and Shan Jiang [1]

[1] School of Management Science and Engineering, Nanjing University of Information Science and Technology, Nanjing 210044, China

[2] School of Management, Shanghai University, Shanghai 200444, China

* Correspondence: jiying_1981@126.com

Abstract: To combat global warming, China proposed the "dual carbon" policy in 2020. In this context, it becomes crucial to improve carbon emissions efficiency. Currently, some scholars have utilized data envelopment analysis (DEA) to study carbon emissions efficiency. However, uncertainty about climate and government economic policy is ignored. This paper establishes a robust DEA model to reduce uncertainty and improve robustness. First, robust optimization theory is combined with DEA to establish the robust DEA model. Second, considering three uncertainty sets (box set, ellipsoid set, and polyhedron set), a robust DEA model for different situations is considered. Finally, to address the problem of over-conservatism in robust optimization, this paper applies the data-driven robust DEA model to further analyze the carbon emissions efficiency of China. The results of the data-driven robust DEA model suggest that the government should focus on coordinated regional development, promote the transformation and upgrading of the energy structure, innovate in green technology, and advocate for people to live a green and low-carbon lifestyle.

Keywords: carbon emissions efficiency; robust optimization; data envelopment analysis (DEA); data-driven; uncertainty set

Citation: Qu, S.; Xu, Y.; Ji, Y.; Feng, C.; Wei, J.; Jiang, S. Data-Driven Robust Data Envelopment Analysis for Evaluating the Carbon Emissions Efficiency of Provinces in China. *Sustainability* **2022**, *14*, 13318. https://doi.org/10.3390/su142013318

Academic Editor: Micro Peron

Received: 4 August 2022
Accepted: 29 September 2022
Published: 17 October 2022

1. Introduction

Climate anomalies and frequent occurrence of extreme weather are global problems faced by mankind. With the continuous emission of carbon dioxide in various countries in the world, greenhouse gases have soared, posing a major threat to life systems [1]. A recent study by Huang et al. [2] found that under a 2 °C global warming scenario, mortality rates associated with extreme temperatures would rise significantly. With the reform and opening up [3], China's economic level has been continuously improving, but at the same time, environmental problems have become more and more prominent. China is currently the largest carbon emitter in Asia and the world, with annual emissions of nearly 10 million tons, more than a quarter of global emissions [4]. In this context, China has proposed the "dual carbon" policy, that is, to achieve a carbon peak by 2030 and carbon neutrality by 2060 [5].

Carbon emissions efficiency indicates the amount of CO_2 that must be emitted to generate each unit of economic output [6]. Yamaji et al. [7] proposed carbon productivity for the first time. In the context of the "dual carbon" policy, improving carbon emissions efficiency becomes crucial. We plan to study carbon emissions efficiency in China in order to find the factors influencing carbon emissions efficiency. Previous scholars have adopted various methods to measure carbon emissions efficiency, and have achieved certain research results. There are two main methods for evaluating carbon emissions efficiency; one is the parametric method and the other is the non-parametric method. The former is represented by stochastic frontier analysis (SFA) and the latter is represented by DEA. SFA was first proposed by Aigner et al. [8] and Meeusen and Broeck [9] in 1977. It takes into account that the production frontier will be affected by random factors, sets the specific function form of the production frontier, and uses statistical methods to estimate the frontier production function. In recent years, some scholars have used SFA to evaluate the carbon emissions efficiency. Sun et al. [10] evaluated greenhouse gas emissions efficiency in China's industry based on SFA. Sun and Huang [11] applied SFA to explore the impact of urbanization on carbon emissions efficiency. Zhang et al. [12] applied SFA to discuss the relationship between economic development and carbon emissions efficiency. However, SFA needs to set the function form of the SFA model and make assumptions about the distribution of the error term. If the assumption is improper, it will cause estimation deviation [13]. In addition, some scholars have applied the DEA method to green development efficiency [14] and to R&D activity efficiency [15].

DEA was first proposed in 1978 by Charnes et al. [16]. The classic DEA models are the CCR model and the BCC model [17]. Zhang et al. [6] applied the three-stage DEA-Tobit model to study carbon emission efficiency in the Chinese construction industry. Meng et al. [18] applied DEA model to measure China's regional energy and carbon emissions efficiency. Cheng et al. [19] used an improved non-radial directional distance function (NDDF) to measure the carbon emissions efficiency of China's provincial industrial sectors. Yan et al. [20] used the SBM model considering undesired output to study the carbon emissions efficiency of China's thermal power industry.

However, previous studies on carbon emissions efficiency have ignored the uncertainty of the climate and governmental economic policies. Ben-Tal [21] pointed out that small perturbations can lead to large errors and even lead to infeasible solutions for the model. It can be seen that it is necessary to consider robust optimization when the environment is uncertain. Robust optimization is now widely used in many fields [22–24], such as the consensus field [25–30], supply chain management [31,32], the energy field [33], etc. Some scholars have combined robust optimization and DEA [34]. Sadjadi and Omrani adopted robust DEA for performance assessment of electricity distribution companies [35]. However, to the best of our knowledge, few scholars have applied robust DEA to the field of carbon emissions efficiency.

In addition, the existing robust optimization models have the drawback of being too conservative. Nowadays, in the era of big data, a data-driven approach can be adopted to alleviate this drawback. On the one hand, the robustness of the original robust model can be retained, and on the other hand, these models are closer to reality and not as conservative. The data-driven approach has been used in some fields, such as consensus aspects [26,36] and energy systems [37]. Some scholars have combined robust DEA and a data-driven approach [38]. However, to the best of our knowledge, few scholars have applied data-driven robust DEA to the field of carbon emissions efficiency.

Although the existing research on carbon emissions efficiency promotes our understanding of this field, there are several deficiencies.

1. Few of the original carbon emissions efficiency studies have considered climate uncertainty and government economic policy uncertainty;
2. The original robust optimization model is too conservative and deviates somewhat from the actual situation.

The main contributions of this paper lie in the following aspects:

1. Due to the uncertainty of climate and government economic policies, robust DEA models are proposed to analyze the treatment;
2. This paper proposes a data-driven robust DEA model to address the inherent over-conservatism of robust optimization;
3. This paper uses real data to study the carbon emissions efficiency of China to verify the validity of the model.

The rest of the paper is organized as follows. Section 2 reviews the literature related to this research. Section 3 introduces the DEA model and builds the robust DEA model. Then, Section 4 presents empirical analysis and discussion. Section 5 applies the data-driven robust DEA model to further analyze the carbon emissions efficiency of China. Finally, Section 6 presents the conclusions. Detailed proof is displayed in Appendix A.

2. Literature Review

2.1. Carbon Emissions Efficiency

Carbon emissions efficiency indicates the amount of carbon dioxide that needs to be emitted when generating each unit of economic output.

Carbon emissions efficiency is the focus of current research by scholars. Carbon emissions efficiency is mainly calculated from two perspectives. One is the single-factor perspective and the other is the full-factor perspective. The single-factor perspective defines carbon emission efficiency as the ratio of two variables. There are two main single-factor carbon emission efficiency indicators. One is carbon productivity, which was first proposed by Kaya et al. in 1993 [7]. Carbon productivity is the ratio of GDP to CO_2 emissions, i.e., GDP per unit of CO_2 emissions. This indicator is based on GDP when measuring carbon efficiency. The second is the carbon index, which was proposed by Mielnik and Goldemberg [39]; its meaning is the carbon emissions per unit of energy consumption, that is, the ratio of total carbon emissions to total energy consumption, to measure carbon emission efficiency.

The above single-factor carbon emission efficiency indicators are simple to calculate and easy to understand, but they are controversial due to the large number of indicators they measure. For this reason, scholars have started to define carbon emission efficiency from the perspective of all factors. The full-factor perspective includes two approaches. One is the parametric method and the other is the non-parametric method. The former is represented by SFA and the latter is represented by DEA.

2.2. The Evolution of the Data-Driven Robust DEA Model

How to calculate carbon emissions efficiency is also a hot topic. Previous researchers have utilized numerous approaches to quantify carbon emissions efficiency and got specific study outcomes. The methods to assess the efficiency of carbon emissions can be divided into parametric and non-parametric methods. The former is represented by stochastic frontier analysis (SFA) [8,9], whereas the latter is represented by discrete event analysis (DEA) [16]. In recent years, several researchers have employed the SFA model to assess carbon emissions efficiency [10–12]. However, the SFA must define the function form of the SFA model and make assumptions about the error term distribution. If the assumption is incorrect, estimate deviation will occur [13].

DEA was first proposed in 1978 by Charnes et al. [16]. Cheng et al. [19] used an improved non-radial directional distance function (NDDF) to estimate the meta-frontier TCEI of China's 30 provincial industrial sectors in 2005–2015 and analyzed their dynamic evolution. Ren et al. (2020) [40] contributed to measuring the energy and carbon emission efficiency (ECEE) of regional transportation systems (RTS) in China considering uncertain carbon emissions.

However, previous studies on carbon emissions efficiency have ignored the uncertainty of climate and government economic policies. Ben-Tal [21] pointed out that small perturbations can lead to large errors, and can even lead to no feasible solution for the model. Robust optimization is now widely used in many fields [22–24], such as the consensus field [25–30], supply chain management [31,32], the energy field [33], etc.

Furthermore, the drawback of present robust optimization models is that they are overly conservative. A data-driven approach can be used to mitigate this disadvantage. Some fields have used the data-driven approach, such as consensus aspects [26,36] and energy systems [37]. Some scholars have combined robust DEA and data-driven approaches [38]. However, to the best of our knowledge, few scholars have applied data-driven robust DEA to the field of carbon emissions efficiency.

2.3. General Comment

In general, the previous studies provide an important reference value for this literature, but there are still some deficiencies. Few of the original carbon emissions efficiency studies considered climate uncertainty and government economic policy uncertainty. The original robust optimization model is too conservative and deviates somewhat from the actual situation. Therefore, this paper constructs robust DEA to study the carbon emissions efficiency problem. Subsequently, in order to overcome the prsoblem of over-conservative robust optimization, this paper constructs a data-driven robust optimization model to further study the carbon emissions efficiency problem.

3. Model Construction

3.1. DEA

Suppose that there are n $DMUs$ (decision making units), and $DMU_j(j = 1, 2, \cdots, n)$ has m inputs $x_{ij}(i = 1, 2, \cdots, m)$ and s outputs $y_{rj}(r = 1, 2, \cdots, s)$. u_r, v_i are the vectors of weight coefficients for the input and output indicators, respectively. We assume that these data are positive numbers. This paper selects labor, energy consumption and capital stock as inputs, and real GDP and carbon emissions as outputs. The DEA model is shown below:

$$\begin{aligned} \max \quad & \theta = \frac{\sum_{r=1}^{s} u_r y_{ro}}{\sum_{i=1}^{m} v_i x_{io}} \\ s.t. \quad & \frac{\sum_{r=1}^{s} u_r y_{rj}}{\sum_{i=1}^{m} v_i x_{ij}} \leq 1, j = 1, \cdots, n \\ & u_r, v_i \geq 0, r = 1, 2, \cdots, s, i = 1, 2, \cdots, m \end{aligned} \tag{1}$$

Through the Charnes–Cooper transformation, this can be transformed into a linear programming model, the objective function of which is a linear function as follows:

$$\begin{aligned} \max \quad & \theta = \sum_{r=1}^{s} u_r y_{ro} \\ s.t. \quad & \sum_{r=1}^{s} u_r y_{rj} - \sum_{i=1}^{m} v_i x_{ij} \leq 0, j = 1, \cdots, n \\ & \sum_{i=1}^{m} v_i x_{ij} = 1 \\ & u_r, v_i \geq 0, r = 1, 2, \cdots, s, i = 1, 2, \cdots, m \end{aligned} \tag{2}$$

In this context, θ refers to the effectiveness of decision-making units. Simultaneously, we use "$\leq$" in our DEA model rather than "=" in the standard DEA model. We make this change to avoid any incompatibility.

Definition 1 ([16]). *If the optimal value of Model* (2) *is 1, the evaluated decision-making unit DMU_j is valid; on the contrary, if the optimal value is less than 1, the evaluated decision-making unit DMU_j is not valid.*

Definition 2 ([16]). *We shall represent the production possibility set as:*

$$T = \{(X, Y) | Y \geq 0 \ can\ be\ produced\ from\ X \geq 0\} \tag{3}$$

According to Banker's research [17], the production possible set (PPS) needs to satisfy the following four assumptions:

Assumption 1. *Convexity: if* $(X_j, Y_j) \in T, j = 1, \cdots, n$ *and* $\lambda_j \geq 0$ *are nonnegative scalars such that* $\sum\limits_{j=1}^{n} \lambda_j = 1$, *then* $(\sum\limits_{j=1}^{n} \lambda_j X_j, \sum\limits_{j=1}^{n} \lambda_j Y_j) \in T$.

Assumption 2. *Inefficiency postulate: (a) if* $(X, Y) \in T$ *and* $\overline{X} \geq X$, *then* $(\overline{X}, Y) \in T$. *(b) If* $(X, Y) \in T$ *and* $\overline{Y} \leq Y$, *then* $(X, \overline{Y}) \in T$.

Assumption 3. *Ray unboundedness: if* $(X, Y) \in T$, *then* $(kX, kY) \in T, \forall k \geq 0$.

Assumption 4. *Minimum extrapolation: T is the intersection set of all satisfying Assumptions 1–3, subject to the condition that each of the observed vectors* $(X, Y) \in \hat{T}, j = 1, \cdots, n$.

In an uncertain economic environment, the outputs of decision-making units are often uncertain. This paper introduces robust optimization into the DEA model. Robust DEA considering uncertainty is proposed.

Next, based on the above model, this paper considers three situations, namely, the box set, ellipsoid set, and polyhedron set. Then, we consider the corresponding uncertainty sets.

$$\begin{aligned} \max \quad & \theta \\ s.t. \quad & \theta - \sum_{r=1}^{s} u_r \left(y_{ro} + \sum_{l=1}^{L} y_{rol}^F \xi_l \right) \leq 0 \\ & \sum_{i=1}^{m} v_i x_{io} \leq 1 \\ & \sum_{r=1}^{s} u_r \left(y_{rj} + \sum_{l=1}^{L} y_{rjl}^F \xi_l \right) - \sum_{i=1}^{m} v_i x_{ij} \leq 0, \forall j \\ & u_r, v_i \geq 0, r = 1, 2, \cdots, s, i = 1, 2, \cdots, m \end{aligned} \tag{4}$$

3.2. RDEA Based on Uncertain Outputs with Box Set

Definition 3. *The box uncertainty set is defined as:*

$$Z^{Box} = \left\{ \xi \in R^L : \|\xi\|_\infty \leq \tau \right\} \tag{5}$$

where τ *is the uncertainty parameter that is used to assess the uncertainty of climate and government economic policies.*

Theorem 1. *The robust DEA model using* Z^{Box} *may then be built as follows:*

$$\begin{aligned} \max \quad & \theta \\ s.t. \quad & \theta + \tau \sum_{l=1}^{L} \sum_{r=1}^{s} u_r y_{rol}^F - \sum_{r=1}^{s} u_r y_{ro} \leq 0 \\ (BRDEA) \quad & \sum_{i=1}^{m} v_i x_{io} \leq 1 \\ & \tau \sum_{l=1}^{L} \sum_{r=1}^{s} u_r y_{rjl}^F + \sum_{r=1}^{s} u_r y_{rj} - \sum_{i=1}^{m} v_i x_{ij} \leq 0, \forall j \\ & u_r, v_i \geq 0, r = 1, 2, \cdots, s, i = 1, 2, \cdots, m \end{aligned} \tag{6}$$

The proof is in Appendix A.1.

3.3. RDEA Based on Uncertain Outputs with Ellipsoid Set

We now examine the model based on the ellipsoid set.

Definition 4. *The ellipsoid uncertainty set is defined as:*

$$Z^{Ellipsoid} = \left\{ \xi \in R^L : \|\xi\|_2 \leq \Omega \right\} \tag{7}$$

where Ω *is the uncertainty parameter that is used to assess the uncertainty of climate and government economic policies.*

Theorem 2. *The robust DEA model using* $Z^{Ellipsoid}$ *may then be built as follows:*

$$\begin{aligned}
& \max \quad \theta \\
& \text{s.t.} \quad \theta + \Omega \sqrt{\sum_{l=1}^{L} \left(\sum_{r=1}^{s} u_r y_{rol}^F \right)^2} - \sum_{r=1}^{s} u_r y_{ro} \leq 0 \\
(ERDEA) \quad & \sum_{i=1}^{m} v_i x_i \leq 1 \\
& \Omega \sqrt{\sum_{l=1}^{L} \left(\sum_{r=1}^{s} u_r y_{rjl}^F \right)^2} + \sum_{r=1}^{s} u_r y_{rj} - \sum_{i=1}^{m} v_i x_{ij} \leq 0, \forall j \\
& u_r, v_i \geq 0, r = 1, 2, \cdots, s, i = 1, 2, \cdots, m
\end{aligned} \tag{8}$$

The proof is in Appendix A.2.

3.4. RDEA Based on Uncertain Outputs with Polyhedron Set

Last, we consider the polyhedron uncertainty set.

Definition 5. *Let us define the polyhedron set as:*

$$Z^{Polyhedron} = \left\{ \xi \in R^L : \|\xi\|_1 \leq \Gamma \right\} \tag{9}$$

where Γ *is the uncertainty parameter that is used to assess the uncertainty of climate and government economic policies.*

Theorem 3. *Then, the robust DEA model with* $Z^{Polyhedron}$ *can be constructed as:*

$$\begin{aligned}
& \max \quad \theta \\
& \text{s.t.} \quad \theta + \Gamma \sum_{l=1}^{L} p_{ol} - \sum_{r=1}^{s} u_r y_{ro} \leq 0 \\
(PRDEA) \quad & \sum_{i=1}^{m} v_i x_{io} \leq 1 \\
& \Gamma \sum_{l=1}^{L} p_{ol} + \sum_{r=1}^{s} u_r y_{rj} - \sum_{i=1}^{m} v_i x_{ij} \leq 0, \forall j \\
& u_r, v_i \geq 0, r = 1, 2, \cdots, s, i = 1, 2, \cdots, m
\end{aligned} \tag{10}$$

The proof is in Appendix A.3.

Based on the above three theorems, the following theorems are proposed.

Theorem 4. *The larger the uncertainty level parameter, the smaller the efficiency value of the robust model.*

The proof is in Appendix A.4.

Theorem 5. *Robust DEA has lower efficiency values than DEA models.*

The proof is in Appendix A.5.

Some economic implications can be seen from the above conclusions; in reality, the decision-making unit may have certain risks when making decisions. For example, when governments make economic decisions, there may be some climate and government economic policy uncertainty. At this time, it can help to improve the robustness of the model. In addition, the government can also choose different parameters according to its own situation to adjust the robustness of the model. In this way, the government can achieve a balance between the economy and the environment under the background of the "dual-carbon" policy.

4. Empirical Analysis and Discussions

4.1. Study Area

In order to scientifically reflect the economic development status of various parts of Chinese society, China's economic regions are divided into four categories, namely eastern, central, western, and northeastern regions. This study will examine the differences in carbon emissions efficiency between different regions from these four economic regions. Table 1 and Figure 1 shows the geographical distribution of these four major regions.

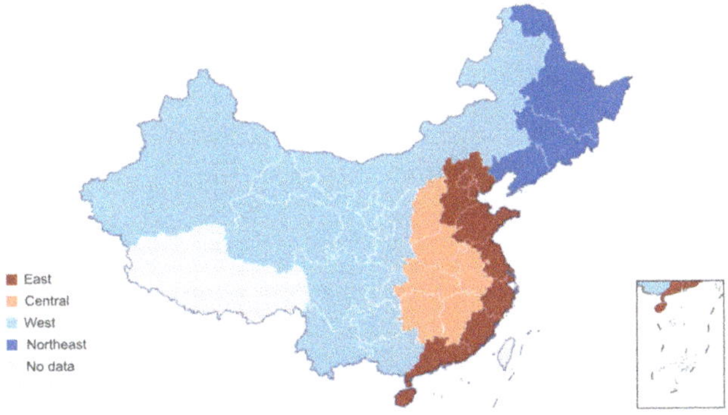

Figure 1. Geographical distribution of the four major regions in China.

Table 1. Geographical distribution of the four major regions in China.

Region	Province
East	Beijing, Tianjin, Hebei, Shandong, Shanghai, Jiangsu, Guangdong, Zhejiang, Fujian, Hainan
Central	Anhui, Henan, Hubei, Hunan, Jiangxi, Shanxi
West	Shaanxi, Ningxia, Sichuan, Qinghai, Gansu, Guizhou, Yunnan, Chongqing, Guangxi, Inner Mongolia, Xinjiang
Northeast	Liaoning, Jilin, Heilongjiang

4.2. Variable Description and Data Sources

4.2.1. Variable Description

To measure the carbon emissions efficiency of 30 provinces in mainland China, this paper selects labor, energy consumption, and capital stock as inputs, and real GDP and carbon emissions as outputs [6]. The details are as follows.

1. Labor. The sum of employees in the three major industries in each region is used as the labor input.
2. Energy consumption. The total energy consumption of each region is used as the energy consumption input.
3. Capital. This study uses the "perpetual inventory method" proposed by Goldsmith [41] in 1951 to calculate the capital stock. The formula for calculating capital stock is
$$K_{i,t} = I_{i,t} + (1 - \delta_{i,t})K_{i,t-1} \tag{11}$$
where K represents the capital stock, I represents the gross capital formation; δ represents the depreciation rate of capital stock, which is set to 9.6% in this research [42]; i represents the province; and t stands for the year.
4. Real GDP. This is calculated by converting nominal GDP and price indices for each provinces.
5. Carbon emissions. The carbon emissions comes from the CEAD database [43,44], which is a carbon accounting database in China. It provides a solid theoretical basis and technical support for China to achieve green and low-carbon development, and contributes to policy design and implementation of greenhouse gas emissions control in China.

4.2.2. Data Sources

This study examined the carbon emissions of 30 provinces in mainland China (all except Tibet, due to the absence of relevant energy data) from 2000 to 2019. The statistical data in this study come from the China Statistical Yearbook, China Energy Statistical Yearbook, China Provincial Statistical Yearbook, and CEAD database [43,44]. Some missing values were filled by interpolation.

4.3. Discussion

In this section, firstly, carbon emissions efficiency evaluation analysis is discussed. Secondly, the influence of some parameters on the model is analyzed, Finally, a comparative analysis is performed with existing DEA methods.

Carbon Emissions Efficiency Evaluation Analysis

This section considers the BRDEA model (τ = 1) to calculate the corresponding values.

As can be seen from Figure 2, the carbon emissions efficiency is generally between 0.6 and 0.9, and the efficiency values are all below 1. It can be seen that most of the efficiency values are low and that the carbon emissions efficiency has not reached the effective level.

From Figure 2, it can be seen that carbon emissions efficiency varies widely among different provinces.

Some places have higher carbon emissions efficiency (greater than 0.79), for example, Shanghai, Zhejiang, Jiangsu, Fujian, and Guangdong. These places have higher levels of economic development, a higher level of urbanization, and a more reasonable industrial structure. In addition, these places have more universities, greater talent and labor forces, higher management levels, and higher science and technology levels. There are also some provinces such as Heilongjiang and Inner Mongolia where the carbon emissions efficiency is also relatively high. Given the geographical location and climatic conditions of these places, tertiary industry in these places does not account for much GDP and there are not many high pollution and high energy consumption enterprises, so their carbon emissions efficiency is also higher.

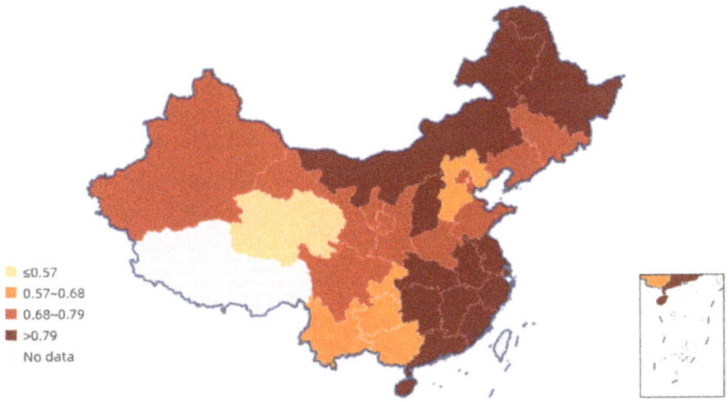

Figure 2. Geographical distribution of mean carbon emissions efficiency of China from 2000 to 2019 in robust DEA model.

Unlike high-efficiency provinces, some places have medium carbon emissions levels, between 0.68–0.79, and these places are widely distributed, including Jilin, Liaoning, Shandong, Xinjiang, etc. Most of these places have medium levels of various indicators.

In addition, some places have a slightly lower level of carbon emissions efficiency, below 0.68, mainly in the western region. These regions have a low level of economic development and a small labor force. With the transfer of industries from the east, a large number of high pollution and high energy consumption enterprises have been transferred to the west, which on the one hand promotes economic development and on the other hand causes environmental pollution. While developing, these places should focus on innovation and combine their own characteristics to develop their economy, develop new energy, and achieve high-quality development.

In order to further analyze the differences in the distribution of carbon emissions efficiency among regions in China, the next step is to analyze the differences in carbon emissions efficiency in China from four regions, one by one.

Figure 3 shows the clear differences between the four regions in recent years. The eastern and central regions are flat, and the northeastern and western regions are lower.

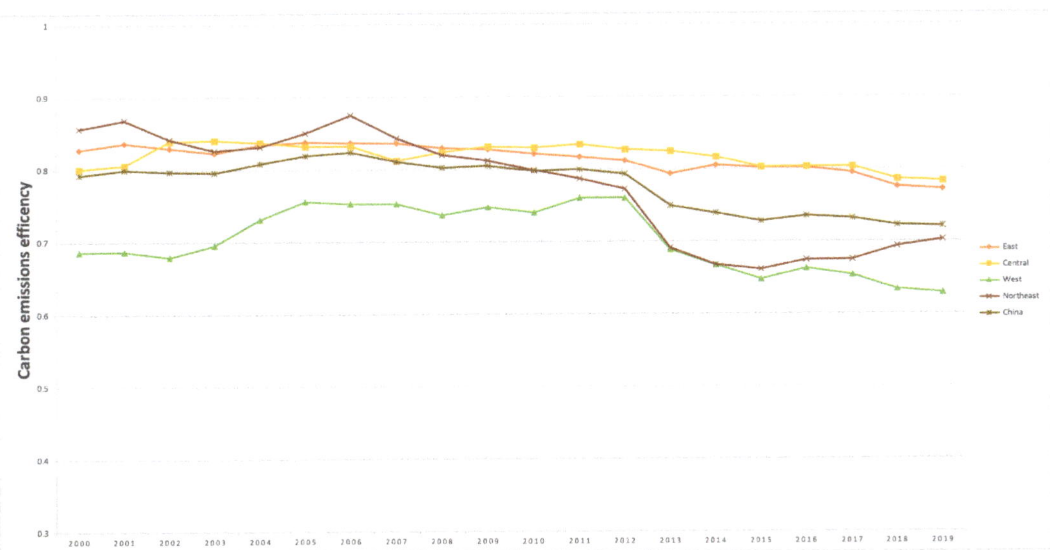

Figure 3. Trends of carbon emissions efficiency for four regions in China from 2000 to 2019.

As a result of the reform and opening up, the eastern region has a higher level of economic development and a higher degree of openness to the outside world, and is, therefore, more efficient in terms of carbon emissions, which is also related to China's current regional policy of "taking the lead in the east".

The carbon emissions efficiency of the central region is around 0.8. Since the central region is located inland, the level of openness needs to be improved. The development of the central region is unbalanced and insufficient, the innovation capability of the manufacturing industry needs to be enhanced, and ecological and green development needs to be improved.

The northeast region belongs to the old industrial region, which is the cradle of China's industrial development and has a long history of industrial development. However, the industrial structure of the eastern region is more homogeneous, and the loss of talent and labor force is more serious.

Due being deep inland, the western region has a low level of openness to the outside world, slow development of transportation, and a low technology level. In addition, in recent years, with the implementation of the western opening strategy, many industries in the eastern coastal areas have been transferred to the western region, but at the same time, this has also brought about environmental problems. Some high pollution, high energy consumption enterprises transferred to the western region have, to a certain extent, caused energy consumption and environmental pollution problems. The self-development ability of the western region is not high. At present, with the development of China's rural revitalization, the western region should combine its characteristics to achieve high-quality development.

4.4. Sensitivity Analysis

Sensitivity analysis is a typical method to quantify the effect of parameter uncertainty on the overall simulation/prediction of uncertainty.

4.4.1. The Effects of Uncertain Parameters in Box Set

This paper sets the uncertainty parameter τ from 1 to 3, and then analyzes the impact of these changes on the RDEA model. As shown in Figure 4 and Table 2, as τ increases, the efficiency value gradually becomes smaller and more conservative, which also confirms Theorem 4. It can be seen that robust optimization considers the best case in the worst case, and is relatively conservative.

Figure 4. The influence of uncertain parameters τ on the model in 2019.

Table 2. The influence of uncertain parameters τ on the model in 2019.

Provinces	Carbon Emissions Efficiency				
	$\tau = 1$	$\tau = 1.5$	$\tau = 2$	$\tau = 2.5$	$\tau = 3$
Beijing	0.778	0.732	0.689	0.648	0.609
Tianjin	0.605	0.570	0.536	0.504	0.474
Hebei	0.596	0.596	0.561	0.528	0.467
Shanxi	0.887	0.835	0.786	0.739	0.695
Inner Mongolia	0.887	0.835	0.786	0.739	0.695
Liaoning	0.736	0.693	0.652	0.614	0.577
Jilin	0.648	0.610	0.574	0.540	0.507
Heilongjiang	0.721	0.679	0.639	0.601	0.565
Shanghai	0.887	0.835	0.786	0.739	0.695
Jiangsu	0.842	0.793	0.746	0.702	0.660
Zhejiang	0.874	0.823	0.774	0.728	0.685
Anhui	0.800	0.753	0.709	0.667	0.627
Fujian	0.887	0.835	0.786	0.739	0.695
Jiangxi	0.774	0.729	0.686	0.645	0.607
Shandong	0.693	0.652	0.614	0.577	0.543
Henan	0.712	0.670	0.631	0.593	0.558
Hubei	0.758	0.714	0.672	0.632	0.594
Hunan	0.764	0.719	0.677	0.637	0.599
Guangdong	0.808	0.761	0.716	0.673	0.633
Guangxi	0.540	0.540	0.479	0.450	0.424
Hainan	0.737	0.694	0.653	0.614	0.578
Chongqing	0.694	0.654	0.615	0.579	0.544
Sichuan	0.795	0.748	0.704	0.662	0.623
Guizhou	0.511	0.481	0.453	0.426	0.401
Yunnan	0.481	0.453	0.426	0.401	0.377
Shaanxi	0.721	0.679	0.639	0.601	0.565
Gansu	0.710	0.669	0.629	0.592	0.557
Qinghai	0.323	0.304	0.287	0.270	0.253
Ningxia	0.679	0.639	0.601	0.566	0.532
Xinjiang	0.566	0.533	0.502	0.472	0.444

4.4.2. The Effects of Uncertain Parameters in the Ellipsoid Set

This paper sets the uncertainty parameter Ω from 1 to 3, and then analyzes the impact of these changes on the RDEA model. As shown in Figure 5 and Table 3, as Ω increases, the efficiency value gradually becomes smaller and more conservative, which also confirms Theorem 4. It can be seen that robust optimization considers the best case in the worst case and is relatively conservative.

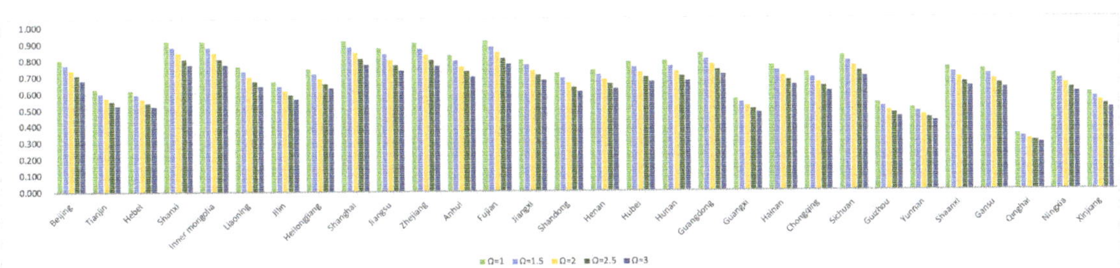

Figure 5. The influence of uncertain parameters Ω on the model in 2019.

Table 3. The influence of uncertain parameters Ω on the model in 2019.

Provinces	Carbon Emissions Efficiency				
	$\Omega = 1$	$\Omega = 1.5$	$\Omega = 2$	$\Omega = 2.5$	$\Omega = 3$
Beijing	0.805	0.772	0.740	0.709	0.679
Tianjin	0.627	0.601	0.576	0.551	0.528
Hebei	0.617	0.592	0.567	0.543	0.520
Shanxi	0.919	0.880	0.844	0.808	0.774
Inner Mongolia	0.919	0.880	0.844	0.808	0.774
Liaoning	0.763	0.731	0.700	0.671	0.643
Jilin	0.671	0.643	0.616	0.590	0.565
Heilongjiang	0.747	0.716	0.686	0.657	0.630
Shanghai	0.919	0.880	0.844	0.808	0.774
Jiangsu	0.872	0.836	0.801	0.768	0.735
Zhejiang	0.905	0.868	0.831	0.796	0.763
Anhui	0.828	0.794	0.761	0.729	0.698
Fujian	0.919	0.880	0.844	0.808	0.774
Jiangxi	0.802	0.768	0.736	0.705	0.676
Shandong	0.718	0.688	0.659	0.631	0.605
Henan	0.737	0.707	0.677	0.649	0.621
Hubei	0.785	0.752	0.721	0.691	0.662
Hunan	0.791	0.758	0.727	0.696	0.667
Guangdong	0.837	0.802	0.769	0.736	0.705
Guangxi	0.560	0.536	0.514	0.493	0.472
Hainan	0.763	0.732	0.701	0.672	0.643
Chongqing	0.719	0.689	0.660	0.633	0.606
Sichuan	0.823	0.789	0.756	0.724	0.694
Guizhou	0.530	0.507	0.486	0.466	0.446
Yunnan	0.498	0.477	0.457	0.438	0.420
Shaanxi	0.747	0.716	0.686	0.657	0.630
Gansu	0.736	0.705	0.676	0.648	0.620
Qinghai	0.335	0.321	0.308	0.295	0.282
Ningxia	0.703	0.674	0.646	0.619	0.593
Xinjiang	0.587	0.562	0.539	0.516	0.494

4.4.3. The Effects of Uncertain Parameters in the Polyhedron Set

This paper sets the uncertainty parameter Γ from 1 to 3, and then analyzes the impact of these changes on the RDEA model. As shown in Figure 6 and Table 4, as Γ increases, the efficiency value gradually becomes smaller and more conservative, which also confirms Theorem 4. It can be seen that robust optimization considers the best case in the worst case and is relatively conservative.

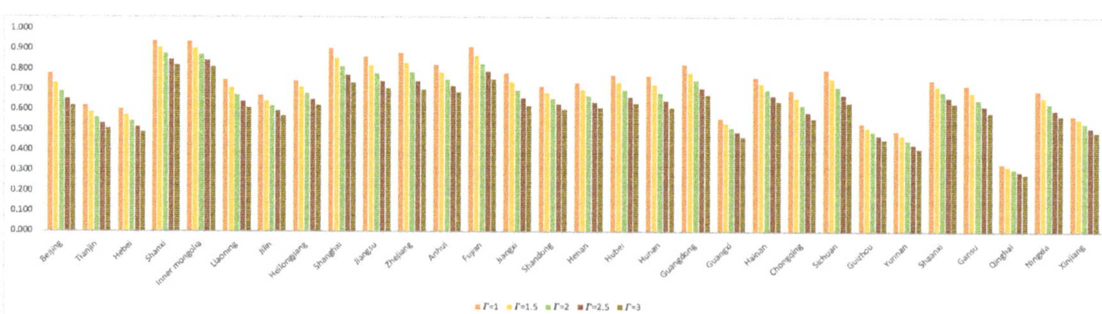

Figure 6. The influence of uncertain parameters Γ on the model in 2019.

Table 4. The influence of uncertain parameters Γ on the model in 2019.

Provinces	Carbon Emissions Efficiency				
	$\Gamma = 1$	$\Gamma = 1.5$	$\Gamma = 2$	$\Gamma = 2.5$	$\Gamma = 3$
Beijing	0.778	0.732	0.692	0.653	0.618
Tianjin	0.618	0.588	0.559	0.532	0.506
Hebei	0.604	0.573	0.543	0.515	0.488
Shanxi	0.938	0.907	0.877	0.847	0.818
Inner Mongolia	0.935	0.903	0.872	0.841	0.811
Liaoning	0.748	0.710	0.674	0.640	0.608
Jilin	0.671	0.644	0.617	0.592	0.568
Heilongjiang	0.744	0.712	0.681	0.652	0.624
Shanghai	0.902	0.857	0.813	0.772	0.732
Jiangsu	0.860	0.818	0.779	0.741	0.705
Zhejiang	0.880	0.831	0.785	0.741	0.700
Anhui	0.822	0.786	0.751	0.718	0.686
Fujian	0.910	0.867	0.827	0.788	0.751
Jiangxi	0.780	0.737	0.696	0.658	0.621
Shandong	0.716	0.685	0.656	0.629	0.602
Henan	0.733	0.700	0.668	0.638	0.610
Hubei	0.773	0.735	0.699	0.665	0.632
Hunan	0.768	0.725	0.685	0.647	0.610
Guangdong	0.824	0.783	0.745	0.708	0.673
Guangxi	0.557	0.533	0.509	0.487	0.466
Hainan	0.761	0.729	0.698	0.669	0.641
Chongqing	0.698	0.659	0.622	0.587	0.558
Sichuan	0.799	0.755	0.713	0.673	0.635
Guizhou	0.531	0.511	0.491	0.472	0.454
Yunnan	0.493	0.470	0.449	0.428	0.408
Shaanxi	0.748	0.717	0.688	0.661	0.634
Gansu	0.722	0.686	0.652	0.619	0.588
Qinghai	0.335	0.321	0.307	0.294	0.282
Ningxia	0.694	0.661	0.630	0.601	0.573
Xinjiang	0.573	0.558	0.533	0.510	0.488

Combining the above three sensitivity analyses, it can be seen that the value of the RDEA model becomes smaller and more conservative as the uncertainty parameter increases, which also verifies Theorem 4.

4.5. Comparison Analysis

In this subsection, the advantages of the proposed methods are described by comparing them with existing methods.

As shown in Figure 7 and Table 5, this paper chooses the data of carbon emissions efficiency in 2019, from which we choose Jiangsu as an example; the value of the DEA model is 0.950. The values of BRDEA ($\tau = 1$), ERDEA ($\Omega = 1$), and PRDEA ($\Gamma = 1$) are 0.842, 0.872, and 0.860, respectively. This shows that the values of the three RDEA models are smaller than the values of the DEA model, meaning that the RDEA model is more conservative and reduces the risk of the model, which also confirms Theorem 5.

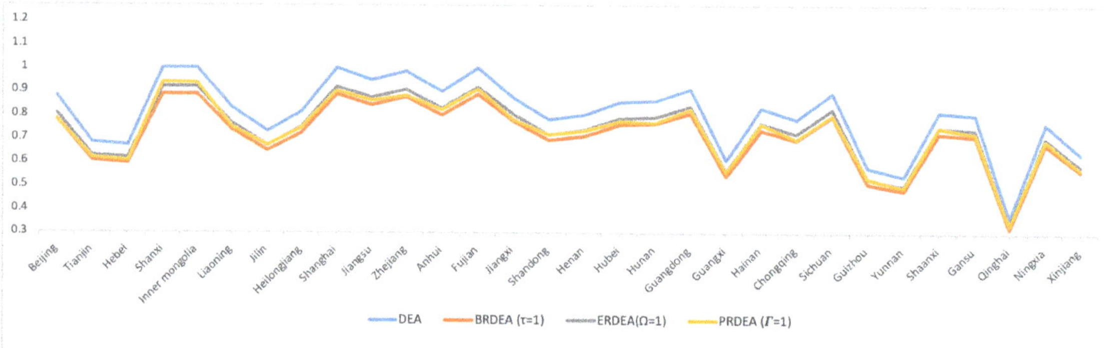

Figure 7. Comparison of the DEA model and robust DEA.

Table 5. Comparison of the DEA model and robust DEA.

Provinces	DEA	BRDEA (τ = 1)	ERDEA (Ω = 1)	PRDEA (Γ = 1)
Beijing	0.877	0.778	0.805	0.778
Tianjin	0.682	0.605	0.627	0.618
Hebei	0.672	0.596	0.617	0.604
Shanxi	1.000	0.887	0.919	0.938
Inner Mongolia	1.000	0.887	0.919	0.935
Liaoning	0.830	0.736	0.763	0.748
Jilin	0.730	0.648	0.671	0.671
Heilongjiang	0.813	0.721	0.747	0.744
Shanghai	1.000	0.887	0.919	0.902
Jiangsu	0.950	0.842	0.872	0.860
Zhejiang	0.985	0.874	0.905	0.880
Anhui	0.902	0.800	0.828	0.822
Fujian	1.000	0.887	0.919	0.910
Jiangxi	0.873	0.774	0.802	0.780
Shandong	0.781	0.693	0.718	0.716
Henan	0.803	0.712	0.737	0.733
Hubei	0.855	0.758	0.785	0.773
Hunan	0.861	0.764	0.791	0.768
Guangdong	0.911	0.808	0.837	0.824
Guangxi	0.609	0.540	0.560	0.557
Hainan	0.831	0.737	0.763	0.761
Chongqing	0.783	0.694	0.719	0.698
Sichuan	0.896	0.795	0.823	0.799
Guizhou	0.576	0.511	0.530	0.531
Yunnan	0.542	0.481	0.498	0.493
Shaanxi	0.813	0.721	0.747	0.748
Gansu	0.801	0.710	0.736	0.722
Qinghai	0.365	0.323	0.335	0.335
Ningxia	0.765	0.679	0.703	0.694
Xinjiang	0.639	0.566	0.587	0.573

5. Data-Driven Robust DEA Model

In Section 3, we show the robust DEA model. The traditional robust model calculates the results based on the uncertainty set, which leads to an overly conservative and somewhat unrealistic model. To deal with this problem, we use the data-driven robust optimization model to improve the DEA model. This approach does not require knowing the exact distribution of carbon emissions output, but only the historical data of carbon emissions output. This makes full use of the historical data and makes it more realistic and economically efficient.

We assume, contrary to the preceding, that the observation sets of the output unit P may be retrieved, where P =

$$y_{r1}, \cdots, y_{rM} \tag{12}$$

and $y_{rq} \in R, q = 1, \cdots, M$. In other words, P represents the initial data that we can obtain. Simultaneously, let y_r represent the mean values of

$$y_{r1}, \cdots, y_{rM}, \tag{13}$$

i.e., $\hat{y_r} = \frac{1}{M} \sum\limits_{q \in M} y_{rq}$, where $[M] = 1, 2, \cdots, M$. As a result, we must identify a decision variable that maximizes worst-case efficiency across all costs in the uncertainty set Z. The robust DEA problem is as follows:

Several approaches for creating Z will be described in the next part [45], with each set having a scaling parameter to determine its size.

$$\begin{aligned} \max \quad & \min_{y_r \in Z} \sum_{r \in [s]} u_r y_r \\ s.t. & \sum_{i=1}^{m} v_i x_i \leq 0 \\ & \sum_{r=1}^{s} u_r y_{rj} - \sum_{i=1}^{m} v_i x_{ij} \leq 0, j = 1, 2, \cdots, n \\ & u_r \geq 0, r = 1, 2, \cdots, s \\ & v_i \geq 0, i = 1, 2, \cdots, m \end{aligned} \tag{14}$$

5.1. Box Uncertainty

We set $y_r^d = \min\limits_{q \in [M]} y_{rq}, y_r^u = \max\limits_{q \in [M]} y_{rq}, \forall \rho \geq 0$; there is $Z_b = \times_{r \in [s]} [\hat{y_r} + \rho(y_r^d - \hat{y_r}), \hat{y_r} + \rho(y_r^a - \hat{y_r})]$, where $\times$ is the Cartesian product and $[s] = 1, 2, \cdots, s$. It is worth noting that $\max\limits_{y_r \in Z} \sum\limits_{r \in [s]} u_r y_r = \sum\limits_{r \in [s]} u_r [\hat{y_r} + \rho(y_r^a - \hat{y_r})]$. Therefore, the obtained data-driven robust DEA model based on the box uncertainty set is as follows:

$$\begin{aligned} \min \quad & -\sum_{r \in [s]} u_r [\hat{y_{ro}} + \rho(y_{ro}^a - \hat{y_{ro}})] \\ s.t. \quad & \sum_{i=1}^{m} v_i x_i \leq 0 \\ & \sum_{r \in [s]} u_r [\hat{y_{rj}} + \rho(y_{rj}^a - \hat{y_{rj}})] \leq \sum_{i=1}^{m} v_i x_{ij}, j = 1, 2, \cdots, n \\ & u_r \geq 0, v_i \geq 0, r = 1, 2, \cdots, s, i = 1, 2, \cdots, m \end{aligned} \tag{15}$$

The proof is in Appendix A.6.

Model (15) is determined by both the uncertain parameters and the observation set, and is different from the previous robust DEA model in Section 3. The proofs for Sections 5.2 and 5.3 are similar to this proof.

5.2. Ellipsoid Uncertainty

The ellipsoid uncertainty set was generated from the fact that the multivariate normal distribution's iso-density locus is an ellipse. Therefore, it usually uses a maximum likelihood fit to create a normal distribution. According to statistical mathematical theory, the maximum likelihood fit of a normal distribution $N(\mu, \sum)$ of data point $\{y_{r1}, \cdots, y_{rm}\}$ is given by $\mu = \hat{y_r}, \sum = \frac{1}{m} \sum (y_{rq} - \mu)(y_{rq} - \mu)^T$. We set an ellipsoid of the form $Z_e = \{y : (y_r - \hat{y_r}) \sum^{-1} (y_r - \hat{y_r} \leq r)\}$ with the scaling parameter and $\kappa \geq 0$ centered on $\hat{y_r}$. Therefore, the obtained data-driven robust DEA model based on the ellipsoid uncertainty set is as follows:

$$\begin{aligned} \min \quad & -\sum_{r\in[s]} \hat{y_{ro}} u_r + \gamma \\ s.t. \quad & \kappa \cdot u_r^T \sum u_r \leq \gamma^2, \gamma \in [s] \\ & \kappa \cdot u_r^T \sum u_r \leq \gamma^2 + \sum_{i=1}^{m} v_i x_{ij}, j = 1,2,\cdots,n \\ & \sum_{i=1}^{m} v_i x_i \leq 1 \\ & u_r \geq 0, v_i \geq 0, r = 1,2,\cdots,s, i = 1,2,\cdots,m \end{aligned} \tag{16}$$

5.3. Polyhedron Uncertainty

A polyhedron defined using linear equations and inequalities is equivalent to a convex hull. We set $Z_p = \left\{ y : \hat{y_r} + (y_r^a - \hat{y_r})\alpha_r, r \in [s], 0 \leq \alpha \leq 1, \sum_{r\in[s]} \alpha_r \leq \rho \right\}$ where the scaling parameter ρ controls the size of the set. Through the duality of the internal maximization problem, the obtained data-driven robust DEA model based on the box uncertainty set is as follows:

$$\begin{aligned} \min \quad & -\sum_{r\in[s]} \hat{y_{ro}} u_r + \rho\gamma + \|t\|_1 \\ s.t. \quad & (y_{ro}^a - \hat{y_{ro}}) u_r \leq u + t_r, r \in [s] \\ & (y_{ro}^a - \hat{y_{rj}}) u_r \leq u + t_r + \sum_{i=1}^{m} v_i x_{ij}, r \in [s], j = 1,2,\cdots,n \\ & \sum_{i=1}^{m} v_i x_i \leq 1 \\ & u_r \geq 0, v_i \geq 0, r = 1,2,\cdots,s, i = 1,2,\cdots,m. \end{aligned} \tag{17}$$

5.4. Case Study

In the previous robust DEA model, although considering the uncertainty of carbon emissions, some data are not taken into account and are not relevant, making the model too conservative. To address this issue, we further analyze the robust DEA in a data-driven context. To unify the comparison, the data in Section 4 are then used to further analyze the carbon emissions efficiency of China. The values will be calculated in turn under the three uncertainty sets.

This section considers the data-driven robust DEA model ($\rho = 1$) to calculate the corresponding values. As can be seen from Figure 8, the carbon emissions efficiency is generally between 0.4 and 0.9, and the efficiency values are all below 1. It can be seen that most of the efficiency values are low and the carbon emissions efficiency has not reached the effective level. Next, the sensitivity analysis of uncertain parameters is studied.

First, when the uncertainty set is the box set, this paper sets the uncertainty parameter ρ from 1 to 3, and then analyzes the impact of these changes on the data-driven robust DEA model. As shown in Figure 9 and Table 6, as ρ increases, the efficiency value gradually becomes smaller. We take Jiangsu as an example. As the parameters ρ become progressively larger, the carbon emissions efficiency of Jiangsu becomes 0.873, 0.830, 0.782, 0.743, and 0.710, respectively.

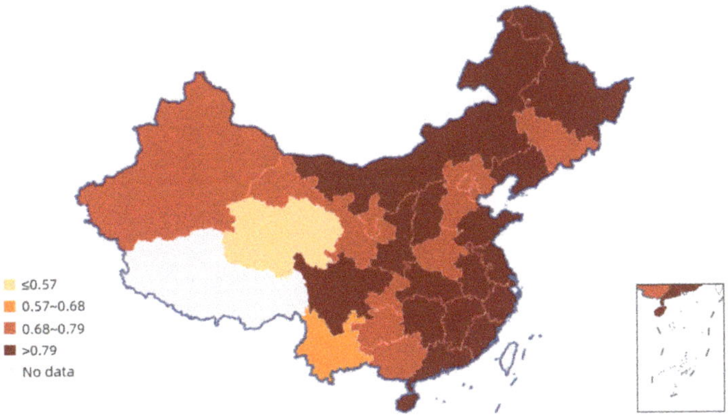

Figure 8. Geographical distribution of mean carbon emissions efficiency of China from 2000 to 2019 in data-driven robust DEA model.

Table 6. Carbon emissions efficiency based on box set in data-driven robust DEA model.

Provinces	Carbon Emissions Efficiency				
	$\rho = 1$	$\rho = 1.5$	$\rho = 2$	$\rho = 2.5$	$\rho = 3$
Beijing	0.783	0.748	0.702	0.663	0.613
Tianjin	0.625	0.581	0.573	0.533	0.502
Hebei	0.633	0.602	0.582	0.543	0.505
Shanxi	0.932	0.862	0.823	0.787	0.723
Inner Mongolia	0.923	0.863	0.822	0.789	0.721
Liaoning	0.783	0.732	0.692	0.663	0.519
Jilin	0.668	0.642	0.602	0.572	0.537
Heilongjiang	0.756	0.721	0.684	0.642	0.612
Shanghai	0.933	0.863	0.822	0.772	0.724
Jiangsu	0.873	0.830	0.782	0.743	0.710
Zhejiang	0.912	0.853	0.812	0.773	0.724
Anhui	0.842	0.783	0.743	0.693	0.652
Fujian	0.921	0.853	0.803	0.763	0.725
Jiangxi	0.824	0.823	0.755	0.683	0.625
Shandong	0.743	0.687	0.642	0.682	0.627
Henan	0.742	0.680	0.653	0.601	0.576
Hubei	0.802	0.762	0.690	0.662	0.621
Hunan	0.805	0.821	0.673	0.643	0.623
Guangdong	0.863	0.823	0.783	0.712	0.678
Guangxi	0.598	0.597	0.583	0.492	0.459
Hainan	0.778	0.712	0.688	0.652	0.612
Chongqing	0.732	0.682	0.644	0.601	0.576
Sichuan	0.862	0.812	0.753	0.682	0.642
Guizhou	0.543	0.504	0.472	0.461	0.412
Yunnan	0.512	0.489	0.443	0.423	0.389
Shaanxi	0.721	0.698	0.647	0.625	0.587
Gansu	0.752	0.714	0.678	0.624	0.589
Qinghai	0.343	0.321	0.293	0.278	0.223
Ningxia	0.732	0.689	0.652	0.621	0.578
Xinjiang	0.601	0.579	0.523	0.487	0.456

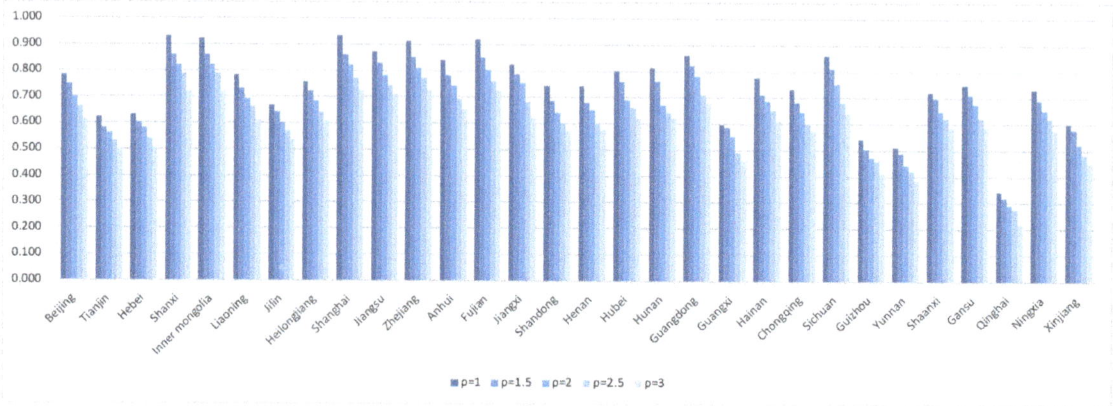

Figure 9. Carbon emissions efficiency based on box set in data-driven robust DEA model.

Secondly, when the uncertainty set is the ellipsoid set, this paper sets the uncertainty parameter κ from 1 to 3 and then analyzes the impact of these changes on the data-driven robust DEA model. As shown in Figure 10 and Table 7, as κ increases, the efficiency value gradually becomes smaller. We take Jiangsu as an example. As the parameters κ become progressively larger, the carbon emissions efficiency of Jiangsu becomes 0.889, 0.854, 0.824, 0.802, and 0.757, respectively.

Table 7. Carbon emissions efficiency based on ellipsoidal set in data-driven robust DEA model.

Provinces	Carbon Emissions Efficiency				
	$\kappa = 1$	$\kappa = 1.5$	$\kappa = 2$	$\kappa = 2.5$	$\kappa = 3$
Beijing	0.824	0.782	0.753	0.725	0.689
Tianjin	0.643	0.612	0.592	0.572	0.542
Hebei	0.638	0.613	0.583	0.562	0.532
Shanxi	0.943	0.902	0.852	0.813	0.793
Inner Mongolia	0.925	0.910	0.873	0.843	0.792
Liaoning	0.792	0.763	0.724	0.682	0.668
Jilin	0.678	0.656	0.634	0.612	0.582
Heilongjiang	0.767	0.743	0.712	0.687	0.663
Shanghai	0.945	0.902	0.862	0.843	0.832
Jiangsu	0.889	0.854	0.824	0.802	0.757
Zhejiang	0.934	0.873	0.845	0.812	0.782
Anhui	0.868	0.793	0.773	0.743	0.702
Fujian	0.943	0.922	0.902	0.843	0.812
Jiangxi	0.842	0.834	0.823	0.743	0.702
Shandong	0.756	0.723	0.702	0.643	0.602
Henan	0.767	0.725	0.701	0.676	0.642
Hubei	0.822	0.802	0.761	0.724	0.682
Hunan	0.833	0.806	0.767	0.726	0.682
Guangdong	0.889	0.843	0.821	0.775	0.726
Guangxi	0.603	0.592	0.565	0.523	0.502
Hainan	0.793	0.763	0.743	0.712	0.672
Chongqing	0.752	0.713	0.682	0.643	0.612
Sichuan	0.872	0.823	0.801	0.763	0.723
Guizhou	0.551	0.523	0.504	0.482	0.475
Yunnan	0.524	0.492	0.471	0.459	0.438
Shaanxi	0.759	0.723	0.702	0.679	0.648
Gansu	0.777	0.743	0.691	0.673	0.654
Qinghai	0.356	0.333	0.313	0.301	0.292
Ningxia	0.743	0.702	0.682	0.652	0.623
Xinjiang	0.613	0.582	0.562	0.535	0.517

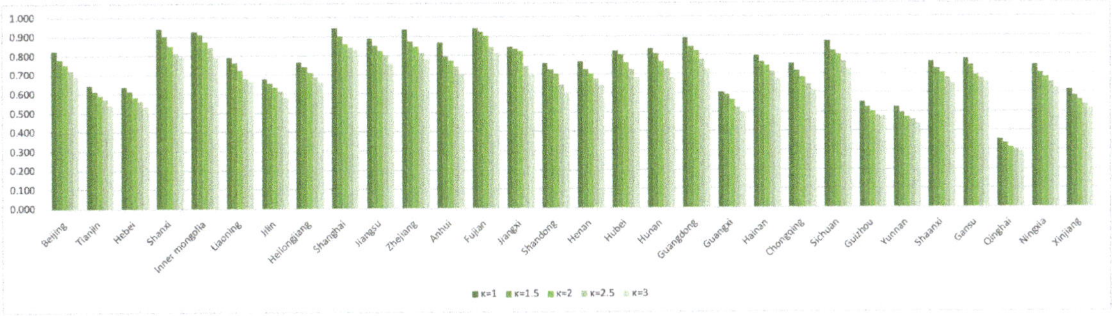

Figure 10. Carbon emissions efficiency based on ellipsoidal set in data-driven robust DEA model.

Finally, when the uncertainty set is the polyhedron set, this paper sets the uncertainty parameter Γ from 1 to 3 and then analyzes the impact of these changes on the data-driven robust DEA model. As shown in Figure 11 and Table 8, as Γ increases, the efficiency value gradually becomes smaller. We take Jiangsu as an example. As the parameters κ become progressively larger, the carbon emissions efficiency of Jiangsu becomes 0.853, 0.832, 0.812, 0.782, and 0.762, respectively.

Table 8. Carbon emissions efficiency based on polyhedron set in data-driven robust DEA model.

Provinces	Carbon Emissions Efficiency				
	Γ = 1	Γ = 1.5	Γ = 2	Γ = 2.5	Γ = 3
Beijing	0.812	0.768	0.742	0.714	0.676
Tianjin	0.632	0.602	0.574	0.563	0.526
Hebei	0.628	0.603	0.562	0.545	0.514
Shanxi	0.939	0.910	0.883	0.852	0.825
Inner Mongolia	0.943	0.893	0.863	0.832	0.788
Liaoning	0.782	0.752	0.713	0.662	0.632
Jilin	0.663	0.642	0.613	0.592	0.582
Heilongjiang	0.753	0.728	0.692	0.652	0.642
Shanghai	0.932	0.902	0.862	0.832	0.824
Jiangsu	0.853	0.832	0.812	0.782	0.762
Zhejiang	0.923	0.863	0.832	0.783	0.762
Anhui	0.878	0.798	0.762	0.725	0.701
Fujian	0.925	0.902	0.885	0.834	0.813
Jiangxi	0.824	0.818	0.802	0.734	0.716
Shandong	0.742	0.713	0.687	0.635	0.614
Henan	0.753	0.714	0.689	0.658	0.625
Hubei	0.808	0.778	0.748	0.713	0.672
Hunan	0.823	0.784	0.746	0.689	0.662
Guangdong	0.878	0.848	0.814	0.735	0.714
Guangxi	0.593	0.573	0.535	0.517	0.491
Hainan	0.783	0.745	0.725	0.682	0.652
Chongqing	0.742	0.683	0.663	0.623	0.582
Sichuan	0.852	0.802	0.763	0.748	0.724
Guizhou	0.542	0.510	0.483	0.474	0.454
Yunnan	0.516	0.482	0.457	0.438	0.415
Shaanxi	0.747	0.712	0.692	0.667	0.647
Gansu	0.756	0.738	0.678	0.646	0.628
Qinghai	0.342	0.331	0.309	0.298	0.272
Ningxia	0.728	0.689	0.663	0.642	0.614
Xinjiang	0.598	0.576	0.543	0.524	0.504

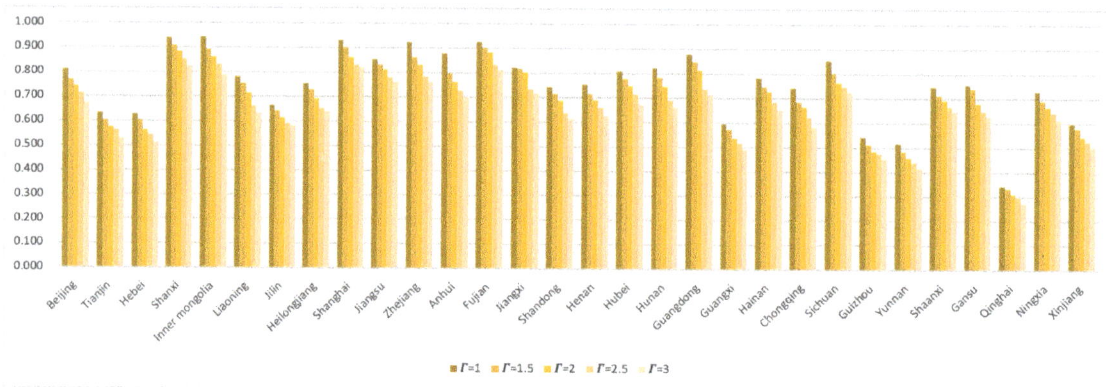

Figure 11. Carbon emissions efficiency based on polyhedron set in data-driven robust DEA model.

Comparison Analysis

Comparing the data-driven robust DEA model (Table 9), the robust DEA model, and the traditional DEA model, it can be seen that the value of the data-driven robust DEA model is between the robust DEA model and the traditional DEA model. In this way, the data-driven robust DEA model overcomes the drawback of over-conservatism of robust optimization and, on the other hand, retains the robustness of robust optimization to some extent. In the case of climate uncertainty and government economic policy uncertainty, the data-driven robust DEA model is more realistic and can be used as a reference for decision makers.

Table 9. Comparison of the DEA model, robust DEA, and data-driven robust DEA.

Provinces	DEA	BRDEA ($\tau = 1$)	ERDEA ($\Omega = 1$)	PRDEA ($\Gamma = 1$)	DRDEA ($\rho = 1$)	DRDEA ($\kappa = 1$)	DRDEA ($\Gamma = 1$)
Beijing	0.877	0.778	0.805	0.778	0.783	0.824	0.812
Tianjin	0.682	0.605	0.627	0.618	0.625	0.643	0.632
Hebei	0.672	0.596	0.617	0.604	0.633	0.638	0.628
Shanxi	1.000	0.887	0.919	0.938	0.932	0.943	0.939
Inner Mongolia	1.000	0.887	0.919	0.935	0.923	0.925	0.943
Liaoning	0.830	0.736	0.763	0.748	0.783	0.792	0.782
Jilin	0.730	0.648	0.671	0.671	0.668	0.678	0.663
Heilongjiang	0.813	0.721	0.747	0.744	0.756	0.767	0.753
Shanghai	1.000	0.887	0.919	0.902	0.933	0.945	0.932
Jiangsu	0.950	0.842	0.872	0.860	0.873	0.889	0.853
Zhejiang	0.985	0.874	0.905	0.880	0.912	0.934	0.923
Anhui	0.902	0.800	0.828	0.822	0.842	0.868	0.878
Fujian	1.000	0.887	0.919	0.910	0.921	0.943	0.925
Jiangxi	0.873	0.774	0.802	0.780	0.824	0.842	0.824
Shandong	0.781	0.693	0.718	0.716	0.743	0.756	0.742
Henan	0.803	0.712	0.737	0.733	0.742	0.767	0.753
Hubei	0.855	0.758	0.785	0.773	0.802	0.822	0.808
Hunan	0.861	0.764	0.791	0.768	0.815	0.833	0.823
Guangdong	0.911	0.808	0.837	0.824	0.863	0.889	0.878
Guangxi	0.609	0.540	0.560	0.557	0.598	0.603	0.593
Hainan	0.831	0.737	0.763	0.761	0.778	0.793	0.783
Chongqing	0.783	0.694	0.719	0.698	0.732	0.752	0.742
Sichuan	0.896	0.795	0.823	0.799	0.862	0.872	0.852
Guizhou	0.576	0.511	0.530	0.531	0.543	0.551	0.542
Yunnan	0.542	0.481	0.498	0.493	0.512	0.524	0.516
Shaanxi	0.813	0.721	0.747	0.748	0.721	0.759	0.747
Gansu	0.801	0.710	0.736	0.722	0.752	0.777	0.756
Qinghai	0.365	0.323	0.335	0.335	0.343	0.356	0.342
Ningxia	0.765	0.679	0.703	0.694	0.732	0.743	0.728
Xinjiang	0.639	0.566	0.587	0.573	0.601	0.613	0.598

From the above comparison with other literature (Table 10), there are regions in the three-stage DEA approach where the efficiency is greater than robust DEA and data-driven robust DEA. It is evident that the three-stage DEA approach is somewhat optimistic and does not take into account climate and policy uncertainties. The method in this paper deals with this problem well. We will further present the corresponding economic and government economic policy implications below.

Table 10. Comparison of the DEA model, robust DEA, data-driven robust DEA, and three-stage DEA.

Provinces	DEA	BRDEA ($\tau = 1$)	DRDEA ($\rho = 1$)	Three-Stage DEA [6]
Beijing	0.877	0.778	0.783	0.987
Tianjin	0.682	0.605	0.625	0.882
Hebei	0.672	0.596	0.633	0.779
Shanxi	1.000	0.887	0.932	0.668
Inner Mongolia	1.000	0.887	0.923	0.564
Liaoning	0.830	0.736	0.783	0.719
Jilin	0.730	0.648	0.668	0.707
Heilongjiang	0.813	0.721	0.756	0.949
Shanghai	1.000	0.887	0.933	0.892
Jiangsu	0.950	0.842	0.873	1.000
Zhejiang	0.985	0.874	0.912	1.000
Anhui	0.902	0.800	0.842	0.791
Fujian	1.000	0.887	0.921	0.853
Jiangxi	0.873	0.774	0.824	0.987
Shandong	0.781	0.693	0.743	0.683
Henan	0.803	0.712	0.742	0.808
Hubei	0.855	0.758	0.802	0.840
Hunan	0.861	0.764	0.815	0.755
Guangdong	0.911	0.808	0.863	0.807
Guangxi	0.609	0.540	0.598	0.948
Hainan	0.831	0.737	0.778	1.000
Chongqing	0.783	0.694	0.732	0.823
Sichuan	0.896	0.795	0.862	0.711
Guizhou	0.576	0.511	0.543	0.813
Yunnan	0.542	0.481	0.512	0.623
Shaanxi	0.813	0.721	0.721	0.810
Gansu	0.801	0.710	0.752	0.612
Qinghai	0.365	0.323	0.343	0.908
Ningxia	0.765	0.679	0.732	0.686
Xinjiang	0.639	0.566	0.601	0.809

5.5. Economic Implications

From the discussion above, it can be seen that the carbon emission efficiency values under different uncertain parameters are different. Governments can formulate policies according to their geographical location and economic development. With a high level of economic development and a relatively high level of carbon emissions, the eastern region can continue to develop its own advantages. In the information age, we will continue to vigorously develop science and technology, closely follow national policies, provide technical support to the central and western regions, and jointly improve carbon emission efficiency. The northeast, central, and western regions have relatively low carbon emissions due to geographical location, resource endowment, and other problems. They need to develop the economy on the one hand and improve the level of science and technology on the other hand, while adjusting the energy structure and industrial structure. To sum up, each region in China can formulate appropriate economic policies related to double carbon according to its own characteristics.

5.6. Policy Implications

To realize the commitment to the global "dual carbon" policy, achieve "high-quality" economic development, and promote the improvement of the ecological environment, several policy recommendations have been put forward.

1. Pay attention to the coordinated development among regions and develop the economy according to local conditions. There are huge differences in regional economic development in China, as well as in resource endowments, industrial advantages, and economic development levels between different regions. The central, western, and northeastern regions should focus on optimizing the energy structure, following the requirements of industrial policies and dual control of energy consumption, promoting the concentration of high-energy-consuming industries in areas with advantages in clean energy, and actively cultivating green development momentum.
2. Promote energy transition and promote green development. Energy is an important material basis for economic and social development, and the most important source of carbon emissions. Continue to reduce the proportion of fossil energy consumption, vigorously develop clean energy, accelerate the construction of a clean, low-carbon, safe and efficient modern energy system, comprehensively promote energy transformation, and take multiple measures to reduce carbon emissions.
3. Innovate green technology and vigorously develop science and technology. Innovation is the first driving force for development. In the new era, energy science and technology must closely follow the national strategic needs, be problem-oriented and goal-oriented, target major frontier areas to speed up development, strengthen the effective matching of systems and mechanisms, and promote the market-oriented allocation of technical elements, thereby promoting the engine of energy science and technology innovation-driven development.
4. Advocate for people to live a green and low-carbon lifestyle and reduce carbon emissions. The sudden outbreak of the COVID-19 epidemic in 2020 reminds us that human beings need a self-revolution, to accelerate the formation of green development methods and lifestyles, and build an ecological civilization and beautiful earth. The key to the implementation of the "dual carbon" policy lies in the awakening and promotion of public awareness. If more and more people can realize the importance of the "dual carbon" policy, reduce or eliminate disposable tools use, and reduce the use of fuel vehicles, etc., it will improve carbon emissions efficiency and promote green economic development.

6. Conclusions

The traditional DEA model focuses on efficiency evaluation, which assumes that the output data are deterministic. However, uncertainty about climate and government economic policies is ignored. Therefore, first we used a robust optimization approach to investigate the uncertainty of traditional DEA. Then, three different types of uncertainty sets were considered, including the box set, ellipsoidal set, and polyhedral set. The model proposed in this paper improves the conservatism of the model and reduces the risk impact of climate and government economic policies. Finally, to verify the validity of the model in this paper, we studied carbon emissions efficiency in China. Some interesting conclusions are described as follows:

1. The higher the level of uncertainty parameters, the less efficient the corresponding robust DEA model is; in other words, more conservative models are more robust. This trend can be reflected in the three robust models in this paper.
2. The values of the robust DEA model proposed in this paper are smaller than those of the original DEA model. This shows that the model in this paper is more robust and conservative. It can be seen that the traditional DEA model is too optimistic and does not take into account the uncertainty caused by various risks. When the effect of uncertain output variables is considered, the efficiency value of the proposed robust model becomes smaller.
3. The carbon emissions efficiency values from the data-driven robust data envelopment analysis are slightly higher than those from the robust optimization, which shows that the data-driven robust DEA model retains the robustness of the model and overcomes the over-conservatism of the robust optimization.

In the future, the government should focus on coordinated regional development, promote the transformation and upgrading of the energy structure, innovate in green technology, and advocate for people to live a green and low-carbon lifestyle.

Author Contributions: Conceptualization, S.Q. and Y.J.; data curation, Y.X. and S.J.; formal analysis, Y.X. and C.F.; investigation, Y.X.; methodology, S.Q. and J.W.; resources, Y.X.; validation, S.Q. and Y.J.; writing—original draft, Y.X.; writing—review and editing, S.Q. and Y.J. All authors have read and agreed to the published version of the manuscript.

Funding: This research was supported by the National Natural Science Foundation of China (No. 72171149, 72171123) and the Shanghai Philosophy and Social Science Foundation (No. 2020BGL010).

Institutional Review Board Statement: Not applicable

Informed Consent Statement: Not applicable

Data Availability Statement: The data presented in this study are available on request from the corresponding author. The data are not publicly available due to privacy considerations.

Acknowledgments: The authors would like to thank the anonymous reviewers for their comments and suggestions.

Conflicts of Interest: The authors declare no conflict of interest.

Appendix A

Appendix A.1. Proof of Theorem 1

Proof of Theorem 1. The first constraint of Model (6) may be rewritten using the form of the box set as:

$$\theta - \sum_{r=1}^{s} u_r y_{ro} - \sum_{l=1}^{L}\sum_{r=1}^{s} u_r y_{rol}^{F} \xi_l, \forall\{\xi : \|\xi\|_\infty \leq \tau\} \tag{A1}$$

This is the same as:

$$\min_{-\gamma \leq \xi_l \leq \tau} \sum_{l=1}^{L}\sum_{r=1}^{s} u_r y_{rol}^{F} \xi_l \geq \theta - \sum_{r=1}^{s} u_r y_{ro}, \forall\{\xi : \|\xi\|_\infty \leq \tau\} \tag{A2}$$

Obviously, the minimal value of Constraint (A2)'s left-hand side is:

$$-\tau \sum_{l=1}^{L}\sum_{r=1}^{s} u_r y_{rjl}^{F} \tag{A3}$$

The explicit version of Model (6)'s initial constraint may then be produced.

$$\theta + \tau \sum_{l=1}^{L}\sum_{r=1}^{s} \mu_r y_{rol}^{F} - \sum_{r=1}^{s} u_r y_{ro} \leq 0 \tag{A4}$$

The same approach may be simply used to produce the explicit version of Model (6)'s third constraint:

$$\tau \sum_{l=1}^{L}\sum_{r=1}^{s} \mu_r y_{rjl}^{F} + \sum_{r=1}^{s} \mu_r y_{rj} - \sum_{i=1}^{m} v_i x_{ij} \leq 0 \tag{A5}$$

The proof is complete. □

Appendix A.2. Proof of Theorem 2

Proof of Theorem 2. The first constraint of Model (8) may be phrased using the ellipsoid set form as:

$$\theta - \sum_{r=1}^{s} u_r y_{ro} - \sum_{l=1}^{L}\sum_{r=1}^{s} u_r y_{rol}^{F} \xi_l, \forall\{\xi : \|\xi\|_2 \leq \Omega\} \tag{A6}$$

This is the same as:

$$\min_{\|\xi\|_2\le\Omega}\sum_{l=1}^{L}\sum_{r=1}^{s}u_r y_{rol}^F\xi_l\ge\theta-\sum_{r=1}^{s}u_r y_{ro},\forall\{\xi:\|\xi\|_2\le\Omega\} \tag{A7}$$

Taking the minimum of the left-hand side of Constraint (A7), we obtain the corresponding inequality:

$$\theta+\Omega\sqrt{\sum_{l=1}^{L}\left(\sum_{r=1}^{s}u_r y_{rol}^F\right)^2}-\sum_{r=1}^{s}u_r y_{ro}\le 0 \tag{A8}$$

Simply repeat the technique for the third constraint of Model (8), and we obtain:

$$\Omega\sqrt{\sum_{l=1}^{L}\left(\sum_{r=1}^{s}u_r y_{rjl}^F\right)^2}+\sum_{r=1}^{s}u_r y_{rj}-\sum_{i=1}^{m}v_i x_{ij}\le 0,\forall j \tag{A9}$$

Finally, the RDEA based on the ellipsoid set may be produced.
The proof is complete. □

Appendix A.3. Proof of Theorem 3

Proof of Theorem 3. In general, the third constraint of Model (10) may be phrased as follows:

$$\max_{\|\xi\|_1\le\Gamma}\sum_{l=1}^{L}\sum_{r=1}^{s}u_r y_{rjl}^F\xi_l\le\sum_{i=1}^{m}v_i x_{io}-\sum_{l=1}^{L}\sum_{r=1}^{s}u_r y_{rjl}^F \tag{A10}$$

We may deduce the explicit form of the dual cone from its properties:

$$\sum_{l=1}^{L}\Gamma p_{jl}+\sum_{r=1}^{s}\mu_r y_{rj}-\sum_{i=1}^{m}x_{ij}v_i\le 0,\quad \textit{where } p_{jl}\ge u_r y_{rjl}^F \tag{A11}$$

Likewise, the precise version of Model (10)'s first constraint may be obtained:

$$\theta+\Gamma\sum_{l=1}^{L}p_{ol}-\sum_{r=1}^{s}u_r y_{ro}\le 0,\quad \textit{where } p_{ol}\ge u_r y_{rol}^F \tag{A12}$$

The proof is complete. □

Appendix A.4. Proof of Theorem 4

Proof of Theorem 4. Let us take BRDEA as an example. The first constraint of Model 9 contains the uncertain variable τ and $\sum_{l=1}^{L}\sum_{r=1}^{s}u_r y_{rol}^F$ is greater than 0. Therefore, we can complete this proof by considering only the first constraint of Model 9. As τ increases, $\tau\sum_{l=1}^{L}\sum_{r=1}^{s}u_r y_{rol}^F$ also increases, and the part of $\sum_{r=1}^{s}u_r y_{ro}$ becomes smaller to satisfy the constraint. That is, Model 9 will obtain a lower efficiency and become more robust.

Similar conclusions can be obtained from other robust optimization models. □

Appendix A.5. Proof of Theorem 5

Proof of Theorem 5. Let us take BRDEA as an example. When the uncertain variables τ are added, the value of $\tau\sum_{l=1}^{L}\sum_{r=1}^{s}u_r y_{rol}^F$ is greater than 0, that is, the value of $\sum_{r=1}^{s}u_r y_{ro}$ Must be smaller to satisfy the constraints. In other words, the efficiency value of Model 9 must be smaller and a little more conservative than the original Model 2.

Similar conclusions can be obtained from other robust optimization models. □

Appendix A.6. Proof of Section Box Uncertainty

Proof of Section 5.1. We simply acquire the largest value of (2)'s goal function:

$$\max_{y_r \in Z} \sum_{r \in [s]} u_r y_r = \sum_{r \in [s]} u_r [\hat{y}_r + \rho(y_r^a - \hat{y}_r)] \tag{A13}$$

It can then be translated into the corresponding form:

$$\min - \sum_{r \in [s]} u_r [\hat{y}_{ro} + \rho(y_{ro}^a - \hat{y}_{ro})] \tag{A14}$$

Likewise, the third inequality of (2) may be expressed as follows:

$$\sum_{r \in [s]} u_r [\hat{y}_{rj} + \rho(y_{rj}^a - \hat{y}_{rj})] \leq \sum_{i=1}^{m} v_i x_{ij}, \quad j = 1, 2, \cdots, n \tag{A15}$$

□

References

1. Jeffry, L.; Mei, Y.O.; Nomanbhay, S.; Mofijur, M.; Show, P.L. Greenhouse gases utilization: A review. *Fuel* **2021**, *301*, 121017. [CrossRef]
2. Huang, W.T.K.; Braithwaite, I.; Charlton-Perez, A.; Sarran, C.; Sun, T. Non-linear response of temperature-related mortality risk to global warming in England and Wales. *Environ. Res. Lett.* **2022**, *17*, 034017. [CrossRef]
3. Lu, Y.; Zhang, Y.; Cao, X.; Wang, C.; Wang, Y.; Zhang, M.; Ferrier, R.C.; Jenkins, A.; Yuan, J.; Bailey, M.J.; et al. Forty years of reform and opening up: China's progress toward a sustainable path. *Sci. Adv.* **2019**, *5*, eaau9413. [CrossRef] [PubMed]
4. Ritchie, H.; Roser, M. CO_2 and Greenhouse Gas Emissions. Our World in Data. 2020. Available online: https://ourworldindata.org/$\protect\protect\unhbox\voidb@x\hbox{CO}_2$-and-other-greenhouse-gas-emissions (accessed on 5 July 2022).
5. Li, Y.; Chen, K.; Zheng, N.; Cai, Q.; Li, Y.; Lin, C. Strategy Research on Accelerating Green and Low-Carbon Development under the Guidance of Carbon Peak and Carbon Neutral Targets. *IOP Conf. Ser. Earth Environ. Sci.* **2021**, *793*, 012009. [CrossRef]
6. Zhang, M.; Li, L.; Cheng, Z. Research on carbon emission efficiency in the Chinese construction industry based on a three-stage DEA-Tobit model. *Environ. Sci. Pollut. Res.* **2021**, *28*, 51120–51136. [CrossRef]
7. Yamaji, K.; Matsuhashi, R.; Nagata, Y.; Kaya, Y. A study on economic measures for CO_2 reduction in Japan. *Energy Policy* **1993**, *21*, 123–132. [CrossRef]
8. Aigner, D.; Lovell, C.K.; Schmidt, P. Formulation and estimation of stochastic frontier production function models. *J. Econom.* **1977**, *6*, 21–37. [CrossRef]
9. Meeusen, W.; van Den Broeck, J. Efficiency Estimation from Cobb-Douglas Production Functions with Composed Error. *Int. Econ. Rev.* **1977**, *18*, 435–444. [CrossRef]
10. Sun, J.; Du, T.; Sun, W.; Na, H.; He, J.; Qiu, Z.; Yuan, Y.; Li, Y. An evaluation of greenhouse gas emission efficiency in China's industry based on SFA. *Sci. Total Environ.* **2019**, *690*, 1190–1202. [CrossRef]
11. Sun, W.; Huang, C. How does urbanization affect carbon emission efficiency? Evidence from China. *J. Clean. Prod.* **2020**, *272*, 122828. [CrossRef]
12. Zhang, C.; Chen, P. Industrialization, urbanization, and carbon emission efficiency of Yangtze River Economic Belt—empirical analysis based on stochastic frontier model. *Environ. Sci. Pollut. Res.* **2021**, *28*, 66914–66929. [CrossRef] [PubMed]
13. Cullinane, K.; Wang, T.F.; Song, D.W.; Ji, P. The technical efficiency of container ports: Comparing data envelopment analysis and stochastic frontier analysis. *Transp. Res. Part A Policy Pract.* **2006**, *40*, 354–374. [CrossRef]
14. Yu, Y.; Yi, Z.; Jia, J. The Efficiency Evolution and Risks of Green Development in the Yangtze River Economic Belt, China. *Sustainability* **2022**, *14*, 10417. [CrossRef]
15. Du, Y.; Seo, W. A Comparative Study on the Efficiency of R&D Activities of Universities in China by Region Using DEA–Malmquist. *Sustainability* **2022**, *14*, 10433.
16. Charnes, A.; Cooper, W.W.; Rhodes, E. Measuring the efficiency of decision making units. *Eur. J. Oper. Res.* **1978**, *2*, 429–444. [CrossRef]
17. Banker, R.D.; Charnes, A.; Cooper, W.W. Some models for estimating technical and scale inefficiencies in data envelopment analysis. *Manag. Sci.* **1984**, *30*, 1078–1092. [CrossRef]
18. Meng, F.; Su, B.; Thomson, E.; Zhou, D.; Zhou, P. Measuring China's regional energy and carbon emission efficiency with DEA models: A survey. *Appl. Energy* **2016**, *183*, 1–21. [CrossRef]

19. Cheng, Z.; Li, L.; Liu, J.; Zhang, H. Total-factor carbon emission efficiency of China's provincial industrial sector and its dynamic evolution. *Renew. Sustain. Energy Rev.* **2018**, *94*, 330–339. [CrossRef]
20. Yan, D.; Lei, Y.; Li, L.; Song, W. Carbon emission efficiency and spatial clustering analyses in China's thermal power industry: Evidence from the provincial level. *J. Clean. Prod.* **2017**, *156*, 518–527. [CrossRef]
21. Ben-Tal, A.; Nemirovski, A. Robust solutions of linear programming problems contaminated with uncertain data. *Math. Program.* **2000**, *88*, 411–424. [CrossRef]
22. Ben-Tal, A.; Nemirovski, A. Robust optimization—Methodology and applications. *Math. Program.* **2002**, *92*, 453–480. [CrossRef]
23. Bertsimas, D.; Sim, M. The price of robustness. *Oper. Res.* **2004**, *52*, 35–53. [CrossRef]
24. Ben-Tal, A.; El Ghaoui, L.; Nemirovski, A. *Robust Optimization*; Princeton University Press: Princeton, NJ, USA, 2009.
25. Han, Y.; Qu, S.; Wu, Z.; Huang, R. Robust consensus models based on minimum cost with an application to marketing plan. *J. Intell. Fuzzy Syst.* **2019**, *37*, 5655–5668. [CrossRef]
26. Qu, S.; Han, Y.; Wu, Z.; Raza, H. Consensus modeling with asymmetric cost based on data-driven robust optimization. *Group Decis. Negot.* **2021**, *30*, 1395–1432. [CrossRef]
27. Lu, Y.; Xu, Y.; Herrera-Viedma, E.; Han, Y. Consensus of large-scale group decision making in social network: The minimum cost model based on robust optimization. *Inf. Sci.* **2021**, *547*, 910–930. [CrossRef]
28. Qu, S.; Li, Y.; Ji, Y. The mixed integer robust maximum expert consensus models for large-scale GDM under uncertainty circumstances. *Appl. Soft Comput.* **2021**, *107*, 107369. [CrossRef]
29. Ji, Y.; Li, H.; Zhang, H. Risk-averse two-stage stochastic minimum cost consensus models with asymmetric adjustment cost. *Group Decis. Negot.* **2022**, *31*, 261–291. [CrossRef]
30. Qu, S.; Wei, J.; Wang, Q.; Li, Y.; Jin, X.; Chaib, L. Robust minimum cost consensus models with various individual preference scenarios under unit adjustment cost uncertainty. *Inf. Fusion* **2022**, *89*, 510–526. [CrossRef]
31. Hosseini-Motlagh, S.M.; Samani, M.R.G.; Homaei, S. Blood supply chain management: Robust optimization, disruption risk, and blood group compatibility (a real-life case). *J. Ambient. Intell. Humaniz. Comput.* **2020**, *11*, 1085–1104. [CrossRef]
32. Cantini, A.; Peron, M.; Carlo, F.D.; Sgarbossa, F. A decision support system for configuring spare parts supply chains considering different manufacturing technologies. *Int. J. Prod. Res.* **2022**, 1–21. doi: 10.1080/00207543.2022.2041757. [CrossRef]
33. Yang, J.; Su, C. Robust optimization of microgrid based on renewable distributed power generation and load demand uncertainty. *Energy* **2021**, *223*, 120043. [CrossRef]
34. Peykani, P.; Mohammadi, E.; Saen, R.F.; Sadjadi, S.J.; Rostamy-Malkhalifeh, M. Data envelopment analysis and robust optimization: A review. *Expert Syst.* **2020**, *37*, e12534. [CrossRef]
35. Sadjadi, S.J.; Omrani, H. Data envelopment analysis with uncertain data: An application for Iranian electricity distribution companies. *Energy Policy* **2008**, *36*, 4247–4254. [CrossRef]
36. Han, Y.; Ji, Y.; Qu, S. A Robust Minimum-Cost Consensus Model With Uncertain Aggregation Weights Based on Data-Driven Method. *IEEE Trans. Comput. Soc. Syst.* **2022**, *9*, 1167–1184. [CrossRef]
37. Shen, F.; Zhao, L.; Du, W.; Zhong, W.; Qian, F. Large-scale industrial energy systems optimization under uncertainty: A data-driven robust optimization approach. *Appl. Energy* **2020**, *259*, 114199. [CrossRef]
38. Qu, S.; Feng, C.; Jiang, S.; Wei, J.; Xu, Y. Data-Driven Robust DEA Models for Measuring Operational Efficiency of Endowment Insurance System of Different Provinces in China. *Sustainability* **2022**, *14*, 9954. [CrossRef]
39. Mielnik, O.; Goldemberg, J. Communication The evolution of the "carbonization index" in developing countries. *Energy Policy* **1999**, *27*, 307–308. [CrossRef]
40. Ren, J.; Gao, B.; Zhang, J.; Chen, C. Measuring the energy and carbon emission efficiency of regional transportation systems in China: Chance-constrained DEA models. *Math. Probl. Eng.* **2020**, *2020*. [CrossRef]
41. Goldsmith, R.W. A perpetual inventory of national wealth. In *Studies in Income and Wealth*; NBER: Cambridge, MA, USA, 1951; Volume 14, pp. 5–73.
42. Jun, Z.; Yuan, Z. reestimate of China's capital stock K. *Econ. Res.* **2003**, *7*, 675–699.
43. Shan, Y.; Guan, D.; Zheng, H.; Ou, J.; Li, Y.; Meng, J.; Mi, Z.; Liu, Z.; Zhang, Q. China CO_2 emission accounts 1997–2015. *Sci. Data* **2018**, *5*, 1–14. [CrossRef]
44. Shan, Y.; Huang, Q.; Guan, D.; Hubacek, K. China CO_2 emission accounts 2016–2017. *Sci. Data* **2020**, *7*, 1–9. [CrossRef] [PubMed]
45. Chassein, A.; Dokka, T.; Goerigk, M. Algorithms and uncertainty sets for data-driven robust shortest path problems. *Eur. J. Oper. Res.* **2019**, *274*, 671–686. [CrossRef]

sustainability

MDPI

Article

Data-Driven Robust DEA Models for Measuring Operational Efficiency of Endowment Insurance System of Different Provinces in China

Shaojian Qu *, Can Feng, Shan Jiang, Jinpeng Wei and Yuting Xu

School of Management Science and Engineering, Nanjing University of Information Science and Technology, Nanjing 210044, China
* Correspondence: qushaojian@nuist.edu.cn

Abstract: China is facing an increasingly serious aging problem, which puts forward higher requirements for the smoothness of the endowment insurance system. Accurate evaluation of the efficiency of the system can help the government to find problems and improve the system. Some scholars have used data envelopment analysis (DEA) method to measure the efficiency of endowment insurance system. However, according to the literature, the impact of government policy adjustment and economic shocks on output of the data was ignored. In this study, a robust optimization method is applied to deal with uncertainty. Robust DEA models proposed in this paper are based on three kinds of uncertainty sets. A data-driven robust optimization method is also applied to resolve the over-conservative problem. Compared with the robust DEA method, based on analysis it is found that the data-driven robust DEA method is more flexible and reliable for efficiency estimating strategies. The results of data-driven robust DEA models illustrate that the government should increase its support for the endowment insurance system, especially for the underdeveloped regions.

Keywords: endowment insurance system; robust optimization; data envelopment analysis (DEA); uncertainty set; data-driven

Citation: Qu, S.; Feng, C.; Jiang, S.; Wei, J.; Xu, Y. Data-Driven Robust DEA Models for Measuring Operational Efficiency of Endowment Insurance System of Different Provinces in China. *Sustainability* **2022**, *14*, 9954. https://doi.org/10.3390/su14169954

Academic Editor: Chia-Lin Chang

Received: 15 July 2022
Accepted: 9 August 2022
Published: 11 August 2022

1. Introduction

In recent years, China's aging process has been accelerating. By October 2020, China's population aged 60 or above reached 264 million, accounting for 18.7% of the total population, of which 191 million will be aged 65 or above, accounting for 13.5% of the total population [1]. The research shows that China will enter a moderately aging society by 2024, with 14% of the population over 65 years old. According to research, China will enter into the serious aging society by 2035 and the proportion of the elderly population over 65 will be increased by more than 20%. In the past few decades, China has moved from the demographic dividend period to the population burden period [2]. The cruel reality of getting old before getting rich will bring a great burden to young people, and will also pose the most severe challenge to the current endowment insurance system. Evaluation of the operation process and effects of the current endowment insurance system objectively and scientifically has become an urgent problem for the further improvement and development of the endowment insurance system. It creates a great significance to grasping the operation status of the endowment insurance system and promoting its sustainable development.

Research on the efficiency of insurance system began as early as the 1950s. Samuelson proposed the Overlapping Generation Model (OLG) to study the performance of endowment insurance [3]. Later, Diamond improved the OLG model, he used the model to study the dynamic impact of two existing financing modes of endowment insurance system on capital [4]. In the 1970s, Aaron proposed the 'Aaron condition', which became a reference standard for the choice of a pension insurance system. He found that the main factors restricting the efficiency of endowment insurance are the growth rate of population and the growth rate of labor production [5]. In recent years, some scholars have studied China's maternity insurance system and basic pension insurance system [6,7]. However, there is little research on endowment insurance systems in China. Charnes et al. proposed that DEA is the most popular non-parametric method to measure the relative efficiency of decision-maker units, and it was developed later on by Banker et al. [8,9]. The DEA method is widely used among various fields: Bank [10–12], energy [13–16], educational system [17,18], entrepreneurship and innovation [19,20], software engineering [21,22]. Andreu proposed four variants of the slacks-based measure of efficiency (SBM) to evaluate the efficiency of the strategic style of pension funds [23]. Further, Hu improved the three-stage DEA model to calculate the operation efficiency of urban and rural residential insurance system in 31 provinces of China from 2012 to 2016, he proposed that different regions should implement different efficiency promotion strategies according to their own problems and situation [24].

With the deepening of population aging, the gap in endowment insurance system is becoming increasingly serious, which means that the consumption of accumulated pension balance will be increased. The change of population structure promotes the change of relevant national policies, it also has an impact on the pension payment rate. Moreover, the stable operation of the economy is very important for the sustainable operations of the endowment insurance system. In case of economic crisis, the accumulated pension balance will be affected. Hence, there are sufficient reasons to believe that the accumulated pension balance and pension expenditures are in an uncertain environment, and this uncertainty is caused by a variety of factors such as economic shock and adjustment of the policy. In previous studies, few scholars consider this uncertainty, likely to lead to wrong ranking results.

Robust Optimization is introduced in this study to resolve the problems mentioned above. It was first proposed by Soyster [25]. El-Ghaoui and Lebret [26] and Ben-Tal and Nemirovski [27–29] extended the RO theory and proposed a new robust model based on ellipsoidal uncertainty sets. Subsequently, Bertsimas and Sim [30–32] and Bertsimas et al. [33] developed a robust optimization approach based on polyhedral uncertainty sets. Sadjadi and Omrani were the first scholars to apply robust optimization for DEA [34]. Kazemia and Haji proposed robust DEA model based on Ben-Tal and Bertsimas approaches to measure the efficiency of high schools [35]. Until now, no scholar has used the robust DEA method to resolve the problems in the field of endowment insurance systems.

However, it is highly subjective that the classical RO methods are mostly based on experience to obtain uncertainty sets. In previous studies, the advantages of big data are ignored, and the results are also too conservative. The data-driven RO method [36,37] uses historical data to construct uncertainty sets, which improves the rationality and economy of the uncertainty sets of the traditional RO method. However, few types of research were conducted previously on introducing data-driven RO to DEA models.

In this paper, a new method is proposed to measure the efficiency of endowment insurance system of 31 provinces in China under uncertainty environment. It is proposed to use a data-driven robust data envelopment analysis (DRDEA) model. It is based on three uncertainty sets: Interval uncertainty set, Ellipsoid uncertainty set, Polyhedron uncertainty set. The determination of the government is to ensure that the operation of the endowment insurance system should be absolutely stable, so the purpose was to resort robust optimization to ensure robustness in the case of data disturbance. The main contributions of this research are as follows: (1) It is proposed that robust DEA models deal with the efficiency measurement in endowment insurance effectively. (2) We deduce the robust counterparts in different deterministic cases. (3) It is proposed that data-driven robust DEA models deal with the inherent defects of robust optimization. (4) This paper uses real numbers to verify the effectiveness of the model.

The structure of this paper is as follows. Section 2 introduces some preliminary concepts. Three robust DEA models are explained in Section 3. Section 4 highlights applicability of models mentioned earlier with endowment insurance system in China along with suggestions and recommendations. Section 5 elaborates the data-driven robust DEA models and its application in endowment insurance system. At the end, Section 6 reflects the conclusion of this paper.

2. Preliminaries

In this section, related knowledge about DEA and Interval DEA theory is presented.

2.1. Data Envelopment Analysis

Data envelopment analysis (DEA) was first proposed by Charnes et al. to measure the relative efficiency [8]. Suppose we have n homogeneous decision-making units (DMU_s). Here, for each DMU_j, it consumes m inputs (fund income, the number of insured and the number of retirees) that are denoted by x_{ij} and produces s outputs (fund expenditure and accumulated fund balance) which are denoted by y_{rj}. Then, we obtain the following model:

$$\begin{aligned} \max\theta = & \frac{\sum\limits_{r=1}^{s} u_r y_{ro}}{\sum\limits_{i=1}^{m} v_i x_{io}} \\ s.t. \quad & \frac{\sum\limits_{r=1}^{s} u_r y_{rj}}{\sum\limits_{i=1}^{m} v_i x_{ij}} \leq 1, \forall_j \\ & u_r, v_i \geq 0. \end{aligned} \tag{1}$$

Obviously, model (1) is a fractional programming. Charnes–Cooper transformation can be used to obtain a linear programming [6]:

$$\begin{aligned} \max\theta = & \sum\limits_{r=1}^{S} u_r y \\ s.t. \quad & \sum\limits_{i=1}^{m} v_i x_{io} \leq 1 \\ & \sum\limits_{r=1}^{S} u_r y_{rj} - \sum\limits_{i=1}^{m} v_i x_{ij} \leq 0, \forall_j \\ & u_r, v_i \geq 0. \end{aligned} \tag{2}$$

Here, θ represents the efficiency of decision-making units. At the same time, we use "$\leq$" in our DEA model instead of "=" in the standard DEA model. The reason why we make this change is to avoid any infeasibility [38].

2.2. Interval DEA Model

This subsection introduces an Interval DEA model, that is, the ratio of all possible comprehensive inputs and comprehensive outputs of the DMU as all possible efficiency values, so the efficiency value obtained is presented as an interval form. We assume that the upper bounds and lower bounds of output are y_{ij}^U and y_{ij}^L, respectively.

Firstly, we consider the most favorable situation for objective DMU. Let the output of objective DMU take the maximum value, and the output of other DMUs take the minimum value, so that the model of the upper bound of interval efficiency value can be obtained.

$$\begin{aligned} \max\theta^U = & \sum\limits_{r=1}^{S} u_r y_{ro}^U \\ s.t. \quad & \sum\limits_{i=1}^{m} v_i x_{io}^L \leq 1 \\ & \sum\limits_{r=1}^{S} u_r y_{rj}^L - \sum\limits_{i=1}^{m} v_i x_{ij}^U \leq 0, \forall_j \\ & u_r, v_i \geq 0. \end{aligned} \tag{3}$$

Similarly, considering the most unfavorable situation, the model of the lower bound of interval efficiency value can be obtained:

$$\begin{array}{ll} \max\theta^L = & \sum\limits_{r=1}^{S} u_r y_{ro}^L \\ s.t. & \sum\limits_{i=1}^{m} v_i x_{io}^U \leq 1 \\ & \sum\limits_{r=1}^{S} u_r y_{rj}^U - \sum\limits_{i=1}^{m} v_i x_{ij}^L \leq 0, \forall_j \\ & u_r, v_i \geq 0. \end{array} \tag{4}$$

When we use the maximum value of input and the minimum value of output in model (3), the result is the lowest value of efficiency. In contrast, we obtain the maximum value of efficiency in model (4), that is to say $\theta \in [\theta^L, \theta^U]$.

3. Robust DEA Model

In this model (2), decision variables and parameters are all deterministic. However, in a real-world scenario, this is likely to lead to errors when uncertain factors exist. In our research, the operational efficiency of the endowment insurance system will be affected by policy adjustments and economic shocks. Therefore, the DEA model under certain circumstances is not suitable for real situations. We must consider the impact of these two uncertainties on output.

Robust optimization is an approach which is seeking the optimal solution in the worst case. In this paper, three uncertainty sets to describe the uncertainty were considered to describe different kinds of uncertainties which may influence the results.

For the output, it consists of two parts. The first part is determinate value, the other part is uncertainty value. We express the uncertainty as follows:

$$U = \left\{ y_{ro}^D = y_{ro} + \sum_{l=1}^{L} y_{rol}^F \xi_l, y_{rj}^D = y_{rj} + \sum_{l=1}^{L} y_{rjl}^F \xi_l, \xi_l \in Z \right\} \tag{5}$$

Thus, model (2) takes the following form

$$\begin{array}{ll} \max\theta = & \sum\limits_{r=1}^{s} u_r y_{ro} \\ s.t. & \sum\limits_{i=1}^{m} v_i x_{io} \leq 1 \\ & \sum\limits_{r=1}^{s} u_r y_{rj} - \sum\limits_{i=1}^{m} v_i x_{io} \leq 0, \forall_j \\ & u_r, v_i \geq 0. \end{array} \tag{6}$$

y_{ro} and y_{rj} denote the outputs in the deterministic situation, y_{rol}^F and y_{rjl}^F are output fluctuation caused by different uncertainty factor. Then, ξ_l represents the uncertainty factor.

Finally, the following Programming can be obtained:

$$\begin{array}{ll} \max & \theta \\ s.t. & \theta - \sum\limits_{r=1}^{S} u_r (y_{ro} + \sum\limits_{l=1}^{L} y_{rol}^F \xi_l) \leq 0 \\ & \sum\limits_{i=1}^{m} v_i x_i \leq 1 \\ & \sum\limits_{r=1}^{S} u_r (\sum\limits_{l=1}^{L} y_{rjl}^F \xi_l) - \sum\limits_{i=1}^{m} v_i x_{ij} \leq 0, \forall_j \\ & u_r, v_i \geq 0. \end{array} \tag{7}$$

3.1. Robust Model Based on Box Uncertainty Set

For the RDEA model, we consider the most simple uncertainty set-box uncertainty set first.

Proposition 1. *The robust data envelopment analysis model based on box uncertainty set can be constructed as:*

$$
\begin{aligned}
\max\quad & \theta \\
\text{s.t.}\quad & \theta + \Phi \sum_{l=1}^{L} \sum_{r=1}^{S} u_r y_{rol}^F - \sum_{r=1}^{S} u_r y_{ro} \le 0 \\
& \sum_{i=1}^{m} v_i x_{io} \le 1 \\
& \Phi \sum_{l=1}^{L} \sum_{r=1}^{S} u_r y_{rjl}^F + \sum_{r=1}^{S} u_r y_{rj} - \sum_{i=1}^{m} v_i x_{ij} \le 0,\ \forall_j \\
& u_r, v_i \ge 0,
\end{aligned}
\tag{8}
$$

where the uncertainty set can be defined as Z^B,$Z^B = \left\{ \xi \in \mathrm{R}^{\mathrm{L}} : ||\xi|| \le \Phi \right\}$. Φ represents the uncertainty parameter, which measures the degree of uncertainty in the case of box uncertainty set. $\sum_{l=1}^{L} \sum_{r=1}^{s} u_r y_{rjl}^F$ represents the disturbance of output data, and measures the disturbance of L uncertain factors to output.

3.2. Robust Model Based on Ellipsoid Uncertainty Set

Now, we consider the model based on the ellipsoid uncertainty set.

Proposition 2. *The robust data envelopment analysis model based on ellipsoid uncertainty set can be constructed as:*

$$
\begin{aligned}
\max\quad & \theta \\
\text{s.t.}\quad & \theta + \Omega \sqrt{\sum_{l=1}^{L} \left(\sum_{r=1}^{S} u_r y_{rol}^F \right)^2} - \sum_{r=1}^{S} u_r y_{ro} \le 0 \\
& \sum_{i=1}^{m} v_i x_{io} \le 1 \\
& \Omega \sqrt{\sum_{l=1}^{L} \left(\sum_{r=1}^{S} u_r y_{rol}^F \right)^2} - \sum_{i=1}^{m} v_i x_{ij} + \sum_{r=1}^{S} u_r y_{rj} \le 0,\ \forall_j \\
& u_r, v_i \ge 0,
\end{aligned}
\tag{9}
$$

where the uncertainty set can be defined as Z^B, $Z^E = \left\{ \xi \in \mathrm{R}^{\mathrm{L}} : ||\xi||_2 \le \Omega \right\}$. Ω represents the uncertainty parameter, which measures the degree of uncertainty in the case of the ellipsoid uncertainty set. $\sum_{l=1}^{L} (\sum_{r=1}^{s} u_r y_{rjl}^F)^2$ represents the disturbance of output data, and measures the disturbance of L uncertain factors to output.

3.3. Robust Model Based on Polyhedron Uncertainty Set

Finally, we consider the polyhedron uncertainty set.

Proposition 3. *The robust data envelopment analysis model based on polyhedron uncertainty set can be constructed as:*

$$
\begin{aligned}
&\max\theta \\
&s.t.\theta + \Gamma \sum_{l=1}^{L} p_{ol} - \sum_{r=1}^{S} y_{ro}u_r \le 0 \\
&\sum_{i=1}^{m} v_i x_{io} \le 1 \\
&\Gamma \sum_{l=1}^{L} p_{jl} + \sum_{r=1}^{S} y_{rj}u_r - \sum_{i=1}^{m} x_{ij}v_i \le 0, \forall_j \\
&u_r, v_i \ge 0,
\end{aligned} \tag{10}
$$

where the uncertainty set can be defined as Z^P, $Z^P = \left\{\xi \in \mathrm{R}^L : ||\xi||_1 \le \Gamma\right\}$. Γ represents the uncertainty parameter, which measures the degree of uncertainty in the case of polyhedron uncertainty set. $\sum_{l=1}^{L} p_{jl}$ represents the disturbance of output data, and measures the disturbance of L uncertain factors to output.

We present proof of Proposition 1, Proposition 2 and Proposition 3 in Appendix A.

4. Simulation Results

4.1. Data and Variable Selections

Although the endowment insurance system was implemented in 2014, the statistical caliber of the data can be traced back to 2012. To ensure the consistency of the data, this paper studies the operation efficiency of the endowment insurance system of 31 provinces in China from 2017 to 2019 (latest available data). The data of input and output indicators are all from China Statistical Yearbook and provincial statistical yearbooks [39]. The data are shown in Table 1.

Table 1. Endowment insurance data of 31 provinces.

Province	Inputs			Outputs	
	Number of Insured Persons	Number of Retirees	Fund Income	Fund Expenditure	Accumulated Fund Balance
Beijing	[1604.5, 1748.2]	[283.1, 302.6]	[2223, 2760.6]	[1394.3, 1698]	[4395, 6018.5]
Tianjin	[655, 695.6]	[213.8, 226.3]	[894.3, 1120.3]	[836.2, 1059.9]	[463.2, 556.5]
Hebei	[1535.8, 1654.5]	[433.8, 466.7]	[1439.2, 2437.4]	[1411.6, 2425]	[735.2, 910]
Shanxi	[798.7, 871.5]	[243.0, 273.6]	[1223.4, 1234.6]	[1082.3, 1168]	[1457.7, 1640]
Neimenggu	[694.3, 763.4]	[257.1, 298.8]	[853.5, 1094.6]	[707.2, 1201.4]	[595.9, 656.5]
Liaoning	[1949.8, 2026.2]	[754.4, 816]	[1863.2, 2486.4]	[2207, 2950]	[303.7, 572.8]
Jilin	[814.6, 882.1]	[332.2, 375.9]	[764.1, 1142.8]	[767, 1263.6]	[340, 504.2]
Heilongjiang	[1206.1, 1364.9]	[523.9, 599.8]	[1240.5, 1785.4]	[1534.2, 2094]	[−557.2, −433.7]
Shanghai	[1548.2, 1589.6]	[489.2, 511.9]	[2767.4, 2933.7]	[2571.1, 2779]	[2069, 2290.3]
Jiangsu	[3034.5, 3417.4]	[796.1, 918.1]	[2885.6, 3923.3]	[2555.3, 3401]	[3731, 4932.4]
Zhejiang	[2712.4, 3031.7]	[747.5, 856.9]	[3011.8, 3052.6]	[2636.7, 3138.]	[3585.4, 3797]
Anhui	[1077.0, 1217.0]	[322.9, 356.7]	[993.3, 2005.6]	[784.6, 1732.7]	[1394, 1909.7]
Fujian	[1022.1, 1137.3]	[119.1, 190.6]	[785.3, 931.8]	[666.5, 782.2]	[820, 976.2]
Jiangxi	[1005.2, 1096.6]	[307.7, 348.4]	[974.1, 1169.8]	[862.6, 1083.9]	[638.1, 824.6]
Shandong	[2660.9, 2868.0]	[638.8, 711.2]	[2289.3, 2784.7]	[2358.7, 2872]	[2217.2, 2387]
Henan	[1897.6, 2133.8]	[460.0, 505.6]	[1521.5, 2053]	[1471.8, 1931]	[1104, 1326.3]
Hubei	[1546.6, 1684.8]	[526.1, 584.3]	[1793.6, 2418]	[1864.2, 2264]	[743.4, 1017.1]
Hunan	[1279.3, 1557.8]	[422.7, 486]	[1448.1, 2129.2]	[1349.1, 1620]	[1104, 1836.7]
Guangdong	[4633.4, 5287.1]	[569.0, 671.2]	[3457.0, 5593.2]	[1898.0, 3761.]	[9245, 12,343.6]
Guangxi	[777.8, 869.5]	[251.9, 268.4]	[977.0, 1248.9]	[881.9, 1126.8]	[556.7, 755.2]
Hainan	[240.9, 281]	[68.9, 72.7]	[271.1, 326.1]	[232, 280]	[173.5, 281.4]
Chongqing	[989.2, 1127.7]	[360.8, 406.6]	[1202.3, 1434.7]	[1093, 1372.4]	[897.1, 1090.1]
Sichuan	[2335.1, 2700.3]	[816.0, 915.7]	[2754.9, 3295.9]	[2276.4, 2764]	[3246, 3759.5]
Guizhou	[588.2, 677.5]	[141.3, 155.8]	[667.1, 799.3]	[575.7, 636.6]	[619.2, 894]
Yunnan	[591.5, 649.9]	[171.3, 181.5]	[878.0, 1096.0]	[690.4, 958.9]	[950.8, 1325.2]
Xizang	[42.9, 48.2]	[9.2, 10]	[110.5, 139.1]	[84.7, 107.4]	[123.6, 171.2]
Shanxi	[953.3, 1080.7]	[246.4, 264.1]	[1049.2, 1254.1]	[961.8, 1187.5]	[566.1, 804.2]
Gansu	[429.8, 469.4]	[141.6, 159.6]	[391.3, 598.5]	[363.5, 599.3]	[403.7, 467]
Qinghai	[138.3, 152.8]	[42.8, 46.7]	[197.6, 300.5]	[205.5, 323.3]	[37, 55.8]
Ningxia	[205.2, 226.6]	[60.2, 66]	[243.0, 269.1]	[221.4, 266.8]	[217.7, 261.5]
Xinjiang	[646.4, 744.2]	[204.3, 219.1]	[1006.1, 1137.1]	[906, 1040.9]	[1074, 1307]

Inputs of endowment insurance mainly include fund income, number of insured and the number of retirees. The first index is fund income. According to the relevant provisions in China, the income of the endowment insurance fund is paid by the payment units. The individuals are included in the scope of the endowment insurance according to the payment base and payment proportion stipulated by the government, as well as the income obtained through other ways to form the source of the fund. It includes the endowment insurance premium paid by the unit and individual employees, the interest income of the endowment insurance fund, the subsidy income of the higher level, the income of the lower level, the transfer income, the financial subsidy, and other income.

The last indicator as an input is the number of retirees. For the service object of endowment insurance, the number of retirees is directly related to the number of services provided by endowment insurance. The more the number of retirees, the higher the expenditure of the endowment insurance fund and the greater the expenditure pressure of the corresponding endowment insurance fund will occur.

In the selection of output indicators, the following two indicators are determined: fund expenditure and accumulated fund balance. The first index is used to measure the number of public services in the operation of endowment insurance system, which is the direct performance of the operations of endowment insurance system. The scope of expenditure mainly includes the pension of retirees, who participate in endowment insurance. The pension of retirees, and the payment of various kinds of stickers, medical expenses, death and funeral subsidies, etc., are also included. Therefore, it can be used as an output indicator. This index refers to the accumulated balance of the endowment insurance fund in a certain period time for the accumulated balance of the fund. It measures the endurance of the endowment insurance system in China, that is, the sustainability of its development. Therefore, it is also an output index that can reflect the operation of the endowment insurance system.

4.2. Interval DEA Results

First of all, this paper uses the average data of input and output as the input and output of DEA model, and obtains the efficiency value and the rank of the comprehensive performance of the operation of endowment insurance in 31 provinces from 2017 to 2019. It should be mentioned that the efficiency value of DEA model does not consider the disturbance of data. Table 2 shows the efficiency values calculated by DEA model. The operation of endowment insurance system in nine provinces is effective: Beijing, Shanxi, Liaoning, Heilongjiang, Shanghai, Zhejiang, Guangdong, Xizang, Qinghai. The efficiency of inefficient securities firms ranged from 0.651 to 0.994.

Table 3 show the results of interval DEA model. The difference between the upper bound and the lower bound is quite different among the 31 DMUs. Among them, the largest value is 0.726, while the smallest is 0.364. It is worth mentioning that all the DMUs have the same upper bound of 1.000, and this does not mean that so many DMUs are efficient. This is because it is hard to find a realistic scenario in which all DMUs are in the most favorable situation at the same time.

In the DEA model, we just need to rank them according to their efficiency value. As mentioned before, the data disturbance is not considered in DEA model. Therefore, the accuracy of the results is too difficult to be assured. In the interval DEA model, the data disturbance is considered. However, it is difficult to rank them accordingly. For DMUs with the same upper bound, ranking them according to their lower bounds is likely to lead to mistakes because we have to know their distribution. In a real-world scenario, these distributions are difficult to describe.

Table 2. Interval DEA efficiency.

Province	DEA Efficiency	Interval DEA Efficiency
Beijing	1.000	[0.663, 1.000]
Tianjin	0.933	[0.523, 1.000]
Hebei	0.935	[0.392, 1.000]
Shanxi	1.000	[0.670, 1.000]
Neimenggu	0.959	[0.452, 1.000]
Liaoning	1.000	[0.532, 1.000]
Jilin	0.910	[0.413, 1.000]
Heilongjiang	1.000	[0.519, 1.000]
Shanghai	1.000	[0.687, 1.000]
Jiangsu	0.963	[0.468, 1.000]
Zhejiang	1.000	[0.562, 1.000]
Anhui	0.906	[0.364, 1.000]
Fujian	0.920	[0.425, 1.000]
Jiangxi	0.889	[0.433, 1.000]
Shandong	0.999	[0.527, 1.000]
Henan	0.921	[0.444, 1.000]
Hubei	0.923	[0.495, 1.000]
Hunan	0.868	[0.431, 1.000]
Guangdong	1.000	[0.636, 1.000]
Guangxi	0.904	[0.475, 1.000]
Hainan	0.857	[0.460, 1.000]
Chongqing	0.923	[0.499, 1.000]
Sichuan	0.922	[0.506, 1.000]
Guizhou	0.902	[0.513, 1.000]
Yunnan	0.938	[0.525, 1.000]
Xizang	1.000	[0.726, 1.000]
Shanxi	0.909	[0.497, 1.000]
Gansu	0.943	[0.420, 1.000]
Qinghai	1.000	[0.557, 1.000]
Ningxia	0.950	[0.535, 1.000]
Xinjiang	0.979	[0.613, 1.000]
Mean	0.947	[0.515, 1.000]

4.3. Robust DEA Results

In the robust DEA model, we consider not only the uncertainty of the output, but also the influence of different uncertain factors on the outputs. In our model, we mainly consider the impact of government policy adjustment and economic shocks on output, thus, we set $l = 2$, which means we consider two uncertain factors that affect the outputs. Here, we suppose $y_{rjl}^F = 0.02y_{rj}, y_{rol}^F = 0.02y_{rj}$, and the uncertainty parameter range from 0 to 5, which represents different degrees of uncertainty. From the previous parameter setting results, we can know the output disturbance value range from 0 to 0.1.

4.3.1. Robust DEA Results Based on Box Set

When the uncertainty set is a box set, the RDEA efficiency values of 31 DMUs are shown in Table 3. As parameter Φ changes, RDEA efficiency changes accordingly. At the beginning, the parameter $\Phi = 0$, box-robust data envelopment analysis model is equivalent to the data envelopment analysis model. In this case, the robust problem here is equal to a nominal problem, which has no disturbance. When parameter Φ increases to 1, the box uncertainty set is equivalent to the interval uncertainty set. The maximum value of the efficiency of endowment insurance system in 31 provinces is 0.923, the minimum value is 0.791, and the average value is 0.874. With the increase of uncertainty parameter Φ, the average efficiency value of the RDEA model decreases from 0.947 to 0.631.

Table 3. RDEA efficiency based on box set.

Province	RDEA Efficiency						Rank
	$\Phi = 0$	$\Phi = 1$	$\Phi = 2$	$\Phi = 3$	$\Phi = 4$	$\Phi = 5$	R
Beijing	1.000	0.923	0.852	0.786	0.724	0.667	01
Tianjin	0.933	0.861	0.795	0.733	0.676	0.622	18
Hebei	0.935	0.863	0.797	0.735	0.677	0.623	17
Shanxi	1.000	0.923	0.852	0.786	0.724	0.667	01
Neimenggu	0.959	0.885	0.817	0.753	0.694	0.639	13
Liaoning	1.000	0.923	0.852	0.786	0.724	0.667	01
Jilin	0.910	0.840	0.775	0.715	0.659	0.607	24
Heilongjiang	1.000	0.923	0.852	0.786	0.724	0.667	01
Shanghai	1.000	0.923	0.852	0.786	0.724	0.667	01
Jiangsu	0.963	0.889	0.820	0.757	0.697	0.642	12
Zhejiang	1.000	0.923	0.852	0.786	0.724	0.667	01
Anhui	0.906	0.836	0.772	0.712	0.656	0.604	26
Fujian	0.920	0.849	0.784	0.723	0.666	0.613	23
Jiangxi	0.889	0.821	0.757	0.698	0.644	0.593	29
Shandong	0.999	0.922	0.851	0.785	0.723	0.666	10
Henan	0.921	0.850	0.785	0.724	0.667	0.614	22
Hubei	0.923	0.852	0.786	0.725	0.668	0.615	19
Hunan	0.868	0.801	0.739	0.682	0.629	0.579	30
Guangdong	1.000	0.923	0.852	0.786	0.724	0.667	01
Guangxi	0.904	0.834	0.770	0.710	0.655	0.603	27
Hainan	0.857	0.791	0.730	0.673	0.621	0.571	31
Chongqing	0.923	0.852	0.786	0.725	0.668	0.615	19
Sichuan	0.922	0.851	0.785	0.724	0.668	0.615	21
Guizhou	0.902	0.833	0.768	0.709	0.653	0.601	28
Yunnan	0.938	0.866	0.799	0.737	0.679	0.625	16
Xizang	1.000	0.923	0.852	0.786	0.724	0.667	01
Shanxi	0.909	0.839	0.774	0.714	0.658	0.606	25
Gansu	0.943	0.870	0.803	0.741	0.683	0.629	15
Qinghai	1.000	0.923	0.852	0.786	0.724	0.667	01
Ningxia	0.950	0.877	0.809	0.746	0.688	0.633	14
Xinjiang	0.979	0.904	0.834	0.769	0.709	0.653	11
Mean	0.947	0.874	0.807	0.744	0.686	0.631	

4.3.2. Robust DEA Results Based on Ellipsoid Set

When the uncertainty set is an ellipsoid set, the RDEA efficiency values of endowment insurance system in 31 provinces is shown in Table 4. Similarly, the problem is equivalent to a nominal problem with the uncertainty parameter Ω. When the parameter increases from 0 to 1, Z^E is the largest ellipsoid contained in the Z^B. In this case, the maximum value of the efficiency of endowment insurance system in 31 provinces is 0.945, the minimum value is 0.616 and the average efficiency is 0.810. As the uncertainty parameter increases, the RDEA efficiency decreases gradually. It is worth noting that the speed of efficiency reduction here is lower than that of the former uncertainty set. When the uncertainty parameter Ω increases to 5, which represents the most conservative situation, the average efficiency drops to 0.712. From the mean results, we can infer that the ellipsoid set has stronger robustness than interval sets.

4.3.3. Robust DEA Results Based on Polyhedron Set

Here, we consider the case that the uncertainty set is a polyhedron uncertainty set. The results of endowment insurance system in 31 provinces are displayed in Table 5. When the uncertainty parameter $\Gamma = 0$, the problem is equivalent to a nominal problem. With the increase of the uncertainty parameter, the RDEA efficiency decrease from 0.947 to 0.686. Different from the previous table, with the change of uncertain parameters, the ranking of DMUs efficiency value is also changing. However, the range of change is small, and it is still relatively stable.

Table 4. RDEA efficiency based on ellipsoid set.

Province	RDEA Efficiency						Rank
	Ω = 0	Ω = 1	Ω = 2	Ω = 3	Ω = 4	Ω = 5	R
Beijing	1.000	0.945	0.893	0.844	0.797	0.752	01
Tianjin	0.933	0.881	0.833	0.787	0.743	0.702	18
Hebei	0.935	0.884	0.835	0.789	0.745	0.703	17
Shanxi	1.000	0.945	0.893	0.844	0.797	0.752	01
Neimenggu	0.959	0.906	0.856	0.809	0.764	0.721	13
Liaoning	1.000	0.945	0.893	0.844	0.797	0.752	01
Jilin	0.910	0.860	0.813	0.768	0.725	0.684	24
Heilongjiang	1.000	0.945	0.893	0.844	0.797	0.752	01
Shanghai	1.000	0.945	0.893	0.844	0.797	0.752	01
Jiangsu	0.963	0.910	0.860	0.812	0.767	0.724	12
Zhejiang	1.000	0.945	0.893	0.844	0.797	0.752	01
Anhui	0.906	0.856	0.809	0.764	0.722	0.681	26
Fujian	0.920	0.869	0.821	0.776	0.733	0.692	23
Jiangxi	0.889	0.840	0.794	0.750	0.708	0.669	29
Shandong	0.999	0.944	0.892	0.843	0.796	0.751	10
Henan	0.921	0.870	0.822	0.777	0.734	0.693	22
Hubei	0.923	0.872	0.824	0.779	0.735	0.694	19
Hunan	0.868	0.820	0.775	0.732	0.692	0.653	30
Guangdong	1.000	0.945	0.893	0.844	0.797	0.752	01
Guangxi	0.904	0.854	0.807	0.763	0.720	0.680	27
Hainan	0.857	0.810	0.765	0.723	0.683	0.645	31
Chongqing	0.923	0.872	0.824	0.779	0.735	0.694	19
Sichuan	0.922	0.871	0.823	0.778	0.735	0.694	21
Guizhou	0.902	0.852	0.805	0.761	0.719	0.678	28
Yunnan	0.938	0.886	0.838	0.791	0.747	0.705	16
Xizang	1.000	0.945	0.893	0.844	0.797	0.752	01
Shanxi	0.909	0.859	0.812	0.767	0.724	0.684	25
Gansu	0.943	0.891	0.842	0.796	0.751	0.709	15
Qinghai	1.000	0.945	0.893	0.844	0.797	0.752	01
Ningxia	0.950	0.898	0.848	0.801	0.757	0.715	14
Xinjiang	0.979	0.925	0.874	0.826	0.780	0.736	11
Mean	0.947	0.895	0.845	0.799	0.754	0.712	

4.4. Comparison between Models Based on a Different Uncertainty Set

Φ, Ω and Γ in the robust DEA model are adjustable parameters to control the size level of an uncertain set, which represents the conservative degree of constraints. They not only describe the fluctuation of the value of the uncertain parameters in the geometry of a certain shape, but also reflect the uncertainty of the outputs. Figure 1 shows three curves of efficiency values varying with uncertainty parameters. With the increase of uncertainty parameters, the efficiency values based on different uncertain sets decrease at different speeds. Intuitively, the results of polyhedron uncertainty model and ellipsoid uncertainty model are better, the efficiency based on box uncertainty model performs the worst, and the results are far from those of the former two models. From the perspective of ranking, the results of the box uncertainty set and the ellipsoid uncertainty set show extremely strong stability. The results of the polyhedron uncertainty set show some fluctuation, but the fluctuation is acceptable. The results of rank are shown in Figures 2–4, respectively. In general, the robust model with ellipsoidal uncertainty shows the highest efficiency and the strongest robustness.

Table 5. RDEA efficiency based on polyhedron set.

Province	RDEA Efficiency (Rank)					
	Γ = 0	Γ = 1	Γ = 2	Γ = 3	Γ = 4	Γ = 5
Beijing	1.000 (01)	0.961 (01)	0.923 (01)	0.886 (03)	0.851 (03)	0.817 (03)
Tianjin	0.933 (18)	0.866 (19)	0.805 (21)	0.748 (21)	0.695 (22)	0.645 (21)
Hebei	0.935 (17)	0.867 (18)	0.805 (21)	0.746 (22)	0.692 (23)	0.641 (23)
Shanxi	1.000 (01)	0.949 (04)	0.900 (04)	0.855 (04)	0.811 (04)	0.770 (04)
Neimenggu	0.959 (13)	0.891 (13)	0.829 (15)	0.771 (17)	0.717 (18)	0.667 (18)
Liaoning	1.000 (01)	0.926 (08)	0.857 (10)	0.793 (10)	0.733 (12)	0.677 (15)
Jilin	0.910 (24)	0.846 (25)	0.787 (26)	0.733 (26)	0.682 (26)	0.634 (26)
Heilongjiang	1.000 (01)	0.923 (10)	0.852 (12)	0.786 (13)	0.724 (16)	0.667 (18)
Shanghai	1.000 (01)	0.933 (06)	0.871 (06)	0.814 (08)	0.760 (09)	0.710 (10)
Jiangsu	0.963 (12)	0.910 (12)	0.862 (09)	0.816 (07)	0.773 (07)	0.732 (07)
Zhejiang	1.000 (01)	0.945 (05)	0.893 (05)	0.844 (05)	0.798 (05)	0.754 (05)
Anhui	0.906 (26)	0.854 (24)	0.807 (19)	0.763 (18)	0.722 (17)	0.684 (13)
Fujian	0.920 (23)	0.861 (21)	0.808 (18)	0.759 (19)	0.714 (19)	0.671 (17)
Jiangxi	0.889 (29)	0.831 (29)	0.777 (29)	0.726 (29)	0.679 (27)	0.636 (24)
Shandong	0.999 (10)	0.933 (06)	0.871 (06)	0.814 (08)	0.761 (08)	0.713 (09)
Henan	0.921 (22)	0.856 (22)	0.797 (23)	0.742 (24)	0.692 (23)	0.645 (21)
Hubei	0.923 (19)	0.856 (22)	0.794 (24)	0.736 (25)	0.683 (25)	0.632 (27)
Hunan	0.868 (30)	0.813 (30)	0.763 (30)	0.716 (30)	0.673 (30)	0.635 (25)
Guangdong	1.000 (01)	0.961 (01)	0.923 (01)	0.887 (01)	0.852 (01)	0.818 (01)
Guangxi	0.904 (27)	0.840 (28)	0.781 (28)	0.726 (28)	0.675 (29)	0.628 (29)
Hainan	0.857 (31)	0.799 (31)	0.744 (31)	0.694 (31)	0.648 (31)	0.606 (30)
Chongqing	0.923 (19)	0.863 (20)	0.807 (19)	0.756 (20)	0.708 (20)	0.563 (31)
Sichuan	0.922 (21)	0.873 (17)	0.826 (16)	0.782 (14)	0.740 (11)	0.701 (11)
Guizhou	0.902 (28)	0.844 (27)	0.792 (25)	0.745 (23)	0.703 (21)	0.663 (20)
Yunnan	0.938 (16)	0.882 (15)	0.835 (13)	0.793 (10)	0.753 (10)	0.715 (08)
Xizang	1.000 (01)	0.961 (01)	0.923 (01)	0.887 (01)	0.852 (01)	0.818 (01)
Shanxi	0.909 (25)	0.845 (26)	0.785 (27)	0.730 (27)	0.679 (27)	0.631 (28)
Gansu	0.943 (15)	0.882 (15)	0.826 (16)	0.775 (16)	0.726 (15)	0.682 (14)
Qinghai	1.000 (01)	0.925 (09)	0.856 (11)	0.792 (12)	0.733 (12)	0.677 (15)
Ningxia	0.950 (14)	0.887 (14)	0.830 (14)	0.777 (15)	0.731 (14)	0.689 (12)
Xinjiang	0.979 (11)	0.916 (11)	0.864 (08)	0.819 (06)	0.776 (06)	0.736 (06)
Mean	0.947	0.887	0.832	0.781	0.733	0.686

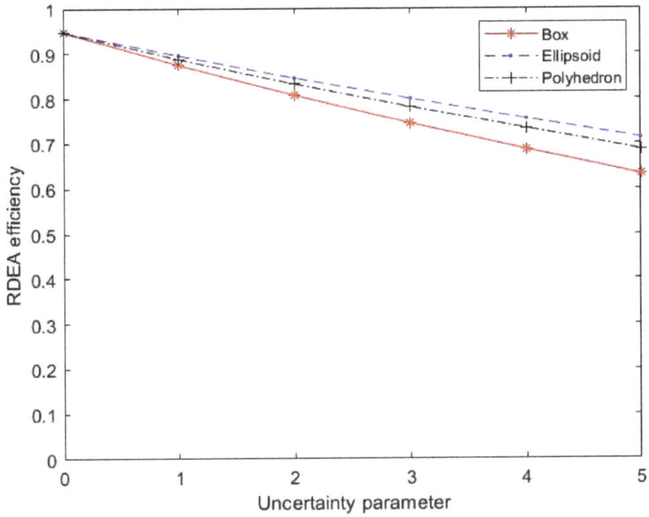

Figure 1. Comparison between different uncertainty sets.

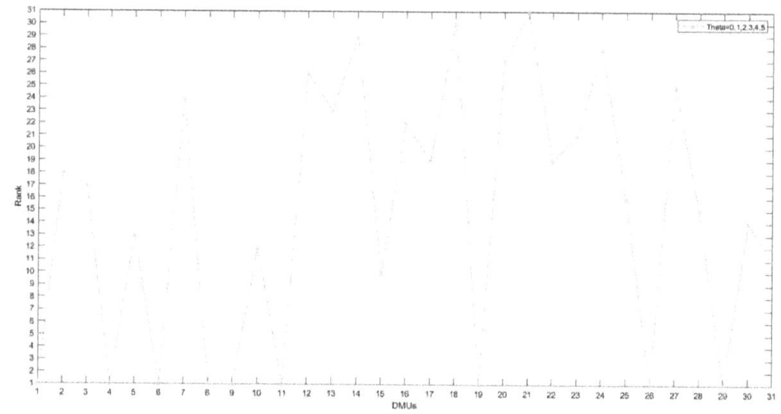

Figure 2. The rank of 31 DMUs in different uncertainty levels (box uncertainty).

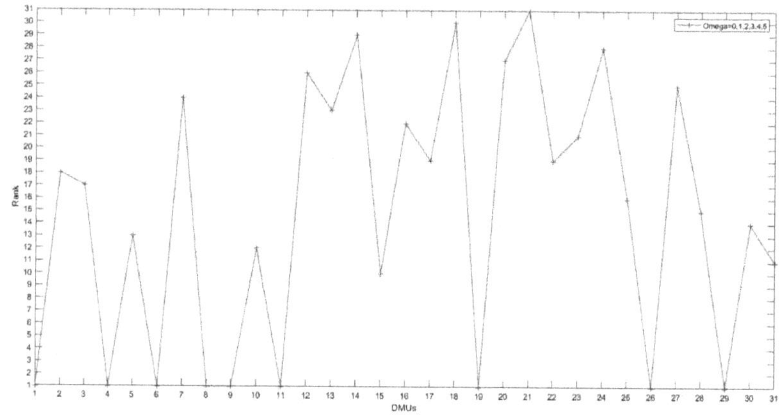

Figure 3. The rank of 31 DMUs in different uncertainty levels (ellipsoid uncertainty).

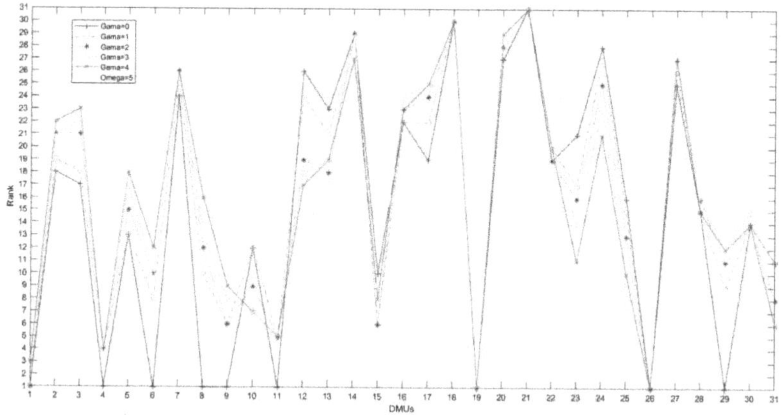

Figure 4. The rank of 31 DMUs in different uncertainty levels (polyhedron uncertainty).

4.5. Comparison between Interval DEA Model and Robust DEA Models

According to the previous subsection, we know that the robust DEA model based on ellipsoid uncertainty set performs the best. In this subsection, we compare the ellipsoid RDEA model with the interval DEA model, and the result are shown in Figure 5.

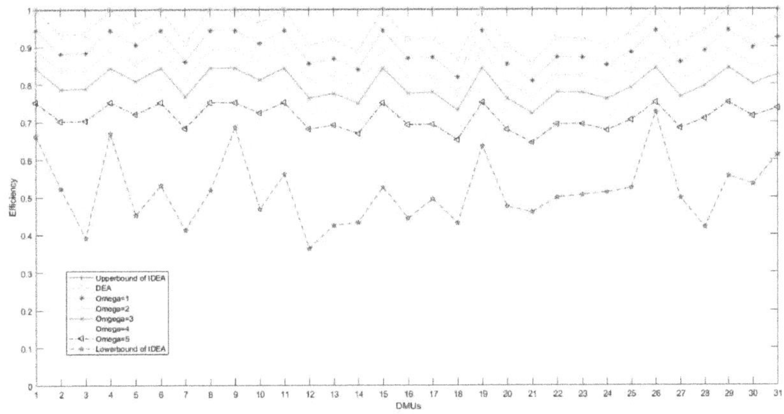

Figure 5. Efficiency of 31 DMUs based on different models.

We compare the two models from two perspectives:

(1) Is the result convenient to compare the efficiency of DMUs?
(2) Is the result close to the real scenario?

According to the results shown in Figure 5, all the upper bounds of the IDEA model are 1. That is to say, all the DMUs can be judged to be effective under extremely favorable conditions. However, in the real world, the probability of this kind of situation is very low because it is difficult to make all units reach their best state. The lower bound of IDEA model ranges from 0.364 to 0.726, which seems to tell us that we can judge the efficiency ranking of DMUs with the same upper bound by comparing the lower bounds of DMUs. Yet this intuition is wrong, we have to know the distribution of efficiency value of each DMU, and this kind of information is often difficult to get in the real world. However, in the robust model, the final result is a certain number, and it is not necessary to determine the distribution of the value. The efficiency value of each DMU can be easily compared. From this point of view, the robust model performs better.

The result of the IDEA model is an interval. Its lower bound shows the lowest value that the DMU may take under the most unfavorable situation, while the upper bound shows the lowest value that the DMU may take under the most unfavorable situation. The IDEA model deals with uncertainty in the form of interval. However, once the interval is calculated, it is a value that cannot be adjusted. In the robust model, the uncertainty parameters describe the size of the uncertainty, we can adjust the size of the parameters according to the actual needs. The results are not only convenient to compare the efficiency of decision-making units, but also fit the real scenario, are more accurate, and provide important information for decision-makers. From the above analysis, a robust model is more suitable to solve our problem.

5. Data-Driven Robust DEA Models

In the Section 3, we assume that the output unit is uncertain because of the needs of the real situation. In the robust optimization method, there is not enough information and we also want to ensure the absolute robustness of the result, which leads to the result being too conservative. In the real situation, the tendency of macro policy of government can be expected approximately. At the same time, we can obtain some useful information from some historical data, rather than having no information available.

In the Section 5, contrary to the above, we assume that the observation sets of the output unit P can be obtained, where $P = \{y_{r1}, \ldots, y_{rM}\}$ and $y_{rq} \in R, q = 1, \ldots, M$. In other words, P is the raw data that we can obtain. At the same time, let y_r represent the average values of $\{y_{r1}, \ldots, y_{rM}\}$, i.e., $\overset{\wedge}{y_r} = \frac{1}{M}\sum_{q \in M} y_{rq}$, where $[M] = \{1, \ldots, M\}$. Therefore, we need to find a decision variable that maximize the worst-case efficiency over all the costs in uncertainty set Z. This is the robust DEA problem:

$$
\begin{aligned}
\max \quad & \{\min_{y_r \in Z} \sum_{r \in [s]} u_r y_r\} \\
s.t. \quad & \sum_{i=1}^{m} v_i x_i \leq 0 \\
& \sum_{r=1}^{s} u_r y_{rj} - \sum_{i=1}^{m} v_i x_{ij} \leq 0, \forall_j \\
& u_r, v_i \geq 0.
\end{aligned} \tag{11}
$$

In the next subsection, several methods for generating Z will be introduced [35], where each set has a scaling parameter to control its size.

5.1. Box Uncertainty

We set $y_r^d = \min_{q \in [M]} y_{rq}, y_r^u = \max_{q \in [M]} y_{rq}$, for any $\lambda \geq 0$, there is

$$
Z_{Int}^{U} = \times_{r \in [s]} [\overset{\wedge}{y_r} + \lambda(y_r^d - \overset{\wedge}{y_r}), \overset{\wedge}{y_r} + \lambda(y_r^a - \overset{\wedge}{y_r})],
$$

where $\times$ is the Cartesian product and $[s] = \{1, \ldots, s\}$. It should be noticed that $\max_{y_r \in Z} \sum_{r \in [s]} u_r y_r = \sum_{r \in [s]} u_r [\overset{\wedge}{y_{ro}} + \lambda(y_{ro}^a - \overset{\wedge}{y_{ro}})]$.

Therefore, the robust problem obtained is

$$
\begin{aligned}
\min - \quad & \sum_{r \in [s]} u_r [\overset{\wedge}{y_{ro}} + \lambda(y_{ro}^a - \overset{\wedge}{y_{ro}})] \\
s.t. \quad & \sum_{i=1}^{m} v_i x_i \leq 0 \\
& \sum_{r \in [s]} u_r [\overset{\wedge}{y_{rj}} + \lambda(y_{rj}^a - \overset{\wedge}{y_{rj}})] \leq \sum_{i=1}^{m} v_i x_{ij}, \forall_j \\
& u_r, v_r \geq 0.
\end{aligned} \tag{12}
$$

Proof. For the objective function of (2), we just obtain its maximum value:

$$
\max_{y_r \in Z} \sum_{r \in [s]} u_r y_r = \sum_{r \in [s]} u_r [\overset{\wedge}{y_{ro}} + \lambda(y_{ro}^a - \overset{\wedge}{y_{ro}})],
$$

Then, it can be transformed to the equivalent form:

$$
\min - \sum_{r \in [s]} u_r [\hat{y_{ro}} + \lambda(y_{ro}^a - \hat{y_{ro}})],
$$

Similarly, the third inequality of (2) can be transformed to the following form:

$$
\sum_{r \in [s]} u_r [\hat{y_{rj}} + \lambda(y_{rj}^a - \hat{y_{rj}})] \leq \sum_{i=1}^{m} v_i x_{ij}, \forall_j.
$$

Here, the size of box uncertainty set in decided by both uncertainty parameter and the observation sets, which is different from the box uncertainty set of Section 3.1. The proof of Sections 5.2 and 5.3 is just similar to the proof above.□

5.2. Ellipsoidal Uncertainty

Ellipsoid uncertainty sets were derived from the observation that the iso-density locus of the multivariate normal distribution is an ellipse. Therefore, the maximum likelihood fit of a normal distribution $N(\mu, \Sigma)$ of data point $\{y_{r1}, \ldots, y_{rm}\}$ is given by $\mu = \hat{y}_r, \Sigma = \frac{1}{M} \sum_{q \in [M]} (y_{rq} - \mu)(y_{rq} - \mu)^T$. We set an ellipsoid of the form $Z_{Elli} = \{y : (y_r - \hat{y}_r)^T \Sigma^{-1}(y_r - \hat{y}_r) \leq r\}$ with the scaling parameter $\omega \geq 0$ and it is centered on $\hat{y}_r$. Following the similar proof of Section 5.1, the robust problem obtained is

$$
\begin{aligned}
\min - \quad & \sum_{r \in [s]} \hat{y}_{ro} u_r + \gamma \\
s.t. \quad & \omega \cdot u_r{}^T \Sigma u_r \leq \gamma^2, \gamma \in [s] \\
& \omega \cdot u_r{}^T \Sigma u_r \leq \gamma^2 + \sum_{i=1}^{m} v_i x_{ij}, \forall_j \\
& \sum_{i=1}^{m} v_i x_i \leq 1 \\
& u_r, v_i \geq 0.
\end{aligned}
\tag{13}
$$

5.3. Polyhedron Uncertainty

A polyhedron defined using linear equations and inequalities is equivalent to a convex hull. We set $Z_{Poly} = \{y : \hat{y}_r + (y_r^a - \hat{y}_r)\alpha_r, r \in [s], 0 \leq \alpha \leq 1, \sum_{r \in [s]} \alpha_r \leq \lambda\}$ where the scaling parameter λ controls the size of the set. Through the duality of the internal maximization problem, we arrive at the robust problem:

$$
\begin{aligned}
\min - \quad & \sum_{r \in [s]} \hat{y}_r u_r + \lambda\gamma + ||t||_1 \\
s.t \quad & (y_{ro}^a - \hat{y}_{ro}) u_r \leq u + t_r, r \in [s] \\
& (y_{rj}^a - \hat{y}_{rj}) u_r \leq u + t_r + \sum_{i=1}^{m} x_{ij} v_i, \forall_j \\
& \sum_{i=1}^{m} v_i x_i \leq 1 \\
& u_r, v_i \geq 0.
\end{aligned}
\tag{14}
$$

5.4. Numerical Analysis

It is obvious that the scaling parameters will affect the size of each uncertainty set. Here, we just follow the analysis of Section 4, comparing the trends of three proposed uncertainty sets under the same scaling parameters. Scaling parameters here are set from 1 to 5 with a step of 1. In this way, we can clearly see the gradual change of efficiency value.

When the uncertainty set is an interval set, the RDEA efficiency values of 31 DMUs are shown in Table 6. When the uncertainty parameter λ equals 1, the maximum value of the efficiency of endowment insurance system in 31 provinces is 0.942, the minimum value is 0.744. Compared with the Robust DEA model, the result has a larger range. The efficiency rank of some provinces has changed greatly under uncertain circumstances. For example, in the robust DEA model, the rank of Hainan is the last one, but it soars up to 21st in a data-driven robust DEA model. When the uncertainty set is an ellipsoidal set, the RDEA efficiency values of 31 DMUs are shown in Table 7. Similar to the interval uncertainty sets of data-driven DEA model, the result shows a larger range and a more stable decreasing with the increasing of the uncertainty level.

Table 6. Data-driven robust DEA efficiency based on interval set.

Province	RDEA Efficiency					Rank
	$\lambda = 1$	$\lambda = 2$	$\lambda = 3$	$\lambda = 4$	$\lambda = 5$	R
Beijing	0.942	0.869	0.804	0.738	0.682	01
Tianjin	0.863	0.795	0.733	0.675	0.621	18
Hebei	0.875	0.797	0.735	0.675	0.617	14
Shanxi	0.939	0.866	0.802	0.736	0.679	05
Neimenggu	0.877	0.799	0.737	0.676	0.619	13
Liaoning	0.899	0.823	0.759	0.695	0.639	10
Jilin	0.804	0.738	0.679	0.621	0.570	25
Heilongjiang	0.881	0.803	0.741	0.679	0.623	12
Shanghai	0.942	0.869	0.804	0.738	0.682	01
Jiangsu	0.942	0.869	0.804	0.738	0.682	01
Zhejiang	0.882	0.805	0.742	0.681	0.625	11
Anhui	0.802	0.736	0.677	0.620	0.568	26
Fujian	0.828	0.761	0.701	0.644	0.592	23
Jiangxi	0.792	0.726	0.668	0.611	0.559	27
Shandong	0.939	0.866	0.802	0.736	0.679	05
Henan	0.823	0.757	0.697	0.639	0.588	24
Hubei	0.862	0.794	0.732	0.674	0.619	19
Hunan	0.744	0.679	0.622	0.566	0.515	31
Guangdong	0.941	0.868	0.803	0.737	0.681	04
Guangxi	0.784	0.718	0.660	0.604	0.552	28
Hainan	0.852	0.784	0.723	0.665	0.613	21
Chongqing	0.856	0.788	0.727	0.669	0.614	20
Sichuan	0.834	0.766	0.706	0.649	0.597	22
Guizhou	0.774	0.708	0.651	0.595	0.544	30
Yunnan	0.866	0.799	0.737	0.679	0.625	17
Xizang	0.917	0.841	0.777	0.711	0.655	07
Shanxi	0.782	0.716	0.658	0.602	0.551	29
Gansu	0.872	0.794	0.732	0.673	0.615	15
Qinghai	0.917	0.841	0.777	0.711	0.655	07
Ningxia	0.872	0.794	0.732	0.673	0.615	15
Xinjiang	0.901	0.825	0.762	0.697	0.642	09
Mean	0.865	0.793	0.732	0.671	0.617	

While the uncertainty set turns out to be a polyhedron set, the rank is quite different with that in robust DEA model. From the Table 8, we can explicitly see that the rank is gradually in a fairly stable state when the uncertainty parameter is greater than 2.

From the analysis above, we can obtain the following conclusion:

(1) In a data-driven robust DEA model, among three uncertainty sets, the ellipsoidal set shows the strongest robustness which is similar to the robust DEA model.

(2) Compared with robust DEA model, the data-driven robust DEA model shows better performance. All three uncertainty sets have greater efficiency than the former. It is worth mentioning that the rank is gradually in a fairly stable state when the uncertainty parameter greater than 2. The reason for these two phenomena is also easy to find. In data-driven methods, more information is provided, so we can describe the uncertainty more accurately, which leads to the more satisfactory result.

Table 7. Data-driven robust DEA efficiency based on ellipsoidal set.

Province	RDEA Efficiency					Rank
	$\omega = 1$	$\omega = 2$	$\omega = 3$	$\omega = 4$	$\omega = 5$	R
Beijing	0.965	0.890	0.823	0.756	0.698	01
Tianjin	0.884	0.814	0.751	0.691	0.636	18
Hebei	0.896	0.816	0.753	0.691	0.632	14
Shanxi	0.962	0.887	0.821	0.754	0.695	05
Neimenggu	0.898	0.818	0.755	0.692	0.634	13
Liaoning	0.921	0.843	0.777	0.712	0.654	10
Jilin	0.823	0.756	0.695	0.636	0.584	25
Heilongjiang	0.902	0.822	0.759	0.695	0.638	12
Shanghai	0.965	0.890	0.823	0.756	0.698	01
Jiangsu	0.965	0.890	0.823	0.756	0.698	01
Zhejiang	0.903	0.824	0.760	0.697	0.640	11
Anhui	0.821	0.754	0.693	0.635	0.582	26
Fujian	0.848	0.779	0.718	0.659	0.606	23
Jiangxi	0.811	0.743	0.684	0.626	0.572	27
Shandong	0.962	0.887	0.821	0.754	0.695	05
Henan	0.843	0.775	0.714	0.654	0.602	24
Hubei	0.883	0.813	0.750	0.690	0.634	19
Hunan	0.762	0.695	0.637	0.580	0.527	31
Guangdong	0.964	0.889	0.822	0.755	0.697	04
Guangxi	0.803	0.735	0.676	0.618	0.565	28
Hainan	0.872	0.803	0.740	0.681	0.628	21
Chongqing	0.877	0.807	0.744	0.685	0.629	20
Sichuan	0.854	0.784	0.723	0.665	0.611	22
Guizhou	0.793	0.725	0.667	0.609	0.557	30
Yunnan	0.887	0.818	0.755	0.695	0.640	17
Xizang	0.939	0.861	0.796	0.728	0.671	07
Shanxi	0.801	0.733	0.674	0.616	0.564	29
Gansu	0.893	0.813	0.750	0.689	0.630	15
Qinghai	0.939	0.861	0.796	0.728	0.671	07
Ningxia	0.893	0.813	0.750	0.689	0.630	15
Xinjiang	0.923	0.845	0.780	0.714	0.657	09
Mean	0.886	0.812	0.749	0.687	0.631	

5.5. Managerial Insights

The only drawback of the robust model is that it may lead to over conservative results. For this reason, we use data-driven method to solve it. According to the results shown in Table 2, the lowest efficiency of endowment insurance system among 31 provinces is 0.857 (Hainan Province). The endowment insurance system efficiency value of 28 provinces is higher than 0.9, which means that China's endowment insurance system is generally in good condition. On the other hand, as shown in Figure 5, there is little difference in the efficiency of endowment insurance among 31 provinces, which means that the difference between regions is not significant. From the results shown in Figure 1, with the increase of uncertainty, the efficiency of endowment insurance system in China's provinces shows a downward trend.

At the same time, the increase of population aging and the improvement of life expectancy are a fixed trend in the next 30 years, which leads to the increase of endowment insurance expenditure. The Chinese government must increase the financial expenditure on endowment insurance to resolve this grim situation. For some provinces with low efficiency, such as Shanxi, Hunan, Hainan, the government should give priority to providing subsidies to promote fair distribution. What is more, local governments should pay more attention to the development of the endowment insurance system, and pay attention to the differences in the operation of the endowment insurance system in different regions. To improve the operation efficiency of endowment insurance, the government of underdeveloped regions can consider learning from the advanced systems and experience of other areas and reasonably guide the input and output of resources.

Table 8. Data-driven robust DEA efficiency based on polyhedron set.

Province	RDEA Efficiency (Rank)				
	Γ = 1	Γ = 2	Γ = 3	Γ = 4	Γ = 5
Beijing	0.957 (01)	0.884 (01)	0.816 (01)	0.754 (01)	0.696 (01)
Tianjin	0.881 (18)	0.807 (18)	0.745 (18)	0.696 (18)	0.643 (18)
Hebei	0.907 (14)	0.833 (14)	0.769 (14)	0.710 (14)	0.656 (14)
Shanxi	0.946 (07)	0.872 (07)	0.805 (07)	0.656 (07)	0.606 (07)
Neimenggu	0.918 (12)	0.844 (12)	0.779 (12)	0.720 (12)	0.664 (12)
Liaoning	0.924 (11)	0.850 (11)	0.784 (11)	0.664 (11)	0.614 (11)
Jilin	0.823 (25)	0.755 (25)	0.697 (25)	0.644 (25)	0.594 (25)
Heilongjiang	0.914 (13)	0.844 (12)	0.779 (12)	0.594 (12)	0.549 (12)
Shanghai	0.951 (05)	0.876 (05)	0.809 (05)	0.747 (05)	0.690 (05)
Jiangsu	0.954 (03)	0.880 (03)	0.813 (03)	0.690 (03)	0.637 (03)
Zhejiang	0.929 (10)	0.854 (10)	0.789 (10)	0.728 (10)	0.672 (10)
Anhui	0.821 (26)	0.753 (26)	0.695 (26)	0.672 (26)	0.621 (26)
Fujian	0.848 (24)	0.777 (24)	0.717 (24)	0.662 (24)	0.612 (24)
Jiangxi	0.815 (27)	0.747 (27)	0.690 (27)	0.612 (27)	0.565 (27)
Shandong	0.948 (06)	0.874 (06)	0.807 (06)	0.745 (06)	0.688 (06)
Henan	0.851 (23)	0.780 (23)	0.714 (23)	0.688 (23)	0.635 (23)
Hubei	0.875 (20)	0.802 (20)	0.720 (20)	0.684 (20)	0.631 (20)
Hunan	0.749 (31)	0.672 (31)	0.684 (31)	0.631 (31)	0.583 (31)
Guangdong	0.954 (03)	0.879 (04)	0.812 (04)	0.749 (04)	0.692 (04)
Guangxi	0.801 (28)	0.722 (28)	0.749 (28)	0.692 (28)	0.639 (28)
Hainan	0.872 (21)	0.799 (21)	0.738 (21)	0.681 (21)	0.629 (21)
Chongqing	0.877 (19)	0.804 (19)	0.681 (19)	0.629 (19)	0.581 (19)
Sichuan	0.858 (22)	0.786 (22)	0.726 (22)	0.670 (22)	0.619 (22)
Guizhou	0.781 (30)	0.703 (30)	0.670 (30)	0.619 (30)	0.571 (30)
Yunnan	0.893 (17)	0.819 (17)	0.756 (17)	0.698 (17)	0.645 (17)
Xizang	0.946 (07)	0.872 (07)	0.698 (07)	0.645 (07)	0.595 (07)
Shanxi	0.795 (29)	0.716 (29)	0.661 (29)	0.610 (29)	0.564 (29)
Gansu	0.907 (14)	0.833 (14)	0.610 (14)	0.564 (14)	0.520 (14)
Qinghai	0.957 (01)	0.883 (02)	0.815 (02)	0.753 (02)	0.695 (02)
Ningxia	0.907 (14)	0.832 (16)	0.753 (16)	0.695 (16)	0.642 (16)
Xinjiang	0.941 (09)	0.866 (09)	0.800 (09)	0.738 (09)	0.682 (09)
Mean	0.887	0.813	0.744	0.678	0.626

6. Conclusions

In this research, robust optimization is applied to a DEA model to deal with data uncertainty. The proposed robust model considers three kinds of uncertainty sets: ellipsoidal uncertainty set, box uncertainty set, polyhedron uncertainty set. At the same time, data-driven robust method was applied to solve its inherent defects, which is over-conservative. The results show that the polyhedron robust model has the worst result, while the ellipsoidal uncertain model has the strongest robustness and the most favorable efficiency. From the analysis of the results, we can describe the uncertainty of market environment by adjusting the size of uncertainty parameters, so as to meet our actual needs. The proposed model was applied to the endowment insurance system of various provinces in China, and it was found that China's endowment insurance system generally operates at a high level and the difference between regions is not significant. In the future, the government should increase financial support for endowment insurance, especially in underdeveloped regions. The government of relatively underdeveloped regions should learn from developed regions to improve the operation efficiency of local pension insurance system.

Author Contributions: S.Q.: Conceptualization, Methodology, Validation, Writing—review and editing. C.F.: Formal analysis, Investigation, Resources, Data curation, Writing—original draft. S.J.: Formal analysis. J.W.: Methodology. Y.X.: Data curation. All authors have read and agreed to the published version of the manuscript.

Funding: This research was supported by the National Natural Science Foundation of China (No. 72171123, 72171149) and the Major Program of National Fund of Philosophy and Social Science of China (No. 21ZDA105).

Informed Consent Statement: All the participants were given an informed consent to be signed upon they consented to participate in this study.

Data Availability Statement: The data presented in this study are available on request from the corresponding author. The data are not publicly available due to privacy considerations.

Conflicts of Interest: The authors declare no conflict of interest.

Appendix A

Proof of Proposition 1: *According to the form of box uncertainty set, we can rewrite the first constraint of model (6) as:*

$$\theta - \sum_{r=1}^{S} u_r y_{ro} - \sum_{l=1}^{L}\sum_{r=1}^{S} u_r y_{rol}^F \xi_l \leq 0, \forall \{\xi : ||\xi|| \leq \Phi\} \tag{A1}$$

It is equivalent to:

$$\min_{-\psi_l \leq \xi_l \leq \psi_l} \sum_{l=1}^{L}\sum_{r=1}^{S} u_r y_{rol}^F \xi_l \geq \theta - \sum_{r=1}^{S} u_r y_{ro}, \forall \{\xi : ||\xi||_\infty \leq \Phi\} \tag{A2}$$

Obviously, the minimum value of the left-hand side of constraint (A1) is:

$$-\Phi \sum_{l=1}^{L}\sum_{r=1}^{S} u_r y_{rol}^F,$$

Then, the explicit form of the first constraint of model (6) can be obtained.

$$\theta + \Phi \sum_{l=1}^{L}\sum_{r=1}^{S} u_r y_{rol}^F - \sum_{r=1}^{S} u_r y_{ro} \leq 0,$$

The same procedure may be easily adapted to obtain the explicit form of the third constraint of model (6):

$$\Phi \sum_{l=1}^{L}\sum_{r=1}^{S} u_r y_{rjl}^F + \sum_{r=1}^{S} u_r y_{rj} - \sum_{i=1}^{m} v_i x_{ij} \leq 0.$$

□

Proof of Proposition 2: *According to the form of box uncertainty set, the first constraint of model (6) can be rewritten as:*

$$\theta - \sum_{r=1}^{S} u_r y_{ro} - \sum_{l=1}^{L}\sum_{r=1}^{S} u_r y_{rol}^F \xi_l \leq 0, \forall \{\xi : ||\xi||_2 \leq \Omega\} \tag{A3}$$

It is equivalent to:

$$\min_{||\xi||_2 \leq \Omega} \sum_{l=1}^{L} \sum_{r=1}^{S} u_r y_{rol}^F \xi_l \geq \theta - \sum_{r=1}^{S} u_r y_{ro}, \forall \{\xi : ||\xi||_2 \leq \Omega\} \tag{A4}$$

Let us take the minimum of the left-hand side of constraint (A3), we obtain the following equivalent inequality:

$$\theta + \Omega \sqrt{\sum_{l=1}^{L} (\sum_{r=1}^{S} u_r y_{rol}^F)^2} - \sum_{r=1}^{S} u_r y_{ro} \leq 0,$$

For the third constraint of model (6), we just follow the same procedure, and we obtain:

$$\Omega \sqrt{\sum_{l=1}^{L} (\sum_{r=1}^{S} u_r y_{rjl}^F)^2} + \sum_{r=1}^{S} u_r y_{rj} - \sum_{i=1}^{m} v_i x_{ij} \leq 0.$$

In conclusion, the RDEA based on ellipsoid uncertainty set can be obtained.□

Proof of Proposition 3: *According to the form of box uncertainty set, the first constraint of model (6) can be rewritten as*:

Generally, the third constraint of model (6) can be equivalently rewritten as:

$$\max_{\xi \in Z^P} \sum_{l=1}^{L} \sum_{r=1}^{S} \xi_l y_{rjl}^F u_r \leq \sum_{i=1}^{m} x_{io} v_i - \sum_{l=1}^{L} \sum_{r=1}^{S} y_{rjl}^F u_r,$$

According to the properties of dual cone, we get its explicit form: $\sum_{l=1}^{L} \Gamma p_{jl} + \sum_{r=1}^{S} y_{rj} u_r - \sum_{i=1}^{m} x_{ij} v_i \leq 0$, where $p_{jl} \geq y_{rjl}^F u_r$.

Similarly, the explicit form of the third constraint of model (6) can be obtained: $\theta + \sum_{l=1}^{L} \Gamma p_{ol} - \sum_{r=1}^{s} y_{ro} u_r \leq 0$, where $p_{ol} \geq y_{rol}^F u_r$.□

References

1. National Bureau of Statistics. Bulletin of the Seventh National Census. 2020. Available online: http://www.stats.gov.cn (accessed on 8 August 2022).
2. Cai, F. Supply- and Demand-side Impacts of Aging on China's Economic Growth. *Econ. Perspect.* **2021**, *1*, 27–34.
3. Samuelson, P.A. The Pure Theory of Public Expenditure. *Rev. Econ. Stat.* **1954**, *36*, 387–389. [CrossRef]
4. Diamond, P. National Debt in a Neoclassical Growth Model. *Am. Econ. Rev.* **1965**, *55*, 1126–1150.
5. Aaron, H.J. The Social Insurance Paradox. *Can. J. Econ. Political Sci.* **1966**, *33*, 371–374. [CrossRef]
6. Zhang, X.; Liu, X.; Wang, Y.; Zhou, L.; Cheng, X. Sustainable Development of China's Maternity Insurance System in the Context of Population Policy Changes: Using a Grounded Theory Approach. *Sustainability* **2022**, *14*, 2138. [CrossRef]
7. Xue, Z.; Ma, Z. Performance Evaluation of China's Basic Pension Insurance Based on a Three-Stage Superefficient SBM-DEA Model. *Discret. Dyn. Nat. Soc.* **2022**, *2022*, 2429927. [CrossRef]
8. Charnes, A.; Cooper, W.W. Measuring the efficiency of decision making units. *Eur. J. Oper. Res.* **1978**, *2*, 429–444. [CrossRef]
9. Banker, R.D.; Charnes, A.; Cooper, W.W. Some models for estimating technical and scale inefficiencies in data envelopment analysis. *Manag. Sci.* **1984**, *30*, 1078–1092. [CrossRef]
10. Chen, Z.F. Chinese bank efficiency during the global financial crisis: A combined approach using satisficing DEA and Support Vector Machines. *N. Am. J. Econ. Financ.* **2018**, *43*, 71–86. [CrossRef]
11. Shaddady, A.; Moore, T. Investigation of the effects of financial regulation and supervision on bank stability: The application of CAMELS-DEA to quantile regressions. *J. Int. Financ. Mark. Inst. Money* **2019**, *58*, 96–116. [CrossRef]
12. Asmild, M.; Zhu, M.Y. Controlling for the use of extreme weights in bank efficiency assessments during the financial crisis. *Eur. J. Oper. Res.* **2016**, *251*, 999–1015. [CrossRef]
13. Geng, Z.Q.; Zeng, R.F. Energy efficiency evaluation and energy saving based on DEA integrated affinity propagation clustering: Case study of complex Petrochemical industries. *Energy* **2016**, *179*, 863–875. [CrossRef]

14. Moon, H.; Min, D. Assessing energy efficiency and the related policy implicatio- ns for energy-intensive firms in Korea: DEA approach. *Energy* **2017**, *133*, 23–34. [CrossRef]
15. Meng, F.Y.; Su, B. Measuring China's regional energy and carbon emission efficiency with DEA models: A survey. *Appl. Energy* **2016**, *183*, 1–21. [CrossRef]
16. Oukil, A.; El-Bouri, A.; Emrouznejad, A. Energy-aware job scheduling in a multi-objective production environment—An integrated DEA-OWA model. *Comput. Ind. Eng.* **2022**, *168*, 108065. [CrossRef]
17. Aparicio, J.; Cordero, J.M. Using non-radial DEA to assess school efficiency in a cross-country perspective: An empirical analysis of OECD countries. *Omega* **2018**, *79*, 9–20. [CrossRef]
18. Huang, M.; Xia, T.; Chen, Z.; Pan, E.; Xi, L. A DEA integrated grey factor analysis approach for efficiency evaluation and ranking in uncertain systems. *Comput. Ind. Eng.* **2021**, *162*, 107681. [CrossRef]
19. Nyamuhirwa, D.M.A.; Awotide, B.A.; Kusinza, D.B.; Bishikwabo, V.K.; Mignouna, J.; Bamba, Z.; Dontsop Nguezet, P.M. A Comparative Analysis of Technical Efficiency and Profitability of Agribusiness and Non-Agribusiness Enterprises in Eastern DRC. *Sustainability* **2022**, *14*, 8384. [CrossRef]
20. Wang, Q.; Chen, Y.; Guan, H.; Lyulyov, O.; Pimonenko, T. Technological Innovation Efficiency in China: Dynamic Evaluation and Driving Factors. *Sustainability* **2022**, *14*, 8321. [CrossRef]
21. Pendharkar, P.C.; Rodger, J.A. Quantitative software project management with mixed data: A comparison of radial, nonradial, and ensemble data envelopment analysis models. *J. Softw. Evol. Process* **2021**, *33*, e2348. [CrossRef]
22. Álvarez, I.; Barbero, J.; Zofío, J.L. A Data Envelopment Analysis Toolbox for MATLAB (No. 2016/03). Ph.D. Thesis, Universidad Autónoma de Madrid (Spain), Department of Economic Analysis (Economic Theory and Economic History), Madrid, Spain, 2016.
23. Andreu, L.; Sarto, J.L.; Vicente, L. Efficiency of the strategic style of pension funds: An application of the variants of the slacks-based measure in DEA. *J. Oper. Res. Soc.* **2014**, *65*, 1886–1895. [CrossRef]
24. Hu, Y.M.; Li, T. Evaluation of the Efficiency of Basic Pension Insurance System Operation of Urban and Rural Residents. *J. Agro-For. Econ. Manag.* **2019**, *18*, 742–751.
25. Soyster, A.L. Convex programming with set-inclusive constraints and applications to inexact linear programming. *Oper. Res.* **1973**, *21*, 1154–1157. [CrossRef]
26. El Ghaoui, L.; Lebret, H. Robust solutions to least-square problems to uncertain data matrices. *SIAM J. Matrix Anal. Appl.* **1997**, *18*, 1035–1064. [CrossRef]
27. Ben-Tal, A.; Nemirovski, A. Robust Convex Optimization. *Math. Oper. Res.* **1998**, *23*, 769–805. [CrossRef]
28. Ben-Tal, A.; Nemirovski, A. Robust solutions of uncertain liner programs. *Oper. Res. Lett.* **1999**, *25*, 1–13. [CrossRef]
29. Ben-Tal, A.; Nemirovski, A. Robust solutions of uncertain liner programming contained data. *Math. Program.* **2000**, *88*, 411–424. [CrossRef]
30. Bertsimas, D.; Sim, M. Robust discrete optimization and network flows. *Math. Program.* **2003**, *98*, 49–71. [CrossRef]
31. Bertsimas, D.; Sim, M. The price of robustness. *Oper. Res.* **2004**, *52*, 35–53. [CrossRef]
32. Bertsimas, D.; Sim, M. Tractable approximations to robust conic optimization problems. *Math. Program.* **2006**, *107*, 5–36. [CrossRef]
33. Bertsimas, D.; Pachamanova, D.; Sim, M. Robust linear optimization under general norms. *Oper. Res. Lett.* **2004**, *32*, 510–516. [CrossRef]
34. Sadjadi, S.J.; Omrani, H. Data envelopment analysis with uncertain data: An application for Iranian electricity distribution companies. *Energy Policy* **2008**, *36*, 4247–4254. [CrossRef]
35. Kazemia, I.; Haji, H.A. A robust DEA model for measuring the relative efficiency of Iranian high schools. *Manag. Sci. Lett.* **2011**, *1*, 389–404.
36. Bertsimas, D.; Gupta, V.; Kallus, N. Data-driven robust optimization. *Math. Program.* **2018**, *167*, 235–292. [CrossRef]
37. Chassein, A.; Dokka, T.; Goerigk, M. Algorithms and uncertainty sets for data-driven robust shortest path problems. *Eur. J. Oper. Res.* **2019**, *274*, 671–686. [CrossRef]
38. Ben-Tal, A.; El Ghaoui, L.; Nemirovski, A. *Robust Optimization*; Princeton University Press: Princeton, NJ, USA, 2009.
39. National Bureau of Statistics. Endowment Insurance for Urban Employees by Region 2019. 2020. Available online: http://www.stats.gov.cn/tjsj/ndsj/2020/indexch.htm (accessed on 8 August 2022).

sustainability

Article

The Pricing Mechanism Analysis of China's Natural Gas Supply Chain under the "Dual Carbon" Target Based on the Perspective of Game Theory

Cheng Che, Xin Geng *, Huixian Zheng, Yi Chen and Xiaoguang Zhang

School of Economics and Management, China University of Petroleum (East China), Qingdao 266580, China; 20000034@upc.edu.cn (C.C.); s21080058@s.upc.edu.cn (H.Z.); s20080061@s.upc.edu.cn (Y.C.); s20080060@s.upc.edu.cn (X.Z.)
* Correspondence: s21080059@s.upc.edu.cn; Tel.: +86-156-9448-8850

Abstract: China is currently the world's largest energy consumer and carbon emitter. In order to reduce the harm of carbon dioxide to the global ecological environment, the use of natural gas instead of coal is a realistic choice for China to achieve the "dual carbon" goal. Opportunities also bring new challenges, and the price of natural gas is an important method of promoting the upstream and downstream industrial chains of natural gas, so it is of great practical significance to study the price of natural gas. This paper builds a three-level supply chain model consisting of suppliers in the natural gas market, city gas companies and consumers in the market and uses the Stackelberg game to study the decision-making models of different subjects under their own dominance and centralized decision-making; it also considers the pricing mechanism and profit situation of stakeholders in the natural gas market under the low-carbon preference of consumers and the level of corporate carbon emission reduction. The research results show that when considering consumers' low-carbon preferences, the sales prices of various stakeholders in the market have increased, which is beneficial for all entities in the natural gas industry chain. At the same time, with the low-carbon transformation of energy companies, the production method drives the price of raw materials to rise in the process of low-carbon innovation, which, in turn, makes the price of various stakeholders in the natural gas market and the level of carbon emission reduction per unit show a positive relationship; in order to maximize the overall profit of the supply chain, the natural gas market should adopt a centralized decision-making method to further promote the reform of China's natural gas marketization.

Keywords: "dual carbon" target; natural gas prices; pricing mechanism; supply chain; Stackelberg game

Citation: Che, C.; Geng, X.; Zheng, H.; Chen, Y.; Zhang, X. The Pricing Mechanism Analysis of China's Natural Gas Supply Chain under the "Dual Carbon" Target Based on the Perspective of Game Theory. *Sustainability* **2022**, *14*, 9510. https://doi.org/10.3390/su14159510

Academic Editor: Roberto Mancinelli

Received: 28 June 2022
Accepted: 28 July 2022
Published: 3 August 2022

1. Introduction

Since human beings entered the era of industrialization, global carbon dioxide emissions have increased year by year, accompanied by global temperature rise, glacier melting, sea level rise and many other environmental problems [1]. According to the survey, more than 85% of carbon emissions come from energy activities, so the pace of the transition to green and low-carbon energy must be accelerated [2]. As the world's largest energy consumer and carbon emitter, China's "carbon peak and carbon neutrality" goal (hereinafter referred to as the "dual carbon" goal) will be the two milestones of China's energy revolution within the next 40 years and an important stage in the construction of a new energy consumption system [3,4]. Using low-carbon or non-fossil energy to replace coal with high carbon emission is an effective way to reduce CO_2. According to the statistics of the International Energy Agency, natural gas is a low-carbon, clean and efficient energy compared with coal and oil [5,6]. As a bridge from high carbon to low-carbon energy, natural gas is a realistic choice for China to achieve the goal of "dual carbon" in the future [7,8]. China's future transformational trend is to take non-fossil energy as the core and natural gas as the transitional energy, so the reform of the system and mechanism of the natural gas market is particularly important [9]. The realization path of the natural gas industry under the "dual carbon" goal mainly involves upstream production, midstream trade and transportation and downstream consumption, which requires us to think about emission reduction from the perspective of the whole industry chain [10]. To be specific, innovation should be carried out in natural gas exploitation technology and natural gas market system and policy; the upstream, middle and downstream of the industrial chain should coordinate with each other; in the process of natural gas exploration and development, environmental damage and exhaust gas emission should be minimized [11]. The upstream exploration and development of the natural gas market should be low-carbon, clean and involve less environmental pollution; the midstream infrastructure is interconnected, economical and efficient; the downstream supply chain channels should be diversified and supply security should be guaranteed [12].

The price of natural gas is an important way to promote the development of upstream and downstream enterprises in the supply chain and this is related to the development of the natural gas market. Therefore, it is of great practical significance to study the price of natural gas. At present, the insufficient supply and high cost of natural gas are the most important problems to be solved, which is also a long-term problem to be faced, namely, to increase supply and reduce cost [13]. For a long time, the natural gas industry has been controlled by the Chinese government in all aspects and its price cannot be adjusted according to the current situation of market supply and demand. Compared with developed countries, the natural gas industry has a low degree of marketization. In terms of short-term goals for the future, the main goal of the Chinese government's price policy is to adjust and optimize the price mechanism of natural gas to make it fluctuate in a reasonable range. Although the state strongly supports the development of natural gas, the development of natural gas still seriously lags behind expectations due to the pricing mechanism [14]. Therefore, it is necessary to continue to deepen the reform of natural gas price marketization and improve the market competitiveness of natural gas [15]. At present, there is still a big gap between the natural gas industry and the complete marketization: First, the pricing of pipeline transportation has not become a system; Second, the upstream exploration and production blocks of natural gas are monopolized by huge oil and gas enterprises, which makes it difficult to form a new situation in which competition is explored; Thirdly, the division of the marketization scope is complicated, which leads to the failure of regulated price policies to a certain extent [9]. To sum up, in order to optimize the natural gas market price mechanism, all enterprises in the industrial chain need to play their respective roles, deepen the division of labor and cooperation, make joint efforts to promote low-carbon emission reduction and establish a qualified price mechanism to promote the development of the natural gas market.

According to the summary of existing research on the natural gas pricing mechanism, the Stackelberg game model has been used to explore many research studies on the natural gas pricing mechanism. For example, some studies discussed the pricing strategy of natural gas, which only considered the game between natural gas production enterprises and industrial users [16] or the construction of two models of competitive pricing of natural gas, in which the advantage orientation of natural gas price in competition was obtained through comparative analysis [17,18]. Studies also analyzed natural gas output under different marginal costs in the energy industry and investigated the influence of a third-party access system on the natural gas pricing mechanism [19,20]. However, the content combining low-carbon emission reduction in previous studies is less. In addition, most previous studies have studied the secondary supply chain [16,21]. On this basis, we studied the three-level supply chain of natural gas, taking into account the upstream, middle and downstream stakeholders, which is more in line with the industrial chain structure of the natural gas market described above. Different from the previous literature, this paper considers consumers' low-carbon preference and the unit carbon emission reduction parameters of various stakeholders in the natural gas market, because we know that consumers' low-carbon preference and enterprises' unit carbon emission reduction level will affect the price of natural gas [22,23]. Finally, the Stackelberg game model is established to analyze the pricing mechanism and profit under different dominant and centralized decision-making conditions, which further expands the relevant research on the pricing mechanism of the natural gas market.

The rest of this paper is organized as follows and relevant literature are listed in Section 2. In Sections 3 and 4, we analyze and describe the problems and models in detail and analyze different situations in detail through the method of game. In Section 5, we assign parameters to the formula obtained after the model calculation and further verify the model through numerical simulation. In Section 6, we summarize the correct conclusions and put forward relevant policy recommendations for the reform of the pricing mechanism of China's natural gas market under the "dual carbon" target. Additionally, we present the direction of future research.

2. Literature Review

The relevant literature have the following three parts: the development status of the natural gas industry under the "dual carbon" goal, the research on the relationship between stakeholders in the natural gas market and the relevant research on the pricing mechanism of stakeholders in the natural gas market. In this section, we review the relevant literature and point out how our study differs from theirs.

2.1. Development Status of the Natural Gas Industry under the "Dual Carbon" Goal

Natural gas is one of the clean energy alternatives to coal and oil. "China's 14th Five-Year Plan" clearly pointed out that in the future, the natural gas industry should constantly improve the production, supply, storage and marketing system based on the "dual carbon" goal and the new economic and social situation to meet the increasing demand for clean energy in social development and build a new pattern of energy development [24]. Based on the experience of Japan, the United States and other developed countries, it has been found that carbon emission reduction can be achieved by controlling the total amount of energy consumption and optimizing the consumption structure. From the perspective of the first change in global energy consumption structure, the proportion of clean energy has increased greatly [25]. Developed countries such as the United States, Japan and the United Kingdom have achieved substantial reductions in carbon emissions by vigorously developing clean energy [26].

In the process of realizing the "dual carbon" goal, how much development space is there for China's natural gas market in the future, whether it can reach the carbon peak at the same time as the world, what is the peak level and what will happen after the peak is reached? These questions will all be discussed. These are issues that directly affect the enthusiasm for exploration, development and infrastructure investment in the natural gas market [2]. There are three challenges facing China's natural gas industry. First, the natural gas infrastructure needs to be further improved. Although China attaches great importance to the construction of natural gas infrastructure, there is still a problem of insufficient pipeline interconnection, which affects the purchase of natural gas by downstream consumers. Secondly, the downstream distribution link of the natural gas market needs to be reformed. The monopoly of city gas companies is not conducive to the marketization of the natural gas industry and the decrease in upstream gas price cannot bring benefits to end consumers directly. Finally, with the expected future increase in natural gas consumption, the capacity of gas storage and peak regulation needs to be improved [27]. In the future, the focus of the natural gas industry is to reduce costs, improve and innovate in technology and further improve infrastructure construction in combination with the goal of low-carbon emission reduction [28]. With the "dual carbon" goal vigorously promoted, the natural gas industry will usher in a prosperous period of development opportunities in the next decade, providing sufficient development space for enterprises in the market, middle and downstream. Therefore, all stakeholders should cooperate to learn from each other and jointly realize low-carbon and sustainable development of the natural gas industry.

2.2. Research on the Relationship between Stakeholders in the Natural Gas Market

In the natural gas market, there are usually multiple stakeholders such as natural gas producers, city gas companies, governments, consumers and environmental protection departments [29]. From the perspective of the supply chain, the natural gas market is composed of the upstream supply market, the downstream consumption market, the pipeline transportation company and the city gas company in the middle. Upstream suppliers are responsible for the exploration, production and processing of natural gas, which is delivered to consumers in the downstream natural gas market through midstream pipeline companies. The pipeline transportation enterprise is responsible for the transportation and dispatching of the long-distance pipeline network, transporting natural gas to gas storage and city gate stations; city gas companies are responsible for buying gas from producers and selling it to the consumer market; consumers in the market generally refer to urban gate stations and downstream consumer markets, including industrial users and residents [30].

The game-based method has been widely used in the optimization of the natural gas supply chain [31]. Most of the existing studies focus on supply reliability, pricing decisions, overall optimization design of supply chain system, etc. We summarize scholars' research on the natural gas supply chain in Table 1:

Table 1. Related research on the natural gas supply chain.

Literature	Model	Research Content and Results
Yan et al. (2017) [32]	Generalized Nash Equilibrium Model	Establish a natural gas supply chain network equilibrium model. Analyze the game process between different levels of natural gas.
Chen et al. (2017) [33]	Stackelberg Model	When the producer is the dominant player in the natural gas supply chain, the impact of the price and profit of the producer and the gas distributor under the condition of fairness concern and the comparison of the price and profit of the two under the condition of fairness and neutrality.
Dong et al. (2018) [34]	Stackelberg Model	Comparative analysis of natural gas markets in China and the United States. Research shows that gas supply chain development levels have increased over time.
Crow et al. (2018) [35]	Cournot Model	A dynamic upstream gas supply model is proposed to simulate investment and operational decisions in the upstream gas industry.
Xu et al. (2020) [36]	Stackelberg Model	Considering the transportation mode, transportation equipment and supply and demand capacity comprehensively, the minimum micro-objective function of the average daily operating cost of the natural gas supply chain was used to optimize the supply chain.
Zhang et al. (2021) [37]	Bertrand Model	The buyer oligopoly model of natural gas importing countries, in addition to considering profits, also takes natural gas reserves into account.
Zhang et al. (2021) [38]	Stackelberg Differential Game Model; Nash Model	Construct a dynamic game model of natural gas trade, incorporate the infrastructure stock into the traditional demand function and consider its impact on natural gas demand and supply security.

In general, many scholars use the game method to analyze and optimize the supply chain of the natural gas industry, laying a certain foundation for further putting forward corresponding policies. Therefore, using the scientific game model and method from the perspective of the natural gas supply chain to explore the reasonable pricing of natural gas has very important guiding significance. In a word, under the "dual carbon" goal, the natural gas industry faces a practical problem in the new market environment: how to coordinate the allocation of upstream resources, improve transportation capacity, optimize the overall value of the sales business chain and on this basis, achieve the goal of maximizing the profit of all stakeholders in the supply chain, which is a considerable problem facing the natural gas industry.

2.3. Research on the Pricing Mechanism of Stakeholders in the Natural Gas Market

Price reform is the core of China's natural gas market reform. With the increasing demand for natural gas, China needs to further improve the pricing mechanism of natural gas. Under the premise of considering the maximum bearing capacity of consumers, the difference in the cost of natural gas from different sources and the appropriate control of the cost of the downstream industry chain, this paper explores how to optimize the benefits and investment returns of all links of the natural gas industry chain. At the same time, the impact of alternative energy prices on the natural gas price system should also be considered and the price should be dynamically adjusted following the changes in market supply and demand [39,40]. According to the structure of the natural gas industry chain, its price involves three aspects: the upstream ex-factory price of natural gas, the midstream pipeline transportation cost and the downstream user price. In the natural gas development plan, it is proposed to give full play to the leverage of the price mechanism in regulating the relationship between supply and demand. Although the measures of natural gas market-oriented reform in different countries in the world are different, most of them are changing from a monopoly market to market pricing [20]. At present, in China's natural gas market, producers monopolize the exploration and production of natural gas and also occupy the midstream transportation through the construction of transportation pipelines with a strong monopoly [41]. The focus of the natural gas market reform is to establish a dedicated pipeline transportation company to separate the upstream and middle reaches of the natural gas industry and to reduce the vertical integration of upstream producers. In the natural gas industry chain, CO_2 emissions run throughout. Therefore, under the "dual carbon" target, the natural gas industry should vigorously develop low-carbon emission reduction technologies to promote sustainable development in the future.

In the study of natural gas pricing mechanisms, some scholars point out the existing problems of natural gas pricing mechanisms and corresponding measures. By learning from the experience of developed countries, building a competitive natural gas market is based on price marketization [42]. Due to problems such as insufficient upstream competitiveness and imperfect infrastructure, China's natural gas market does not have the conditions to completely liberalize prices. To sum up, the current pricing mechanism has the following problems: the permitted rate of return of the natural gas pipeline is too high, the price management method of the "base price + floating range" still has room for optimization and there is still no influential benchmark price [43,44]. Therefore, the natural gas price reform needs to straighten out the original price, formulate reasonable prices for transportation, storage and gas transmission and distribution and accelerate the construction of a market trading center to form an influential market benchmark price [45]. Other scholars have discussed the role of speculation in natural gas pricing [46,47]. Manera et al. used the DCC multivariate GARCH model to analyze the future prices of energy commodities (including natural gas) and agricultural commodities; the results showed a spillover effect between commodities and a surge in the correlation between energy and agricultural commodities. In addition, they considered alternative measures of speculative activities in their research to assess the role of speculation in simulating commodity futures price volatility [48,49]. The research of comprehensive scholars has found that in recent years, energy and agricultural product markets have become the most concerning issues in various countries and the relationship between the two has become closer. Factors such as macroeconomic uncertainty, emergencies and speculation have a significant impact on natural gas prices. The quantitative easing monetary policies successively introduced in Europe and the United States have intensified global excess liquidity and the emergence of some crises has severely impacted the supply chains of commodities such as food and energy, which has further promoted global inflation. However, the green and low-carbon transformation provides a strong basis for the future market-oriented reform of domestic energy. In addition, some scholars have studied the emission reduction of the supply chain and analyzed the optimal emission reduction cost input and optimal subsidy rate of enterprises by constructing models of government subsidies for cooperative emission reduction between manufacturers and retailers under different game relationships [50]. Zhao considered relevant research on emission reduction investment strategies of the supply chain under equity concern and low-carbon preference [51]. Fan constructed a two-stage decision-making model between the government and supply chain enterprises and explored the similarities and differences between government subsidy strategies and enterprises' carbon emission reduction decisions under different circumstances [52].

Through reading literature related to the natural gas supply chain and pricing mechanism, most scholars have pointed out the market-oriented direction of natural gas pricing reform in their studies but have not formed a systematic market-oriented pricing reform strategy. At the same time, the pricing reform schemes proposed in literature are mostly qualitative research, lacking quantitative analysis and data model verification. Moreover, considering the current "dual carbon" goal of China, the content of the natural gas supply chain combined with low-carbon emission reduction is reduced. Therefore, this study assigns relevant stakeholders in China's natural gas market as the research object and builds a game model considering the carbon emission reduction of enterprises and consumers' low-carbon preferences so as to further improve China's natural gas price mechanism.

3. Problem Description and Model Assumptions

3.1. Problem Description

In order to facilitate the "dual carbon" goal as soon as possible and further improve the price system of the natural gas market, we will solve this problem from the perspective of the supply chain. This paper establishes a three-level supply chain model consisting of suppliers in the natural gas market, city gas companies and consumers in the market. The natural gas supplier is responsible for the exploration and exploitation of natural gas and sells the natural gas to the city gas company at a certain price P_i. When the supplied natural gas is transported to the city gas company through the pipeline, certain pipeline transportation costs are incurred. The city gas company also sells to the natural gas market at a certain price P_t and finally, the natural gas market sells the natural gas to the consumers in the market at a certain price P_r.

By constructing this model, this paper focuses on the impact of consumers' low-carbon preferences η and corporate low-carbon emission reduction levels e on the prices and profits of various stakeholders in the natural gas market. The three-level supply chain model of the natural gas market is shown in Figure 1:

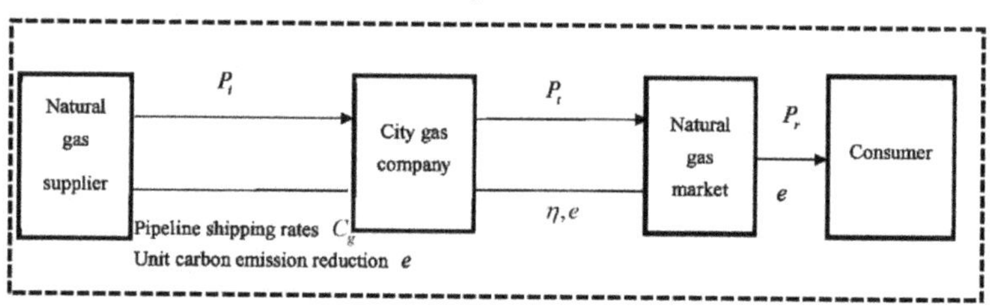

Figure 1. Natural gas market supply chain model.

3.2. Model Assumptions

According to the natural gas three-tier supply chain model, the following are the assumptions:

Hypothesis 1 (H1). *Under the natural gas market conditions, all stakeholders are pursuing the maximization of their own interests. According to the "individual behavior rationality" criterion, there is bound to be strong non-cooperative competition and each supply chain member has decision-making power. In the context of China's current "dual carbon" goal, the power of various stakeholders in the natural gas market varies; it is predicted that competition will be more significant under the pricing mechanism of the natural gas market in the future; that is, it is assumed that the pricing process is sequential and a dynamic process. Additionally, we assume that all stakeholders in the natural gas market are in a complete information environment and according to the information* *they have, they set the price of natural gas that already includes reasonable profits. This study does not consider the impact of the international market, domestic economic environment, related taxes and government regulations on the game model.*

Hypothesis 2 (H2). *All stakeholders in the natural gas market produce according to the quantity determined by the market demand function. The natural gas market demand function is:* $d = D - kP_r + \eta e (k > 0)$ *[20,53]. Among them, D represents the market size of the natural gas market (that is, the largest possible demand in the market), k is the price sensitivity coefficient and* P_r *is the market value of natural gas. Consumers are sensitive to the purchase of low-carbon products, and the inverse demand function of price has a linear relationship with the level of emission reduction. In the demand function,* η *represents the low-carbon preference coefficient of consumers and e represents the unit emission reduction.*

Hypothesis 3 (H3). *Natural gas suppliers will adopt corresponding low-carbon emission reduction methods so that the carbon emission reduction per unit of natural gas is e. Referring to various literature, it is concluded that the carbon emission reduction cost of natural gas suppliers is* $\frac{1}{2}e^2m$ *[54,55], the investment cost coefficient of carbon emission reduction is m. That is to say, as the level of emission reduction increases, the cost of emission reduction for suppliers also increases. In order to encourage natural gas suppliers to reduce carbon emissions in the production process, city gas companies share the carbon emission reduction costs of suppliers. Therefore, the carbon emission reduction costs of city gas companies are* $\frac{1}{2}e^2(1-a)m$ *and a is the share of carbon emission reduction costs.*

Hypothesis 4 (H4). *The rate at which a pipeline company transports natural gas per unit of production is* C_g *[33]. The natural gas pipeline company operates as an independent third party and only bears the transportation cost of natural gas in the entire natural gas supply chain and the established unit transportation rate already includes reasonable profits.*

3.3. Model Parameters

All parameters involved in this paper and their definitions are summarized in Table 2.

Table 2. Symbol parameters.

Symbolic Parameters	Parameter Meaning
D	Market size
k	Consumer sensitivity to price
P_r	Market value of natural gas
P_i	The price that a gas supplier sells to a city gas company
P_t	The price that the city gas company sells to market consumers
C_i	Marginal cost of natural gas suppliers
C_{fi}	Fixed costs for natural gas suppliers
C_t	Marginal cost of city gas company
C_{ft}	Fixed costs for city gas companies
C_g	The rate of transportation per unit of output by a pipeline company
η	Consumer's low-carbon preference coefficient
e	Unit carbon emission reduction
a	Proportion of emission reduction cost sharing between natural gas suppliers and city gas companies
m	Carbon emission reduction investment cost factor
π_i	Profits of natural gas suppliers
π_t	Profits of city gas companies
π_r	Market consumer profit
π_c	Total profit of the natural gas supply chain

4. The Decision Model of Various Stakeholders in the Natural Gas Market

In this section, we make mathematical derivations based on the Stackelberg game model. According to the above, we have established a three-tier supply chain model consisting of suppliers in the natural gas market, city gas companies and consumers in the market. The Stackelberg model is a price leadership model and there are differences in the order of actions of each subject in the supply chain. We explore decision models for decentralized as well as centralized decision-making below. Under decentralized decision-making, we explored the situation of leadership by different leaders. The leader first decides one of its own outputs and then the rest of the followers decide their own output according to the observed output. In this model of the game, leaders determine their own output with full knowledge of how followers in the market will behave. This means that the leader can know the reaction function of the follower and then decide how to act on the premise of maximizing his own profit. When different stakeholders are dominant, we use backward induction to deduce, from the follower side to the leader side, how to get the relevant function. Under the centralized decision-making model, the main bodies of the natural gas supply chain make joint decisions and they take the maximization of the interests of the entire supply chain as the ultimate goal and then finally determine the market price P_r of natural gas.

4.1. *Game Research of Stakeholders in the Natural Gas Market under Decentralized Decision-Making*

4.1.1. The Situation of the Natural Gas Supplier Is the Dominant Player

According to this assumption, city gas companies and natural gas suppliers will also produce according to the demand function under the condition of maximizing their own interests; then we solve the equilibrium solution of the Stackelberg model according to the idea of reverse induction.

According to the parameters in the table, the profit function of each stakeholder in the natural gas market can be obtained as follows:

Supplier's profit:

$$\pi_i = (P_i - C_i)(D - kP_r + \eta e) - C_{fi} - \frac{1}{2}e^2 m \tag{1}$$

Profits of city gas companies:

$$\pi_t = (P_t - P_i - C_t - C_g)(D - kP_r + \eta e) - C_{ft} - \frac{1}{2}(1-a)e^2 m \tag{2}$$

Consumer profits:

$$\pi_r = (P_r - P_t)(D - kP_r + \eta e) \tag{3}$$

The first stage: Consumers in the natural gas market based on the price given by the city gas company is P_t and to determine the price P_r with the goal of maximizing Formula (3), obviously, π_r is a concave function of P_r and there is a maximum value. Letting $\frac{\partial \pi_r}{\partial P_r} = 0$, we can get:

$$\frac{\partial \pi_r}{\partial P_r} = D - 2kP_r + \eta e + kP_t \tag{4}$$

$$P_r = \frac{D + \eta e + kP_t}{2k} \tag{5}$$

The second stage: the city gas company based on the price P_i given by the natural gas and the P_r determined in the first stage, determined the price P_t with the goal of maximizing Formula (2), and there is a maximum value. Letting $\frac{\partial \pi_t}{\partial P_t} = 0$, we can get:

$$\frac{\partial \pi_t}{\partial P_t} = \frac{D}{2} + \frac{\eta e}{2} - kP_t + \frac{kP_i}{2} + \frac{kC_t}{2} + \frac{kC_g}{2} \tag{6}$$

$$P_t = \frac{D + \eta e + kP_i + kC_t + kC_g}{2k} \tag{7}$$

The third stage: The gas supplier decides the price P_i based on the P_r and P_t obtained in the first two stages with the goal of maximizing Formula (1). Similarly, π_i is a concave function of P_i and there is a maximum value. Letting $\frac{\partial \pi_i}{\partial P_i} = 0$, we can get:

$$\frac{\partial \pi_i}{\partial P_i} = \frac{1}{4}D + \frac{1}{4}\eta e - \frac{1}{4}kC_t - \frac{1}{4}kC_g + \frac{1}{4}kC_i - \frac{1}{2}kP_i \tag{8}$$

$$P_{i1} = \frac{D + \eta e - kC_t - kC_g + kC_i}{2k} \tag{9}$$

By substituting P_{i1} into P_t and P_r, respectively, we get:

$$P_{t1} = \frac{3D + 3\eta e + kC_i + kC_t + kC_g}{4k} \tag{10}$$

$$P_{r1} = \frac{7D + 7\eta e + kC_t + kC_g + kC_i}{8k} \tag{11}$$

Substitute P_{i1}, P_{t1} and P_{r1} into π_{i1}, π_{t1} and π_{r1}, respectively, to get:

$$\pi_{i1} = \frac{(D + \eta e - kC_t - kC_g - kC_i)^2}{16k} - C_{fi} - \frac{1}{2}e^2 m \tag{12}$$

$$\pi_{t1} = \frac{(D + \eta e - kC_t - kC_g - kC_i)^2}{32k} - C_{ft} - \frac{1}{2}(1-a)e^2 m \tag{13}$$

$$\pi_{r1} = \frac{(D + \eta e - kC_t - kC_g - kC_i)^2}{64k} \tag{14}$$

In summary, the total profit function of the natural gas supply chain is:

$$\pi_{c1} = \pi_{i1} + \pi_{t1} + \pi_{r1} \tag{15}$$

$$\pi_{c1} = \frac{7(D + \eta e - kC_t - kC_g - kC_i)^2}{64k} - C_{fi} - C_{ft} - \frac{1}{2}e^2 m - \frac{1}{2}(1-a)e^2 m \tag{16}$$

4.1.2. The Situation When the City Gas Company Is in a Dominant Position

In Sections 4.1.2 and 4.1.3, in order to make the formulas look simpler and more intuitive, we added three new parameters (w, q, v), which are the unit product margins of each stakeholder in the three-tier supply chain profit. Although in Sections 4.1.2 and 4.1.3, we use different parameters than those in Section 4.1.1. However, what we want to explain here is that no matter whether the unit product profit or the price of each stakeholder is used for derivation, it will not affect the final derivation result and the deduced profit of the natural gas supplier is the same. Because as we all know, marginal profit = sales revenue − marginal cost.

The natural gas supplier determines the selling price P_i based on the unit product profit margin w,

$$P_i = C_i + w \tag{17}$$

The city gas company determines the selling price P_t according to the unit product profit margin q,

$$P_t = C_i + w + C_g + C_t + q \tag{18}$$

The consumer determines the acceptable price P_r based on the profit margin v per unit of product,

$$P_r = C_i + w + C_t + C_g + q + v \tag{19}$$

The profit of the natural gas supplier is:

$$\pi_i = w(D - kP_r + \eta e) - Cf_i - \frac{1}{2}e^2 m \tag{20}$$

$$\pi_i = w(D - k(C_i + w + C_t + C_g + q + v) + \eta e) - Cf_i - \frac{1}{2}e^2 m \tag{21}$$

The profit of the city gas company is:

$$\pi_t = q(D - kP_r + \eta e) - C_{ft} - \frac{1}{2}(1-a)e^2 m \tag{22}$$

$$\pi_t = q(D - k(C_i + w + C_t + C_g + q + v) + \eta e) - Cf_t - \frac{1}{2}(1-a)e^2 m \tag{23}$$

The consumer's profit is:

$$\pi_r = v(D - kP_r + \eta e) \tag{24}$$

$$\pi_r = v(D - k(C_i + w + C_t + C_g + q + v) + \eta e) \tag{25}$$

The first stage: According to the q given by the city gas company, the natural gas supplier determines its own marginal profit w with the goal of maximizing its own profit, that is, Formula (21) maximization. It can be seen from the above formula that π_i is a concave function of w, so there is a maximum value. Letting $\frac{\partial \pi_i}{\partial w} = 0$, we can get: $0 = D - kC_i - 2wk - kC_t - kC_g - kq - kv + \eta e$

$$w = \frac{D - k(C_i + C_t + C_g + q + v) + \eta e}{2k} \tag{26}$$

The second stage: According to the q given by the city gas company, consumers also aim to maximize their own profit, that is, Formula (25) maximization in order to determine their own marginal profit v. From the above formula, we can see that π_r is a concave function of v, so there is a maximum value. Letting $\frac{\partial \pi_r}{\partial v} = 0$, we can get:

$$0 = D - kC_i - kw - kC_t - kC_g - kq - 2kv + \eta e \tag{27}$$

$$v = \frac{D - k(C_i + C_t + C_g + q + w) + \eta e}{2k} \tag{28}$$

The third stage: Since natural gas suppliers and consumers have unknowns when making decisions, they restrict each other, so for the simultaneous equations, let $w = v$:

$$\frac{D - k(C_i + C_t + C_g + q + v) + \eta e}{2k} = \frac{D - k(C_i + C_t + C_g + q + w) + \eta e}{2k} \tag{29}$$

According to their interrelationships, the final simplification is:

$$w = v = \frac{D - k(C_i + C_t + C_g + q) + \eta e}{3k} \tag{30}$$

The fourth stage: The city gas company aims at maximizing its own profit according to the above obtained w and v by maximizing the Equation (23) to determine the profit margin q; it can be seen from the above Formula (23) that π_t is a concave function of q, so there is a maximum. Letting $\frac{\partial \pi_t}{\partial q} = 0$, we can get:

$$q = \frac{D - k(C_i + C_t + C_g) + \eta e}{2k} \tag{31}$$

By substituting q into $w = v = \frac{D-k(C_i+C_t+C_g+q)+\eta e}{3k}$, we can get:

$$w' = v' = \frac{D - k(C_i + C_t + C_g) + \eta e}{6k} \tag{32}$$

By substituting q, w and v into the following three formulas: $P_i = C_i + w'$, $P_t = C_i + w' + C_g + C_t + q$ and $P_r = C_i + w' + C_t + C_g + q + v$, we can get:

$$p_{i2} = C_i + w' = \frac{D + 5C_ik - C_tk - C_gk + \eta e}{6k} \tag{33}$$

$$p_{t2} = C_i + w' + C_g + C_t + q = \frac{2D + C_ik + C_gk + C_tk + \eta e}{3k} \tag{34}$$

$$p_{r2} = C_i + w' + C_t + C_g + q + v' = \frac{5D + C_ik + C_gk + C_tk + \eta e}{6k} \tag{35}$$

Then, by substituting p_{i2}, p_{t2} and p_{r2} into π_{i2}, π_{t2} and π_{r2}, we can get:

$$\pi_{i2} = \frac{(D - C_ik - C_gk - C_tk)^2 + 6\eta e(D - C_ik - C_gk - C_tk) + 5(\eta e)^2}{36k} - C_{fi} - \frac{1}{2}e^2m \tag{36}$$

$$\pi_{t2} = \frac{(D - C_ik - C_gk - C_tk)^2 + 6\eta e(D - C_ik - C_gk - C_tk) + 5(\eta e)^2}{12k} - C_{ft} - \frac{1}{2}(1-a)e^2m \tag{37}$$

$$\pi_{r2} = \frac{(D - C_ik - C_gk - C_tk)^2 + 6\eta e(D - C_ik - C_gk - C_tk) + 5(\eta e)^2}{36k} \tag{38}$$

In summary, the total profit function of the natural gas supply chain is:

$$\pi_{c2} = \pi_{i2} + \pi_{t2} + \pi_{r2} \tag{39}$$

$$\pi_{c2} = \frac{5(D - C_ik - C_gk - C_tk)^2 + 30\eta e(D - C_ik - C_gk - C_tk) + 15(\eta e)^2}{36k} - C_{fi} - C_{ft} - \frac{1}{2}e^2m - \frac{1}{2}(1-a)e^2m \tag{40}$$

4.1.3. The Situation When the Consumer Is in the Main Position

In this section, the derivation content before the second stage is the same as the Formulas (17)–(26) in Section 4.1.2, so it is omitted here.

The second stage: According to the v given by the consumer and the w obtained in the first stage, the city gas company seeks q with the goal of maximizing the Formula (23). From the above formula, we can see that π_t is a concave function of q and there is a maximum value; therefore, letting $\frac{\partial \pi_t}{\partial q} = 0$, we get:

$$q = \frac{D - K(C_i + C_t + C_g + v) + \eta e}{2k} \tag{41}$$

The third stage: According to the w and q obtained in the first two stages, consumers aim to maximize their own profits, so they can find v by maximizing Formula (25). It can be seen from the above equation that π_r is a concave function of v and has a maximum value. Therefore, letting $\frac{\partial \pi_r}{\partial v} = 0$, we get:

$$v = v^* = \frac{D - k(C_i + C_t + C_g) - 3\eta e}{2k} \tag{42}$$

By substituting v^* into q, we get:

$$q^* = \frac{D - k(C_i + C_t + C_g) + 5\eta e}{8k} \tag{43}$$

Then, by substituting v^* and q^* into w, we get:

$$w^* = \frac{D - k(C_i + C_t + C_g) + 5\eta e}{8k} \tag{44}$$

Then, by substituting v^*, w^*, q^* into $P_i = C_i + w$, $P_t = C_i + w + C_g + C_t + q$, $P_r = C_i + w + C_t + C_g + q + v$ from the three formulas, we can get:

$$p_{i3} = \frac{D + 7kC_i - kC_t - kC_g + 5\eta e}{8k} \tag{45}$$

$$p_{t3} = \frac{3D + 5kC_i + 5kC_g + 5kC_t + 15\eta e}{8k} \tag{46}$$

$$p_{r3} = \frac{7D + kC_i + kC_t + kC_g + 3\eta e}{8k} \tag{47}$$

By substituting p_{i3}, p_{t3} and p_{r3} into π_{i3}, π_{t3} and π_{r3}, respectively, we can get:

$$\pi_{r3} = \frac{[D - k(C_i + C_t + C_g)]^2 + 2\eta e[D - k(C_i + C_t + C_g)] - 15(\eta e)^2}{16k} \tag{48}$$

$$\pi_{t3} = \frac{[D - k(C_i + C_t + C_g)]^2 + 10\eta e[D - k(C_i + C_t + C_g)] + 25(\eta e)^2}{32k} - C_{ft} - \frac{1}{2}e^2(1-a)m \tag{49}$$

$$\pi_{i3} = \frac{[D - k(C_i + C_t + C_g)]^2 + 10\eta e[D - k(C_i + C_t + C_g)] + 25(\eta e)^2}{64k} - Cf_i - \frac{1}{2}e^2 m \tag{50}$$

In summary, the total profit function of the natural gas supply chain is:

$$\pi_{c3} = \pi_{r3} + \pi_{t3} + \pi_{i3} \tag{51}$$

$$\pi_{c3} = \frac{7[D - k(C_i + C_t + C_g)]^2 + 38\eta e[D - k(C_i + C_t + C_g)] + 15(\eta e)^2}{64k} - C_{fi} - C_{ft} - \frac{1}{2}e^2 m - \frac{1}{2}e^2(1-a)m \tag{52}$$

4.2. Game Research of Natural Gas Stakeholders under Centralized Decision-Making

Under the centralized decision-making model, the main bodies of the supply chain cooperate closely to make decisions together. Taking the profit maximization of the entire supply chain as the ultimate goal of decision-making, the natural gas market price P_r is jointly determined. Therefore, when natural gas suppliers, city gas companies and consumers in the natural gas market make joint decisions in the natural gas supply chain, the profit function of the entire natural gas market supply chain can be obtained as:

$$\pi = (P_r - C_i - C_t - C_g)(D - kP_r + \eta e) - C_{ft} - C_{fi} - \frac{1}{2}e^2 m - \frac{1}{2}e^2(1-a)m \tag{53}$$

It can be seen from the above formula that π is a concave function of P_r and there is a maximum value, so let $\frac{\partial \pi}{\partial P_r} = 0$. We can obtain the equilibrium price under centralized decision-making as:

$$0 = D - 2kP_r + \eta e + kC_i + kC_t + kC_g \tag{54}$$

$$P_r^* = \frac{D + \eta e + k(C_i + C_t + C_g)}{2k} \tag{55}$$

By substituting $P_r{}^*$ into the supply chain profit function, the maximum total profit of the natural gas market supply chain can be obtained as:

$$\pi^* = \frac{[D + \eta e - k(C_i + C_t + C_g)]^2}{4k} - C_{ft} - C_{fi} - \frac{1}{2}e^2 m - \frac{1}{2}e^2(1-a)m \quad (56)$$

5. Numerical Simulation

In order to better verify the changing relationship between the parameters, this section will compare and analyze the parameters by assigning them. First, we describe and explain the parameters involved in the article in the third part of the problem description and model assumptions. Then, based on these descriptions, we refer to the previous relevant literature and set the values of the fixed parameters [52,56,57]. The values are explained as follows:D= 45, $k = 0.1$, $C_t = 0.9$, $C_i = 1.2$, $C_g = 0.15$, $\eta = 3$, $m = 0.4$, $a = 0.6$, $C_{fi} = 0.3$, $C_{ft} = 0.2$ and $e = 0.3$. In the drawing in this section, under the decentralized decision-making model, the natural gas supplier-led model is represented by Model 1, the city gas company-led model is represented by Model 2 and the market consumer-led model is represented by Model 3. When the natural gas market is in centralized decision-making mode, it is represented by Model 4.

5.1. The Impact of Consumers' Low-Carbon Preferences on the Prices of Various Stakeholders in the Natural Gas Market

Through numerical simulation, the impact of consumers' low-carbon preference on the prices of various stakeholders can be more clearly demonstrated. Under the "dual carbon" target, these values are substituted into Formulas (9), (33), (45) and (55) to obtain the change in the price of natural gas suppliers considering the dominance of different stakeholders and the centralized decision-making of the natural gas supply chain condition. As can be seen from Figure 2, with the increase in consumers' low-carbon preference, the price of natural gas suppliers also shows an increasing trend and when using the centralized decision model, the ex-factory price of natural gas is the highest. Similarly, the values are substituted into Formulas (10), (34), (46) and (55) to obtain the price changes of the city gas companies in Figure 3. When natural gas suppliers dominate, city gas companies sell at higher prices than when city gas companies dominate the supply chain. In addition, the selling price of natural gas in the above-mentioned case is also higher than the price in the other two cases, namely the centralized decision-making and the market consumer-led situation. In a supplier-dominated and market consumer-dominated situation, the magnitude of changes in the price of city gas companies is larger than in the other two situations, which is the dominant and centralized decision-making model of the city gas company. As shown in Figure 4, by substituting the values into Formulas (11), (35), (47) and (55), it can be seen that with the increase in consumers' low-carbon preference, the price of natural gas in the consumer market presents an increasing trend. When the natural gas supplier is in the dominant position, the market price is higher than the price in other situations. Under the background of the "dual carbon" target, natural gas suppliers carry out low-carbon innovation; the price of natural gas will rise as production technology is innovated, which will also run up prices for other stakeholders. Therefore, this also reflects the dominant position of natural gas suppliers in the pricing mechanism. The low-carbon innovation of suppliers also satisfies consumers' low-carbon preference, which is conducive to the further development of the downstream consumer market of the natural gas industry and the improvement of consumers' acceptable price.

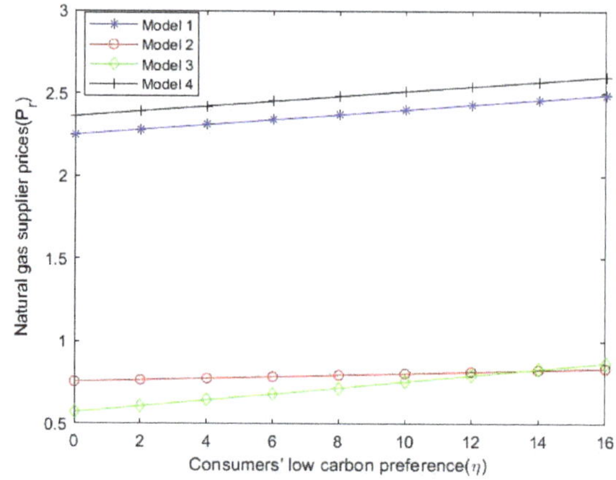

Figure 2. P_r with the changing trend of η.

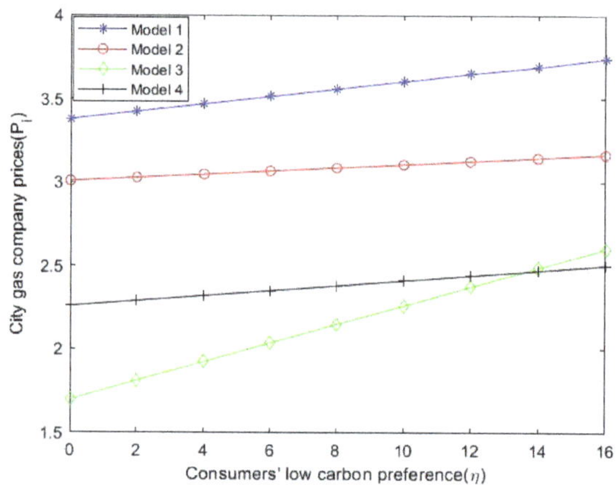

Figure 3. P_i with the changing trend of η.

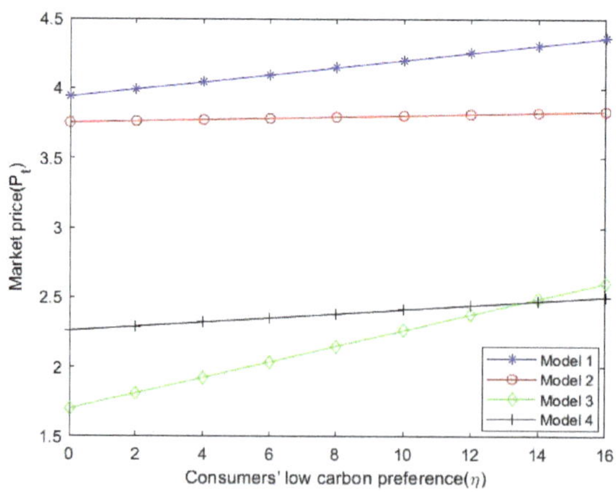

Figure 4. P_t with the changing trend of η.

5.2. The Impact of Unit Carbon Emission Reduction Level on the Price of Various Stakeholders in the Natural Gas Market

As shown in Figure 5, by substituting numerical values, it can be seen that by comparing Formulas (10), (35), (55) and (64), gas suppliers have the highest pricing levels when city gas companies dominate. Additionally, with the increase in unit carbon emission reduction level, the price of each stakeholder shows an increasing trend. As can be seen in Figure 6, city gas companies have the highest pricing levels when natural gas suppliers dominate the supply chain. At the same time, when the market is dominated by consumers, with the increase in the level of carbon emission reduction per unit, the price of city gas companies fluctuates the most. Figure 7 shows that when suppliers are dominant, the natural gas market price increases with the level of unit carbon emission reduction and the market price at this time is greater than that when other stakeholders are dominant and supply chain decision-making is centralized. The total profit of the natural gas supply chain can be obtained by substituting the value into Formulas (15), (39) and (51).

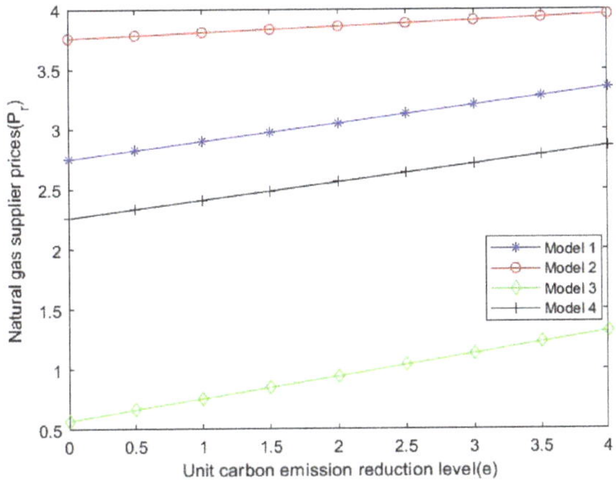

Figure 5. P_r with the changing trend of e.

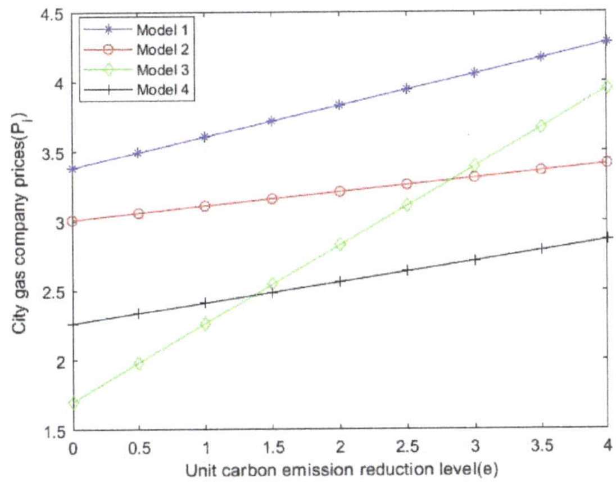

Figure 6. P_i with the changing trend of e.

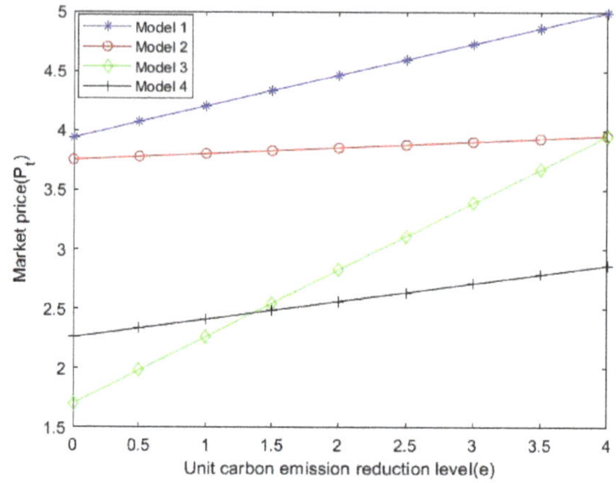

Figure 7. P_t with the changing trend of e.

5.3. *The Impact of Consumers' Low-Carbon Preference and Unit Carbon Emission Reduction Level on the Total Profit of the Natural Gas Supply Chain*

As shown in Figures 8 and 9, the total profit of the supply chain increases with the increase in consumers' low-carbon preference and the total profit level of the supply chain is positively related to the unit carbon reduction. Under the centralized decision model, the total profit of the natural gas supply chain is the largest. To sum up, in the natural gas market, the prices of various stakeholders in the natural gas market increase with the increase in consumers' low-carbon preference; that is, the price of stakeholders in the natural gas market is proportional to consumers' low-carbon preference. For the natural gas supply chain, natural gas suppliers are responsible for natural gas production, exploration, development and other activities, occupy a core position in the overall supply chain and have absolute dominance in the game with related enterprises in the middle and lower reaches. Therefore, it also reflects that the market price of city gas companies and consumers is often determined by supplier pricing.

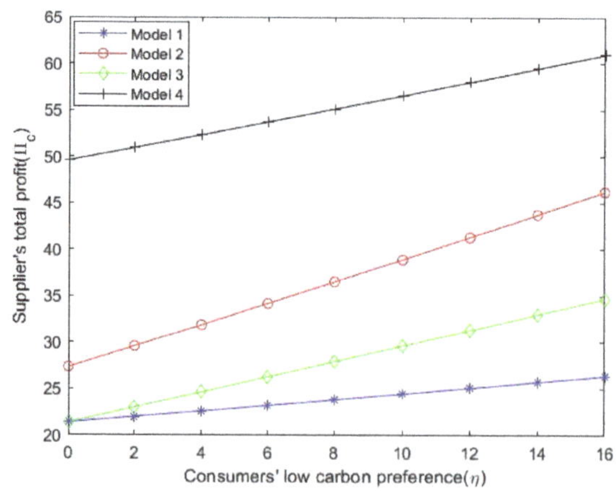

Figure 8. π_c with the changing trend of η.

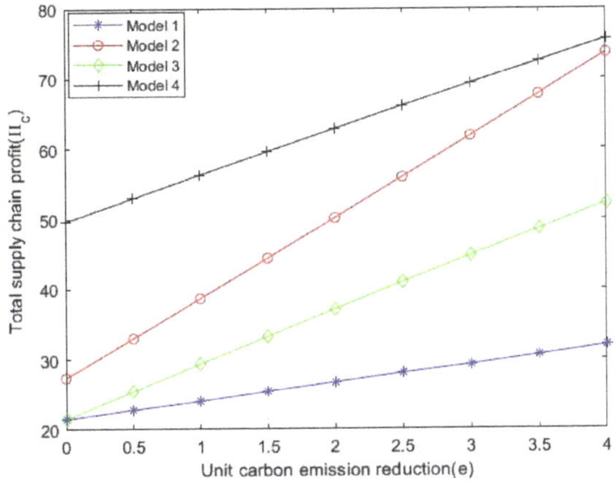

Figure 9. π_c with the changing trend of e.

At present, the academic circle pays more and more attention to the low-carbon transformation of energy enterprises under the background of the "dual carbon" goal. With the adjustment of energy structure and the continuous promulgation of relevant policies for decarbonization in various countries, energy prices have been rising for a long time. Since the signing of the Paris Climate Agreement, all economies around the world will face the issue of green transition and the transition from fossil energy to non-fossil energy is a medium and long-term process. The transformation of energy companies to low-carbon production methods will drive up the prices of some raw materials. This also explains the reason for the consequent increase in selling prices for various stakeholders in the natural gas supply chain under the pricing model considering the unit carbon emission reduction level. China's 2021 Sustainable Consumption Report focuses on low-carbon consumption trends in the context of "carbon peaking" and "carbon neutrality" goals. At present, China's "dual carbon" goal and the concept of low-carbon development have become the consensus for all sectors of society and low-carbon consumption is also becoming the daily action for many people. The green and low-carbon products provided by energy companies have enriched the choices of consumers and the increasingly low-carbon preference of consumers in the market has also forced the low-carbon transformation of the production field. The two promote each other at both ends of the industrial chain. It is precisely because consumers' increasing low-carbon consumption preference encourages enterprises to improve carbon emission reduction levels, the price of various stakeholders in the natural gas market and low-carbon preference show a positive relationship. In the future, low-carbon consumption will be the general trend. Only by taking the market as the guide and promoting the green and low-carbon transformation of products and value chains in the whole cycle can enterprises gain sustained growth and momentum in the low-carbon era.

6. Conclusions and Future Directions

6.1. Conclusions

Based on the current situation of natural gas pricing mechanisms at home and abroad, this paper analyzes the current natural gas pricing model in China, finds out the existing problems and provides relevant policies and suggestions for the market-oriented reform of natural gas pricing. This paper applies the Stackelberg game model under the "dual carbon" goal, aiming at the natural gas suppliers, city gas companies and consumers in the natural gas market, considering the decision-making model dominated by each stakeholder and the main body of the supply chain. Considering the impact of consumers' low-carbon preference level and carbon emission reduction level on the three-level supply chain, numerical simulation and analysis of the model are carried out to study the impact on the price and profit of the natural gas market and the following conclusions are drawn:

1. In the pricing model considering consumers' low-carbon preference, when the natural gas supplier dominates, the selling price of each stakeholder in the natural gas supply chain reaches the maximum value. In the natural gas market, the prices of various stakeholders in the natural gas market increase with the increase in consumers' low-carbon preference; that is, the price of stakeholders in the natural gas market is proportional to consumers' low-carbon preference. However, compared with other stakeholders, the increase rate of city gas companies tends to be more stable.
2. In the pricing model considering the unit carbon emission reduction level, the sales price of each stakeholder in the natural gas market increases with the increase in the unit carbon emission reduction level. When the city gas company is in the dominant position, the supplier sells the highest price at this time. When the natural gas supplier is in a dominant position, the city gas company and the sales price in the market are maximized.
3. For the natural gas industry, all stakeholders in the supply chain will gain more benefits under the centralized decision-making model. In the case considering consumers' low-carbon preference and the unit carbon emission reduction level, the total profit of the natural gas supply chain is positively proportional to the low-carbon preference and the unit carbon emission reduction level; that is, it increases with the increase in low-carbon preference and unit carbon emission. Moreover, the income of the natural gas industry chain under the centralized decision model is greater than that under the dominance of city gas companies and greater than that under the market consumer and natural gas suppliers.

Based on the above conclusions and corresponding analysis, with the vigorous advancement of the "dual carbon" goal, the natural gas suppliers responsible for exploration and production and the city gas companies responsible for transporting and storing natural gas will vigorously promote low-carbon emission reductions. With the improvement in the level of carbon emission reduction per unit and the increase in low-carbon preference due to the improvement of consumers' awareness of environmental protection, the sales prices of various players in the natural gas market will further increase. On the whole, in order to maximize the overall profit of the supply chain, the natural gas market should adopt a centralized decision-making method to ensure the interests of all parties and further promote the reform of China's natural gas marketization. On the basis of the above research, the following suggestions are put forward, hoping to provide some help for the formulation of relevant price policies for China's natural gas industry: First, the natural gas industry should further strengthen the internal management of the market and promote the implementation of price; secondly, under the condition of ensuring the steady supply of our country's natural gas market, the natural gas price should be in line with international standards as soon as possible; finally, the natural gas industry should improve and optimize the user market structure and strengthen price management innovation research.

6.2. Future Directions

Limitations and future prospects resulting from this study: First, future research can explore how to maximize the benefits among the main bodies of the supply chain and further improve the innovation level of low-carbon emission reduction under the encouragement of various preferential policies of the government for the natural gas industry. Second, future research should focus on solving the problem of upstream oligopoly and the lack of a benchmark, step-by-step in order to solve control pricing to market pricing, introducing a competitive mechanism to form a diversified market competition structure, translating dynamic price adjustment mechanisms into practice, strengthening the construction of supporting facilities, all of which will lead to the steady and healthy development of China's natural gas industry. Finally, with the continuous development of the global natural gas trade, in the future, the pricing mechanism of natural gas may tend to be globally integrated. Looking at the future of the natural gas industry, in areas such as equipment development, energy consumption and environmental protection, low-carbon methods and technologies and investment profits are the key points and difficulties that the natural gas industry needs to solve in the future.

Author Contributions: Conceptualization, C.C. and X.G.; methodology, X.G.; software, X.G. and Y.C.; validation, X.G., X.Z. and H.Z.; formal analysis, C.C., X.G. and Y.C.; investigation, X.Z. and H.Z.; resources, C.C.; data curation, X.G. and Y.C.; writing—original draft preparation, C.C. and X.G.; writing—review and editing, C.C. and X.G.; visualization, Y.C., X.Z. and H.Z.; supervision, C.C.; project administration, H.Z.; funding acquisition, C.C. All authors have read and agreed to the published version of the manuscript.

Funding: This research was funded by the Shandong Provincial Key Research and Development Program (Soft Science) (2020RKE28013) and the Qingdao Social Science Planning Research Project (QDSKL2101039).

Data Availability Statement: The original contributions presented in the study are included in the article. Further inquiries can be directed to the corresponding author.

Acknowledgments: We would like to thank the reviewers and the editor-in-charge for their valuable time spent on this article. We are also grateful to all the foundations that support us.

Conflicts of Interest: The authors declare that there are no conflict of interest regarding the publication of this paper.

References

1. Zou, C.N.; Xiong, B.; Xue, H.Q. The role and role of new energy in carbon neutrality. *PED* **2021**, *48*, 411–420.
2. Zhou, S.H.; Wang, J.; Liang, Y. Development of China's "14th Five-Year Plan" natural gas industry under the background of carbon neutrality. *NGI* **2021**, *41*, 171–182.
3. Xi, J.P. Delivers an Important Speech at the General Debate of the 75th Session of the United Nations (UN) General Assembly (22 September 2020, Beijing). People's Daily Overseas Edition. Available online: https://baijiahao.baidu.com/s?id=1678590270030743941&wfr=spider&for=pc (accessed on 22 September 2020).
4. Zhou, S.W.; Zhu, J.L. Exploring the path to help the strategy of "carbon peaking and carbon neutrality". *NGI* **2021**, *41*, 1–8.
5. IEA. Global CO_2 Emissions in 2019 [EB/OL]. (11 February 2020). Available online: https://wwwjea.org/articles/global-co2-emissions-in-2019 (accessed on 19 February 2021).
6. IEA. CO_2 Emissions Statistics [EB/OL]. (16 November 2020). Available online: https://wwwjea.org/subscribe-to-data-services/co2-emissions-statistics (accessed on 19 February 2021).
7. Luo, Z.X. Carbon neutralization activates the potential of natural gas demand in many fields. *Energy* **2020**, *11*, 30–32.
8. Li, S.O.; Zhang, B.S.; Xu, T. Forecasting of China's natural gas production and its policy implications. *Pet. Sci.* **2016**, *13*, 593–603. [CrossRef]
9. Fan, Y.; Yi, B.W. Laws of Energy Transformation, Driving Mechanisms and China's Path. *Manag. Wld.* **2021**, *37*, 95–105. [CrossRef]
10. Zhou, D. Prospects of the international natural gas market in the post-epidemic era. *Sino-Foreign Energy* **2021**, *26*, 79.
11. Xiao, J.Z.; Xiao, Y.T.; Li, J.Y. Evaluation of high-quality development of my country's natural gas market based on principal component analysis. *J. CHA Univ. Petro.* **2021**, *37*, 20–27. [CrossRef]
12. Wang, X.Y. China's natural gas supply security assessment and countermeasures. *J. CHA Univ. Petro.* **2015**, *31*, 6–10. [CrossRef]
13. Bai, J. City gas companies should do a good job in expanding natural gas consumption. *Sinopec* **2019**, *10*, 21–24.
14. Shi, D. Deepening the reform of natural gas prices and promoting the development of clean energy in China. *CHN Petro. Chem. Ind.* **2017**, *1*, 27.
15. Wang, Z. Some thoughts on building my country's 5A-level natural gas industry chain. *NET Econ.* **2018**, *12*, 11–15.
16. Zhou, G.D.; Kuang, J.C. Research on Natural Gas Pricing Strategy Based on Game Theory. *N G Econ.* **2004**, *3*, 17–19.
17. Feng, S. Game Analysis of Natural Gas Competitive Strategy. *J. Xihua Univ.* **2004**, 136–137. [CrossRef]
18. Chang, Y.; Trang, D.T.Q.; Tan, T.S. Competition and cooperation in the natural gas market: A game-theoretic demand-base analysis. *AEJ* **2021**, *19*, 21–49. [CrossRef] [PubMed]
19. Qian, Y. Game Analysis of Natural Gas Industry Competitive Strategy. *Mark. MOD* **2009**, *563*, 256–257.
20. Nie, G.H. Research on Natural Gas Pricing Mechanism Based on Stackelberg Game. *J. CHA Univ. Petro.* **2013**, *37*, 184–188.
21. Bazyar, A.; Zarrinpoor, N.; Safavian, A. Optimal design of a sustainable natural gas supply chain network under uncertainty. *Chem. Eng. Res. Des.* **2021**, *176*, 60–88. [CrossRef]

22. Qin, Y.; Tong, F.; Yang, G. Challenges of using natural gas as a carbon mitigation option in China. *Energy Policy* **2018**, *117*, 457–462. [CrossRef]
23. Luo, L.J.; Zhu, S.L.; Fu, S.X. The impact of low-carbon development on China's energy price reform. *Teach. Res.* **2015**, *5*, 14–23.
24. Tian, G. "China Natural Gas Development Report (2021)" released. *NGI* **2021**, *41*, 68.
25. Yang, Y.; Yu, H.Y.; Lu, G. Centennial change of world energy and national energy security. *J. Nat. Res.* **2020**, *35*, 2803–2820.
26. Wang, L.N.; Peng, T.D.; Xiang, Z.J. Analysis of China's energy transition path under the carbon neutrality goal. *Int. Petro Econ.* **2021**, *29*, 2–8.
27. Wang, X.Q.; Liu, W.; Luo, H.H. Development Opportunities and Challenges of China's Natural Gas Industry under the "Dual Carbon" Target. *Int. Petro Econ.* **2021**, *29*, 35–42.
28. Jin, Y.N.; Ni, Z.; Tian, Z. Challenges and Opportunities of Carbon Neutral Vision to Oil and Gas Industry. *Petro Tech. Econ.* **2021**, *37*, 1–6.
29. Wang, X.Y.; Guo, J.; Zheng, Z. Development and suggestion of natural gas market under the independent opening of pipeline network. *Macroecon. Manag.* **2021**, 76–83. [CrossRef]
30. Gong, C.Z.; Wu, D.S.; Gong, N.J. The potential impact and path of China's upstream natural gas market-oriented reform. *Environ. Econ. Res.* **2019**, *4*, 8–27. [CrossRef]
31. Chen, Y.; Li, J.; Lu, H.; Yang, Y. Impact of unconventional natural gas development on regional water resources and market supply in China from the perspective of game analysis. *Energy Policy* **2020**, *145*, 111750. [CrossRef]
32. Yan, Z.C.; Yuan, Y.H. Research on natural gas supply chain network problems from the perspective of supply side. *Sta. Dec.* **2017**, 58–62. [CrossRef]
33. Chen, Y.J.; Liu, Z.H. Analysis of natural gas pricing decision considering fairness concerns. *Nat. Gas. Tech. Econ.* **2017**, *11*, 78–80.
34. Dong, K.; Sun, R.; Wu, J. The growth and development of natural gas supply chains: The case of China and the US. *Energy Policy* **2018**, *123*, 64–71. [CrossRef]
35. Crow, D.J.G.; Giarola, S.; Hawkes, A.D. A dynamic model of global natural gas supply. *Appl Energy* **2018**, *218*, 452–469. [CrossRef]
36. Xu, N.; Zhang, X.Q.; Zhang, H.R. Overall Optimization of Natural Gas Supply Chain. *Oil Gas. Storage Trans.* **2020**, *39*, 1250–1261.
37. Zhang, R.; Li, S. The natural gas import game between China and the EU under the influence of reserve preference. *J. Syst. Manag.* **2021**, *30*, 729–742.
38. Zhang, H.; Zhou, B.; Zeng, X.Y. The impact of supply security on the natural gas game from the perspective of infrastructure. *CHA J. Sys. Engr.* **2021**, *36*, 314–329. [CrossRef]
39. Gu, S.; Dang, L.R.; Du, X.T. Sustainable Development Mechanism of Natural Gas Industry. *NGI* **2019**, *39*, 117–123.
40. Yin, J.P. Perfecting the natural gas pricing mechanism must take into account the interests of multiple parties. *Price Theory Pract.* **2016**, *380*, 46–48.
41. Yu, L.; Xiao, X.Z.; Jiang, C.H. The "Trinity" Theory of Natural Monopoly. *Contemp. Financ.* **2004**, *8*, 5–11.
42. Zeng, M.; Liu, R.; Wang, S.Y. Research on Competitive Natural Gas Market Construction and Price Formation Mechanism—and Analysis of Main Experience and Practice of Natural Gas Market Construction in Typical Countries. *Price Theory Pract.* **2020**, 56–59. [CrossRef]
43. Li, T.D.; He, C.L.; Dong, Z.Y. Paths and policies to improve my country's natural gas pricing mechanism. *Nat. Gas. Tech. Econ.* **2021**, *15*, 68–75.
44. Fu, S.; Zhang, P.C.; Dong, Z.Y. The key direction of my country's natural gas price reform during the "14th Five-Year Plan" period and enterprise countermeasures. *NGI* **2020**, *40*, 159–167.
45. Dong, B.G.; He, C.L.; Zhang, Y. Reconstructing the price formation mechanism of China's natural gas production, supply, storage and sales—and on the central task of China's "14th Five-Year Plan" natural gas price reform. *NGI* **2020**, *40*, 126–133.
46. Sifat, I.; Ghafoor, A.; Mand, A.A. The COVID-19 pandemic and speculation in energy, precious metals, and agricultural futures. *J. Behav Exp. Financ.* **2021**, *30*, 100498. [CrossRef]
47. Xiao, J.; Wang, Y. Macroeconomic uncertainty, speculation, and energy futures returns: Evidence from a quantile regression. *Energy* **2022**, *241*, 122517. [CrossRef]
48. Manera, M.; Nicolini, M.; Vignati, I. Financial speculation in energy and agriculture futures markets: A multivariate GARCH approach. *Energy J.* **2013**, *34*. [CrossRef]
49. Manera, M.; Nicolini, M.; Vignati, I. Modelling futures price volatility in energy markets: Is there a role for financial speculation? *Energy Econ.* **2016**, *53*, 220–229. [CrossRef]
50. Li, Y.D.; Zhao, D.Z.; Xia, L.J. Government subsidy strategy under vertical emission reduction cooperation in low carbon supply chain. *Ops. Res. Mgt* **2014**, *23*, 1–11.
51. Zhao, D.; Jian, S.F. Research on investment strategies for supply chain emission reduction under fairness concerns and low-carbon preference. *Ind. Tech. Econ.* **2020**, *39*, 94–104.
52. Fan, D.D.; Xu, Q. Analysis of corporate carbon emission reduction and government subsidy decision-making under different power structures. *Soft Sci.* **2018**, *32*, 64–70. [CrossRef]
53. Liang, X.; Zhang, Y.T. Low-carbon dual-channel supply chain pr icing and emission reduction strategies based on consumer preferences. *Ops. Res. Manag.* **2020**, *29*, 107–117.
54. Wang, K.; Qian, Q.H.; Zhou, S.F. Research on supply chain low-carbon emission reduction mechanism based on green financial loans and cost sharing. *Financ. Theory Pract.* **2019**, *1*, 84–92.

55. Che, C.; Chen, Y.; Zhang, X.G. Study on Emission Reduction Strategies of Dual-Channel Supply Chain Considering Green Finance. *Front. Environ. Sci.* **2021**, *9*, 197. [CrossRef]
56. Yao, L.M.; He, L.H.; Chen, X.D. A study on the profit distribution mechanism of the resource-Based supply chain considering low-carbon constraints and ecological restoration. *Resour. Por.* **2021**, *74*, 101539. [CrossRef]
57. Che, C.; Chen, Y.; Zhang, X.G. The impact of different government subsidy methods on low-carbon emission reduction strategies in dual-channel supply chain. *Complexity* **2021**. [CrossRef]

sustainability

Article

Procurement Strategies and Auction Mechanism for Heterogeneous Service Providers in a Service Supply Chain

Jifeng Cao and Cheng Ma *

School of Business, Qingdao University, Qingdao 266071, China; gltjx2018@163.com
* Correspondence: mc_0812@163.com

Abstract: This paper examines competition between two heterogeneous service providers (SPs) in a service procurement market. We investigate a service supply chain consisting of one service integrator (SI), and two SPs who compete the SI's order with their own reservation profits. To select the favorable SP and centralize the total system, we propose an auction mechanism in symmetric and asymmetric information scenarios and find that channel coordination can be achieved by means of an efficient auction mechanism, hence, Pareto improvement conditions can be clearly illustrated. In addition, we characterize the effects of information asymmetry on optimal decisions and derive some important managerial insights into the value of information.

Keywords: service supply chain management; auction mechanism; asymmetric information; value of information

Citation: Cao, J.; Ma, C. Procurement Strategies and Auction Mechanism for Heterogeneous Service Providers in a Service Supply Chain. *Sustainability* **2022**, *14*, 9201. https://doi.org/10.3390/su14159201

Academic Editors: Shaojian Qu, Qingguo Bai, Ying Ji and Congjun Rao

Received: 8 June 2022
Accepted: 25 July 2022
Published: 27 July 2022

1. Introduction

In recent years, service supply chain management has received increasing attention in the operations management and operations research literature as the world economy has grown increasingly service oriented. Due to fierce market competition, many manufacturing enterprises have gradually expanded their products range from physical products to value-added services. This trend is called "product servitization" in [1]. Service supply chain systems can be subdivided into two categories based on the specific form of the product, the service-only supply chain, and the product service supply chain. For example, in many well-established service industries such as healthcare body checking, psychology advice, and financial consultancy, the respective supply chains are service-only supply chain [2]. Likewise, we can find the product service supply chain in restaurant and food retail supply chain, product design, and logistics supply chain.

To benefit the channel members, cooperation is one of the most effective means to improve channel members' or a supply chain's operational performance, and a service supply chain is no exception. So far, most studies on supply chain coordination mechanisms have investigated the issue of how to establish game theory models and design contracts among supply chain members [3–5]. By contrast, little work was done on auctions for supply chain coordination. Actually, it is known that an auction can provide an ideal and important market mechanism for the competitive allocation of the price for goods or services. In a service supply chain, an auction offers direct access to numerous competitive SPs at a relatively low cost. For the SP, the auction offers a transparent form of competition that depends on quantitatively defined terms such as price, delivery time, and service quality. Meanwhile, the auction could facilitate supply chain coordination and has a significant influence on the supply chain performance. In order to minimize procurement costs, many companies purchase via reverse auctions or employ online bidding systems instead of traditional procurement contracts. For instance, Wal-Mart uses a reverse auction mechanism to choose competitive manufacturers so as to save purchasing costs, and hence increase Wal-Mart's profit. In another instance, eBay, the largest consumer-oriented auction site, attains nearly half of the gross merchandise via auction transactions.

When it comes to demand uncertainty, the supply chain often inevitably suffers from the risk of the trade-off between overstocks and stock-outs. Moreover, there are many other factors inducing supply failures such as earthquakes, floods, and climate environment. Therefore, in the face of uncertain demand, it is necessary to take participants' risk preferences into consideration. Nowadays, risk analysis and management have drawn much attention from both the industry and academia in uncertain environments (Choi and Andrzej [6], Choi et al. [7], Choi and Andrzej [8]). At present, based on probability and statistics, value at risk (VaR) is one of the most widely and frequently employed models for risk measure in risk management (Olson and Wu [9], Paul et al. [10]). VaR can be characterized as a maximum expected loss, given some time horizon and within a given confidence interval, which is suitable for our model setting. Inspired by real business problems, this study considers a service supply chain consisting of two heterogeneous SPs, with different reservation profits, who competitively provide service to a single risk-averse SI measured by VaR. Under such a context, the contract must be designed to offer the SPs no less than their own reservation profits. Specifically, the research questions are: whether or not an appropriate auction mechanism can be regarded as a service supply chain coordination mechanism? How do the critical system factors, especially reservation profits and information asymmetry, affect the optimal outcomes and contract decisions?

Our specific research questions related to the impacts of information asymmetry and auction schemes on service competition in a service supply chain in which SI chooses the favorable SP and coordinates the whole service supply chain. Toward that end, we develop a duopoly competition model in an asymmetric information case to examine whether auction and contracting mechanisms can be served as a coordination mechanism in a service supply chain considering participants' different risk preferences and individual reservation profit. Several managerial insights about how participants' risk preferences and cost variation influence their own optimal decisions and system optimums are presented. The main contributions of this paper can be summarized as follows:

1. For the purpose of encouraging the SP to cooperate fully with the SI, and satisfying their reservation profits at the same time, proper auction and contracting mechanisms are established to ensure that the SP and the SI are inclined to get involved.
2. In certain circumstances, the optimal decisions deduced by the auction and contracting mechanisms are exactly the same as channel coordination optimums. Furthermore, Pareto improvement conditions are also clearly illustrated.
3. We characterize the impact of the SP's cost variation on the optimal decisions and total supply chain profits. We find that the high-cost SP is always motivated to truthfully announce rather than distort exact cost information with the SI. This conclusion is consistent with the outcome of separate equilibrium in a signal game.
4. We derive more practical insights into the relationship between the decision makers' risk preferences and policies. We also perform some comprehensive comparisons of the channel coordination optimums obtained in the risk-neutral system and the risk-averse one.

In addition to offering insights into the service competition in a service supply chain, our results extend the results of the literature that examines firm competition using service price. First, proper auction contracting, as opposed to a traditional coordination contract, achieves service supply chain coordination. In particular, Pareto improvement conditions ensure every member is a voluntary participant in auction contracting. Second, as compared with a risk-neutral case in which the SI is risk-neutral and maximizes one's own expected profit, we find that the risk-averse SI has a lower service price and higher order quantity. Additionally, in the presence of information asymmetry, we find that the high-cost SP may be worse off if encountered with a high cost variation.

The rest of the paper is organized as follows. We briefly present the relevant literature in Section 2. In Section 3, information symmetric case is studied. Section 4 presents the optimal auction mechanism in asymmetric information cases and explores the value of information. In Section 5, the comparative analyses among different risk preferences of the SI are conducted. Numerical experiments and sensitivity analyses are performed to validate and enrich our analytical results in Section 6. Section 7 concludes this study. Detail proof is displayed in Appendix A.

2. Literature Review

The body of research closely related to this work can be divided into three broad sets. The first set studies service supply chain coordination. The second set refers to the effects of information asymmetry on the decision maker's policies in supply chain management. The third set includes the literature on auction mechanisms in operations management.

There is considerable literature devoted to service supply chain coordination. Here, we define coordination as the mechanism used to achieve the best performance in a service supply chain system, which is consistent with the existing literature [11–15]. A widely accepted structure of a service supply chain is: "Service Provider (SP)-Service Integrator (SI)-Customers" [16]. Usually, the SI has stronger control power and outsources the functional service to SPs in order to maintain a competitive advantage. Bernstein and Federgruen [17] examined supply chain coordination with price and service competition. They found that with an exogenous service level, a simple linear wholesale pricing contract can be designed so as to achieve service supply chain coordination. However, with the endogenous service level, they proved that a simple linear wholesale pricing contract cannot achieve coordination. Sethi et al. [18] examined a two-stage service supply chain with information updating. They identified the optimal order quantity and studied the effect of order cancellations in such a service supply chain. They employed the buyback contract to coordinate the supply chain. Sieke et al. [19] proposed several service-level-based supply contracts to achieve supply chain coordination. They identified the optimal service level contracts. Xiao and Xu [20] studied the service level in a supply chain under the vendor-managed inventory. They identified the equilibrium prices and service quality under both centralized and decentralized cases and found that a revenue-sharing contract can achieve supply chain coordination. Heydari [21] investigated a coordination mechanism in a supply chain taking customer service level into consideration and found that stochastic lead time may harm the customer service level. However, different from the above-mentioned literature, this study designed auction contracting as a coordination mechanism.

In addition, we are aware of several studies in the supply chain literature that explore the impacts of cost/demand information asymmetry on supply chain performance. Cakanyildirim et al. [22] concluded that information asymmetry alone does not necessarily induce loss in channel efficiency and designed the optimal contract for coordination. Chiu et al. [4] proposed a PRR (price, rebate, and returns) contract to coordinate a decentralized supply chain with both additive and multiplicative price-dependent demands and presented the fact that multiple equilibrium policies for channel coordination existed. Ha [23] designed a contract to maximize the supplier's profit with asymmetric cost information in a one-supplier-one-buyer relationship for a short-life-cycle product. Ha et al. [24] investigated contracting and information sharing in two competing supply chains, in which each consisting of one manufacturer and one retailer. They evaluated the value of information sharing and fully characterized the equilibrium information sharing decisions under different investment costs. In this study, we investigate the impact of service cost variation on the optimal service price, optimal order quantity, and participants' profits.

This study specifically relates to the literature on auction mechanisms in operations management in a competitive context. Previous studies have examined how an auction is utilized to award supply contracts to a selected group of suppliers. Auctions find successful applications in the electricity procurement market, which attracts a great number of researchers to make further investigation into the market mechanism. Ryzin and Vulcano [25] studied the inventory management for items sold through forward (selling) auctions. Abhishek et al. [26] studied the problem of selling a resource through an auction mechanism. Jain et al. [27] proposed the multi-stage auction mechanism by analyzing two-way competitions, a Bertrand and Cournot competition. Huang et al. [28] proved that the PT-BOCR method is a useful tool for risk aversion buyers to avoid losses and for suppliers to win the bids. Chen et al. [29] investigated keyword auctions for advertising. Another research stream in procurement auctions analyzed auctions in a more complicated supply chain setting. Ivengar and Kumar [30] proposed an optimal auction mechanism for a buyer that sources from suppliers with different capacity limits and production costs. Chen and Vulcano [31] compared the first- and the second-price auctions as mechanisms for two competing retailers to procure capacity from a supplier. Multi-attribute procurement auction mechanism was also extensively studied. Che [32] studied the two-attribute auction in which price and quality are the two attributes under consideration. Branco [33] extended Che's model to allow a correlation in the suppliers' costs. Parkes and Kalagnanam [34] provided an iterative auction design for an important special case of the multi-attribute allocation problem. Recently, there is also some research on auction coordination mechanisms for procurement. Chen [35] investigated the auction coordination mechanism for the supply chain with one manufacturer and multiple competing suppliers in the electronic market.

In this study, we consider the channel participants' risk preferences and reservation profits, and the main differences between our paper and the literature mentioned above can be summarized as follows. Firstly, in consideration of individual reservation profit, we employ an auction and contracting mechanism for channel coordination for a service supply chain in which traditional contracts are generally used such as two-part contracts, cost-sharing contracts, risk-sharing contracts, etc. Our findings also provide a useful tool for firms to deal with supplier selection and channel coordination. Secondly, we investigate the impacts of production cost variation and the value of information on the participants' profits and preferences in contrast to associated optimal policies in symmetric and asymmetric situations, which rarely appeared in the existing literature.

3. Symmetric Information Case

In this section, we consider a supply chain consisting of one SI and two competitive SPs providing services to the SI with different costs. We begin this section with some notations used throughout this paper in Table 1.

Table 1. List of Notations.

Notation	Definition
r:	service price;
q:	order quantity;
c:	the SI's holding cost;
L_i:	side payment of SP i, for $i = l, h$;
w_i:	the SP i's wholesale price;
s_i:	the SP i's service cost, s_l stands for low cost, s_h stands for high cost, $s_l \leq s_h$;
Π_i :	the SI's profit when $i = R$; the entire supply chain's profit when $i = SC$; the low-(high-)cost SP's profit when $i = S_{l(h)}$;
α :	the confidence level for the SI's risk measure;
k_{S_i}:	the SP i's reservation profit, $i = l, h$.

For the demand in the system, we assume a stochastic and price-dependent demand function (Arcelus et al. [36], Leng et al. [37])

$$D_d = gr^{-b}x, \quad \text{for } g > 0, b > 1,$$

where x denotes a random variable with a finite mean μ that follows a density function $f(\cdot)$ and a cumulative distribution function $F(\cdot)$. Denote the inverse function of $F(\cdot)$ by $F^{-1}(\cdot)$. Then, the SI's profit can be expressed as

$$\Pi_R(r,q) = \begin{cases} gr^{-b+1}x - (c+w)q, & \text{if } x \leq \frac{qr^b}{g}, \\ (r-c-w)q, & \text{if } x > \frac{qr^b}{g}. \end{cases}$$

Consider the case that the SI follows the VaR criterion to determine the optimal order quantity and optimal service price. Then, the SI's profit with VaR is defined as

$$\max_{r,q\geq 0}\{T \mid P(\Pi_R(r,q) \leq T) \leq \alpha\} \tag{1}$$

where T is the SI's reservation profit, and $0 < \alpha < 1$ denotes the confidence level which reflects the degree of risk-aversion for the SI. The optimization problem (1) with a chance constraint that maximizes the SI's profit turns out to be the main challenge.

Lemma 1. *For the VaR model, the optimal price, optimal order quantity, and optimal profit of the SI are*

$$\begin{cases} r^\star = \frac{b}{b-1}(c+w), \\ q^\star = \Phi_\alpha g b^{-b}(b-1)^b(c+w)^{-b}, \\ T^\star = \Phi_\alpha g b^{-b}(b-1)^{b-1}(c+w)^{-b+1}, \end{cases} \tag{2}$$

respectively.

From
$$\frac{d\Phi_\alpha}{d_\alpha} = \frac{dF^{-1}(1-\alpha)}{d_\alpha} = -\frac{1}{\frac{dF(\Phi_\alpha)}{d\Phi_\alpha}} = -\frac{1}{f(F^{-1}(1-\alpha))} < 0,$$

we have $\frac{dq^\star}{d\alpha} < 0$ and $\frac{dT^\star}{d\alpha} < 0$, then $q^\star$ and $T^\star$ decrease with respect to α. The lower the risk tolerance of managers, the less order quantity as a security strategy, hence, the SI's profit decreases accordingly.

Now, consider the situation where the symmetric information case including the SPs' service cost (s_l, s_h), reservation profit (k_l, k_h). Without loss of generality, we assume that $s_l < s_h$, and the market imposes a competitive requirement $s_h \leq \frac{c+bs_l}{b-1}$ on the costs of the two SPs. There is a twofold explanation for imposing this assumption. Firstly, the high-cost SP's cost should not be too high compared to the low-cost SP's cost in order to make the market more competitive. Secondly, this requirement is aimed at making both the SI and SPs have incentives to join in the two-part contract auction.

3.1. Supply Chain Coordination Model with Symmetric Information

For channel coordination issues, we can obtain the following optimization problem

$$\begin{aligned} \max_{w} \pi_{SC,\alpha} &= \pi_{R,\alpha} + \pi_{R,\alpha} \\ &= \Phi_\alpha g b^{-b}(b-1)^{b-1}(c+w_i)^{-b+1} + (w_i - s_i)\Phi_\alpha g b^{-b}(b-1)^b(c+w_i)^{-b}. \end{aligned}$$

For this problem, by Equation (2), we have the following conclusion.

Lemma 2. *For the channel, the optimal order quantity, optimal service price, and coordinated supply chain profit are respectively*

$$\begin{cases} r^\star_{R,\alpha} = \frac{b}{b-1}(c+s_i), \\ q^\star_{R,\alpha} = \Phi_\alpha g b^{-b}(b-1)^b(c+s_i)^{-b}, \\ \pi^\star_{R,\alpha} = \Phi_\alpha g b^{-b}(b-1)^{b-1}(c+s_i)^{-b+1}. \end{cases}$$

Next, we consider the decentralized case where the SPs and the SI are independent of each other, that is, both of the two SPs determine the optimal wholesale price from their viewpoints. Therefore, the SI will set the service price and order quantity to maximize its profit. For the SP in the decentralized case, the optimal wholesale price $w^\star$ can be derived from the following optimization problem

$$\max_{w_i} \pi_{S_i,\alpha} = (w_i - s_i)\Phi_\alpha g b^{-b}(b-1)^b(c+w_i)^{-b},\ i = l,h$$

Lemma 3. *For the decentralized case, the optimal wholesale price of the SP is*

$$w_i^{\star ind} = \frac{c+bs_i}{b-1},\ i = l,h.$$

From Lemma 3, one has

$$\begin{cases} \pi^{\star ind}_{S_i,\alpha} = \Phi_\alpha g b^{-2b}(b-1)^{2b-1}(c+s_i)^{-b+1}, \\ \pi^{\star ind}_{R,\alpha} = \Phi_\alpha g b^{-2b+1}(b-1)^{2b-2}(c+s_i)^{-b+1}, \\ \pi^{\star ind}_{SC,\alpha} = \Phi_\alpha g b^{-2b}(b-1)^{2b-2}(2b-1)(c+s_i)^{-b+1}, \end{cases}$$

where $\pi^{\star ind}_{S_i,\alpha}$, $\pi^{\star ind}_{R,\alpha}$, $\pi^{\star ind}_{SC,\alpha}$ are respectively, the SP's optimal profit, the SI's optimal profit, and the total supply chain profit for the decentralized case.

By comparing $\pi^\star_{SC,\alpha}$ and $\pi^{\star ind}_{SC,\alpha}$, we find that $\frac{\pi^\star_{SC,\alpha}}{\pi^{\star ind}_{SC,\alpha}} = \frac{b^b}{(b-1)^{b-1}(2b-1)} > 1$. As a consequence of Lemma 3, the total supply chain profit cannot achieve the system optimums. Therefore, designing the supply chain coordination mechanism becomes particularly important. On the other hand, it should be noted that, in the channel coordination, the SI possesses the total profit of the supply chain system–in other words, the SP has no profit. Although the supply chain optimum is achieved, the SP has no incentive to cooperate with the SI because of no profit. Consequently, it is quite necessary to design an effective supply chain coordination mechanism to ensure the efficient operation of the entire supply chain.

3.2. Two-Part Contract Auction with Symmetric Information

In this subsection, we define a two-part contract auction as follows, also illustrated in Figure 1.

(1) The SI announces an order quantity function and a service price function

$$q = \Phi_\alpha g(c+w)^{-b}, r = c+w. \tag{3}$$

(2) Given that reservation profit k_{S_i} respectively, each SP in the two-part contract proposes a wholesale price $w_i^\star$ and a side payment $L_i^\star$ to the SI.

(3) The SI selects the more profitable contract and then makes decisions on its order quantity and service price.

Likewise, we use $w_i^{\star tca}$, $r^{\star tca}_{R,\alpha}$, $q^{\star tca}_{R,\alpha}$, $\pi^{\star tca}_{R,\alpha}$, $\pi^{\star tca}_{S_i,\alpha}$, $\pi^{\star tca}_{SC,\alpha}$ to denote the optimal wholesale price of the SP i, $i = l,h$, the optimal service price, the optimal quantity, the SI's optimal profit, the SP's optimal profit and the total supply chain profit in the two-part contract auction, respectively. Notice that both the service price function and order quantity function of the SI are related to the SP's wholesale price. In this setting, it offers the SI more flexibility to make decisions. From the definition of q and r in Equation (3), the two SPs can determine the wholesale price by solving the problem as follows

$$\max_{w} \pi_{S_i,\alpha} = (w_i - s_i)\Phi_\alpha g(c+w_i)^{-b},\ i = l,h.$$

As mentioned above, the optimal wholesale price of the SP is

$$w_i^{\star tca} = \frac{c+bs_i}{b-1}, i = l,h$$

For the SP's reservation profit $k_{S_i}(i = l, h)$, only when the profit that the SP gets from the contract is greater than its reservation profit, the SP will accept the contract and cooperate with the SI, otherwise, they will not.

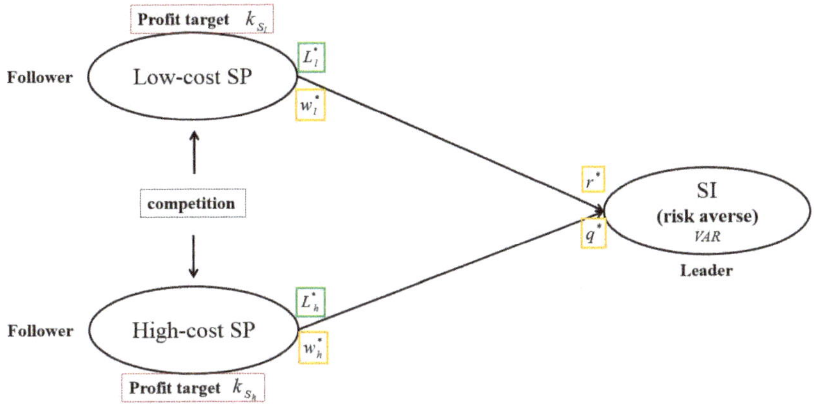

Figure 1. Process flow diagram for two-part contract auction.

Proposition 1. *For the symmetric information case, the profit of the total supply chain under a two-part contract auction is the same as the one in channel coordination optimum.*

A nice consequence of Proposition 1 is that the two-part contract auction guarantees the channel coordination optimum in a symmetric information scenario.

Proposition 2. *For any given $k_{S_i} = \Phi_\alpha g b^{-2b}(b-1)^{2b-1}(c+s_i)^{-b+1}$, which is the profit of the SP $i = l, h$ in decentralized case, each SP's profit under the two-part contract auction $\pi^{\star tca}_{S_i,\alpha}$ is always greater than that in decentralized case.*

From Proposition 2, compared with the decentralized case, the SP's profit in the two-part contract auction can be improved. Therefore, both of them have an incentive to participate in the designed auction mechanism. Combining Propositions 1 and 2 yields the following Parato improvement conditions.

Proposition 3. *For the SI's profit under the two-part contract auction, it holds that*

(1) *If the SI cooperates with the high-cost SP, then the SI's profit under two-part contract auction is greater than that in the decentralized case provided that*

$$k_{S_l} < \Phi_\alpha g b^{-b}(b-1)^{b-1}[(c+s_l)^{-b+1} - (\frac{b-1}{b})^{b-1}(c+s_h)^{-b+1}];$$

(2) *If the SI cooperates with the low-cost SP, then the SI's profit under two-part contract auction is greater than that in the decentralized case provided that*

$$k_{S_h} < \Phi_\alpha g b^{-b}(b-1)^{b-1}[(c+s_h)^{-b+1} - (\frac{b-1}{b})^{b-1}(c+s_l)^{-b+1}];$$

(3) *For the case*

$$k_{S_i} = \Phi_\alpha g b^{-2b}(b-1)^{2b-1}(c+s_i)^{-b+1}, i = l, h$$

if the SI cooperates with the low-cost SP, the SI's profit under two-part contract auction is greater than that in the decentralized case provided that

$$\frac{(c+s_h)^{-b+1}}{(c+s_l)^{-b+1}} < \frac{1-(\frac{b-1}{b})^b}{(\frac{b-1}{b})^{b-1}};$$

if the SI cooperates with the low-cost SP, the SI's profit under two-part contract auction is greater than that in the decentralized case provided that

$$\frac{(c+s_l)^{-b+1}}{(c+s_h)^{-b+1}} < \frac{1-(\frac{b-1}{b})^b}{(\frac{b-1}{b})^{b-1}}.$$

Proposition 3 clearly characterizes Pareto improvement conditions in which every participant's profit can be improved under the two-part contract auction strategy in comparison with that in the decentralized case. Therefore, both of them have an incentive to voluntarily participate in this auction mechanism.

4. Supply Chain System with Asymmetric Information

4.1. Supply Chain Coordination Model with Asymmetric Information

In this section, we focus on the case where the SPs hold asymmetric cost s_i. The SI assumes that the two SPs' marginal costs both follow a uniform distribution on the interval $[\bar{s}-\varepsilon, \bar{s}+\varepsilon]$ and have a prior probability density function $g(s)$ and a cumulative distribution function $G(s)$, where ε denotes cost variation.

Similar to the discussion in Section 3, let $\tilde{q}^{\star}_{R,\alpha}$, $\tilde{r}^{\star}_{R,\alpha}$ and $\tilde{\pi}^{\star}_{SC,\alpha}$ be the optimal order quantity, optimal service price and optimal expected profit of the supply chain system in the asymmetric case.

Lemma 4. *In the asymmetric information case,*

(1) *when the low-cost SP wins the contract, the expected optimal service price, order quantity, and profit of the entire supply chain are*

$$E[\tilde{r}^{\star}_{R,\alpha}] = \frac{b}{b-1}[c+2\int_{\bar{s}-\varepsilon}^{\bar{s}+\varepsilon} s(1-G(s))g(s)ds] \tag{4}$$

$$E[\tilde{q}^{\star}_{R,\alpha}] = 2\Phi_\alpha g b^{-b-1}(b-1)^b \int_{\bar{s}-\varepsilon}^{\bar{s}+\varepsilon} s^{-\frac{1}{b}}(1-G(s^{-\frac{1}{b}}-c))g(s^{-\frac{1}{b}}-c)ds \tag{5}$$

$$E[\tilde{\pi}^{\star}_{SC,\alpha}] = 2\Phi_\alpha g b^{-b-1}(b-1)^{b-2} \int_{\bar{s}-\varepsilon}^{\bar{s}+\varepsilon} s^{-\frac{1}{b-1}}(1-G(s^{-\frac{1}{b-1}}-c))g(s^{-\frac{1}{b-1}}-c)ds \tag{6}$$

(2) *when the high-cost SP wins the contract, the expected optimal service price, order quantity, and profit of the entire supply chain are*

$$E[\tilde{r}^{\star}_{R,\alpha}] = \frac{b}{b-1}[c+2\int_{\bar{s}-\varepsilon}^{\bar{s}+\varepsilon} sG(s)g(s)ds] \tag{7}$$

$$E[\tilde{q}^{\star}_{R,\alpha}] = 2\Phi_\alpha g b^{-b-1}(b-1)^b \int_{\bar{s}-\varepsilon}^{\bar{s}+\varepsilon} s^{-\frac{1}{b}}G(s^{-\frac{1}{b}}-c)g(s^{-\frac{1}{b}}-c)ds \tag{8}$$

$$E[\tilde{\pi}^{\star}_{SC,\alpha}] = 2\Phi_\alpha g b^{-b-1}(b-1)^{b-2} \int_{\bar{s}-\varepsilon}^{\bar{s}+\varepsilon} s^{-\frac{1}{b-1}}G(s^{-\frac{1}{b-1}}-c)g(s^{-\frac{1}{b-1}}-c)ds \tag{9}$$

4.2. Two-Part Contract Auction with Asymmetric Information Case

Now, we introduce the two-part auction mechanism for asymmetric information cases so as to achieve channel coordination optimums.

(1) An order quantity function and a service price function w.r.t. wholesale price w announced by the SI as follows:

$$q = \Phi_\alpha g(\hat{k} + w)^{-b}, r = \hat{k} + w,$$

where $\hat{k}$ is an unknown parameter decided by the channel system.

(2) The SPs compete in this two-part contract auction, in which each of them proposes a wholesale price of $w_i^\star$ and a side payment of $L_i^\star$ to the SI.

(3) The SI chooses the more profitable contract and determines the order quantity and service price on $w_i^\star$.

If the low-cost SP wins the auction, one needs to determine the optimal wholesale price by solving the following optimization

$$\max_w \pi_{S_l,\alpha} = (w - \min(s_l, s_h))\Phi_\alpha g(\hat{k} + w)^{-b}.$$

There exists a unique optimal wholesale price

$$w_l^{\star tca} = \frac{\hat{k} + b\min(s_l, s_h)}{b-1},$$

and hence

$$\begin{aligned} \hat{q}_{R,\alpha}^{\star tca} &= \Phi_\alpha g b^{-b}(b-1)^b(\hat{k} + \min(s_l, s_h))^{-b}, \\ \hat{r}_{R,\alpha}^{\star tca} &= \tfrac{b}{b-1}(\hat{k} + \min(s_l, s_h)), \end{aligned}$$

where $\tilde{r}_{R,\alpha}^\star$, $\tilde{q}_{R,\alpha}^\star$ and $\tilde{\pi}_{SC,\alpha}^\star$ denote the optimal service price, order quantity and profit of the supply chain under two-part asymmetric information case.

Lemma 5. *There may not exist any other values but $\hat{k} = c$ which can improve the total supply chain profit.*

Lemma 5 implies that the best choice of the parameter $\hat{k} = c$, from the viewpoint of improving the system efficiency, which shows that, under the two-part auction setting, the SI's profit completely comes from the SP's side payment.

Proposition 4. *For asymmetric information cases, the supply chain can achieve coordination in a two-part contract auction.*

Likewise, Proposition 4 shows that for asymmetric service cost information, channel coordination can be achieved in the two-part contract auction as well.

Proposition 5. *For any given $k_{S_i} = \Phi_\alpha g b^{-2b}(b-1)^{2b-1}(c + E[s_i])^{-b+1}, i = l, h$, which is the profit of the SP obtained in the decentralized case, the SP's expected profit in this two-part contract auction is more than that in decentralized case.*

Proposition 5 shows us, in contrast to the decentralized case, the SP has an incentive to join the two-part contracting auction. In practice, to stimulate every participant's incentive to join the two-part contract, we need to explicitly clarify the Pareto improvement conditions.

Proposition 6. *For the SI's profit in a two-part contract auction, it holds that*

(1) *If the SI cooperates with the high-cost SP, then the SI's profit under two-part contract auction is greater than that in the decentralized case provided that*

$$\begin{aligned} k_{S_l} < \; & 2\Phi_\alpha g b^{-b}(b-1)^{b-2}[\textstyle\int_{\bar{s}-\varepsilon}^{\bar{s}+\varepsilon} s^{-\frac{1}{b-1}}(1 - G(s^{-\frac{1}{b-1}} - c))g(s^{-\frac{1}{b-1}} - c)ds - \\ & (\tfrac{b-1}{b})^{b-1}\textstyle\int_{\bar{s}-\varepsilon}^{\bar{s}+\varepsilon} s^{-\frac{1}{b-1}} G(s^{-\frac{1}{b-1}} - c)g(s^{-\frac{1}{b-1}} - c)ds]. \end{aligned}$$

(2) *If the SI cooperates with the low-cost SP, then the SI's profit under two-part contract auction is greater than that in the decentralized case provided that*

$$k_{S_h} < 2\Phi_\alpha g b^{-b}(b-1)^{2b-2}[\int_{\bar{s}-\varepsilon}^{\bar{s}+\varepsilon} s^{-\frac{1}{b-1}}(1-G(s^{-\frac{1}{b-1}}-c))g(s^{-\frac{1}{b-1}}-c)ds - (\tfrac{b-1}{b})^{b-1}\int_{\bar{s}-\varepsilon}^{\bar{s}+\varepsilon} s^{-\frac{1}{b-1}}G(s^{-\frac{1}{b-1}}-c)g(s^{-\frac{1}{b-1}}-c)ds].$$

(3) *Specially, for the case that*

$$k_{S_l} = 2\Phi_\alpha g b^{-2b}(b-1)^{2b-2}\int_{\bar{s}-\varepsilon}^{\bar{s}+\varepsilon} s^{-\frac{1}{b-1}}(1-G(s^{-\frac{1}{b-1}}-c))g(s^{-\frac{1}{b-1}}-c)ds,$$

which is the profit of the low-cost SP in the decentralized case. The SI's profit under two-part contract auction is greater than that in the decentralized case if and only if

$$\frac{\int_{\bar{s}-\varepsilon}^{\bar{s}+\varepsilon} s^{-\frac{1}{b-1}}G(s^{-\frac{1}{b-1}}-c)g(s^{-\frac{1}{b-1}}-c)ds}{\int_{\bar{s}-\varepsilon}^{\bar{s}+\varepsilon} s^{-\frac{1}{b-1}}(1-G(s^{-\frac{1}{b-1}}-c))g(s^{-\frac{1}{b-1}}-c)ds} < \frac{1-(\frac{b-1}{b})^b}{(\frac{b-1}{b})^{b-1}}.$$

For the case that

$$k_{S_h} = 2\Phi_\alpha g b^{-2b}(b-1)^{2b-2}\int_{\bar{s}-\varepsilon}^{\bar{s}+\varepsilon} s^{-\frac{1}{b-1}}G(s^{-\frac{1}{b-1}}-c)g(s^{-\frac{1}{b-1}}-c)ds,$$

which is the profit of the high-cost SP in the decentralized case. The SI's profit under two-part contract auction is greater than that in the decentralized case if and only if

$$\frac{\int_{\bar{s}-\varepsilon}^{\bar{s}+\varepsilon} s^{-\frac{1}{b-1}}(1-G(s^{-\frac{1}{b-1}}-c))g(s^{-\frac{1}{b-1}}-c)ds}{\int_{\bar{s}-\varepsilon}^{\bar{s}+\varepsilon} s^{-\frac{1}{b-1}}G(s^{-\frac{1}{b-1}}-c)g(s^{-\frac{1}{b-1}}-c)ds} < \frac{1-(\frac{b-1}{b})^b}{(\frac{b-1}{b})^{b-1}}.$$

Therefore, when the above conditions hold, both the SP's and the SI's profit can be improved in this two-part contract auction strategy compared with the decentralized case. Therefore, both of them have an incentive to join in this contracting and auction mechanism. Note that, the formulations under the symmetric and asymmetric cases are quite different. Therefore, it is necessary to explore the impacts of the SP's service cost variation on optimal decisions.

4.3. Value of Information

In order to explore the value of information, we compare participants' optimal decisions and profits under both symmetric and asymmetric information cases and analyze the difference between different cases.

Proposition 7. *Suppose the low-cost SP wins the contract, then*

(a) *for the service's optimal decisions, it holds that*
   - (a.1) *the optimal service price decreases w.r.t. the low-cost SP's cost variation (ε);*
   - (a.2) *the optimal order quantity increases w.r.t. the low-cost SP's cost variation (ε);*
   - (a.3) *the greater the level of the low-cost SP's cost variation, the larger the difference of service prices under information symmetry and information asymmetry cases;*
   - (a.4) *the greater the level of the low-cost SP's cost variation, the smaller the difference of the optimal order quantity under both symmetric and asymmetric information cases.*

(b) *For the SI's profit, it holds that*
   - (b.1) *the SI's profit decreases with the low-cost SP's cost variation (ε);*
   - (b.2) *the greater the level of the low-cost SP's cost variation, the larger the difference of the SI's profits under both symmetric and asymmetric information cases.*

(c) *For the low-cost SP's profit, it holds that*

(c.1) the low-cost SP's profit increases with its own cost variation (ε);

(c.2) the greater the level of the low-cost SP's cost variation, the smaller the difference between the low-cost SP's profit under both symmetric and asymmetric information cases;

(d) and for the supply chain's profit, it holds that

(d.1) the supply chain's profit increases with the low-cost SP's cost variation (ε);

(d.2) the greater the level of the low-cost SP's cost variation, the smaller the difference in the supply chain's profit under both symmetric and asymmetric information cases.

For ease of exposition, we can use a table to clearly illustrate the results of Proposition 7.

Proposition 8. *When the high-cost SP wins the contract,*

(a) for the service's optimal decisions we have that

(a.1) the optimal service price increases with the high-cost SP's cost variation (ε);

(a.2) the optimal order quantity decreases with the high-cost SP's cost variation (ε);

(a.3) the greater the level of the high-cost SP's cost variation (large ε), the smaller the difference of the service price under both symmetric and asymmetric information cases (two cases for short).

(a.4) the greater the level of the high-cost SP's cost variation (large ε), the larger the difference of the optimal order quantity under two cases.

(b) for the SI's profit we have that

(b.1) the SI's profit increases with the high-cost SP's cost variation (ε);

(b.2) the greater the level of the high-cost SP's cost variation (large ε), the smaller the difference of the SI's profit under two cases.

(c) for the low-cost SP's profit we have that

(c.1) the low-cost SP's profit decreases with its own cost variation (ε);

(c.2) the greater the level of the high-cost SP's cost variation (large ε), the larger the difference of the low-cost SP's profit under two cases.

(d) for the supply chain's profit we have that

(d.1) the supply chain's profit decreases with the high-cost SP's cost variation (ε);

(d.2) the greater the level of the high-cost SP's cost variation (large ε), the larger the difference of the supply chain's profit under two cases.

Likewise, for ease of exposition, we can use a table to clearly illustrate the results of Proposition 8.

Observing from Tables 2 and 3, we find that information asymmetry benefits the low-cost SP, however, the SI may be worse off without the SP's full cost information. On the contrary, information asymmetry benefits SI when the high-cost SP wins the contract, but the high-cost SP may be worse off if encountered with high cost variation. Therefore, the high-cost SP is always motivated to truthfully announce rather than distort his exact cost information. This conclusion is somewhat similar to separate equilibrium in a signal game.

Table 2. Results of Proposition 7 "↑" increasing; "↓" decreasing.

	$\tilde{r}^{\star tca}_{R,\alpha}$	$\tilde{q}^{\star tca}_{R,\alpha}$	$E[\tilde{\pi}^{\star tca}_{R,\alpha}]$	$E[\tilde{\pi}^{\star tca}_{S_i,\alpha}]$	$E[\tilde{\pi}^{\star tca}_{SC,\alpha}]$
ε ↑	↓	↑	↓	↑	↑
	$\Delta\tilde{r}^{\star tca}_{R,\alpha}$	$\Delta\tilde{q}^{\star tca}_{R,\alpha}$	$\Delta E[\tilde{\pi}^{\star tca}_{R,\alpha}]$	$\Delta E[\tilde{\pi}^{\star tca}_{S_i,\alpha}]$	$\Delta E[\tilde{\pi}^{\star tca}_{SC,\alpha}]$
ε ↑	↑	↓	↑	↓	↓

Table 3. Results of Proposition 8 "↑" increasing; "↓" decreasing.

	$\bar{r}^{\star tca}_{R,\alpha}$	$\bar{q}^{\star tca}_{R,\alpha}$	$E[\bar{\pi}^{\star tca}_{R,\alpha}]$	$E[\bar{\pi}^{\star tca}_{S_i,\alpha}]$	$E[\bar{\pi}^{\star tca}_{SC,\alpha}]$
$\varepsilon\uparrow$	↑	↓	↑	↓	↓
	$\Delta\bar{r}^{\star tca}_{R,\alpha}$	$\Delta\bar{q}^{\star tca}_{R,\alpha}$	$\Delta E[\bar{\pi}^{\star tca}_{R,\alpha}]$	$\Delta E[\bar{\pi}^{\star tca}_{S_i,\alpha}]$	$\Delta E[\bar{\pi}^{\star tca}_{SC,\alpha}]$
$\varepsilon\uparrow$	↓	↑	↓	↑	↑

5. The Case with a Risk Neutral SI

In this section, we consider the optimal service price, the optimal order quantity, and supply chain profits when the participants are all risk-neutral. The relationship between the risk-neutral agent case and at least one risk-averse agent case will also be explored.

In the risk-neutral agent system, facing the announced wholesale price, the SI needs to solve the following optimization to determine his optimal service price and order quantity

$$\max_{r>c,q\geq 0} E[\Pi_R(r,q)]$$

where,

$$\begin{aligned} E[\Pi_R(r,q)] &= rE[\min(gr^{-b}x,q)] - (c+w)q \\ &= r[q(1-F(\tfrac{q}{g}r^b)) + \textstyle\int_0^{\frac{q}{g}r^b}(gr^{-b}x)f(x)dx] - (c+w)q \\ &= r[q(1-F(\tfrac{q}{g}r^b)) + gr^{-b}(\tfrac{q}{g}r^bF(\tfrac{q}{g}r^b)) - \textstyle\int_0^{\frac{q}{g}r^b}F(x)dx] - (c+w)q \\ &= r(q - gr^{-b}\textstyle\int_0^{\frac{q}{g}r^b}F(x)dx) - (c+w)q. \end{aligned}$$

Thus

$$E[\Pi_{SC}(r,q)] = r(q - gr^{-b}\int_0^{\frac{q}{g}r^b}F(x)dx) - (c+s)q.$$

Differentiate $E[\Pi_{SC}(r,q)]$ w.r.t. q and r, and denote $\frac{dE[\Pi_{SC}(r,q)]}{dq} = \frac{dE[\Pi_{SC}(r,q)]}{dr} = 0$, which yields the optimal values $q^\star_N$ and $r^\star_N$, where

$$q^\star_N = g(r^\star_N)^{-b}F^{-1}(\frac{r^\star_N - c - s}{r^\star_N}) \tag{10}$$

$$(b-1)g(r^\star_N)^{-b}\int_0^{F^{-1}(\frac{r^\star_N-c-s}{r^\star_N})}F(x)dx + q^\star_N(1 - b(\frac{r^\star_N - c - s}{r^\star_N})) = 0, \tag{11}$$

From which we can obtain

$$(b-1)\int_0^{F^{-1}(\frac{r^\star_N-c-s}{r^\star_N})}F(x)dx + F^{-1}(\frac{r^\star_N - c - s}{r^\star_N})(1 - b(\frac{r^\star_N - c - s}{r^\star_N})) = 0.$$

Let $SL(r) = \frac{r-c-s}{r}, \alpha = 1 - SL(r^\star_N)$. Then

$$F^{-1}(\frac{r^\star_N - c - s}{r^\star_N}) = F^{-1}(1-\alpha) = \Phi_\alpha.$$

The following theorem illustrates the relationship between service prices and order quantities in agents' risk-neutral cases and risk-averse cases.

Proposition 9. *For any $0 < \alpha < 1$, it holds that $r^\star_\alpha \leq r^\star_N$ and $q^\star_\alpha \geq q^\star_N$.*

Note that even though we are unable to obtain a closed-form expression of the optimal order quantity and optimal service price in the risk-neutral SI case, Proposition 9 theoretically proves the relationship between the optimal service price and optimal order quantity between the cases with a risk-neutral and a risk-averse SI, respectively. To be specific, Proposition 9 shows that a risk-averse SI tends to set a service price higher than that set by a risk-neutral SI and induce a lower order quantity than that induced by a risk-neutral SI, which accounts for how decisions might adapt with respect to the degree of risk aversion. This is an important and neat analytical result.

6. Numerical Examples and Illustrations

In order to verify the efficiency of the obtained results, we take a numerical experiment on some practical industrial productions. We assume that Foxconn and Quanta are two competing manufacturing service SPs for the SI Apply Company, which plays the role of an SI. As stated in literature (Mathewson and Winter [38], Gans and King [39]), pricing power endows the SI first mover advantage. The exponential and stochastic demand function is shown as follows:

$$D_d = gr^{-b}x, g > 0, b > 1.$$

For the parameters involved in the model, we take $g = 10{,}000$, $b = 2$, $\varepsilon = 5$, low-cost and high-cost are denoted as s_l and s_h, respectively, $10 \leq s_l, s_h \leq 20$. The reservation profits of the SPs are k_{S_l}, k_{S_h} respectively, $200 \leq k_{S_l} \leq 500$, $100 \leq k_{S_h} \leq 400$.

To test the effect of the SP costs on the optimal order quantity, the optimal service price, and the profits of the supply chain under a different situation, $\lambda = 1$. the probability density function and the cumulative distribution function of x are respectively

$$f_1(x) = e^{-x}, x \geq 0, \quad F_1(x) = 1 - e^{-x}, x \geq 0.$$

Hence $F_1^{-1}(\alpha) = -\ln(1-x)$, and

$$\Phi_{1,\alpha} = F_1^{-1}(\alpha) = -\ln(1-\alpha)$$

Next, we assume that the variable representing the disturbance in the demand D_d follows a normal distribution with $\mu = 4$ and variance $\sigma^2 = 1$. In other words, the probability density function and the cumulative distribution function of x are

$$f_2(x) = \frac{1}{\sqrt{2\pi}}e^{-\frac{(x-4)^2}{2}}$$

$$F_2(x) = \int_{-\infty}^{x} \frac{1}{\sqrt{2\pi}}e^{-\frac{(y-4)^2}{2}}dy = \int_{-\infty}^{x-4} \frac{1}{\sqrt{2\pi}}e^{-\frac{y^2}{2}}dy$$

respectively. Hence $\Phi_{2,\alpha}$ is defined by

$$\Phi_{2,\alpha} = F_2^{-1}(1-\alpha)$$

Thus $\int_{-\infty}^{\Phi_{2,\alpha}} \frac{1}{\sqrt{2\pi}}e^{-\frac{y^2}{2}}dy = \alpha$. We divide the SI's risk preferences into three cases (i) $\alpha = 0.9$, (ii) $\alpha = 0.95$, (iii) $\alpha = 0.99$.

6.1. Comparisons of Optimal Decisions under the Symmetric and Asymmetric Information Cases

In the case of symmetric information, we assume that the service costs of the SPs $s_i, i = l, h$ is either 10 or 15, the SI knows that the service costs of the SPs follow a uniform distribution on the interval $[10, 20]$ and thus, the expected cost $E[s_i]$ is 15. The probability density function and the cumulative distribution function of s_i are

$$g(s) = \begin{cases} 0.1, \ 10 \le s \le 20, \\ 0, \ \text{otherwise}, \end{cases} \quad G(s) = \begin{cases} 0, \ 0 \le s \le 10 \\ 0.1(s-10), 10 \le s \le 20 \\ 1, \ s > 20 \end{cases}$$

respectively. Thus,

$$f_{\min}(s) = F'_{\min}(s) = 2(1 - G(s))g(s) = \begin{cases} 0.02(20-s), & 10 \le s \le 20 \\ 0, & \text{otherwise} \end{cases}$$

$$f_{\max}(s) = F'_{\max}(s) = 2G(s)g(s) = \begin{cases} 0.02(s-10), & 10 \le s \le 20 \\ 0, & \text{otherwise} \end{cases}$$

Let $q^{\star}_{R,\alpha}$ and $r^{\star}_{R,\alpha}$, $\tilde{q}^{\star}_{R,\alpha}$ and $\tilde{r}^{\star}_{R,\alpha}$ denote the optimal order quantity and service price under the symmetric information and asymmetric information cases with risk preference. We consider two cases where the demand disturbance follows the exponential distribution and the normal distribution, respectively. When the SI cooperates with the low-cost SP, we have

$$\begin{aligned} E[\tilde{q}^{\star}_{R,\alpha}] &= \Phi_{\alpha} g b^{-b}(b-1)^{b} E[(c + \min(s_l, s_h))^{-b}] \\ &= \Phi_{\alpha} \times 10000 \times 0.25 \times 0.00317 \\ &= 7.7925\Phi_{\alpha} \end{aligned}$$

$$E[\tilde{r}^{\star}_{R,\alpha}] = \frac{110}{3} = 36.67.$$

When the SI cooperates with the low-cost SP, we have

$$\begin{aligned} E[\tilde{q}^{\star}_{R,\alpha}] &= \Phi_{\alpha} g b^{-b}(b-1)^{b} E[(c + \max(s_l, s_h))^{-b}] \\ &= \Phi_{\alpha} \times 10000 \times 0.25 \times 0.002217 \\ &= 5.5425\Phi_{\alpha} \end{aligned}$$

$$E[\tilde{r}^{\star}_{R,\alpha}] = \frac{130}{3} = 43.33.$$

Detail results are illustrated in Tables 4 and 5.

Table 4. Comparison of the optimal ordering quantities and optimal service prices obtained under the complete information and asymmetric information situations when the SI cooperates with low-cost SP (g = 10,000, $b = 2$, $c = 5$).

Cooperation with the Low-Cost SP		Demand Disturbance ∼ Exponential				Demand Disturbance ∼ Normal			
		Symmetric Information		Asymmetric Information		Symmetric Information		Asymmetric Information	
α	s_1	$q^{\star}_{R,\alpha}$	$r^{\star}_{R,\alpha}$	$\tilde{q}^{\star}_{R,\alpha}$	$\tilde{r}^{\star}_{R,\alpha}$	$q^{\star}_{R,\alpha}$	$r^{\star}_{R,\alpha}$	$\tilde{q}^{\star}_{R,\alpha}$	$\tilde{r}^{\star}_{R,\alpha}$
	10	25.58	30.00			58.67	30.00		
0.9	12	19.92	34.00	17.94		45.67	34.00	41.14	
	14	15.95	38.00			36.57	38.00		
	10	33.29	30.00			62.67	30.00		
0.95	12	25.91	34.00	23.34	36.67	48.79	34.00	43.95	36.67
	14	20.75	38.00			39.06	38.00		
	10	51.17	30.00			70.33	30.00		
0.99	12	39.84	34.00	35.89		54.76	34.00	49.33	
	14	31.89	38.00			43.84	38.00		

Table 5. Comparison of the optimal ordering quantities and optimal service prices obtained under the complete information and asymmetric information situations when the SI cooperates with high-cost SP (g = 10,000, $b = 2$, $c = 5$).

Cooperation with the High-Cost SP		Demand Disturbance ~ Exponential				Demand Disturbance ~ Normal			
		Symmetric Information		Asymmetric Information		Symmetric Information		Asymmetric Information	
α	s_1	$q^{\star}_{R,\alpha}$	$r^{\star}_{R,\alpha}$	$\bar{q}^{\star}_{R,\alpha}$	$\bar{r}^{\star}_{R,\alpha}$	$q^{\star}_{R,\alpha}$	$r^{\star}_{R,\alpha}$	$\bar{q}^{\star}_{R,\alpha}$	$\bar{r}^{\star}_{R,\alpha}$
	16	13.05	42.00			29.93	42.00		
0.9	18	10.88	46.00	12.76	43.33	24.95	46.00	29.26	43.33
	20	9.21	50.00			21.12	50.00		
	16	16.98	42.00			31.97	42.00		
0.95	18	14.16	46.00	16.60		26.65	46.00	31.26	
	20	11.98	50.00		43.33	22.56	50.00		43.33
	16	26.11	42.00			35.88	42.00		
0.99	18	21.76	46.00	25.52		29.91	46.00	35.08	
	20	18.42	50.00			25.32	50.00		

6.2. *The Impact of Service Cost Variation*

In this section, we use a numerical example to explore the service cost variation's effect on the SI's optimal decisions and profit, SP's profit, and supply chain's profit. Thus, to compare the performances of service cost variation in this section, we use normal distribution density functions to present the demand disturbances with $\mu = 4$ and variance $\sigma^2 = 1$. For the SI's risk preference, we let $\alpha = 0.9$. Then Figures 2–5 depict the impact of parameter ε on the SI and the SP under the information asymmetry.

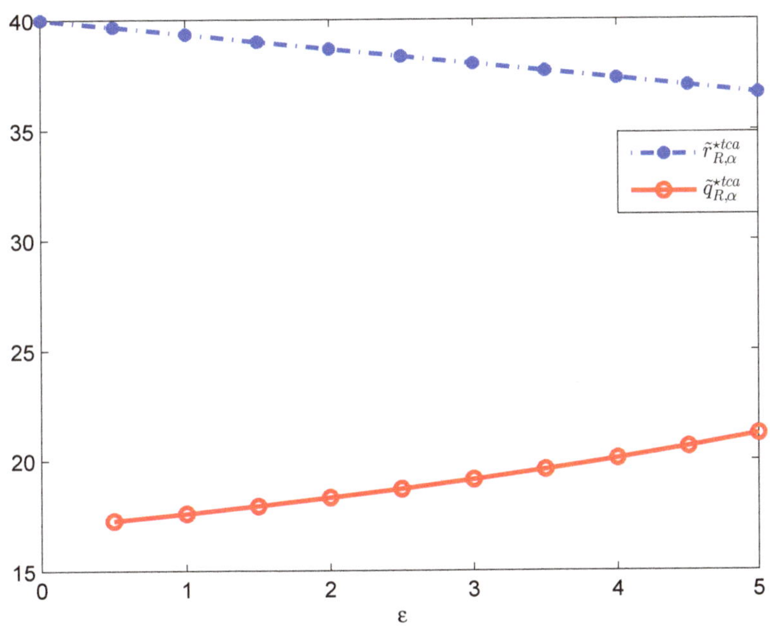

Figure 2. The impact of ε on $r^{\star tca}_{R,\alpha}$ and $q^{\star tca}_{R,\alpha}$ when low-cost SP wins the contract ($b = 2, g$ = 10,000, $c = 5, \bar{s} = 15, k_l = 300, k_h = 200$).

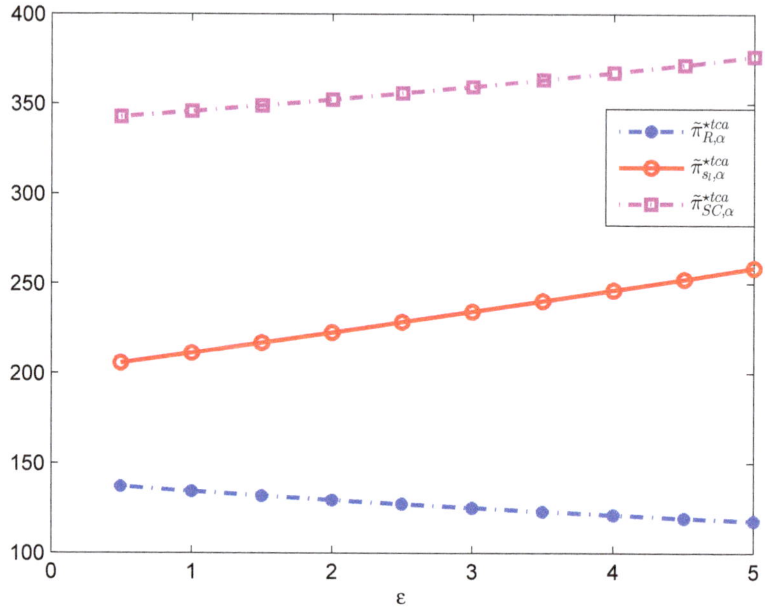

Figure 3. The impact of ε on $\pi^{\star tca}_{R,\alpha}$,$\pi^{\star tca}_{S_l,\alpha}$ and $\pi^{\star tca}_{S_l,\alpha}$ when low-cost SP wins the contract ($b = 2$, g = 10,000, $c = 5, \bar{s} = 15, k_l = 300, k_h = 200$).

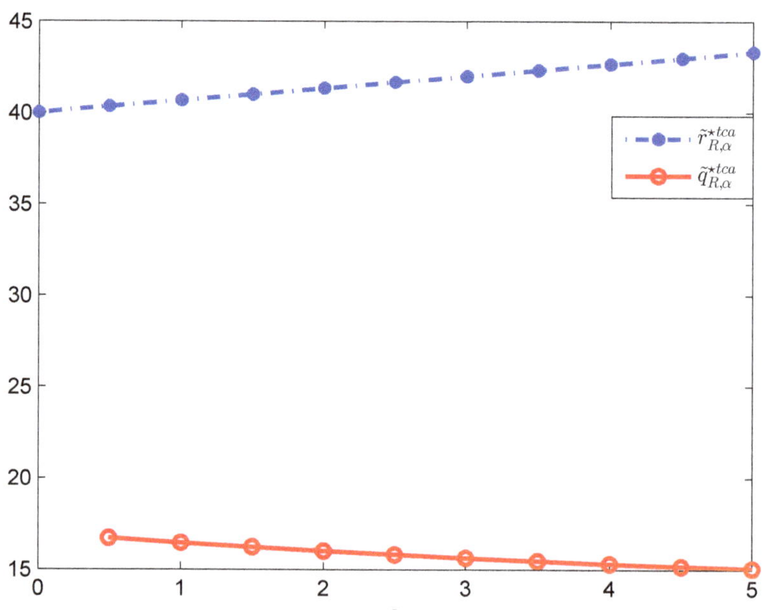

Figure 4. The impact of ε on $r^{\star tca}_{R,\alpha}$ and $q^{\star tca}_{R,\alpha}$ when high-cost SP wins the contract ($b = 2, g$ = 10,000, $c = 5, \bar{s} = 15, k_l = 300, k_h = 200$).

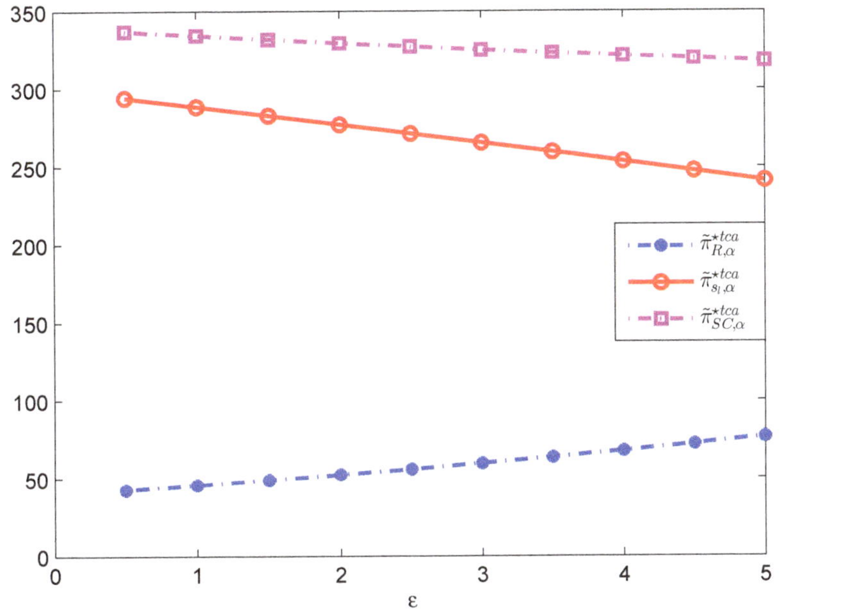

Figure 5. The impact of ε on $\pi^{\star tca}_{R,\alpha}$,$\pi^{\star tca}_{S_l,\alpha}$ and $\pi^{\star tca}_{S_l,\alpha}$ when high-cost SP wins the contract ($b = 2$, g = 10,000, $c = 5, \bar{s} = 15, k_l = 300, k_h = 200$).

From Figures 2–5, information asymmetry benefits SP when the low-cost SP wins the contract, but the SI may be worse off without the SP's full cost information. To be specific, the optimal service price and the SI's profit decrease with the low-cost SP's cost variation, while the optimal order quantity, the low-cost SP's profit, and the supply chain's profit increase with the low-cost SP's cost variation. When the high-cost SP wins the contract, the result is opposite to that when the low-cost SP wins the contract.

7. Conclusions, Practical Implications and Future Research Directions

In this study, we design the procurement auction and contracting mechanism and demonstrate it would be served as a channel coordination mechanism incorporating parties' risk preferences and individual reservation profit and further, analytically provide new insights on the bidding and auction strategies in both symmetric information and asymmetric information, respectively. We characterize the Pareto improvement conditions and find that the high-cost SP may be worse off if encountered with high cost variation. Therefore, the high-cost SP is always motivated to truthfully announce his cost information. With risk-averse agents, we show that the obtained results have lower service prices and higher order quantity when compared with the supply chain with risk-neutral agents case.

Several important practical implications are derived from this study. (1) When the SI practitioner is engaged in service procurement from heterogeneous SPs, an appropriate auction and contracting mechanism can be employed for SP selection. Our study underscores the importance of the auction mechanism in the competitive bidding environment, as shown in the main body, which can be taken as a successful channel coordination strategy. (2) In practice, the SIs with various risk preferences have different decision-making abilities. Notably, the agents' risk preferences are therefore important and non-negligible influencing factors in a service procurement supply chain. (3) In an asymmetric information environment, relative high-cost SP should strive for avoiding cost variation, i.e., to truthfully share exact cost information with the SI.

Our work leaves many interesting directions for future research. It would be interesting to develop a multi-period auction and contracting mechanism for multi-product service supply chain coordination. It is also an intriguing question to understand how much the production deviation cost of the two SPs differs in optimal decision-making and policy.

Author Contributions: J.C.: Conceptualization, Methodology, Formal analysis, Investigation, Writing—original draft, Writing review & editing, Funding acquisition. C.M.: Conceptualization, Methodology, Supervision, Visualization, Resources, Data curation, Writing review & editing, Funding acquisition. All authors have read and agreed to the published version of the manuscript.

Funding: This research was supported by the National Natural Science Foundation of China (No. 71991474; 71721001; 11401331), Humanities and social sciences research projects of the Ministry of Education of China (No. 18YJC630119), the Natural Science Foundation of Shandong Province, China (No. ZR2020MA024) and digital Shandong research project of Social Science Planning fund program of Shandong province (No. 20CSDJ16).

Informed Consent Statement: Informed consent was obtained from all subjects involved in the study.

Data Availability Statement: Not applicable.

Conflicts of Interest: The authors declare that they have no known competing financial interests or personal relationships that could have appeared to influence the work reported in this paper.

Appendix A

Appendix A.1. Proof of Lemma 1

According to the SI's profit, when $x \leq \frac{qr^b}{g}$, that is, $gr^{-b+1}x - (c+w)q \leq (r-c-w)q$, then

$$\pi(r,q) \leq (r-c-w)q.$$

Further, when $T > (r-c-w)q$, one has

$$P(\pi(r,q) \leq T) = 1;$$

and when $T \leq (r-c-w)q$, i.e., $\pi(r,q) = gr^{-b+1}x - (c+w)q \leq T$, one has

$$x \leq \frac{T+(c+w)q}{gr^{-b+1}}.$$

Let $\beta = \frac{T+(c+w)q}{gr^{-b+1}}, \Phi_\alpha = F^{-1}(1-\alpha)$. Then using the fact that $F(\cdot)$ is increasing, we have

$$P(\pi(r,q) \leq T) = P(x \leq \beta) = F(\beta) \leq 1-\alpha.$$

Hence, $\beta \leq \Phi_\alpha$. Thus, $T \leq (r-c-w)q$ can be transformed into

$$q \geq \frac{T}{r-c-w}.$$

Then

$$\frac{T}{(r-c-w)r^{-b}} = \frac{T+(c+w)\frac{T}{r-c-w}}{gr^{-b+1}} \leq \frac{T+(c+w)q}{gr^{-b+1}} = \beta.$$

Therefore,

$$\frac{T}{(r-c-w)r^{-b}} \leq \beta \leq \Phi_\alpha \Rightarrow \Phi_\alpha g r^{-b}(r-c-w)$$

and optimization problem Equation (1) can be written as

$$\max_{r \geq 0} T = \Phi_\alpha g r^{-b}(r-c-w).$$

Letting

$$\begin{cases} \frac{dT}{dr} = \Phi_\alpha g r^{-b-1}[(-b+1)r + b(c+w)] = 0, \\ \frac{d^2T}{dr^2} = \Phi_\alpha g r^{-b-2} b[(b-1)r - (b+1)(c+w)] = 0 \end{cases}$$

yields the desired results.

Appendix A.2. Proof of Lemma 2

It follows from

$$\begin{aligned} \frac{d\pi_{s_i}}{dw_i} &= \Phi_\alpha g(c+w_i)^{-b} + \Phi_\alpha g(-b)(c+w_i)^{-b-1}(w_i - s_i) \\ &= \Phi_\alpha g(c+w_i)^{-b}(1 - b(c+w_i)^{-1}(w_i - s_i)) = 0 \end{aligned}$$

that

$$1 - b(c+w_i)^{-1}(w_i - s_i) = 0.$$

Therefore,

$$w_i^{\star ind} = \frac{c + bs_i}{b-1},\ i = l, h. \qquad \square$$

Appendix A.3. Proof of Proposition 1

The maximum side payment that SP can provide to the SI is $L_i = \pi_{S_i,\alpha}^{\star tca} - k_{S_i}$, where $\pi_{S_i,\alpha}^{\star tca} = \Phi_\alpha g b^{-b}(b-1)^{b-1}(c+s_i)^{-b+1}, i = l, h$. However, in the two-part auction, the SP who wins the contract would offer a side payment that is equal to the maximum side payment the other SP can provide. We break the discussion into different cases.

(1) If $L_l = \pi_{S_l,\alpha}^{\star tca} - k_{S_l} \le 0, L_h = \pi_{S_h,\alpha}^{\star tca} - k_{S_h} \le 0$, i.e.,

$$\Phi_\alpha g b^{-b}(b-1)^{b-1}(c+s_l)^{-b+1} - k_{S_l} \le 0, \Phi_\alpha g b^{-b}(b-1)^{b-1}(c+s_h)^{-b+1} - k_{S_h} \le 0,$$

then, neither of the two SPs can get their target profit, and thus this is beyond our consideration.

(2) If $L_l \ge L_h \ge 0$, we have $\pi_{S_l,\alpha}^{\star tca} - k_{S_l} \ge \pi_{S_h,\alpha}^{\star tca} - k_{S_h} \ge 0$, i.e.,

$$\Phi_\alpha g b^{-b}(b-1)^{b-1}(c+s_l)^{-b+1} - k_{S_l} \ge \Phi_\alpha g b^{-b}(b-1)^{b-1}(c+s_h)^{-b+1} - k_{S_h} \ge 0,$$

based on the service price setting, if the low-cost SP wants to win the contract, he should set side payment L_l which is not less than the side payment L_h that the high-cost SP can provide, where L_h is the maximal profit of the SI if the contract is won by high-cost SP. This means that $L_l^\star = \Phi_\alpha g b^{-b}(b-1)^{b-1}(c+s_h)^{-b+1} - k_{S_h}$, the SI may extract the profit by side payment $L_l^\star$. Then we can derive the optimal wholesale price $w_l^{\star tca} = \frac{c+bs_l}{b-1}$, and the profit of the low-cost SP is $\pi_{S_l,\alpha}^{\star tca} = \Phi_\alpha g b^{-b}(b-1)^{b-1}(c+s_l)^{-b+1} - L_l^\star$. Accordingly, the service price and the order quantity are

$$r_{R,\alpha}^{\star tca} = \frac{b}{b-1}(c+s_l) \text{ and } q_{R,\alpha}^{\star tca} = \Phi_\alpha g b^{-b}(b-1)^b(c+s_l)^{-b}.$$

Therefore, in this two-part contract auction, the profits of the SP, the SI, and the total supply chain can be expressed as follows

$$\begin{aligned} \pi_{S_l,\alpha}^{\star tca} &= \Phi_\alpha g b^{-b}(b-1)^{b-1}(c+s_l)^{-b+1} - L_l^\star \\ &= \Phi_\alpha g b^{-b}(b-1)^{b-1}(c+s_l)^{-b+1} - \Phi_\alpha g b^{-b}(b-1)^{b-1}(c+s_h)^{-b+1} + k_{S_h}, \\ \pi_{R,\alpha}^{\star tca} &= L_l^\star = \Phi_\alpha g b^{-b}(b-1)^{b-1}(c+s_h)^{-b+1} - k_{S_h}, \\ \pi_{SC,\alpha}^{\star tca} &= \Phi_\alpha g b^{-b}(b-1)^{b-1}(c+s_l)^{-b+1}. \end{aligned}$$

By comparing the optimums given in Equation (1), when the SI cooperates with the low-cost SP, channel coordination can be achieved in the two-part contract auction.

(3) If $L_h > L_l$, then $\pi^{\star tca}_{S_h,\alpha} - k_{S_h} > \pi^{\star tca}_{S_l,\alpha} - k_{S_l} \geq 0$, i.e.,

$$\Phi_\alpha g b^{-b}(b-1)^{b-1}(c+s_h)^{-b+1} - k_{S_h} \geq \Phi_\alpha g b^{-b}(b-1)^{b-1}(c+s_l)^{-b+1} - k_{S_l} \geq 0.$$

Using a similar discussion for case (2), we know that when the high-cost SP wants to win the contract, he should set a side payment equal to the maximum side payment L_l that the low-cost SP can provide. Thus $L^\star_h = \Phi_\alpha g b^{-b}(b-1)^{b-1}(c+s_l)^{-b+1} - k_{S_l}$, and the profit of the high-cost SP is $\pi^{\star tca}_{S_h,\alpha} = \Phi_\alpha g b^{-b}(b-1)^{b-1}(c+s_h)^{-b+1} - L^\star_h$. The SI can extract the profit from side payments. We can derive the optimal wholesale price from analysis above that $w^{\star tca}_h = \frac{c+bs_h}{b-1}$. Accordingly, the service price and the order quantity are

$$r^{\star tca}_{R,\alpha} = \frac{b}{b-1}(c+s_h) \text{ and } q^{\star tca}_{R,\alpha} = \Phi_\alpha g b^{-b}(b-1)^b(c+s_h)^{-b}.$$

Therefore, in this two-part contract auction, the profits of the SP, the SI, and the total supply chain can be expressed as follows

$$\begin{aligned}
\pi^{\star tca}_{S_h,\alpha} &= \Phi_\alpha g b^{-b}(b-1)^{b-1}(c+s_h)^{-b+1} - L^\star_h \\
&= \Phi_\alpha g b^{-b}(b-1)^{b-1}(c+s_h)^{-b+1} - \Phi_\alpha g b^{-b}(b-1)^{b-1}(c+s_h)^{-b+1} + k_{S_l}, \\
\pi^{\star tca}_{R,\alpha} &= L^\star_h = \Phi_\alpha g b^{-b}(b-1)^{b-1}(c+s_l)^{-b+1} - k_{S_l}, \\
\pi^{\star tca}_{SC,\alpha} &= \Phi_\alpha g b^{-b}(b-1)^{b-1}(c+s_h)^{-b+1}.
\end{aligned}$$

This concludes the proof.

Appendix A.4. Proof of Proposition 2

For $k_{S_i} = \Phi_\alpha g b^{-2b}(b-1)^{2b-1}(c+s_i)^{-b+1}, i = l, h$, we have

$$\pi^{\star tca}_{S_i,\alpha} - k_{S_i} = \phi_\alpha g(b^{-b}(b-1)^{b-1} - b^{-2b}(b-1)^{2b-1})(c+s_i)^{-b+1}.$$

Since $b^{-b}(b-1)^{b-1} - b^{-2b}(b-1)^{2b-1} > 0$, we conclude that $\pi^{\star tca}_{S_i,\alpha} - k_{S_i} > 0$.

Appendix A.5. Proof of Proposition 3

Without loss of generality, in the two-part contract auction, when choosing the high-cost SP, we have

$$\pi^{\star tca}_{R,\alpha} = L^\star_h = \Phi_\alpha g b^{-b}(b-1)^{b-1}(c+s_l)^{-b+1} - k_{S_l}.$$

In the decentralized case, we have

$$\pi^{\star ind}_{R,\alpha} = \Phi_\alpha g b^{-2b+1}(b-1)^{2b-2}(c+s_h)^{-b+1}$$

To ensure that the SI chooses the two-part contract auction, we derive the following conditions:

$$\Phi_\alpha g b^{-b}(b-1)^{b-1}(c+s_l)^{-b+1} - k_{S_l} - \Phi_\alpha g b^{-2b+1}(b-1)^{2b-2}(c+s_h)^{-b+1} > 0.$$

Thus

$$k_{S_l} < \Phi_\alpha g b^{-b}(b-1)^{b-1}[(c+s_l)^{-b+1} - (\frac{b-1}{b})^{b-1}(c+s_h)^{-b+1}].$$

If

$$k_{S_l} = \Phi_\alpha g b^{-2b}(b-1)^{2b-1}(c+s_l)^{-b+1},$$

we have

$$\Phi_\alpha g b^{-2b}(b-1)^{2b-1}(c+s_l)^{-b+1} < \Phi_\alpha g b^{-b}(b-1)^{b-1}[(c+s_l)^{-b+1} - (\frac{b-1}{b})^{b-1}(c+s_h)^{-b+1}]$$

Therefore,

$$\frac{(c+s_h)^{-b+1}}{(c+s_l)^{-b+1}} < \frac{1-(\frac{b-1}{b})^{b-1}}{(\frac{b-1}{b})^{b-1}}$$

Likewise, in the low-cost SP case, we can get a corresponding conclusion. Therefore, both the SP and the SI will choose to join in the two-part contract auction model. The channel coordination can be achieved and the supply chain can achieve the maximum profits.

Appendix A.6. Proof of Lemma 4

(1) We first consider the case that low-cost SP wins the contract. Similar to the symmetric information case, the expected optimal service price, order quantity, and profit of the entire supply chain are

$$\begin{aligned} E[\tilde{r}^\star_{R,\alpha}] &= \frac{b}{b-1}(c+E[\min(s_l,s_h)]), \\ E[\tilde{q}^\star_{R,\alpha}] &= \Phi_\alpha g b^{-b}(b-1)^b E[(c+\min(s_l,s_h))^{-b}], \\ E[\tilde{\pi}^\star_{SC,\alpha}] &= \Phi_\alpha g b^{-b}(b-1)^{b-1} E[(c+\min(s_l,s_h))^{-b+1}], \end{aligned}$$

Let $E(s) = \int_{\bar{s}-\varepsilon}^{\bar{s}+\varepsilon} s g(s) ds$, and assume that s_l and s_h are independent and follow the uniform distribution $G(s)$. Therefore, the cumulative distribution function and the probability density function of $\min(s_l, s_h)$ are

$$\begin{aligned} F_{\min}(s) &= 1-(1-G(s))^2, \\ f_{\min}(s) &= F'_{\min} = 2(1-G(s))g(s), \end{aligned}$$

respectively. Hence

$$E[\min(s_l,s_h)] = \int_{\bar{s}-\varepsilon}^{\bar{s}+\varepsilon} s f_{\min}(s) ds.$$

Then, the probability density function is

$$f_{c+\min(s_l,s_h)}(s) = 2(1-G(s-c))g(s-c).$$

Moreover,

$$f_{(c+\min(s_1,s_2))^{-b}}(s) = \frac{1}{b}s^{-\frac{1+b}{b}} f_{c+\min(s_l,s_h)}(s^{-\frac{1}{b}}) = \frac{2}{b}s^{-\frac{1+b}{b}}(1-G(s^{-\frac{1}{b}}-c))g(s^{-\frac{1}{b}}-c).$$

and thus

$$E[(c+\min(s_l,s_h))^{-b}] = \int_{\bar{s}-\varepsilon}^{\bar{s}+\varepsilon} s f_{(c+\min(s_1,s_2))^{-b}}(s) ds.$$

If the contract is obtained by the low-cost SP, we have the following optimums:

$$\begin{aligned} E[\tilde{r}^\star_{R,\alpha}] &= \frac{b}{b-1}(c+2\int_{\bar{s}-\varepsilon}^{\bar{s}+\varepsilon} s(1-G(s))g(s)ds), \\ E[\tilde{q}^\star_{R,\alpha}] &= 2\Phi_\alpha g b^{-b-1}(b-1)^b \int_{\bar{s}-\varepsilon}^{\bar{s}+\varepsilon} s^{-\frac{1}{b}}(1-G(s^{-\frac{1}{b}}-c))g(s^{-\frac{1}{b}}-c)ds, \\ E[\tilde{\pi}^\star_{SC,\alpha}] &= 2\Phi_\alpha g b^{-b-1}(b-1)^{b-2} \int_{\bar{s}-\varepsilon}^{\bar{s}+\varepsilon} s^{-\frac{1}{b-1}}(1-G(s^{-\frac{1}{b-1}}-c))g(s^{-\frac{1}{b-1}}-c)ds. \end{aligned}$$

which yields Equations (4)–(6).

(2) The cumulative distribution function and the probability density function of $\max(s_l, s_h)$ are:

$$F_{\max}(s) = G(s)^2, \quad f_{\max}(s) = F_{\max}^{'} = 2G(s)g(s)$$

respectively. Hence

$$E[\max(s_l, s_h)] = \int_{\bar{s}-\varepsilon}^{\bar{s}+\varepsilon} s f_{\max}(s) ds.$$

Then, the probability density function is

$$f_{c+\max(s_l,s_h)}(s) = 2G(s-c)g(s-c).$$

Furthermore,

$$f_{(c+\max(s_1,s_h))^{-b}}(s) = \frac{1}{b} s^{-\frac{1+b}{b}} f_{c+\max(s_l,s_h)}(s^{-\frac{1}{b}}) = \frac{2}{b} s^{-\frac{1+b}{b}} G(s^{-\frac{1}{b}} - c) g(s^{-\frac{1}{b}} - c)$$

and thus

$$E[(c+\max(s_l, s_h))^{-b}] = \int_{\bar{s}-\varepsilon}^{\bar{s}+\varepsilon} s f_{(c+\max(s_1,s_2))^{-b}}(s) ds.$$

If the contract is obtained by the low-cost SP, we have the following optimums:

$$E[\tilde{r}_{R,\alpha}^{\star}] = \frac{b}{b-1}(c + 2\int_{\bar{s}-\varepsilon}^{\bar{s}+\varepsilon} sG(s)g(s)ds),$$
$$E[\tilde{q}_{R,\alpha}^{\star}] = 2\Phi_{\alpha} g b^{-b-1}(b-1)^{b} \int_{\bar{s}-\varepsilon}^{\bar{s}+\varepsilon} s^{-\frac{1}{b}} G(s^{-\frac{1}{b}} - c) g(s^{-\frac{1}{b}} - c) ds,$$
$$E[\tilde{\pi}_{SC,\alpha}^{\star}] = 2\Phi_{\alpha} g b^{-b-1}(b-1)^{b-2} \int_{\bar{s}-\varepsilon}^{\bar{s}+\varepsilon} s^{-\frac{1}{b-1}} G(s^{-\frac{1}{b-1}} - c) g(s^{-\frac{1}{b-1}} - c) ds.$$

Appendix A.7. Proof of Lemma 5

The total profit of the supply chain is

$$\begin{aligned} \max_{\hat{k}} \pi_{SC,\alpha} &= \pi_{R,\alpha} + \pi_{S,\alpha} \\ &= \Phi_{\alpha} g b^{-b}(b-1)^{b}[\hat{k} + \min(s_l), s_h]^{-b}[\tfrac{b}{b-1}(\hat{k} + \min(s_l, s_h)) - c - \min(s_l, s_h))] \end{aligned}$$

Differentiating $\pi_{SC,\alpha}$ with respect to the parameter $\hat{k}$ yields

$$\frac{d\pi_{SC,\alpha}}{d\hat{k}} = \Phi_{\alpha} g b^{-b+1}(b-1)^{b}[\hat{k} + \min(s_l, s_h)]^{-b-1}(c - \hat{k}).$$

Solving the equation $\frac{\pi_{SC,\alpha}}{d\hat{k}} = 0$ yields that $\hat{k} = c$. Putting $\hat{k} = c$ into the second derivative of $\pi_{SC,\alpha}$ yields that

$$\frac{d^2\pi_{SC,\alpha}}{d\hat{k}^2}|_{\hat{k}=c} = -\Phi_{\alpha} g b^{-b+1}(b-1)^{b}[c + \min(s_l, s_h)]^{-b-1} < 0.$$

Appendix A.8. Proof of Proposition 4

We break the discussion into two cases.

(1) $\tilde{L}_l \geq \tilde{L}_h \geq 0$, i.e.,

$$\Phi_{\alpha} g b^{-b}(b-1)^{b-1}(c + \min(s_l, s_h))^{-b+1} - k_{S_l} \geq \Phi_{\alpha} g b^{-b}(b-1)^{b-1}(c + \max(s_l, s_h))^{-b+1} - k_{S_h} \geq 0.$$

If the low-cost SP wants to win the contract, he should set side payment $\tilde{L}_l$ which is not less than the side payment $\tilde{L}_h$, where $\tilde{L}_h$ is the maximal profit offered by the high-cost SP. This means that $\tilde{L}_l^{\star} = \Phi_{\alpha} g b^{-b}(b-1)^{b-1}(c + \max(s_l, s_h))^{-b+1} - k_{S_h}$, the

SI can extract the profit by side payment $\tilde{L}_l^\star$. We can derive the optimal wholesale price, the low-cost SP's profit, and the SI's profit as follows:

$$\begin{aligned}
\tilde{w}_l^{\star tca} &= \frac{c + b\min(s_l, s_h)}{b-1}, \\
\tilde{\pi}_{S_l,\alpha}^{\star tca} &= \Phi_\alpha g b^{-b}(b-1)^{b-1}(c+\min(s_l,s_h))^{-b+1} - \tilde{L}_l^\star \\
&= \Phi_\alpha g b^{-b}(b-1)^{b-1}[(c+\min(s_l,s_h))^{-b+1} - (c+\max(s_l,s_h))^{-b+1}] + k_{S_h}, \\
\tilde{\pi}_{R,\alpha}^{\star tca} &= \Phi_\alpha g b^{-b}(b-1)^{b-1}(c+\max(s_l,s_h))^{-b+1} - k_{S_h}.
\end{aligned}$$

Therefore, the expected optimal service price, order quantity, and supply chain profit are

$$E[\tilde{r}_{R,\alpha}^{\star tca}] = \frac{b}{b-1}(c + 2\int_{\bar{s}-\varepsilon}^{\bar{s}+\varepsilon} s(1-G(s))g(s)ds) \tag{A1}$$

$$E[\tilde{q}_{R,\alpha}^{\star tca}] = 2\Phi_\alpha g b^{-b-1}(b-1)^b \int_{\bar{s}-\varepsilon}^{\bar{s}+\varepsilon} s^{-\frac{1}{b}}(1-G(s^{-\frac{1}{b}}-c))g(s^{-\frac{1}{b}}-c)ds \tag{A2}$$

$$E[\tilde{\pi}_{SC,\alpha}^{\star tca}] = 2\Phi_\alpha g b^{-b-1}(b-1)^{b-2} \int_{\bar{s}-\varepsilon}^{\bar{s}+\varepsilon} s^{-\frac{1}{b-1}}(1-G(s^{-\frac{1}{b-1}}-c))g(s^{-\frac{1}{b-1}}-c)ds \tag{A3}$$

respectively, where $E[\max(s_l,s_h)] = 2\int_{\bar{s}-\varepsilon}^{\bar{s}+\varepsilon} sG(s)g(s)ds$, $E[\min(s_l,s_h)] = 2\int_{\bar{s}-\varepsilon}^{\bar{s}+\varepsilon} s(1-G(s))g(s)ds$. By comparing the optimums Equations (4)–(6) with Equations (A1)–(A3), we have $E[\tilde{r}_{R,\alpha}^\star] = E[\tilde{r}_{R,\alpha}^{\star tca}]$, $E[\tilde{q}_{R,\alpha}^\star] = E[\tilde{q}_{R,\alpha}^{\star tca}]$, $E[\tilde{\pi}_{SC,\alpha}^\star] = E[\tilde{\pi}_{SC,\alpha}^{\star tca}]$. Therefore, channel coordination can be achieved in the two-part contract auction.

(2) $\tilde{L}_h > \tilde{L}_l$, i.e.,

$$\Phi_\alpha g b^{-b}(b-1)^{b-1}(c+\max(s_l,s_h))^{-b+1} - k_{S_h} \geq \Phi_\alpha g b^{-b}(b-1)^{b-1}(c+\min(s_l,s_h))^{-b+1} - k_{S_l} \geq 0.$$

In this case, similar to the argument for case (1), when the high-cost SP wants to win the contract, one needs to submit a side payment equal to the maximum side payment $\tilde{L}_l$ that the low-cost SP can provide.
This means that $\tilde{L}_h^\star = \Phi_\alpha g b^{-b}(b-1)^{b-1}(c+\min(s_l,s_h))^{-b+1} - k_{S_l}$, the SI can extract the profit from side payment $\tilde{L}_l^\star$. We can derive the optimal wholesale price, the low-cost SP's profit, and the SI's profit as follows:

$$\begin{aligned}
\tilde{w}_h^{\star tca} &= \frac{c + b\max(s_l, s_h)}{b-1}, \\
\tilde{\pi}_{S_h,\alpha}^{\star tca} &= \Phi_\alpha g b^{-b}(b-1)^{b-1}(c+\max(s_l,s_h))^{-b+1} - \tilde{L}_h^\star \\
&= \Phi_\alpha g b^{-b}(b-1)^{b-1}[(c+\max(s_l,s_h))^{-b+1} - (c+\min(s_l,s_h))^{-b+1}] + k_{S_l}, \\
\tilde{\pi}_{R,\alpha}^{\star tca} &= \Phi_\alpha g b^{-b}(b-1)^{b-1}(c+\min(s_l,s_h))^{-b+1} - k_{S_l}.
\end{aligned}$$

Therefore, the expected optimal service price, order quantity, and supply chain profit are

$$E[\tilde{r}_{R,\alpha}^{\star tca}] = \frac{b}{b-1}(c + 2\int_{\bar{s}-\varepsilon}^{\bar{s}+\varepsilon} sG(s)g(s)ds) \tag{A4}$$

$$E[\tilde{q}_{R,\alpha}^{\star tca}] = 2\Phi_\alpha g b^{-b-1}(b-1)^b \int_{\bar{s}-\varepsilon}^{\bar{s}+\varepsilon} s^{-\frac{1}{b}}G(s^{-\frac{1}{b}}-c)g(s^{-\frac{1}{b}}-c)ds \tag{A5}$$

$$E[\tilde{\pi}_{SC,\alpha}^{\star tca}] = 2\Phi_\alpha g b^{-b-1}(b-1)^{b-2} \int_{\bar{s}-\varepsilon}^{\bar{s}+\varepsilon} s^{-\frac{1}{b-1}}G(s^{-\frac{1}{b-1}}-c)g(s^{-\frac{1}{b-1}}-c)ds \tag{A6}$$

where $E[\max(s_l,s_h)] = 2\int_{\bar{s}-\varepsilon}^{\bar{s}+\varepsilon} sG(s)g(s)ds$, $E[\min(s_l,s_h)] = 2\int_{\bar{s}-\varepsilon}^{\bar{s}+\varepsilon} s(1-G(s))g(s)ds$. By comparing the optimums (7)–(9) with Equations (A4)–(A6), when the SI cooperates with the low-cost SP, we have $E[\tilde{r}_{R,\alpha}^\star] = E[\tilde{r}_{R,\alpha}^{\star tca}]$, $E[\tilde{q}_{R,\alpha}^\star] = E[\tilde{q}_{R,\alpha}^{\star tca}]$, $E[\tilde{\pi}_{SC,\alpha}^\star] = E[\tilde{\pi}_{SC,\alpha}^{\star tca}]$. Therefore, channel coordination can be achieved in the two-part contract auction.

Appendix A.9. Proof of Proposition 5

For

$$k_{S_i} = \Phi_\alpha g b^{-2b}(b-1)^{2b-1}(c+E[s_i])^{-b+1}, i = l, h,$$

we have

$$E[\tilde{\pi}^{\star tca}_{S_l,\alpha}] - k_{S_i} = \Phi_\alpha g(b^{-b}(b-1)^{b-1} - b^{-2b}(b-1)^{2b-1})(c+E[s_i])^{-b+1}.$$

It's easy to observe that $b^{-b}(b-1)^{b-1} - b^{-2b}(b-1)^{2b-1} > 0$, then $E[\tilde{\pi}^{\star tca}_{S_l,\alpha}] - k_{S_i} > 0, i = l, h$.

Appendix A.10

The proof is broken into two cases.

Case I: The high-cost SP wins the contract.

In the two-part contract auction, the optimal profit of the SI is

$$\tilde{\pi}^{\star tca}_{R,\alpha} = L^\star_h = \Phi_\alpha g b^{-b}(b-1)^{b-1}(c+\min(s_l,s_h))^{-b+1} - k_{S_l}.$$

Under the decentralized case, the optimal profit of the SI is

$$\pi^{\star ind}_{R,\alpha} = \Phi_\alpha g b^{-2b+1}(b-1)^{2b-2}(c+\max(s_l,s_h))^{-b+1}.$$

To ensure that the SI chooses the two-part contract auction, we derive the following conditions

$$\Phi_\alpha g b^{-b}(b-1)^{b-1}(c+\min(s_l,s_h))^{-b+1} - k_{S_l} - \Phi_\alpha g b^{-2b+1}(b-1)^{2b-2}(c+\max(s_l,s_h))^{-b+1} > 0.$$

Therefore,

$$\begin{aligned} k_{S_l} < \; & 2\Phi_\alpha g b^{-b}(b-1)^{b-2}\Big[\int_{\bar{s}-\varepsilon}^{\bar{s}+\varepsilon} s^{-\frac{1}{b-1}}(1-G(s^{-\frac{1}{b-1}}))g(s^{-\frac{1}{b-1}})ds \\ & -(\tfrac{b-1}{b})^{b-1}\int_{\bar{s}-\varepsilon}^{\bar{s}+\varepsilon} s^{-\frac{1}{b-1}}G(s^{-\frac{1}{b-1}})g(s^{-\frac{1}{b-1}})ds\Big]. \end{aligned}$$

Especially, when

$$k_{S_l} = 2\Phi_\alpha g b^{-2b}(b-1)^{2b-2}\int_{\bar{s}-\varepsilon}^{\bar{s}+\varepsilon} s^{-\frac{1}{b-1}}(1-G(s^{-\frac{1}{b-1}}-c))g(s^{-\frac{1}{b-1}}-c)ds,$$

we have

$$\frac{\int_{\bar{s}-\varepsilon}^{\bar{s}+\varepsilon} s^{-\frac{1}{b-1}}G(s^{-\frac{1}{b-1}}-c)g(s^{-\frac{1}{b-1}}-c)ds}{\int_{\bar{s}-\varepsilon}^{\bar{s}+\varepsilon} s^{-\frac{1}{b-1}}(1-G(s^{-\frac{1}{b-1}}-c))g(s^{-\frac{1}{b-1}}-c)ds} < \frac{1-(\frac{b-1}{b})^b}{(\frac{b-1}{b})^{b-1}}.$$

Case II: The low-cost SP wins the contract.

Under the two-part contract auction, the optimal profit of the SI is

$$\tilde{\pi}^{\star tca}_{R,\alpha} = L^\star_l = \Phi_\alpha g b^{-b}(b-1)^{b-1}(c+\max(s_l,s_h))^{-b+1} - k_{S_h}.$$

In the decentralized case, the optimal profit of the SI is

$$\pi^{\star ind}_{R,\alpha} = \Phi_\alpha g b^{-2b+1}(b-1)^{2b-2}(c+\min(s_l,s_h))^{-b+1},$$

To ensure that the SI chooses the two-part contract auction, we derive the following conditions

$$\Phi_\alpha g b^{-b}(b-1)^{b-1}(c+\min(s_l,s_h))^{-b+1} - k_{S_h} - \Phi_\alpha g b^{-2b+1}(b-1)^{2b-2}(c+\min(s_l,s_h))^{-b+1} > 0.$$

Therefore,

$$k_{S_h} < 2\Phi_\alpha g b^{-b}(b-1)^{b-2}[\int_{\bar{s}-\varepsilon}^{\bar{s}+\varepsilon} s^{-\frac{1}{b-1}}(1-G(s^{-\frac{1}{b-1}}))g(s^{-\frac{1}{b-1}})ds - (\frac{b-1}{b})^{b-1}\int_{\bar{s}-\varepsilon}^{\bar{s}+\varepsilon} s^{-\frac{1}{b-1}}G(s^{-\frac{1}{b-1}})g(s^{-\frac{1}{b-1}})ds].$$

Especially, when

$$k_{s_h} = 2\Phi_\alpha g b^{-2b}(b-1)^{2b-2}\int_{\bar{s}-\varepsilon}^{\bar{s}+\varepsilon} s^{-\frac{1}{b-1}}(1-G(s^{-\frac{1}{b-1}}-c))g(s^{-\frac{1}{b-1}}-c)ds,$$

we have

$$\frac{\int_{\bar{s}-\varepsilon}^{\bar{s}+\varepsilon} s^{-\frac{1}{b-1}}(1-G(s^{-\frac{1}{b-1}}-c))g(s^{-\frac{1}{b-1}}-c)ds}{\int_{\bar{s}-\varepsilon}^{\bar{s}+\varepsilon} s^{-\frac{1}{b-1}}G(s^{-\frac{1}{b-1}}-c)g(s^{-\frac{1}{b-1}}-c)ds} < \frac{1-(\frac{b-1}{b})^b}{(\frac{b-1}{b})^{b-1}}.$$

□

Appendix A.11. Proof of Proposition 7

(a) Let $X = s^{-\frac{1}{b}} - c$. When the low-cost SP wins the contract, the expected optimal service price and order quantity are

$$\begin{aligned} E[\tilde{r}^{\star tca}_{R,\alpha}] &= \frac{b}{b-1}(c + 2\int_{\bar{s}-\varepsilon}^{\bar{s}+\varepsilon} s(1-\frac{s-\bar{s}+\varepsilon}{2\varepsilon})\frac{1}{2\varepsilon}ds \\ &= \frac{b}{b-1}(c+\frac{3\bar{s}-\varepsilon}{3}), \\ E[\tilde{q}^{\star tca}_{R,\alpha}] &= 2\Phi_\alpha g b^{-b-1}(b-1)^b\int_{\bar{s}-\varepsilon}^{\bar{s}+\varepsilon}(X+c)\frac{\bar{s}+\varepsilon-X}{2\varepsilon}\frac{b(X+c)^{-1-b}}{2\varepsilon}dX \\ &= \frac{\Phi_\alpha g b^{-b-1}(b-1)^b}{2\varepsilon^2}\int_{\bar{s}-\varepsilon}^{\bar{s}+\varepsilon}((\bar{s}+\varepsilon+c)(X+c)^{-b}-(X+c)^{1-b})dX, \end{aligned}$$

respectively. For ease of exposition, let $m = \bar{s}+\varepsilon+c, n = \bar{s}-\varepsilon+c, M = (\bar{s}+\varepsilon+c)^b, N = (\bar{s}-\varepsilon+c)^b$. Derivating $E[\tilde{r}^{\star tca}_{R,\alpha}]$ with ε, we find that

$$\frac{\partial E[\tilde{r}^{\star tca}_{R,\alpha}]}{\partial\varepsilon} = -\frac{b}{3(b-1)}.$$

Since $b > 1$, we have $\frac{\partial E[\tilde{r}^{\star tca}_{R,\alpha}]}{\partial\varepsilon} < 0$, which yields (a.1). For the optimal order quantity, we divide into two cases for further discussion.

Case I: $b = 2$.

$$E[\tilde{q}^{\star tca}_{R,\alpha}] = \frac{\Phi_\alpha g}{8\varepsilon^2}(\frac{2\varepsilon}{n} - \ln\frac{m}{n}).$$

Case II: $b > 1$ and $b \neq 2$. The expected optimal order quantity is

$$E[\tilde{q}^{\star tca}_{R,\alpha}] = \frac{\Phi_\alpha g b^{-b}(b-1)^b}{2\varepsilon^2}\frac{m^{2-b}-n^{1-b}[m+2\varepsilon(1-b)]}{(1-b)(2-b)}.$$

Derivating $E[\tilde{q}^{\star tca}_{R,\alpha}]$ with respect to ε, we find that
In **Case I**,

$$\frac{\partial E[\tilde{q}^{\star tca}_{R,\alpha}]}{\partial\varepsilon} = \frac{\Phi_\alpha g(-\frac{4\varepsilon}{n}+\frac{4(c+\bar{s})\varepsilon^2}{n^2 m}+2\ln\frac{m}{n})}{8\varepsilon^3}.$$

Let $h(\varepsilon) = -\frac{4\varepsilon}{n}+\frac{4(c+\bar{s})\varepsilon^2}{n^2 m}+2\ln\frac{m}{n}$. Then $h(0) = 0$ and

$$h'(\varepsilon) = \frac{4(c+\bar{s})\varepsilon^2(m+2\varepsilon)}{n^3 m^2} > 0.$$

Therefore, $\frac{\partial E[\tilde{q}^{\star tca}_{R,\alpha}]}{\partial\varepsilon} > 0$.

In **Case II**, we have

$$\frac{\partial E[\tilde{q}^{\star tca}_{R,\alpha}]}{\partial \varepsilon} = \Phi_\alpha g b^{-b}(b-1)^b \frac{-\frac{m}{M}(2(c+\bar{s})+b\varepsilon)+\frac{2(c+\bar{s}^2)-(3b-2)(c+\bar{s})\varepsilon+b\varepsilon^2(2b-3)}{N}}{2\varepsilon^3(b-2)(b-1)}.$$

Let $k_1(\varepsilon) = -\frac{m}{M}(2(c+\bar{s})+b\varepsilon)+\frac{2(c+\bar{s}^2)-(3b-2)(c+\bar{s})\varepsilon+b\varepsilon^2(2b-3)}{N}$, then

$$k_1'(\varepsilon) = (b-2)\frac{(c+\bar{s})^2(N-M)-b\varepsilon^2(N+(3-2b)M)+\varepsilon(c+\bar{s})(M-N+b(M+N))}{nNM}$$

and

$$k_1''(\varepsilon) = b\varepsilon(b-2)(b-1)\frac{(c+\bar{s})^2(5M-N)-\varepsilon^2(N+(3-2b)M)+2\varepsilon(c+\bar{s})(N+b(M+1))}{nNM}$$

Now we divide the discussion into two cases.

Case II.1 $b>2$. We have $k_1''(\varepsilon)>0$. Since $k_1'(0)=0$, it's easy to see that $k_1'(\varepsilon)>0$ and hence $k_1(\varepsilon)>0$. Using the fact that $b>2$, we can obtain $\frac{\partial E[\tilde{q}^{\star tca}_{R,\alpha}]}{\partial \varepsilon}>0$.

Case II.2 $1<b<2$. $k_1''(\varepsilon)<0$. Since $k_1'(0)=0$, it's easy to see that $k_1'(\varepsilon)<0$ and hence $k_1(\varepsilon)<0$. Using the fact that $1<b<2$, we can obtain $\frac{\partial E[\tilde{q}^{\star tca}_{R,\alpha}]}{\partial \varepsilon}>0$, which yields (a.2).

To find the difference between $r^{\star tca}_{R,\alpha}$ and $E[\tilde{r}^{\star tca}_{R,\alpha}]$, we have $\triangle E[\tilde{r}^{\star tca}_{R,\alpha}] = r^{\star tca}_{R,\alpha} - E[\tilde{r}^{\star tca}_{R,\alpha}]$. Similarly, we have $\triangle E[\tilde{r}^{\star tca}_{R,\alpha}] = r^{\star tca}_{R,\alpha} - E[\tilde{r}^{\star tca}_{R,\alpha}]$. Taking the derivative of $\triangle E[\tilde{r}^{\star tca}_{R,\alpha}]$ and $\triangle E[\tilde{r}^{\star tca}_{R,\alpha}]$ with respect to ε yields

$$\frac{\partial \triangle E[\tilde{r}^{\star tca}_{R,\alpha}]}{\partial \varepsilon} = \frac{b}{3(b-1)} > 0$$

and

$$\frac{\partial \triangle E[\tilde{q}^{\star tca}_{R,\alpha}]}{\partial \varepsilon} = \begin{cases} -\frac{\Phi_\alpha g h(\varepsilon)}{8\varepsilon^3} < 0, & b=2, \\ -\Phi_\alpha g b^{-b}(b-1)^b \frac{k_1(\varepsilon)}{2\varepsilon^3(b-2)(b-1)} < 0, & b \neq 2. \end{cases}$$

Therefore, $\frac{\partial \triangle E[\tilde{r}^{\star tca}_{R,\alpha}]}{\partial \varepsilon}>0$ and $\frac{\partial \triangle E[\tilde{q}^{\star tca}_{R,\alpha}]}{\partial \varepsilon}<0$, which yields (a.3) and (a.4).

(b) We now analyze the SI's profit.

Let $Y = s^{-\frac{1}{b-1}} - c$, then the SI's profit can be rewritten as follows:

$$\begin{aligned} E[\tilde{\pi}^{\star tca}_{R,\alpha}] &= 2\Phi_\alpha g b^{-b}(b-1)^{b-1}\int_{\bar{s}-\varepsilon}^{\bar{s}+\varepsilon}(Y+c)\frac{Y-\bar{s}+\varepsilon}{2\varepsilon}\frac{(b-1)(Y+c)^{-b}}{2\varepsilon}dY - k_{S_h} \\ &= \frac{2\Phi_\alpha g b^{-b}(b-1)^{b-1}}{2\varepsilon^2}\int_{\bar{s}-\varepsilon}^{\bar{s}+\varepsilon}[(Y+c)^{2-b}-\bar{s}-\varepsilon+c(Y+c)^{1-b}]dY - k_{S_h}. \end{aligned}$$

Case I: $b=2$. We can obtain the expected optimal profit for SI

$$E[\tilde{\pi}^{\star tca}_{R,\alpha}] = \frac{\Phi_\alpha g}{8\varepsilon^2}(2\varepsilon - n\ln\frac{m}{n}) - k_{S_h}.$$

Taking the derivative of $E[\tilde{\pi}^{\star tca}_{R,\alpha}]$ w.r.t. ε yields

$$\frac{\partial E[\tilde{\pi}^{\star tca}_{R,\alpha}]}{\partial \varepsilon} = \Phi_\alpha g \frac{-\frac{2\varepsilon(m+c+\bar{s})}{m}+(n+c+\bar{s})\ln\frac{m}{n}}{8\varepsilon^3}$$

Let $f(\varepsilon) = -\frac{2\varepsilon(m+c+\bar{s})}{m}+(n+c+\bar{s})\ln\frac{m}{n}$. Then $f(0)=0$ and

$$f'(\varepsilon) = \frac{2\varepsilon((c+\bar{s})^2+\varepsilon^2)}{nm^2} - \ln\frac{m}{n}$$

$$f^{''}(\varepsilon) = -\frac{4(c+\bar{s})^2(n-2\varepsilon)\varepsilon}{n^2m^3} - \ln\frac{m}{n}$$

It is easy to find that $f^{''}(0) = 0$ and $f^{'}(0) = 0$. Then

$$f^{'''}(\varepsilon) = -\frac{4(c+\bar{s})^2(n-2\varepsilon)\varepsilon}{n^2m^3} - \ln\frac{m}{n}$$

Since $f^{'''}(0) < 0$, it holds that $f^{'}(\varepsilon) < f^{'}(0) = 0$. Therefore, $f(\varepsilon) < 0$, and hence $\frac{\partial E[\tilde{\pi}^{\star tca}_{R,\alpha}]}{\partial\varepsilon} < 0$.

Case II: $b = 3$. We can obtain the SI's expected optimal profit

$$E[\tilde{\pi}^{\star tca}_{R,\alpha}] = \frac{2\Phi_\alpha g}{27\varepsilon^2}(\frac{-2\varepsilon}{m} + \ln\frac{m}{n}) - k_{S_h}$$

Taking the derivative of $E[\tilde{\pi}^{\star tca}_{R,\alpha}]$ w.r.t. ε yields

$$\frac{\partial E[\tilde{\pi}^{\star tca}_{R,\alpha}]}{\partial\varepsilon} = 2\Phi_\alpha g\frac{\frac{4\varepsilon^2(c+\bar{s})}{m^2n} + \frac{4\varepsilon}{m} - 2\ln\frac{m}{n}}{27\varepsilon^3}$$

Let $l(\varepsilon) = \frac{4\varepsilon^2(c+\bar{s})}{m^2n} + \frac{4\varepsilon}{m} - 2\ln\frac{m}{n}$, then $f(0) = 0$ and

$$l^{'}(\varepsilon) = -\frac{4\varepsilon^2(c+\bar{s})(n-2\varepsilon)}{n^2m^3}$$

$$l^{''}(\varepsilon) = -\frac{8(c+\bar{s})\varepsilon((c+\bar{s})^3 - 5(c+\bar{s})^2\varepsilon + 3(c+\bar{s}) - 3\epsilon^3)}{n^3m^4}$$

Since $l^{''}(0) = 0$ and $l^{'}(0) = 0$, one has

$$l^{'''}(\varepsilon) = -\frac{8(c+\bar{s})\varepsilon((c+\bar{s})^5 - 11(c+\bar{s})^4\varepsilon + 20(c+\bar{s})^3\varepsilon^2 - 40(c+\bar{s})^2\varepsilon^3 + 15(c+\bar{s})\varepsilon^4 - 9\varepsilon^5}{n^4m^5}$$

Since $l^{'''}(0) < 0$, then we have $l^{'}(\varepsilon) < l^{'}(0) = 0$. Therefore, we can find $l(\varepsilon) < 0$, then $\frac{\partial E[\tilde{\pi}^{\star tca}_{R,\alpha}]}{\partial\varepsilon} < 0$.

Case III: $b \neq 2$ and $b \neq 3$, we can obtain the expected optimal profit of SI as follows:

$$E[\tilde{\pi}^{\star tca}_{R,\alpha}] = \Phi_\alpha g b^{-b}(b-1)^{b-1}\frac{n^{3-b} - m^{2-b}(\bar{s} - \varepsilon(5-2b) + c)}{2\varepsilon^2(2-b)(3-b)} - k_{S_h}$$

Derivating $E[\tilde{\pi}^{\star tca}_{R,\alpha}]$ with ε, one has

$$\frac{\partial E[\tilde{\pi}^{\star tca}_{R,\alpha}]}{\partial\varepsilon} = \Phi_\alpha g b^{-b}(b-1)^b\frac{-\frac{n^2}{N}(2m-(1+b)\varepsilon) + \frac{m}{M}(2(c+\bar{s})^2 + (-5+3b)(c+\bar{s})\varepsilon + (b-1)(2b-5)\varepsilon^2)}{2\varepsilon^3(b-2)(b-3)}$$

Let $t(\varepsilon) = -\frac{n^2}{N}(2m-(1+b)\varepsilon) + \frac{m}{M}(2(c+\bar{s})^2 + (-5+3b)(c+\bar{s})\varepsilon + (b-1)(2b-5)\varepsilon^2)$, then we have $t(0) = 0$ and

$$t^{'}(\varepsilon) = (b-3)[-\frac{n}{N}(m-b\varepsilon) + \frac{1}{M}((c+\bar{s})^2 + b(c+\bar{s})\varepsilon - (b-1)(2b-5)\varepsilon^2)]$$

$$t^{''}(\varepsilon) = (b-3)(b-1)(b-2)\varepsilon[\frac{1}{N} + \frac{-5(c+\bar{s}) + (-5+2b)\varepsilon}{mM}]$$

It is easy to find that $t^{''}(0) = 0$ and $t^{'}(0) = 0$. Then

$$t^{'''}(\varepsilon) = (b-3)(b-1)(b-2)[\frac{n+b\varepsilon}{nN} + \frac{-5(c+\bar{s})^2 + (-10+9b)(c+\bar{s})\varepsilon - (b-1)(2b-5)\varepsilon^2}{m^2M}]$$

Case III.1 $b > 3$. We have $t'''(0) < 0$, then $t'(\varepsilon) < t'(0) = 0$. Therefore, $t(\varepsilon) < 0$. Since $3 < b$ then $\frac{\partial E[\tilde{\pi}^{\star tca}_{R,\alpha}]}{\partial \varepsilon} < 0$.

Case III.2 $2 < b < 3$. We have $t'''(0) > 0$, then $t'(\varepsilon) > t'(0) = 0$. Therefore, $t(\varepsilon) > 0$. Since $2 < b < 3$, then $\frac{\partial E[\tilde{\pi}^{\star tca}_{R,\alpha}]}{\partial \varepsilon} < 0$.

Case III.3 $1 < b < 2$. We have $t'''(0) < 0$, then $t'(\varepsilon) < t'(0) = 0$. Therefore, we can find $t(\varepsilon) < 0$. Since $1 < b < 2$ then $\frac{\partial E[\tilde{\pi}^{\star tca}_{R,\alpha}]}{\partial \varepsilon} < 0$, which yields (b.1) as desired.

It suffices to show the difference between $\pi^{\star tca}_{R,\alpha}$ and $E[\tilde{\pi}^{\star tca}_{R,\alpha}]$, i.e., $\triangle E[\tilde{\pi}^{\star tca}_{R,\alpha}] = \pi^{\star tca}_{R,\alpha} - E[\tilde{\pi}^{\star tca}_{R,\alpha}]$, we take the derivative of $\triangle E[\tilde{\pi}^{\star tca}_{R,\alpha}]$ with ε to obtain

$$\frac{\partial \Delta E[\tilde{\pi}^{\star tca}_{R,\alpha}]}{\partial \varepsilon} = \begin{cases} -\frac{\Phi_\alpha g f(\varepsilon)}{8\varepsilon^3} > 0, & b = 2 \\ -\frac{2\Phi_\alpha g l(\varepsilon)}{27\varepsilon^3} > 0, & b = 3 \\ -\Phi_\alpha g b^{-b}(b-1)^b \frac{t(\varepsilon)}{2\varepsilon^3(b-2)(b-3)} > 0, & b \neq 2 \text{ and } b \neq 3, \end{cases}$$

that means $\frac{\partial \Delta E[\tilde{\pi}^{\star tca}_{R,\alpha}]}{\partial \varepsilon} > 0$, which yields (b.2) as desired.

(c) The low-cost SP's profit can be rewritten as

$$\begin{aligned} E[\tilde{\pi}^{\star tca}_{S_l,\alpha}] &= 2\Phi_\alpha g b^{-b}(b-1)^{b-1} \int_{\bar{s}-\varepsilon}^{\bar{s}+\varepsilon} (Y+c)(1-2G(Y))g(Y)(Y+c)^{-b} dY + k_{S_h} \\ &= \frac{\Phi_\alpha g b^{-b}(b-1)^{b-1}}{\varepsilon^2} \int_{\bar{s}-\varepsilon}^{\bar{s}+\varepsilon} [(Y+c)^{1-b}(\bar{s}+\varepsilon-c) - (Y+c)^{2-b}] dY + k_{S_h} \end{aligned}$$

Case I: When $b = 2$, we can obtain the expected optimal profit of low-cost SP

$$E[\tilde{\pi}^{\star tca}_{S_l,\alpha}] = \frac{\Phi_\alpha g}{4\varepsilon^2}[-2\varepsilon + (\bar{s}+c)\ln\frac{m}{n}] + k_{S_h}$$

Derivating $E[\tilde{\pi}^{\star tca}_{S_l,\alpha}]$ w.r.t. ε yields

$$\frac{\partial E[\tilde{\pi}^{\star tca}_{S_l,\alpha}]}{\partial \varepsilon} = \Phi_\alpha g \frac{2\varepsilon + \frac{\varepsilon^3}{mn} - (c+\bar{s})\ln\frac{m}{n}}{2\varepsilon^3}$$

Let $u(\varepsilon) = 2\varepsilon + \frac{\varepsilon^3}{mn} - (c+\bar{s})\ln\frac{m}{n}$, then $u(0) = 0$ and $u'(\varepsilon) = \frac{(c+\bar{s})^2\varepsilon^2+\varepsilon^4}{n^2m^2} > 0$. It is easy to see that $u(\varepsilon) > 0$, i.e., $\frac{\partial E[\tilde{\pi}^{\star tca}_{S_l,\alpha}]}{\partial \varepsilon} > 0$.

Case II: When $b = 3$, we can obtain the expected optimal profit of low-cost SP

$$E[\tilde{\pi}^{\star tca}_{S_l,\alpha}] = \frac{2\Phi_\alpha g}{27\varepsilon^2}\left(\frac{4\varepsilon(\bar{s}+c)}{(\bar{s}+c)^2-\varepsilon^2} - 2\ln\frac{m}{n}\right) + k_{S_h}$$

Derivating $E[\tilde{\pi}^{\star tca}_{S_l,\alpha}]$ w.r.t. ε yields that

$$\frac{\partial E[\tilde{\pi}^{\star tca}_{S_l,\alpha}]}{\partial \varepsilon} = 8\Phi_\alpha g \frac{\frac{4(c+\bar{s})\varepsilon^3 - 2\varepsilon(c+\bar{s})^3}{m^2n^2} + \ln\frac{m}{n}}{27\varepsilon^3}$$

Let $v(\varepsilon) = \frac{4(c+\bar{s})\varepsilon^3 - 2\varepsilon(c+\bar{s})^3}{m^2n^2} + \ln\frac{m}{n}$, then we have $v(0) = 0$ and $v'(\varepsilon) = \frac{2\varepsilon^2(c+\bar{s})[(c+\bar{s})^2+3\varepsilon^2]}{n^3m^3} > 0$. It is easy to find that $v(\varepsilon) > 0$, then $\frac{\partial E[\tilde{\pi}^{\star tca}_{S_l,\alpha}]}{\partial \varepsilon} > 0$.

Case III: When $b \neq 2$ and $b \neq 3$, the expected optimal profit of low-cost SP is as follows

$$E[\tilde{\pi}^{\star tca}_{S_l,\alpha}] = \Phi_\alpha g b^{-b}(b-1)^{b-1} \frac{\frac{m^2}{M}[n+\varepsilon(b-1)] - \frac{n^2}{N}[n-\varepsilon(1-b)]}{\varepsilon^2(2-b)(3-b)} + k_{S_h}$$

Derivating $E[\tilde{\pi}^{\star tca}_{S_l,\alpha}]$ w.r.t. ε yields that

$$\frac{\partial E[\tilde{\pi}^{\star tca}_{S_l,\alpha}]}{\partial \varepsilon} = \Phi_\alpha g b^{-b}(b-1)^b \frac{(\frac{n}{N}-\frac{m}{M})[2(c+\bar{s})^2(b-2)(b-1)\varepsilon^2]-2(b-1)(c+\bar{s})\varepsilon(\frac{n}{N}+\frac{m}{M})}{\varepsilon^3(2-b)(3-b)}$$

Let $x(\varepsilon)=(\frac{n}{N}-\frac{m}{M})[2(c+\bar{s})^2(b-2)(b-1)\varepsilon^2]-2(b-1)(c+\bar{s})\varepsilon(\frac{n}{N}+\frac{m}{M})$, then $x(0)=0$ and

$$x'(\varepsilon)=\frac{(b-3)(b-2)(b-1)\varepsilon^2(M+N)}{NM}$$

Case III.1 when $b>3$, we have $x'(\varepsilon)>0$, then we can obtain $x(\varepsilon)>0$ and $\frac{\partial E[\tilde{\pi}^{\star tca}_{S_l,\alpha}]}{\partial \varepsilon}>0$.

Case III.2 when $2<b<3$, we have $x'(\varepsilon)<0$, then we can obtain $x(\varepsilon)<0$ and $\frac{\partial E[\tilde{\pi}^{\star tca}_{S_l,\alpha}]}{\partial \varepsilon}>0$.

Case III.3 when $1<b<2$, we have $x'(\varepsilon)>0$, then we can obtain $x(\varepsilon)>0$ and $\frac{\partial E[\tilde{\pi}^{\star tca}_{S_l,\alpha}]}{\partial \varepsilon}>0$.

Combination of Cases I, II, and III yields (c.1) as desired.

For $\triangle E[\tilde{\pi}^{\star tca}_{S_l,\alpha}]=\pi^{\star tca}_{S_l,\alpha}-E[\tilde{\pi}^{\star tca}_{S_l,\alpha}]$, taking the derivative yields

$$\frac{\partial \Delta E[\tilde{\pi}^{\star tca}_{S_l,\alpha}]}{\partial \varepsilon}=\begin{cases}-\frac{\Phi_\alpha g u(\varepsilon)}{2\varepsilon^3}, & b=2\\ -\frac{2\Phi_\alpha g v(\varepsilon)}{27\varepsilon^3}, & b=3\\ -\Phi_\alpha g b^{-b}(b-1)^b\frac{t(\varepsilon)}{\varepsilon^3(2-b)(3-b)}, & b\neq 2 \text{ and } b\neq 3\end{cases}$$

which means $\frac{\partial \Delta E[\tilde{\pi}^{\star tca}_{S_l,\alpha}]}{\partial \varepsilon}<0$ as desired.

(d) The supply chain's profit can be rewritten as follows:

$$\begin{aligned}E[\tilde{\pi}^{\star tca}_{SC,\alpha}] &= 2\Phi_\alpha g b^{-b}(b-1)^{b-1}\int_{\bar{s}-\varepsilon}^{\bar{s}+\varepsilon}(Y+c)(1-G(Y))g(Y)(Y+c)^{-b}dY\\ &= \frac{\Phi_\alpha g b^{-b}(b-1)^{b-1}}{\varepsilon^2}\int_{\bar{s}-\varepsilon}^{\bar{s}+\varepsilon}[(Y+c)^{1-b}(\bar{s}+\varepsilon+c)-(Y+c)^{2-b}]dY\end{aligned}$$

Case I: when $b=2$, we can obtain the expected optimal profit of the supply chain

$$E[\tilde{\pi}^{\star tca}_{SC,\alpha}]=\frac{\Phi_\alpha g}{8\varepsilon^2}[-2\varepsilon+m\ln\frac{m}{n}]$$

Derivating $E[\tilde{\pi}^{\star tca}_{SC,\alpha}]$ w.r.t. ε yields that

$$\frac{\partial E[\tilde{\pi}^{\star tca}_{SC,\alpha}]}{\partial \varepsilon}=\Phi_\alpha g\frac{2\varepsilon[-2(c+\bar{s})+\varepsilon]+n[2(c+\bar{s})+\varepsilon]\ln\frac{m}{n}}{-8n\varepsilon^3}$$

Let $y(\varepsilon)=2\varepsilon[-2(c+\bar{s})+\varepsilon]+n[2(c+\bar{s})+\varepsilon]\ln\frac{m}{n}$, then $y(0)=0$ and $y'(\varepsilon)=\frac{(m+\varepsilon)(2\varepsilon-m\ln\frac{m}{n})}{m}<0$. It is easy to find that $y(\varepsilon)<0$, and hence $\frac{\partial E[\tilde{\pi}^{\star tca}_{SC,\alpha}]}{\partial \varepsilon}>0$.

Case II: when $b=3$, we can obtain the expected optimal profit of the supply chain

$$E[\tilde{\pi}^{\star tca}_{SC,\alpha}]=\frac{2\Phi_\alpha g}{27\varepsilon^2}(\frac{2\varepsilon}{n}-\ln\frac{m}{n})$$

Derivating $E[\tilde{\pi}^{\star tca}_{SC,\alpha}]$ w.r.t. ε yields

$$\frac{\partial E[\tilde{\pi}^{\star tca}_{SC,\alpha}]}{\partial \varepsilon}=4\Phi_\alpha g\frac{\frac{2\varepsilon[-(c+\bar{s})^2+(c+\bar{s})\varepsilon+\varepsilon^2]}{n^2m}+\ln\frac{m}{n}}{27\varepsilon^3}$$

Let $z(\varepsilon) = \frac{2\varepsilon[-(c+\bar{s})^2+(c+\bar{s})\varepsilon+\varepsilon^2]}{n^2 m} + \ln\frac{m}{n}$, then $z(0) = 0$ and $z'(\varepsilon) = \frac{4\varepsilon^2(c+\bar{s})(m+2\varepsilon)}{n^3 m^2} > 0$. It is easy to find that $z(\varepsilon) > 0$, then $\frac{\partial E[\tilde{\pi}^{\star tca}_{SC,\alpha}]}{\partial\varepsilon} > 0$.

Case III: when $b \neq 2$ and $b \neq 3$, we can obtain the expected optimal profit of supply chain

$$E[\tilde{\pi}^{\star tca}_{SC,\alpha}] = \Phi_\alpha g b^{-b}(b-1)^{b-1}\frac{n^{3-b} - m^{2-b}(\bar{s} - \varepsilon(5-2b) + c)}{2\varepsilon^2(2-b)(3-b)}$$

Derivating $E[\tilde{\pi}^{\star tca}_{SC,\alpha}]$ w.r.t. ε yields

$$\frac{\partial E[\tilde{\pi}^{\star tca}_{SC,\alpha}]}{\partial\varepsilon} = \Phi_\alpha g b^{-b}(b-1)^b\frac{\frac{m^2}{M}[-2(c+\bar{s}) + \varepsilon(1-b)] + \frac{n}{N}[2(c+\bar{s})^2 - (3b-5)(c+\bar{s})\varepsilon + (b-1)(2b-5)\varepsilon^2]}{2\varepsilon^3(b-2)(b-3)}$$

Let $\varphi(\varepsilon) = \frac{m^2}{M}[-2(c+\bar{s}) + \varepsilon(1-b)] + \frac{n}{N}[2(c+\bar{s})^2 - (3b-5)(c+\bar{s})\varepsilon + (b-1)(2b-5)\varepsilon^2]$, then $\varphi(0) = 0$ and

$$\varphi'(\varepsilon) = (b-3)[\frac{m}{M}(c+b\varepsilon) + \frac{-(c+\bar{s})^2 + b\varepsilon(c+\bar{s}) + (b-1)(2b-5)\varepsilon^2}{N}]$$

$$\varphi''(\varepsilon) = (b-3)(b-1)(b-2)\varepsilon[-\frac{1}{M} + \frac{5(c+\bar{s}) + (-5+2b)\varepsilon}{nN}]$$

Case III.1: when $b > 3$, we have $\varphi''(\varepsilon) > 0$, then $\varphi'(\varepsilon) > 0$, and hence $\varphi(\varepsilon) > 0$. Therefore, $\frac{\partial E[\tilde{\pi}^{\star tca}_{SC,\alpha}]}{\partial\varepsilon} > 0$.

Case III.2: when $2 < b < 3$, we have $\varphi''(\varepsilon) < 0$, then $\varphi'(\varepsilon) < 0$, and hence $\varphi(\varepsilon) < 0$. Therefore, $\frac{\partial E[\tilde{\pi}^{\star tca}_{SC,\alpha}]}{\partial\varepsilon} > 0$.

Case III.3: when $1 < b < 2$, we have $\varphi''(\varepsilon) > 0$, then $\varphi'(\varepsilon) > 0$, and hence $\varphi(\varepsilon) > 0$. Therefore, $\frac{\partial E[\tilde{\pi}^{\star tca}_{SC,\alpha}]}{\partial\varepsilon} > 0$, which yields (d.1).

To consider $\triangle E[\tilde{\pi}^{\star tca}_{SC,\alpha}] = \pi^{\star tca}_{SC,\alpha} - E[\tilde{\pi}^{\star tca}_{SC,\alpha}]$, since

$$\frac{\partial\triangle E[\tilde{\pi}^{\star tca}_{S_I,\alpha}]}{\partial\varepsilon} = \begin{cases} -\frac{\Phi_\alpha g y(\varepsilon)}{-8n\varepsilon^3} < 0, b = 2 \\ -\frac{2\Phi_\alpha g z(\varepsilon)}{27\varepsilon^3} < 0, b = 3 \\ -\Phi_\alpha g b^{-b}(b-1)^b\frac{\varphi(\varepsilon)}{\varepsilon^3(b-2)(b-3)} < 0,\ b \neq 2\ and\ b \neq 3 \end{cases}$$

we have $\frac{\partial\triangle E[\tilde{\pi}^{\star tca}_{SC,\alpha}]}{\partial\varepsilon} < 0$, and hence (d.2) holds as desired.

Appendix A.12. Proof of Proposition 8

(a) When the high-cost SP wins the contract, Let $X = s^{-\frac{1}{b}} - c$, then the expected optimal service price and the expected optimal order quantity can be written as

$$\begin{aligned} E[\tilde{r}^{\star tca}_{R,\alpha}] &= \frac{b}{b-1}(c + 2\int_{\bar{s}-\varepsilon}^{\bar{s}+\varepsilon} s\frac{s-\bar{s}+\varepsilon}{2\varepsilon}\frac{1}{2\varepsilon}ds, \\ &= \frac{b}{b-1}(c + \frac{3\bar{s}+\varepsilon}{3}), \\ E[\tilde{q}^{\star tca}_{R,\alpha}] &= 2\Phi_\alpha g b^{-b-1}(b-1)^b\int_{\bar{s}-\varepsilon}^{\bar{s}+\varepsilon}(X+c)\frac{X-\bar{s}+\varepsilon}{2\varepsilon}\frac{b(X+c)^{-1-b}}{2\varepsilon}dX, \\ &= \frac{\Phi_\alpha g b^{-b-1}(b-1)^b}{2\varepsilon^2}\int_{\bar{s}-\varepsilon}^{\bar{s}+\varepsilon}[(X+c)^{1-b} - (\bar{s}-\varepsilon+c)(X+c)^{-b}]dX. \end{aligned}$$

Case I: When $b = 2$, we can obtain the expected optimal order quantity

$$E[\tilde{q}^{\star tca}_{R,\alpha}] = \frac{\Phi_\alpha g}{8\varepsilon^2}(\frac{-2\varepsilon}{m} + \ln\frac{m}{n})$$

Case II: When $b \neq 2$,we can obtain the expected optimal order quantity

$$E[\tilde{q}^{\star tca}_{R,\alpha}] = \frac{\Phi_\alpha g b^{-b}(b-1)^b}{2\varepsilon^2}\frac{n^{2-b} - m^{1-b}[n - 2\varepsilon(1-b)]}{(1-b)(2-b)}$$

Derivating $E[\tilde{r}^{\star tca}_{R,\alpha}]$ with ε, one has

$$\frac{\partial E[\tilde{r}^{\star tca}_{R,\alpha}]}{\partial \varepsilon} = \frac{b}{3(b-1)}.$$

Since $b > 1$, $\frac{\partial E[\tilde{r}^{\star tca}_{R,\alpha}]}{\partial \varepsilon} > 0$ yields (a.1) as desired. For the optimal order quantity, when $b = 2$, taking the derivative of $E[\tilde{q}^{\star tca}_{R,\alpha}]$ w.r.t. ε yields

$$\frac{\partial E[\tilde{q}^{\star tca}_{R,\alpha}]}{\partial \varepsilon} = \frac{\Phi_\alpha g(\frac{4\varepsilon}{m} + \frac{4(c+\bar{s})\varepsilon^2}{m^2 n} - 2\ln\frac{m}{n})}{8\varepsilon^3} = \frac{\Phi_\alpha g l(\varepsilon)}{8\varepsilon^3} < 0$$

Therefore, we have $\frac{\partial E[\tilde{q}^{\star tca}_{R,\alpha}]}{\partial \varepsilon} < 0$. When $b \neq 2$, we have

$$\frac{\partial E[\tilde{q}^{\star tca}_{R,\alpha}]}{\partial \varepsilon} = \Phi_\alpha g b^{-b}(b-1)^b \frac{\frac{n}{N}(-2(c+\bar{s}) + b\varepsilon) + \frac{2(c+\bar{s}^2)+(3b-2)(c+\bar{s})\varepsilon + b\varepsilon^2(2b-3)}{M}}{2\varepsilon^3(b-2)(b-1)}$$

Let $k_2(\varepsilon) = \frac{n}{N}(-2(c+\bar{s}) + b\varepsilon) + \frac{2(c+\bar{s}^2)+(3b-2)(c+\bar{s})\varepsilon+b\varepsilon^2(2b-3)}{M}$, then

$$k_2^{'}(\varepsilon) = (b-2)\frac{(c+\bar{s})^2(N-M) - b\varepsilon^2(N+(3-2b)M) + \varepsilon(c+\bar{s})(M-N+b(M+N))}{nNM}$$

$$k_2^{'''}(\varepsilon) = b\varepsilon(b-2)(b-1)[\frac{1}{nN} + \frac{-5(c+\bar{s}) - (3-2b)\varepsilon}{Mm^2}]$$

Then, $k_2^{'}(0) = 0$ and $k_2^{''}(0) = 0$. Furthermore, we can obtain that

$$k_2^{'''}(\varepsilon) = b(b-2)(b-1)[\frac{c+\bar{s}+b\varepsilon}{n^2 N} + \frac{-5(c+\bar{s})^2 - (1+9b)(c+\bar{s})\varepsilon}{+}(3-2b)b\varepsilon^2]$$

Case II.1: when $b > 2$, then $k_2^{'''}(0) < 0$, then $k_2^{'}(\varepsilon) < k_2^{'}(0) = 0$. Therefore, we can find $k_2(\varepsilon) < 0$. Since $2 < b$ then $\frac{\partial E[\tilde{\pi}^{\star tca}_{q,\alpha}]}{\partial \varepsilon} < 0$.

Case II.2: when $1 < b < 2$, then $k_2^{'''}(0) > 0$, then $k_2^{'}(\varepsilon) > k_2^{'}(0) = 0$. Therefore, we can find $k_2(\varepsilon) > 0$. Since $1 < b < 2$ then $\frac{\partial E[\tilde{\pi}^{\star tca}_{q,\alpha}]}{\partial \varepsilon} < 0$.

Combination of discussions in Cases I and II yields (a.2).

For $\triangle E[\tilde{r}^{\star tca}_{R,\alpha}] = r^{\star tca}_{R,\alpha} - E[\tilde{r}^{\star tca}_{R,\alpha}]$, taking the derivative of $\triangle E[\tilde{r}^{\star tca}_{R,\alpha}]$ and $\triangle E[\tilde{r}^{\star tca}_{R,\alpha}]$ w.r.t. ε yields

$$\frac{\partial \Delta E[\tilde{r}^{\star tca}_{R,\alpha}]}{\partial \varepsilon} = -\frac{b}{3(b-1)} < 0$$

$$\frac{\partial \Delta E[\tilde{q}^{\star tca}_{R,\alpha}]}{\partial \varepsilon} = \begin{cases} -\frac{\Phi_\alpha g l(\varepsilon)}{8\varepsilon^3} > 0, b = 2 \\ -\Phi_\alpha g b^{-b}(b-1)^b \frac{k_2(\varepsilon)}{2\varepsilon^3(b-2)(b-1)} > 0,\ b \neq 2 \end{cases}$$

Therefore, $\frac{\partial \Delta E[\tilde{r}^{\star tca}_{R,\alpha}]}{\partial \varepsilon} < 0$ (a.3) and $\frac{\partial \Delta E[\tilde{q}^{\star tca}_{R,\alpha}]}{\partial \varepsilon} > 0$. Hence, (a.3) and (a.4) hold as required.

(b) Let $Y = s^{-\frac{1}{b-1}} - c$, then the SI's profit is

$$E[\tilde{\pi}^{\star tca}_{R,\alpha}] = 2\Phi_\alpha g b^{-b}(b-1)^{b-2}\int_{\bar{s}-\varepsilon}^{\bar{s}+\varepsilon}(Y+c)\frac{\bar{s}+\varepsilon-Y}{2\varepsilon}\frac{(b-1)(Y+c)^{-b}}{2\varepsilon}dY - k_{S_I}$$
$$= \frac{2\Phi_\alpha g b^{-b}(b-1)^{b-1}}{2\varepsilon^2}\int_{\bar{s}-\varepsilon}^{\bar{s}+\varepsilon}[\bar{s}+\varepsilon+c(Y+c)^{1-b}-(Y+c)^{2-b}]dY - k_{S_I}$$

Case I: when $b = 2$, then SI's expected optimal profit is

$$E[\tilde{\pi}^{\star tca}_{R,\alpha}] = \frac{\Phi_\alpha g}{8\varepsilon^2}(-2\varepsilon + m\ln\frac{m}{n}) - k_{S_I}$$

Taking the derivative of $E[\tilde{\pi}^{\star tca}_{R,\alpha}]$ w.r.t. ε yields

$$\frac{\partial E[\tilde{\pi}^{\star tca}_{R,\alpha}]}{\partial\varepsilon} = \Phi_\alpha g\frac{y(\varepsilon)}{-8n\varepsilon^3}$$

Since $y(\varepsilon) < 0$, it holds that $\frac{\partial E[\tilde{\pi}^{\star tca}_{R,\alpha}]}{\partial\varepsilon} > 0$.
Case II: when $b = 3$, we can obtain the expected optimal profit of SI :

$$E[\tilde{\pi}^{\star tca}_{R,\alpha}] = \frac{2\Phi_\alpha g}{27\varepsilon^2}(\frac{2\varepsilon}{n} - \ln\frac{m}{n}) - k_{S_I}$$

Taking the derivative of $E[\tilde{\pi}^{\star tca}_{R,\alpha}]$ w.r.t. ε yields

$$\frac{\partial E[\tilde{\pi}^{\star tca}_{R,\alpha}]}{\partial\varepsilon} = \Phi_\alpha g\frac{z(\varepsilon)}{-8n\varepsilon^3}$$

Since $z(\varepsilon) > 0$, it holds that $\frac{\partial E[\tilde{\pi}^{\star tca}_{R,\alpha}]}{\partial\varepsilon} > 0$.
Case III: when $b \neq 2$ and $b \neq 3$, the SI's expected optimal profit is

$$E[\tilde{\pi}^{\star tca}_{R,\alpha}] = \Phi_\alpha g b^{-b}(b-1)^{b-1}\frac{\varphi(\varepsilon)}{2\varepsilon^2(b-2)(b-3)} - k_{S_I}$$

Taking the derivative of $E[\tilde{\pi}^{\star tca}_{R,\alpha}]$ w.r.t. ε yields

$$\frac{\partial E[\tilde{\pi}^{\star tca}_{R,\alpha}]}{\partial\varepsilon} = \Phi_\alpha g b^{-b}(b-1)^b\frac{\varphi(\varepsilon)}{2\varepsilon^3(b-2)(b-3)}$$

Case III.1: when $b > 3$, then $\varphi(\varepsilon) > 0$. Therefore, $\frac{\partial E[\tilde{\pi}^{\star tca}_{R,\alpha}]}{\partial\varepsilon} > 0$.
Case III.2: when $2 < b < 3$, then $\varphi(\varepsilon) < 0$. Therefore, $\frac{\partial E[\tilde{\pi}^{\star tca}_{R,\alpha}]}{\partial\varepsilon} > 0$.
Case III.3: when $1 < b < 2$, then $\varphi(\varepsilon) > 0$. Therefore, $\frac{\partial E[\tilde{\pi}^{\star tca}_{R,\alpha}]}{\partial\varepsilon} > 0$.
Therefore, combination of Cases I, II and III yields (b.1). For $\triangle E[\tilde{\pi}^{\star tca}_{R,\alpha}] = \pi^{\star tca}_{R,\alpha} - E[\tilde{\pi}^{\star tca}_{R,\alpha}]$, since

$$\frac{\partial\Delta E[\tilde{\pi}^{\star tca}_{S_I,\alpha}]}{\partial\varepsilon} = \begin{cases} -\frac{\Phi_\alpha g y(\varepsilon)}{-8n\varepsilon^3} < 0, b = 2 \\ -\frac{2\Phi_\alpha g z(\varepsilon)}{27\varepsilon^3} < 0, b = 3 \\ -\Phi_\alpha g b^{-b}(b-1)^b\frac{\varphi(\varepsilon)}{\varepsilon^3(b-2)(b-3)} < 0,\ b \neq 2 \text{ and } b \neq 3 \end{cases}$$

one has that $\frac{\partial\Delta E[\tilde{\pi}^{\star tca}_{R,\alpha}]}{\partial\varepsilon} < 0$, and thus, (b.2) holds as desired.

(c) The high-cost SP's profit can be rewritten :

$$E[\tilde{\pi}^{\star tca}_{S_h,\alpha}] = 2\Phi_\alpha g b^{-b}(b-1)^{b-1}\int_{\bar{s}-\varepsilon}^{\bar{s}+\varepsilon}(Y+c)(2G(Y)-1)g(Y)(Y+c)^{-b}dY + k_{S_l}$$
$$= \frac{\Phi_\alpha g b^{-b}(b-1)^{b-1}}{\varepsilon^2}\int_{\bar{s}-\varepsilon}^{\bar{s}+\varepsilon}[(Y+c)^{2-b}(\bar{s}+\varepsilon-c)-(Y+c)^{1-b}]dY + k_{S_l}$$

Case I: when $b=2$, we can obtain the expected optimal profit of low-cost SP :

$$E[\tilde{\pi}^{\star tca}_{S_h,\alpha}] = \frac{\Phi_\alpha g}{4\varepsilon^2}[2\varepsilon - (\bar{s}+c)\ln\frac{m}{n}] + k_{S_l}$$

Taking the derivative of $E[\tilde{\pi}^{\star tca}_{S_h,\alpha}]$ w.r.t. ε yields

$$\frac{\partial E[\tilde{\pi}^{\star tca}_{S_h,\alpha}]}{\partial\varepsilon} = -\Phi_\alpha g\frac{u(\varepsilon)}{2\varepsilon^3}$$

Since $u(\varepsilon)>0$, one has $\frac{\partial E[\tilde{\pi}^{\star tca}_{S_h,\alpha}]}{\partial\varepsilon}<0$.

Case II: when $b=3$, we can obtain the expected optimal profit of low-cost SP

$$E[\tilde{\pi}^{\star tca}_{S_h,\alpha}] = \frac{2\Phi_\alpha g}{27\varepsilon^2}(-\frac{4\varepsilon(\bar{s}+c)}{(\bar{s}+c)^2-\varepsilon^2}+2\ln\frac{m}{n}) + k_{S_l}.$$

Taking the derivative of $E[\tilde{\pi}^{\star tca}_{S_l,\alpha}]$ w.r.t. ε yields

$$\frac{\partial E[\tilde{\pi}^{\star tca}_{S_h,\alpha}]}{\partial\varepsilon} = -8\Phi_\alpha g\frac{v(\varepsilon)}{27\varepsilon^3}$$

Since $v(\varepsilon)>0$, one has $\frac{\partial E[\tilde{\pi}^{\star tca}_{S_h,\alpha}]}{\partial\varepsilon}<0$.

Case III: when $b\neq 2$ and $b\neq 3$,we can obtain the expected optimal profit of low-cost SP

$$E[\tilde{\pi}^{\star tca}_{S_h,\alpha}] = -\Phi_\alpha g b^{-b}(b-1)^{b-1}\frac{-\frac{m^2}{M}[n+\varepsilon(b-1)]+\frac{n^2}{N}[n-\varepsilon(1-b)]}{\varepsilon^2(2-b)(3-b)} + k_{S_l}.$$

Taking the derivative of $E[\tilde{\pi}^{\star tca}_{S_h,\alpha}]$ w.r.t. ε yields

$$\frac{\partial E[\tilde{\pi}^{\star tca}_{S_h,\alpha}]}{\partial\varepsilon} = -\Phi_\alpha g b^{-b}(b-1)^b\frac{x(\varepsilon)}{\varepsilon^3(b-2)(b-3)}.$$

Case III.1: when $b>3$, then $x(\varepsilon)>0$ and $\frac{\partial E[\tilde{\pi}^{\star tca}_{S_h,\alpha}]}{\partial\varepsilon}<0$.

Case III.2: when $2<b<3$, then $x(\varepsilon)<0$ and $\frac{\partial E[\tilde{\pi}^{\star tca}_{S_h,\alpha}]}{\partial\varepsilon}<0$.

Case III.3: when $1<b<2$, then $x(\varepsilon)>0$ and $\frac{\partial E[\tilde{\pi}^{\star tca}_{S_h,\alpha}]}{\partial\varepsilon}<0$.

Hence, (c.1) holds as desired. For $\triangle E[\tilde{\pi}^{\star tca}_{S_h,\alpha}] = \pi^{\star tca}_{S_h,\alpha} - E[\tilde{\pi}^{\star tca}_{S_h,\alpha}]$, taking the derivative $\triangle E[\tilde{\pi}^{\star tca}_{S_l,\alpha}]$ w.r.t. ε yields

$$\frac{\partial\Delta E[\tilde{\pi}^{\star tca}_{S_h,\alpha}]}{\partial\varepsilon} = \begin{cases} \frac{\Phi_\alpha g u(\varepsilon)}{2\varepsilon^3} > 0, b=2 \\ \frac{2\Phi_\alpha g v(\varepsilon)}{27\varepsilon^3} > 0, b=3 \\ \Phi_\alpha g b^{-b}(b-1)^b\frac{t(\varepsilon)}{\varepsilon^3(2-b)(3-b)} > 0,\ b\neq 2 \text{ and } b\neq 3 \end{cases}$$

Therefore, $\frac{\partial\Delta E[\tilde{\pi}^{\star tca}_{S_h,\alpha}]}{\partial\varepsilon}>0$, which yields (c.2) as desired.

(d) The supply chain's profit can be rewritten

$$E[\tilde{\pi}^{\star tca}_{SC,\alpha}] = 2\Phi_\alpha g b^{-b}(b-1)^{b-1}\int_{\bar{s}-\varepsilon}^{\bar{s}+\varepsilon}(Y+c)G(Y)g(Y)(Y+c)^{-b}dY$$
$$= \frac{\Phi_\alpha g b^{-b}(b-1)^{b-1}}{\varepsilon^2}\int_{\bar{s}-\varepsilon}^{\bar{s}+\varepsilon}[(Y+c)^{2-b}(\bar{s}-\varepsilon+c)-(Y+c)^{1-b}]dY$$

Case I: when $b=2$, the expected optimal profit of the supply chain is

$$E[\tilde{\pi}^{\star tca}_{SC,\alpha}] = \frac{\Phi_\alpha g}{8\varepsilon^2}[2\varepsilon - n\ln\frac{m}{n}]$$

Taking the derivative of $E[\tilde{\pi}^{\star tca}_{SC,\alpha}]$ w.r.t. ε yields

$$\frac{\partial E[\tilde{\pi}^{\star tca}_{SC,\alpha}]}{\partial \varepsilon} = \Phi_\alpha g\frac{f(\varepsilon)}{8\varepsilon^3}$$

Since $f(\varepsilon)<0$, one has $\frac{\partial E[\tilde{\pi}^{\star tca}_{SC,\alpha}]}{\partial \varepsilon}<0$.
Case II: when $b=3$, the expected optimal profit of the supply chain is

$$E[\tilde{\pi}^{\star tca}_{SC,\alpha}] = \frac{2\Phi_\alpha g}{27\varepsilon^2}(\frac{2\varepsilon}{m} - \ln\frac{m}{n})$$

Taking the derivative of $E[\tilde{\pi}^{\star tca}_{SC,\alpha}]$ w.r.t. ε yields

$$\frac{\partial E[\tilde{\pi}^{\star tca}_{SC,\alpha}]}{\partial \varepsilon} = 2\Phi_\alpha g\frac{l(\varepsilon)}{27\varepsilon^3}$$

Since $l(\varepsilon)<0$, one has $\frac{\partial E[\tilde{\pi}^{\star tca}_{SC,\alpha}]}{\partial \varepsilon}<0$.
Case III: when $b\neq 2$ and $b\neq 3$, the expected optimal profit of supply chain is

$$E[\tilde{\pi}^{\star tca}_{SC,\alpha}] = \Phi_\alpha g b^{-b}(b-1)^{b-1}\frac{n^{3-b}-m^{2-b}(\bar{s}-\varepsilon(5-2b)+c)}{2\varepsilon^2(2-b)(3-b)}$$

Taking the derivative of $E[\tilde{\pi}^{\star tca}_{SC,\alpha}]$ w.r.t. ε yields

$$\frac{\partial E[\tilde{\pi}^{\star tca}_{SC,\alpha}]}{\partial \varepsilon} = \Phi_\alpha g b^{-b}(b-1)^{b}\frac{t(\varepsilon)}{2\varepsilon^2(b-2)(b-3)}$$

Case III.1: when $b>3$, one has $t(\varepsilon)<0$ and hence $\frac{\partial E[\tilde{\pi}^{\star tca}_{SC,\alpha}]}{\partial \varepsilon}<0$.
Case III.2: when $2<b<3$, one has $t(\varepsilon)>0$ and hence $\frac{\partial E[\tilde{\pi}^{\star tca}_{SC,\alpha}]}{\partial \varepsilon}<0$.
Case III.3: when $1<b<2$, one has $t(\varepsilon)<0$ and hence $\frac{\partial E[\tilde{\pi}^{\star tca}_{SC,\alpha}]}{\partial \varepsilon}<0$.
Hence, (d.1) holds as desired. For $\triangle E[\tilde{\pi}^{\star tca}_{SC,\alpha}] = \pi^{\star tca}_{SC,\alpha} - E[\tilde{\pi}^{\star tca}_{SC,\alpha}]$, since

$$\frac{\partial \triangle E[\tilde{\pi}^{\star tca}_{S_I,\alpha}]}{\partial \varepsilon} = \begin{cases} -\frac{\Phi_\alpha g f(\varepsilon)}{-8\varepsilon^3} > 0, b=2 \\ -\frac{2\Phi_\alpha g l(\varepsilon)}{27\varepsilon^3} > 0, b=3 \\ -\Phi_\alpha g b^{-b}(b-1)^b\frac{t(\varepsilon)}{\varepsilon^3(b-2)(b-3)} > 0,\ b\neq 2 \text{ and } b\neq 3 \end{cases}$$

one has $\frac{\partial \triangle E[\tilde{\pi}^{\star tca}_{SC,\alpha}]}{\partial \varepsilon}>0$, which yields (d.2) as desired.

Appendix A.13. Proof of Proposition 9

(a) We outline a proof by contradiction. Suppose that $r^\star_\alpha > r^\star_N$. Then using the fact that $E[\pi_{SC}(r,q)]$ is unimodal in r for $r>c$, one has $\frac{dE[\Pi_{SC}(r,q)]}{dr}|_{r=r^\star_\alpha}<0$. Therefore,

$$\frac{dE[\Pi_{SC}(r,q)]}{dr}|_{r=r_\alpha^\star} = (b-1)\int_0^{F^{-1}(\frac{r_\alpha^\star-c-s}{r_\alpha^\star})} F(x)dx + F^{-1}(\frac{r_\alpha^\star-c-s}{r_\alpha^\star})(1-b(\frac{r_\alpha^\star-c-s}{r_\alpha^\star})) < 0.$$

Since $r_\alpha^\star = \frac{b}{b-1}(c+s)$, it follows that

$$(b-1)\int_0^{F^{-1}(\frac{r_\alpha^\star-c-s}{r_\alpha^\star})} F(x)dx + F^{-1}(\frac{r_\alpha^\star-c-s}{r_\alpha^\star})(1-b(\frac{r_\alpha^\star-c-s}{r_\alpha^\star})) < 0$$

and thus

$$\int_0^{F^{-1}(SL(r_\alpha^\star))} F(x)dx < 0.$$

This leads to a contradiction. Therefore $r_\alpha^\star < r_N^\star$.

(b) It holds that

$$\begin{aligned} q_N^\star - q_\alpha^\star &= g(r_N^\star)^{-b}F^{-1}(\frac{r_\alpha^\star-c-s}{r_\alpha^\star}) - \Phi_\alpha g b^{-b}(b-1)^b(c+s)^{-b} \\ &= g\Phi_\alpha[(r_N^\star)^{-b} - b^{-b}(b-1)^b(c+s)^{-b}] \\ &= g\Phi_\alpha[(r_N^\star)^{-b} - (\frac{b}{b-1})^{-b}(c+s)^{-b}] \\ &= g\Phi_\alpha[(r_N^\star)^{-b} - ((\frac{b}{b-1})(c+s))^{-b}] \\ &= g\Phi_\alpha[(r_N^\star)^{-b} - (r_\alpha^\star)^{-b}] \le 0, \end{aligned}$$

where the second equality uses the fact $F^{-1}(\frac{r_\alpha^\star-c-s}{r_\alpha^\star}) = \Phi_\alpha$, and the fifth equality follows from $r_\alpha^\star = \frac{b}{b-1}(c+s)$, and the last inequality uses the result of (a). Thus, $q_\alpha^\star \ge q_N^\star$ and the proof is completed.

References

1. Sousa, R.; da Silveira, G.J. The relationship between servitization and product customization strategies. *Int. J. Oper. Prod. Manag.* **2019**, *39*, 454–474. [CrossRef]
2. Wang, Y.; Wallace, S.W.; Shen, B.; Choi, T.M. Service supply chain management: A review of operational models. *Eur. J. Oper. Res.* **2015**, *247*, 685–698. [CrossRef]
3. Cachon, G. Supply chain coordination with contracts. In *Handbooks in Operations Research and Management Science: Supply Chain Management*; de Kok, A.G., Graves, S.C., Eds.; North-Holland: Amsterdam, The Netherlands, 2003; pp. 229–339.
4. Chiu, C.H.; Choi, T.M.; Tang, C.S. Price, rebate, and returns supply contracts for coordinating supply chains with price-dependent demand. *Prod. Oper.* **2011**, *20*, 81–91. [CrossRef]
5. Tsay, A.A.; Lovejoy, W.S. Quantity flexibility contracts and supply chain performance. *Manuf. Serv. Oper. Manag.* **1999**, *1*, 89–111. [CrossRef]
6. Choi, S.; Andrzej, R. A risk-averse newsvendor with law invariant coherent measures of risk. *Oper. Res. Lett.* **2008**, *36*, 77–82. [CrossRef]
7. Choi, S.; Andrzej, R.; Zhao, Y. A multiproduct risk-averse newsvendor with law-invariant coherent measures of risk. *Oper. Res.* **2001**, *59*, 346–364. [CrossRef]
8. Choi, S.; Andrzej, R. A multi-product risk-averse newsvendor with exponential utility function. *Eur. J. Oper. Res.* **2011**, *214*, 78–84. [CrossRef]
9. Olson, D.L.; Wu, D. Value at Risk Models. In *Enterprise Risk Management Models. Springer Texts in Business and Economics*; Springer: Berlin/Heidelberg, Germany, 2017.
10. Paul, E.; Liu, H.; Wang, R. Quantile-Based Risk Sharing. *Oper. Res.* **2018**, *66*, 936–949.
11. Corbett, C.J.; Zhou, D.; Tang, C.S. Designing supply contracts: Contract type and information asymmetry. *Manag. Sci.* **2004**, *50*, 550–559. [CrossRef]
12. Chung, W.; Talluri, S.; Narasimhan, R. Quantity flexibility contract in the presence of discount incentive. *Decis. Sci.* **2014**, *45*, 49–79.
13. Gan, X.; Sethi, S.P.; Yan, H. Channel coordination with a risk-neutral supplier and a downside-risk-averse retailer. *Prod. Oper. Manag.* **2005**, *14*, 80–89. [CrossRef]

14. Jin, M.; Wu, S.D. *Supply Chain Coordination in Electronic Markets Auction and Contracting Mechanisms*; Department of Industrial and System Engineering, Lehigh University: Bethlehem, PA, USA, 2001.
15. Jin, M.; Yu, A. Procurement auctions and supply chain performance. *Int. J. Prod. Econ.* **2015**, *162*, 192–200. [CrossRef]
16. Liu, W.; Shen, X.; Wang, D. The impacts of dual overconfidence behavior and demand updating on the decisions of port service supply chain: A real case study from China. *Ann. Oper. Res.* **2020**, *291*, 591–604. [CrossRef]
17. Bernstein, F.; Federgruen, A. Coordination mechanisms for supply chains under price and service competition. *Manuf. Serv. Oper. Manag.* **2007**, *9*, 242–262. [CrossRef]
18. Sethi, S.P.; Yan, H.; Zhang, H.; Zhou, J. A supply chain with a service requirement for each market signal. *Prod. Oper. Manag.* **2007**, *16*, 322–342. [CrossRef]
19. Sieke, M.A.; Seifert, R.W.; Thonemann, U.W. Designing service level contracts for supply chain coordination. *Prod. Oper. Manag.* **2012**, *21*, 698–714.
20. Xiao, T.; Xu, T. Coordinating price and service level decisions for a supply chain with deteriorating item under vendor managed inventory. *Int. J. Prod. Econ.* **2013**, *145*, 743–752. [CrossRef]
21. Heydari, J. Coordinating supplier's reorder point: A coordination mechanism for supply chains with long supplier lead time. *Comput. Oper. Res.* **2014**, *48*, 89–101. [CrossRef]
22. Cakanyildirim, M.; Gan, X.; Sethi, S. Contracting and coordination under asymmetric production cost information. *Prod. Oper. Manag.* **2012**, *21*, 345–360. [CrossRef]
23. Ha, A.Y. service provider-buyer contracting: Asymmetric cost information and cutoff level policy for buyer participation. *Nav. Res. Logist.* **2001**, *48*, 41–64. [CrossRef]
24. Ha, A.Y.; Tong, S. Contracting and information sharing under supply chain competition. *Manag. Sci.* **2008**, *54*, 701–715. [CrossRef]
25. Van Ryzin, G.; Vulcano, G. Optimal auctioning and ordering in an infinite horizon inventory-pricing system. *Oper. Res.* **2004**, *52*, 346–367. [CrossRef]
26. Abhishek, V.; Hajek, B.; Williams, S.R. Auctions with a profit sharing contract. *Games Econ. Behav.* **2013**, *77*, 247–270. [CrossRef]
27. Jain, V.; Panchal, G.; Kumar, S. Universal service provider selection via multi-dimensional auction mechanisms for two-way competition in oligopoly market of supply chain. *Omega* **2014**, *47*, 127–137. [CrossRef]
28. Huang, M.; Qian, X.; Fang, S.C.; Wang, X. Winner determination for risk aversion buyers in multi-attribute reverse auction. *Omega* **2016**, *592*, 184–200. [CrossRef]
29. Chen, J.; Feng, J.; Whinston, A.B. Keyword auctions, unit-price contracts, and the role of commitment. *Prod. Oper. Manag.* **2007**, *19*, 30–321. [CrossRef]
30. Iyengar, G.; Kumar, A. Optimal procurement mechanisms for divisible goods with capacitated suppliers. *Rev. Econ. Des.* **2008**, *12*, 129–154. [CrossRef]
31. Chen, Y.J.; Vulcano, G. Effects of information disclosure under first- and second-price auctions in a supply chain setting. *Manuf. Serv. Oper. Manag.* **2008**, *11*, 299–316. [CrossRef]
32. Che, Y.K. Design competition through multi-dimensional auctions. *RAND J. Econ.* **1993**, *24*, 668–680. [CrossRef]
33. Branco, F. The design of multidimensional auctions. *RAND J. Econ.* **1997**, *28*, 63–81. [CrossRef]
34. Parkes, D.; Kalagnanam, J. Models for iterative multiattribute procurement auctions. *Manag. Sci.* **2005**, *51*, 435–451. [CrossRef]
35. Chen, K. Procurement strategies and coordination mechanism of the supply chain with one manufacturer and multiple suppliers. *Int. J. Prod. Econ.* **2012**, *138*, 125–135. [CrossRef]
36. Arcelus, F.J.; Satyendra, K.; Srinivasan, G. Inventory model with stochastic lead-time and price dependent demand incorporating advance payment. *Int. Trans. Oper. Res.* **2006**, *13*, 209–227. [CrossRef]
37. Leng, M.; Parlar, M. Game-Theoretic Analyses of Decentralized Assembly Supply Chains: Non-Cooperative Equilibria vs. Coordination with Cost-Sharing Contracts. *Eur. J. Oper. Res.* **2010**, *204*, 96–104. [CrossRef]
38. Mathewson, F.; Winter, R.A. Buyer groups. *Int. J. Ind. Organ.* **1997**, *15*, 137–164. [CrossRef]
39. Gans, J.S.; King, S.P. Exclusionary contracts and competition for large buyers. *Int. J. Ind. Organ.* **2002**, *20*, 1363–1381. [CrossRef]

sustainability

MDPI

Article

Impact of Information Asymmetry on the Operation of Green Closed-Loop Supply Chain under Government Regulation

Jianteng Xu [1], Peng Wang [1] and Qi Xu [2,*]

1 School of Management, Qufu Normal University, Rizhao 276826, China; jiantengxu@qfnu.edu.cn (J.X.); w_peng233@163.com (P.W.)
2 School of Medical Information Engineering, Jining Medical University, Rizhao 276826, China
* Correspondence: xuqi8079@163.com

Abstract: Recycling subsidy and carbon tax policies are ways to achieve energy and environmental sustainability. The implementation of these policies has changed the operating environment of traditional closed-loop supply chains, while the privacy of relevant information increases the difficulty of decision-making. Under the background, this paper considers the green closed-loop supply chain (GCLSC) under the hybrid policy of recycling subsidy and carbon tax where the manufacturer is in charge of recycling and the retailer invests in green marketing. Taking green marketing cost coefficient as the retailer's private information, this paper explores the influence of information asymmetry on optimal decisions and performance of the GCLSC. By constructing game models of information symmetry and asymmetry, the optimal decisions, economic and environmental performance, and social welfare are provided. Combined with numerical analysis, the influence of uncertainty of the manufacturer's estimation, subsidies and carbon tax on the GCLSC is proposed. The results indicate that the uncertainty in the manufacturer's estimation can improve the social welfare under certain conditions, but it cannot reduce carbon emissions. Recycling subsidy and carbon tax policies oppositely affect the manufacturer's optimal decisions and carbon emissions. Information asymmetry is beneficial to the retailer. However, less uncertainty in estimation is not always better for the manufacturer. The manufacturer needs to proactively adopt strategies to stimulate the retailer's information sharing.

Keywords: closed-loop supply chain; information asymmetry; carbon tax; government subsidy; social welfare

Citation: Xu, J.; Wang, P.; Xu, Q. Impact of Information Asymmetry on the Operation of Green Closed-Loop Supply Chain under Government Regulation. *Sustainability* **2022**, *14*, 7999. https://doi.org/10.3390/su14137999

Academic Editor: Andrea Trianni

Received: 16 May 2022
Accepted: 28 June 2022
Published: 30 June 2022

1. Introduction

With the booming economy, people have gradually realized the limitation of natural resources and have begun to pay attention to conservation. A closed-loop supply chain (CLSC) is designed to collect used-products from consumers by reverse logistics and remanufactures the entire product or parts to create new value [1]. In April 2019, Apple announced the expansion of it's recycling program by quadrupling the number of recycling locations available to consumers in America. In 2020, Huawei processed more than 4500 tons of e-waste through its own recycling channels. It contributes to the carbon peak and carbon neutrality targets of China. By improving energy efficiency and reducing waste, CLSC extends traditional supply chains to the green supply chain.

In addition to improving energy efficiency, environmental sustainability has received increasing attention in recent years [2–4]. Green closed-loop supply chains (GCLSCs) have become a major trend in energy and environmental sustainability, as it reduces the use of raw materials and decreases energy consumption and associated carbon emissions [5]. Implementations of emission reduction regulations (such as carbon tax, cap-and-trade, mandatory cap) have enriched the significance of GCLSCs. Assuming that the information between CLSC members is symmetric, some literature concentrates on the optimal operations decisions for the CLSC under certain carbon emission regulation [6–9]. Other literature interests in comparing the impact of different carbon emission policies on CLSC [10–12]. In order to encourage the remanufacturing and recycling of products, many governments subsidize recycling programs during implementing carbon reduction policies. For example, in 2018, after Shanghai launched the carbon trading scheme in November 2013, the government gave enterprises that recycled batteries a subsidy of 1000 RMB per set. Under the carbon tax regulation, Japan spent approximately 100 billion yen on subsidies to support battery-related industries in 2021, such as the sorting and recycling of renewable battery materials. The interaction of carbon reduction policy and recycling subsidy policy challenges the GCLSC's operational decisions. Since the government charges taxes for each unit of carbon emissions emitted by enterprises under carbon tax policy [13,14]. For the CLSC, carbon tax increases the environmental cost of manufacturers while the subsidy policy reduces the costs of recycling and remanufacturing. The manufacturer has to balance the environmental cost and remanufacturing cost. Therefore, scientific guidance, with regard to the operation and decision-making, is needed for members of the GCLSC. However, there is little literature concentrates on it. Dou and Choi [15] compared the green investment and recycling decisions of CLSC under the subsidy for trade-in program. They thought that carbon tax and subsidy policies motivate the GCLSC and consumers to accept the trade-in program. Shang et al. [16] analyzed optimal operational decisions of CLSCs when the government subsidize manufacturer's emission reduction, recycling and the retailer's advertising investment. All above literature is studied based on the assumption of information symmetry among GCLSC members.

In reality, information asymmetry among members is also an important factor affecting the operations of GCLSCs apart from the operating environment and policy regulations. Every enterprise holds private information that may significantly affect supply chain operations, such as market demand, recovery and green marketing efforts. Information asymmetry means that one participant with information advantages does not share his private information with the others. As opposed to the uncertainty of information, the information asymmetry may result in different status for participants in GCLSC. It will further affect their operational decisions, economic profits, and environmental impacts. There is quite a lot of literature on optimal operation decisions of CLSCs under uncertainty market demand [17,18], recovery product quality [10,19], and carbon price [20]. All these literature assumes that the information among CLSC participants is symmetrical.

On the basis of such background, this paper studies the influence of the information asymmetry on a GCLSC under the hybrid policy of carbon tax and recycling subsidy. In the considered GCLSC, the retailer carries out green marketing promotion and the manufacturer is responsible for recycling. The manufacturer has to estimate the parameter of green marketing efforts because it is the retailer's private information. This paper constructs decision optimization models for both cases of information symmetry and asymmetry. The main contributions are three-fold: (1) This paper takes green marketing cost coefficient as retailer's private information, provides the closed-form solutions of the optimal pricing, recycling rate, and marketing promotion decisions of the GCLSC under the interaction of subsidy and carbon tax policies. (2) Via analyzing game behaviors, the difference of the optimal decisions and GCLSC's performance between information symmetry and asymmetry are displayed under the hybrid policy of carbon tax and recycling subsidy. (3) The impact of the hybrid policy and the uncertainty in the manufacturer's estimation on operation decisions, the economic and environmental performance of a GCLSC is revealed. The interesting results show that when the retailer keeps marketing cost coefficient as the private information under the hybrid policy of carbon tax and subsidy, it is not always adverse to the manufacturer. Under certain conditions, it is beneficial to the manufacturer.

The following sections are organized as follows. The related literature is reviewed in Section 2. The problem description and notations that will be used in the rest of this paper are described in Section 3. The optimal decisions and characters under both information symmetry and asymmetry scenarios are analyzed in Section 4. Numerical analysis is presented in Section 5 to complement theoretical results. Section 6 concludes the findings, managerial insights and the direction for further research.

2. Literature Review

This paper contributes to the following two research themes: CLSC management under the constraint of carbon emissions reduction and supply chain management with information asymmetry.

2.1. CLSC Management under the Constraint of Carbon Emissions Reduction

With increasing global attention to green supply chains, lots of scholars have studied how carbon emissions reduction affects the operational strategies of the supply chain from different angles [6,21–23]. The influence of emissions reduction on CLSCs has also received considerable attention because the remanufacturing process generates carbon emissions [24,25]. Growing literature studies the optimal remanufacturing and recycling strategies of an enterprise under carbon emission reduction regulations. For example, Chai et al. [16] explain how the adoption of carbon trading policy influences the optimal remanufacturing decisions of an enterprise. Chen et al. [26] focus on the optimal collection and remanufacturing decisions for a remanufacturing system under carbon cap and take-back policies. Bai et al. [27] employ a distributionally robust newsboy approach to propose the optimal production and collection decisions under a cap-and-trade policy. Other literature has studied the operational strategies of multi-echelon CLSCs under the constraint of carbon emissions reduction [28,29]. Recently, Dou and Cao [5] investigate the optimal operational strategies of the CLSC in two operational periods by considering three product collection channels under a carbon tax regulation. Yang et al. [30] study how the implementation of the cap-and-trade policy affects the collection model selection of a two-echelon CLSC. Jauhari et al. [31] consider two recovery processes in a three-echelon CLSC under a carbon trading regulation, and concentrate on the optimal decisions including green technology investment, product quality and selling price under five different scenarios. Shekarian et al. [32] explore the effect of remanufacturing and emissions on a dual-channel CLSC with competitive collection. Wang and Wu [33] study the recycling and carbon reduction investment decisions for two types of CLSCs under cap-and-trade regulation.

All above literature studies the deterministic environment in which the market demand and parameter information are known. Taking into account the universal existence of uncertainty in reality, some scholars investigate the low-carbon operation strategies of CLSCs in an uncertain scenario. For example, Jauhari et al. [7] reveal the operation decisions of CLSC with stochastic demand and return rate under carbon tax regulation. Xu et al. [20] formulate a stochastic model for a CLSC facing uncertain demand and carbon price under the carbon trading scenario to find the optimal operation decisions in a multi-period planning horizon. Guo et al. [19] consider a remanufacturing enterprise that faces uncertain recycled product quality and demand under subsidy and carbon tax policies. They employ heuristic and intelligent methods to find the approximate solutions of the provided discrete optimization models.

The main characteristics of above literature on CLSCs are that they study the impact of uncertainty on operation strategies by assuming that the relevant information is symmetrical among CLSC members. In contrast, this paper considers a GCLSC with information asymmetry under a hybrid policy of carbon emission reduction and recycling subsidy. When the green marketing cost coefficient is the retailer's private information, the manufacturer has to estimate it before making decision. Hence, the effect of information asymmetry and the uncertainty of estimation on optimal joint remanufacturing and carbon abatement strategies are studied.

2.2. Supply Chain Management with Information Asymmetry

In reality, it is difficult to realize information sharing among supply chain participants. A growing number of scholars studied the operational strategies for supply chains with information asymmetry [34–38]. Recently, some scholars pay close attention to the influence of information asymmetry on CLSCs [39–42]. Wang et al. [43] consider a dual-channel CLSC consisted of one retailer and one third party recycling institution under a government reward-penalty mechanism. By taking recycling efforts as private information, they design contracts for the manufacturer to obtain real information. Via analyzing 288 articles, Chen and Huang [44] deem that the information asymmetry in CLSC remains be solved. Wu et al. [45] explore how the government incentivizes the retailer to report his recovery information when the recovery technology-type is taken as the retailer's private information. Wang et al. [46] reveal the effect of information asymmetry and fairness concerns on the performance of CLSCs by taking fairness concerns as the manufacturer's private information.

Above literature on CLSC studies the impacts of different types of information asymmetry without considering the external factors. On the contrary, this paper concentrates on the comprehensive influence of both the information asymmetry and the hybrid policy of carbon reduction and subsidy on the GCLSC. Considering the green marketing cost coefficient as the retailer's private information, it compares the optimal equilibrium strategies of the GCLSC with information symmetry and asymmetry under carbon tax and subsidy policies, and provides some managerial insights. The differences of models between relevant literature and present work are summarized in Table 1.

Table 1. Comparison of models between relevant literature and present work.

literature	Echelons of CLSC	Carbon Policy	Recycling Subsidy	Uncertainty	Information Asymmetry
Dou and Cao [5]	Two	Carbon tax	×	×	×
Jauhari et al. [31]	Three	Cap-and-trade	×	×	×
Wang and Wu [33]	Three	Cap-and-trade	×	×	×
Xu et al. [20]	Two	Cap-and-trade	×	Demand and carbon price	×
Jauhari et al. [7]	Two	Carbon tax	×	Demand and return	×
Guo et al. [19]	One	Carbon tax	✓	Demand and recycled products quality	×
Gao et al. [42]	Two	×	×	Demand	✓
Wang et al. [43]	Three	×	×	Demand	✓
Wang et al. [46]	Two	×	×	Fairness concern	✓
Present work	Two	Carbon tax	✓	Marketing cost coefficient	✓

3. Problem Descriptions and Notations

3.1. Problem Description

This paper considers a two-echelon CLSC under both carbon tax and recycling subsidy regulations. The diagram of the considered GCLSC is shown in Figure 1. In this GCLSC, the manufacturer produces new products at a unit manufacturing cost c_m and collects used-products with collection rate τ from customers at a unit collection price f. The used-products are remanufactured at the same cost c_n. The remanufactured product does not differ from the new product in appearance and function. The retailer buys products from the manufacturer at unit price w, and sells them with green marketing efforts A_r at unit price p. The manufacturer's production process is the main source of carbon emissions. The unit carbon emissions generated during manufacturing and remanufacturing is e_0 and e_1, respectively. Under the carbon tax regulation, the government announces a tax to the manufacturer, who pays tax for the unit carbon emissions at price r. In addition, to encourage the remanufacturing of used-products, the government subsidizes G for used-products. The research approach of this paper is provided in Figure 2.

3.2. Notations and Assumtions

The symbols and notations that will be used in the rest of this paper are summarized in Table 2.

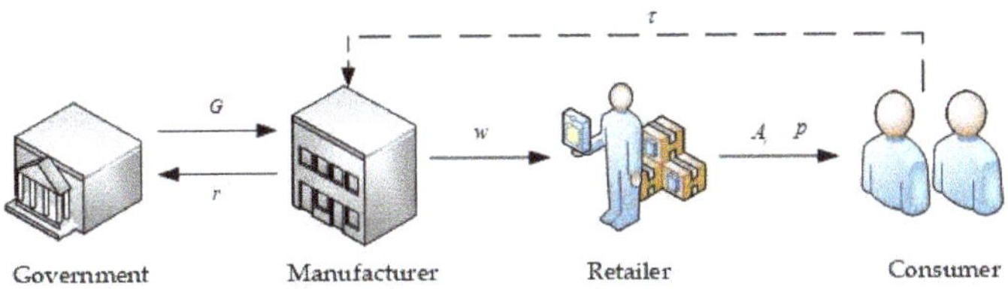

Figure 1. The diagram of GCLSC.

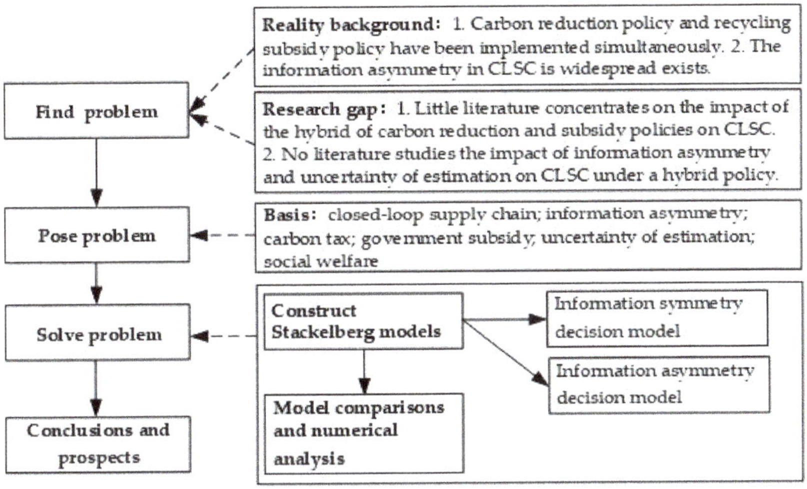

Figure 2. Flowchart of the research approach.

Table 2. The main parameters and notations.

Parameters	
D	The total market demand (unit)
ϕ	Basic market scale of products (unit)
β	Price-sensitive parameter of demand (unit/$)
b	Elasticity coefficient of the demand to green marketing efforts (unit/unit effort)
c_m	manufacturing cost per unit new product ($/unit)
c_n	remanufacturing cost per unit used-product ($/unit)
μ_r	Coefficient of the retailer's green marketing effort cost ($)
c_l	Coefficient of the manufacturer's collection cost ($)
f	Unit collection price of the manufacturer ($/unit)
G	Unit government subsidy for used-products ($/unit)
r	Unit carbon tax price ($/unit emission)
e_0	Carbon emissions generated during manufacturing one product (kg/unit)
e_1	Carbon emissions generated during remanufacturing one used-product (kg/unit)
Decision variables	
w	Wholesale price charged by the manufacturer ($/unit)
τ	Collection rate of the manufacturer
p	Selling price of retailer ($/unit)
A_r	Green marketing efforts of the retailer (index of efforts level)
Objective functions	
Π_R^{sym}, Π_M^{sym}	Profit of the retailer and manufacturer under information symmetry ($)
$E(\Pi_R^{asy}), E(\Pi_M^{asy})$	Expected profit of the retailer and manufacturer under information asymmetry ($)

The following assumptions are used to make this research closer to reality and concentrate on the key points.

Assumption 1. *The corresponding collection cost and green marketing efforts cost is* $\frac{1}{2}c_l\tau^2$ *and* $\frac{1}{2}\mu_r A_r^2$*, where* c_l *is the collection cost coefficient, and* μ_r *is the green marketing cost coefficient [13,35]. It accords with the economic principle of increasing marginal cost.*

Assumption 2. *The demand for products is linearly influenced by both the retail price and green marketing efforts, i.e.,* $D = \phi - \beta p + bA_r$*, where* ϕ *is the basic market scale,* β *is the price-sensitive parameter, and b is the elastic coefficient of demand to the green marketing efforts [35].*

Assumption 3. $c_m > c_n + f + re_1$*. It means that the manufacturer pays more for manufacturing than remanufacturing. It ensures the sustainability of the collection activity [26].*

Assumption 4. $c_l > \frac{\beta^2\mu_r(\Delta - re_1)^2}{2(2\beta\mu_r - b^2)} > 0$*, where* $\Delta = c_m - c_n + G - f$*. It indicates a higher cost for collection used-products, and guarantees the feasibility of the developed models [13].*

According to above notations and assumptions, we can formulate the profits, carbon emissions and social welfare as follows.

The profit functions of the retailer and the manufacturer are expressed as follows:

$$\prod\nolimits_R = (p - w)D - \frac{1}{2}\mu_r A_r^2 \tag{1}$$

$$\prod\nolimits_M = (w - c_m)D + (c_m - c_n)\tau D + (G - f)\tau D - \frac{1}{2}c_l\tau^2 - r(e_0 D + e_1\tau D) \tag{2}$$

Equation (1) is composed by the profit on sale and the green marketing effort cost. In Equation (2), the first two terms are the profit on sale, the third and fourth terms are the profit earned from collecting used-products, and the fifth term is the carbon tax, where $e_0 D$ is the carbon emissions from manufacturing new products and $e_1\tau D$ is the carbon emissions from remanufacturing used-products. To facilitate the analysis, the total carbon emissions are expressed as

$$J = e_0 D + e_1\tau D \tag{3}$$

The government subsidizes the manufacturer's collection activity from a social welfare perspective. Referring to previous studies [2,47], social welfare (SW) consists of three components: total profits of GCLSC participants, consumer surplus (CS) and government subsidy. CS is the difference between the maximum unit price and the actually unit paid price. Then, CS and SW can be expressed as follows:

$$CS = \int_{\frac{\phi + bA_r - D}{\beta}}^{\frac{\phi + bA_r}{\beta}} (\phi - \beta p + bA_r)dp = \frac{D^2}{2\beta} \tag{4}$$

$$SW = \Pi_M + \Pi_R + CS - G\tau D \tag{5}$$

4. Problem Formulation and Solutions

Considering the power of the retailer and the manufacturer, the Stackelberg game is used to establish a decision model in which the manufacturer first announces the wholesale price and the collection rate as the leader, then the retailer decides the selling price and the green marketing efforts. Under this decision model, the decision-making of supply chain members in the cases of information symmetry and information asymmetry are considered.

4.1. Information Symmetry Decision Model

In this model (represented as the "sym" model), all information of the retailer is shared with the manufacturer. The retailer's profit is first analyzed, and the manufacturer makes decisions based on the response function of the retailer, solving Equations (1) and (2) leads to the following conclusions.

Theorem 1. *There exist optimal solutions that maximize the GCLSC members' profits under information symmetry.*

(1) The retailer's optimal retail price and optimal green marketing efforts are

$$p^{sym*} = \frac{\mu_r \phi}{2\beta\mu_r - b^2} + \frac{\beta\mu_r - b^2}{2\beta\mu_r - b^2} \frac{c_l(2\beta\mu_r - b^2)[\phi + \beta(c_m + re_0)] - \beta^2\mu_r\phi(\Delta - re_1)^2}{2\beta c_l(2\beta\mu_r - b^2) - \beta^3\mu_r(\Delta - re_1)^2},$$

$$A_r^{sym*} = \frac{bc_l[\phi - \beta(c_m + re_0)]}{2c_l(2\beta\mu_r - b^2) - \beta^2\mu_r(\Delta - re_1)^2}$$

(2) The manufacturer's optimal wholesale price and collection rate are

$$w^{sym*} = \frac{c_l(2\beta\mu_r - b^2)[\phi + \beta(c_m + re_0)] - \beta^2\mu_r\phi(\Delta - re_1)^2}{2\beta c_l(2\beta\mu_r - b^2) - \beta^3\mu_r(\Delta - re_1)^2},$$

$$\tau^{sym*} = \frac{\beta\mu_r(\Delta - re_1)[\phi - \beta(c_m + re_0)]}{2c_l(2\beta\mu_r - b^2) - \beta^2\mu_r(\Delta - re_1)^2}.$$

The proof is shown in Appendix A.

Theorem 1 shows that there are optimal solutions for the manufacturer and retailer when information is symmetric. Substituting p^{sym*}, A_r^{sym*}, w^{sym*} and τ^{sym*} into the relevant functions, the optimal profits of both retailer and manufacturer, total carbon emissions and social welfare under information symmetry are given as follows:

$$\Pi_R^{sym*} = \frac{\mu_r c_l^2[\phi - \beta(c_m + re_0)]^2(2\beta\mu_r - b^2)}{2[2c_l(2\beta\mu_r - b^2) - \beta^2\mu_r(\Delta - re_1)^2]^2}, \tag{6}$$

$$\Pi_M^{sym*} = \frac{\mu_r c_l[\phi - \beta(c_m + re_0)]^2}{2[2c_l(2\beta\mu_r - b^2) - \beta^2\mu_r(\Delta - re_1)^2]}, \tag{7}$$

$$J^{sym*} = \frac{e_0\beta\mu_r c_l[\phi - \beta(c_m + re_0)]}{2c_l(2\beta\mu_r - b^2) - \beta^2\mu_r(\Delta - re_1)^2} + \frac{e_1\beta^2\mu_r^2 c_l(\Delta - re_1)[\phi - \beta(c_m + re_0)]^2}{[2c_l(2\beta\mu_r - b^2) - \beta^2\mu_r(\Delta - re_1)^2]^2}, \tag{8}$$

$$SW^{sym*} = \frac{\mu_r c_l[\phi - \beta(c_m + re_0)]^2[3c_l(2\beta\mu_r - b^2) - \beta^2\mu_r(\Delta - re_1)^2 + 2\beta\mu_r(c_l - g\beta(\Delta - re_1))]}{2[2c_l(2\beta\mu_r - b^2) - \beta^2\mu_r(\Delta - re_1)^2]^2} \tag{9}$$

4.2. Information Asymmetry Decision Model

Since the participants in the decentralized GCLSC make decisions independently, the retailer may not share all information with other participants that results in information asymmetry. In this section, the green marketing effort cost coefficient μ_r is the retailer's private information, which the manufacturer lacks full information about it (denoted as model "asy"). It is assumed that μ_r is uniformly distributed, that is, $\mu_r \sim U[\overline{\mu}_r - \varepsilon, \overline{\mu}_r + \varepsilon]$, where ε, $0 < \varepsilon < \overline{\mu}_r$, denotes the degree of information uncertainty that reflects the uncertainty in the manufacturer's estimation of the green marketing cost coefficient. When the value of ε increases, the uncertainty in manufacturer's estimation on the green marketing cost coefficient increases.

The retailer has the same information under asymmetric information and keeps the private information about the green marketing effort cost coefficient μ_r. Given the retailer's decisions, the manufacturer decides the optimal collection rate and wholesale price by maximizing its expected profit $E(\prod_M^{asy})$ which is given as follows:

$$\begin{aligned} E(\prod_M^{asy}) &= \int_{\overline{\mu_r}-\varepsilon}^{\overline{\mu_r}+\varepsilon}[(w-c_m+\Delta\tau)\frac{\beta\mu_r(\phi-\beta w)}{2\beta\mu_r-b^2}-\frac{1}{2}c_l\tau^2-r(e_0+e_1\tau)\frac{\beta\mu_r(\phi-\beta w)}{2\beta\mu_r-b^2}]\frac{1}{2\varepsilon}d\mu_r \\ &= (\phi-\beta w)[w-c_m+\Delta\tau-r(e_0+e_1\tau)][\frac{1}{2}+\frac{b^2}{8\beta\varepsilon}\ln\frac{2\beta(\overline{\mu_r}+\varepsilon)-b^2}{2\beta(\overline{\mu_r}-\varepsilon)-b^2}]-\frac{1}{2}c_l\tau^2. \end{aligned} \quad (10)$$

Let $h(\varepsilon)=\frac{1}{2}+\frac{b^2}{8\beta\varepsilon}\ln\frac{2\beta(\overline{\mu_r}+\varepsilon)-b^2}{2\beta(\overline{\mu_r}-\varepsilon)-b^2}$. Solving Equation (10) leads to the following conclusion.

Theorem 2. *There exist optimal solutions that maximize the GCLSC participants' profits under information asymmetry.*

(1) The retailer's optimal selling price and green marketing efforts are

$$p^{asy*}=\frac{\mu_r\phi}{2\beta\mu_r-b^2}+\frac{\beta\mu_r-b^2}{2\beta\mu_r-b^2}\frac{c_l[\phi+\beta(c_m+re_0)]-\beta\phi h(\varepsilon)(\Delta-re_1)^2}{\beta[2c_l-\beta h(\varepsilon)(\Delta-re_1)^2]},$$

$$A_r^{asy*}=\frac{bc_l[\phi-\beta(c_m+re_0)]}{(2\beta\mu_r-b^2)\left[2c_l-\beta h(\varepsilon)(\Delta-re_1)^2\right]}$$

(2) The manufacturer's optimal collection rate and wholesale price are

$$\tau^{asy*}=\frac{h(\varepsilon)(\Delta-re_1)[\phi-\beta(c_m+re_0)]}{2c_l-\beta h(\varepsilon)(\Delta-re_1)^2},$$

$$w^{asy*}=\frac{c_l[\phi+\beta(c_m+re_0)]-\beta\phi h(\varepsilon)(\Delta-re_1)^2}{\beta[2c_l-\beta h(\varepsilon)(\Delta-re_1)^2]}.$$

The proof is shown in Appendix A.

The above process proves the existence of optimal solutions for GCLSC participants when information is asymmetric. Substituting the optimal decisions into the correlation function yields the optimal expected profits of the GCLSC participants, total carbon emissions and social welfare as follows:

$$E(\prod_R^{asy*})=\frac{\mu_r c_l^2[\phi-\beta(c_m+re_0)]^2}{2(2\beta\mu_r-b^2)[2c_l-\beta h(\varepsilon)(\Delta-re_1)^2]^2}, \quad (11)$$

$$E(\prod_M^{asy*})=\frac{c_l[\phi-\beta(c_m+re_0)]^2[2\mu_r c_l-h^2(\varepsilon)(\Delta-re_1)^2(2\beta\mu_r-b^2)]}{2(2\beta\mu_r-b^2)[2c_l-\beta h(\varepsilon)(\Delta-re_1)^2]^2}, \quad (12)$$

$$J^{asy*}=\frac{e_0\beta\mu_r c_l[\phi-\beta(c_m+re_0)]}{(2\beta\mu_r-b^2)[2c_l-\beta h(\varepsilon)(\Delta-re_1)^2]}+\frac{e_1\beta\mu_r c_l h(\varepsilon)(\Delta-re_1)[\phi-\beta(c_m+re_0)]^2}{(2\beta\mu_r-b^2)[2c_l-\beta h(\varepsilon)(\Delta-re_1)^2]^2}, \quad (13)$$

$$\begin{aligned} SW^{asy*} &= \frac{c_l[\phi-\beta(c_m+re_0)]^2[3\mu_r c_l-h^2(\varepsilon)(\Delta-re_1)^2(2\beta\mu_r-b^2)]}{2(2\beta\mu_r-b^2)[2c_l-\beta h(\varepsilon)(\Delta-re_1)^2]^2} \\ &+ \frac{\beta\mu_r c_l[\phi-\beta(c_m+re_0)]^2[\mu_r c_l+2gh(\varepsilon)(\Delta-re_1)(2\beta\mu_r-b^2)]}{2(2\beta\mu_r-b^2)^2[2c_l-\beta h(\varepsilon)(\Delta-re_1)^2]^2}. \end{aligned} \quad (14)$$

4.3. Model Comparisons and Analysis

This section reveals the effect of ε, r and G on optimal decisions, profits, carbon emissions, and social welfare, and further compares the optimal decisions between the models with information symmetry and asymmetry.

Corollary 1. *Uncertainty in estimation and government policies affect the manufacturer's decisions, and the following conclusions hold when $\phi>\beta(c_m+re_0)$ is satisfied.*

(1) As ε and G increase, the optimal wholesale price w^{asy*} decreases while the optimal collection rate τ^{asy*} increases.

(2) As r increases, the optimal wholesale price w^{asy*} increases while the optimal collection rate τ^{asy*} decreases.

The proof is shown in Appendix A.

As seen from Corollary 1, when the basic market scale is large, the manufacturer's uncertainty about the green marketing effect coefficient increases with the increases of ε, and the manufacturer's dominant position in the game weakens to reduce the wholesale price and improve the collection rate. The optimal wholesale price decreases with G while it increases with r. However, the optimal collection rate increases with G while it decreases with r. It implicates that the impact of recycling subsidy and carbon tax on the manufacturer's optimal decisions is opposite. The carbon tax and subsidy policies are complementary to each other.

Corollary 2. *Uncertainty in estimation affects the manufacturer's profit, and the following conclusions hold.*

(1) If $\frac{\beta\mu_r}{2\beta\mu_r-b^2} < h(\varepsilon) < \frac{2c_l}{\beta(\Delta-re_1)^2}$, $E(\prod_M^{asy*})$ decreases while $\Pi_M^{sym*} - E(\prod_M^{asy*})$ increases with ε.

(2) If $\frac{1}{2} < h(\varepsilon) < \frac{\beta\mu_r}{2\beta\mu_r-b^2}$, $E(\prod_M^{asy*})$ increases while $\Pi_M^{sym*} - E(\prod_M^{asy*})$ decreases with ε.

Corollary 3. *Both $E(\prod_R^{asy*}$ and $E(\prod_R^{asy*}) - \Pi_R^{sym*}$ increase as ε increases.*

The proofs of Corollaries 2 and 3 are shown in Appendix A.

Corollaries 2 and 3 show that the retailer's expected profit increases with ε, indicating that keeping the green marketing effort cost coefficient as private information facilitates the increase of profit and improves the disadvantageous position in the game. In contrast, the uncertainty of estimation has more complex effect on the manufacturer's profit. When $h(\varepsilon) < \frac{2c_l}{\beta(\Delta-re_1)^2}$, as the uncertainty in estimation increases, the manufacturer's expected profit first increases and then decreases. The difference of the manufacturer's profit in the cases of information symmetry and asymmetry first decreases and then increases. It indicates that the retailer prefers a large uncertainty in the manufacturer's estimation. From the aspect of the manufacturer, it is not true that the less uncertainty in estimation is the better.

Corollary 4. *If $\frac{1}{2} < h(\varepsilon) < \frac{\beta\mu_r}{2\beta\mu_r-b^2}$, the social welfare SW^{asy*} increases with ε.*

Corollary 5. *When $\phi > \beta(c_m + re_0)$ holds, carbon emissions J^{asy*} increase with ε and G, and decrease with r.*

The proofs of Corollaries 4–5 are shown in Appendix A.

Corollaries 4–5 provide the conditions in which the social welfare and carbon emissions increase with respect to the uncertainty in the manufacturer's estimation. Corollary 5 further reveals that when the basic market scale is large, the carbon tax and subsidies oppositely affect carbon emissions. It indicates that the uncertainty in the manufacturer's estimation can improve the social welfare under certain conditions, but it cannot reduce carbon emissions.

Corollary 6. *The optimal decisions of the GCLSC are affected by information asymmetry. When $\phi > \beta(c_m + re_0)$, the following conclusions hold.*

(1) If $h(\varepsilon) < \frac{\beta\mu_r}{2\beta\mu_r-b^2}$, $\tau^{sym*} > \tau^{asy*}$ and $A_r^{sym*} > A_r^{asy*}$ hold.

(2) *If* $\frac{2c_l\left[1-(2\beta\mu_r-b^2)\right]}{\beta(\Delta-re_1)^2}+\beta\mu_r<h(\varepsilon)<\frac{c_l\left[\phi+\beta(c_m+re_0)\right]\left[1-(2\beta\mu_r-b^2)\right]}{\beta\phi(\Delta-re_1)^2}+\beta\mu_r$, $w^{sym*}<w^{asy*}$ *and* $p^{sym*}<p^{asy*}$ *hold.*

The proof is shown in Appendix A.

Corollary 6 compares the optimal decisions between the model with information symmetry and asymmetry. Conclusion (1) proposes the condition under which the collection rate and green marketing efforts in the information symmetry scenario are greater than those in the information asymmetry scenario. Conclusion (2) proposes the condition under which the wholesale price and selling price in the information asymmetry scenario are greater than those under information symmetry. It indicates that the influence of information asymmetry on the optimal decisions depends on the uncertainty in the manufacturer's estimation.

5. Numerical Analysis

This section illustrates the theoretical results and draws several managerial insights by the numerical analysis. Referring to the data in literature [35,48], we use the following values of parameters: $c_m=15$, $c_n=10$, $c_l=20$, $f=2.5$, $r=1.1$, $G=0.5$, $e_0=5$, $e_1=2$, $\phi=55$, $b=1$, $\beta=1.2$, $\mu_r=2$. When information is symmetric, the optimal decisions and profits of GCLSC members, social welfare and carbon emissions are as follows: $w^{sym*}=33.0112$, $\tau^{sym*}=0.3887$, $p^{sym*}=41.1094$, $A_r^{sym*}=4.0491$, $\Pi_R^{sym*}=62.3018$, $\Pi_M^{sym*}=123.0927$, $SW^{sym*}=222.8543$, $J^{sym*}=56.1441$. Let $\overline{\mu}_r=2$ and $\varepsilon=1$. When information is asymmetric, the optimal decisions and profits of GCLSC members, social welfare and carbon emissions are as follows: $w^{asy*}=33.0053$, $\tau^{asy*}=0.4033$, $p^{asy*}=41.1072$, $A_r^{asy*}=4.0509$, $E(\prod_R^{asy*})=62.3586$, $E(\prod_M^{asy*})=127.7171$, $SW^{asy*}=227.4995$, $J^{asy*}=56.4537$. The comparison reveals that information asymmetry has small impacts on the retailer's optimal decisions and profit, while it has significant impacts on the manufacturer's collection decision and profit, carbon emissions and social welfare. In this case, information asymmetry is not conducive to reducing carbon emissions but has positive impacts on GCLSC's economic performance and social welfare.

Next, the effect of the main parameters on the GCLSC is examined. Figure 3 proposes the effect of ε on GCLSC members' decisions and profits, carbon emissions, and social welfare.

Figure 3 shows that for the manufacturer, the collection rate and the expected profit increases while the wholesale price decreases as ε increases. For the retailer, the green marketing efforts and the expected profit increases while the retail price decreases as ε increases, but the trend is not significant. Carbon emissions and social welfare increase as ε increases. The reasons are as follows. The manufacturer's dominant position is weakened under asymmetric green marketing effort cost information. As the uncertainty grows, the manufacturer reduces the wholesale price and increases the collection rate, leading to an increase in the expected profit. According to the manufacturer's decisions, the retailer reduces the retail price and increases green marketing efforts. The retailer's expected profit increases under information asymmetry, but the private information does not fundamentally change the retailer's disadvantageous position as a follower. So the changes in the retailer's decisions and profit are not significant. On the other hand, lower selling price and higher green marketing efforts lead to higher product demand, and a higher collection rate leads to higher collection quantities. Therefore, the number of products increases, as well as carbon emissions. Meanwhile, the increase in product demand and supply chain members' profits lead to an increase in social welfare.

Based on the information asymmetry model, the influence of carbon tax price r and government subsidy G on the GCLSC is further analyzed. Figure 4 shows the influence of r and G on GCLSC members' decisions, profits, carbon emissions, and social welfare.

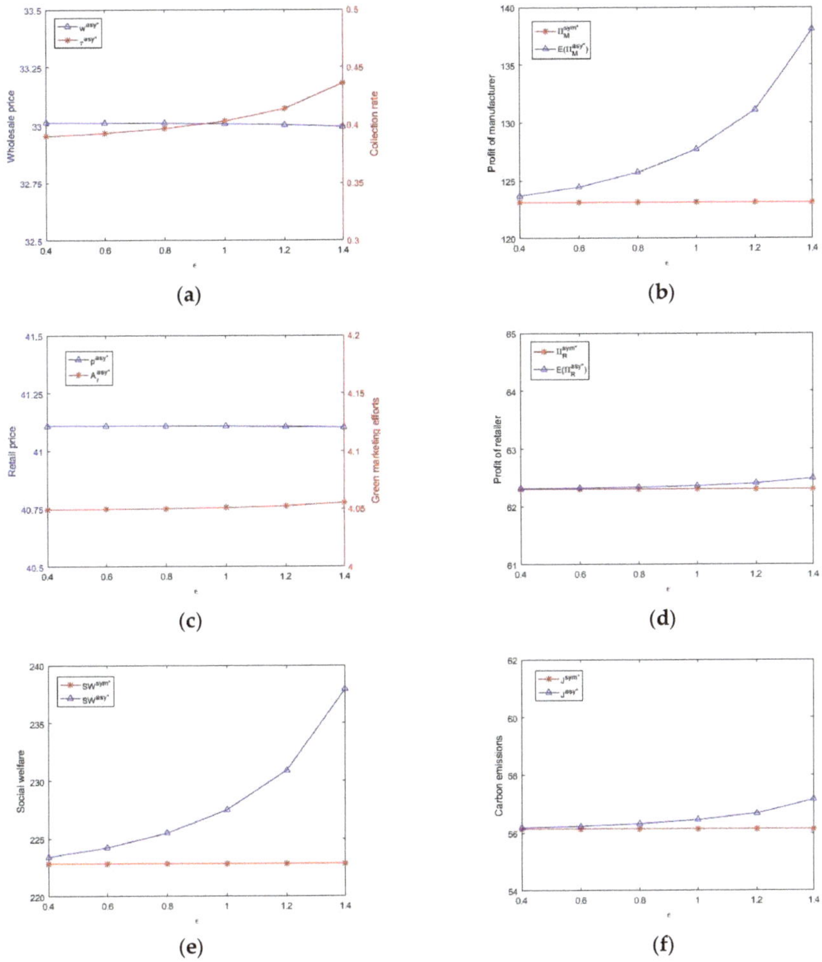

Figure 3. The effect of ε on the GCLSC. (**a**) The effect on w^{asy*} and τ^{asy*}, (**b**) The effect on Π_M^{sym*} and $E(\prod_M^{asy}*)$, (**c**) The effect on p^{asy*} and A_r^{asy*}, (**d**) The effect on Π_R^{sym*} and $E(\prod_R^{asy}*)$, (**e**) The effect on SW^{sym*} and SW^{asy*}, (**f**) The effect on J^{sym*} and J^{asy*}.

As seen in Figure 4, for the manufacturer, the collection rate and the expected profit decrease while the wholesale price increases as r increases. For the retailer, the green marketing efforts and the expected profit decrease while the retail price increases as r increases. Carbon emissions and social welfare decrease as r increases. The reasons are as follows. The increase in the carbon tax price means an increase in costs for the manufacturer. Hence, the manufacturer compensates for the loss by raising the wholesale price and reduces collection cost by decreasing the collection rate. Specifically, when $r = 1.5$, $\tau = 0$ holds and the manufacturer no longer collects used-products. The increase in wholesale price causes a higher selling price and fewer green marketing efforts. Subsequently, the market demand and carbon emissions of manufacturing new products decrease. The decrease in collection rate leads to fewer remanufactured products, so the carbon emissions of remanufactured products decrease. As market demand and supply chain members' profits decline, social welfare decreases as well.

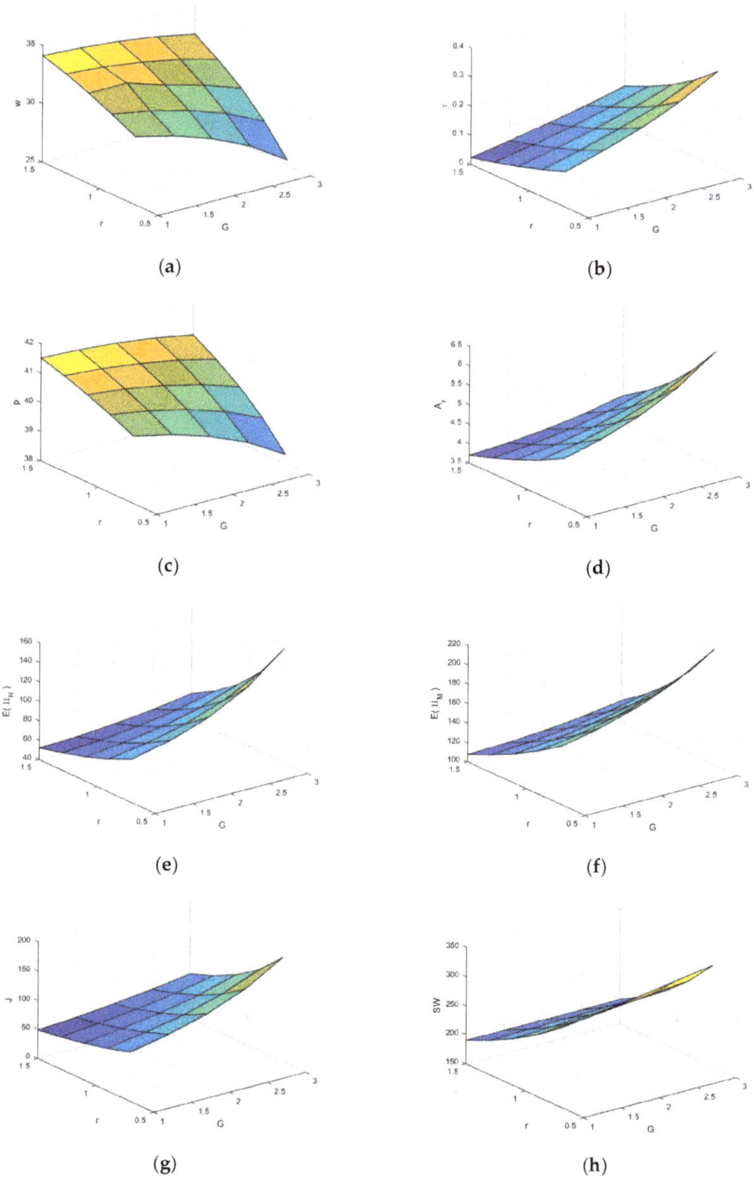

Figure 4. The effect of r and G on the GCLSC. (**a**) The effect on w^{asy*}, (**b**) The effect on τ^{asy*}, (**c**) The effect on p^{asy*}, (**d**) The effect on A_r^{asy*}, (**e**) The effect on $E(\prod_R^{asy}*)$. (**f**) The effect on $E(\prod_M^{asy}*)$, (**g**) The effect on J^{asy*}, (**h**) The effect on SW^{asy*}.

From Figure 4, we can also observe that for the manufacturer, the expected profit and the collection rate increase while the wholesale price decreases as G increases. For the retailer, the expected profit and the green marketing efforts increase while the retail price decreases as G increases. Carbon emissions and social welfare increase as G increases. The reasons are as follows. Because of the government subsidies, the manufacturer obtains more subsidies by reducing wholesale price and increasing collection rate. A decrease in wholesale price causes a lower retail price and higher green marketing efforts. The market demand and the carbon emissions of manufacturing new products increase. The increase in the collection rate rises the quantity of remanufactured products, which causes an increase in the carbon emissions from remanufacturing process. As market demand and total profits increase, social welfare increases as well.

According to the above analysis, it can be seen that information asymmetry does not give the retailer greater advantages since the increase in the retailer's profit is not significant when the green marketing effort cost information is asymmetric. But the manufacturer's decisions are greatly affected by information asymmetry. Under certain conditions, the expected profit of the manufacturer increases with uncertainty ε. Otherwise, the manufacturer's expected profit decreases. Overall, information asymmetry brings great uncertainty to the supply chain, which has negative impacts on social welfare and environmental benefits. The collection rate increases when the green marketing effort cost coefficient is asymmetric, but with increasing uncertainty, carbon emissions increase subsequently, which is not conducive to the environmental protection and emission reduction policies. As the price of carbon tax increases, carbon emissions and the collection rate decrease together with the social welfare. However, carbon emissions and social welfare increase as G increases. Hence, determining the carbon tax price and subsidies that will both control carbon emissions and effectively collect products is a matter for careful decision-making.

6. Conclusions

This paper considers a two-echelon GCLSC under a hybrid of carbon tax and subsidy regulations, where the manufacturer is in charge of production and collection, and the retailer sells products and makes green marketing efforts. Viewing the marketing effort cost coefficient as the private information of the retailer, we derive optimal decisions, profits, carbon emissions and social welfare under the game models with information symmetry and asymmetry. The results obtained by theoretical and numerical analysis indicate that: (1) The influence of information asymmetry on the optimal decisions relates to the uncertainty in the manufacturer's estimation. Information asymmetry is beneficial to the retailer. But from the aspect of the manufacturer, it is not true that the less uncertainty in estimation is the better. (2) The uncertainty in the manufacturer's estimation can improve the social welfare under certain conditions, but it cannot reduce carbon emissions. (3) Recycling subsidy and carbon tax policy oppositely affect the manufacturer's optimal decisions and carbon emissions.

The following suggestions are putting forward to governments and closed-loop supply chains. (1) The government needs to determine a reasonable carbon tax price and subsidies for the GCLSC to tradeoff the economic and environmental sustainability although subsidy and carbon tax policies are complementary to each other. (2) From the perspective of the CLSC participants, the manufacturer needs to proactively adopt strategies to stimulate the retailer's information sharing, which is beneficial to improve the accuracy of decision-making and reduce profit loss caused by the information asymmetry. Because less uncertainty in estimation is not always better for the manufacturer when the retailer has the private information. Besides, measures such as collecting used-products, introducing green technology and improving green marketing efforts should be taken to reduce carbon emissions and realize sustainable development.

In this paper, optimal decisions for the GCLSC with asymmetric information are explored by taking the green marketing cost coefficient as the retailer's private information. It assumes that the manufacturer's estimation of the retailer's green marketing cost coefficient follows uniform distribution. However, another distribution may result in different conclusion. Hence, whether the probability distribution of the estimation affects the research findings is another interesting issue. The analysis results of this paper indicate that the manufacturer should proactively adopt strategies to stimulate the retailer's information sharing. Then, the optimal decisions under other information asymmetry cases and how to encourage information sharing among GCLSC participants will be interesting research topics.

Author Contributions: Conceptualization, J.X. and Q.X.; formal analysis, J.X. and P.W; methodology, J.X. and P.W.; software P.W. and Q.X.; visualization, P.W. and Q.X.; supervision, J.X.; writing—original draft, P.W.; writing—review and editing, J.X. and Q.X.; funding acquisition, J.X. All authors have read and agreed to the published version of the manuscript.

Funding: This research was funded by the National Natural Science Foundation of China, grant number 71702087, the Youth Innovation Science and Technology Support Program of Shandong Province Higher Education, grant number 2021RW024, and the Special funds for Taishan Scholars, Shandong, grant number tsqn202103063.

Institutional Review Board Statement: Not applicable.

Informed Consent Statement: Not applicable.

Data Availability Statement: Not applicable.

Acknowledgments: We greatly appreciate the editor and the anonymous reviewers for their insightful comments and suggestions, which have greatly helped to improve the research.

Conflicts of Interest: The authors declare no conflict of interest.

Appendix A

Proof of Theorem 1. Simplifying Equation (1) and taking the first-order and second-order derivatives of Π_R, we can obtain the Hessian matrix of $\prod_R$ on variables p and A_r as follows:

$$H_1 = \begin{pmatrix} \frac{\partial^2 \Pi_R}{\partial p^2} & \frac{\partial^2 \Pi_R}{\partial p \partial A_r} \\ \frac{\partial^2 \Pi_R}{\partial A_r \partial p} & \frac{\partial^2 \Pi_R}{\partial A_r^2} \end{pmatrix} = \begin{pmatrix} -2\beta & b \\ b & -\mu_r \end{pmatrix}$$

Hence, when $\beta > \frac{b^2}{2\mu_r}$, Π_R is a concave function of p and A_r. According to the first-order condition, that is, $\frac{\partial \Pi_R}{\partial p} = -2\beta p + \phi + bA_r + \beta w = 0$ and $\frac{\partial \Pi_R}{\partial A_r} = b(p-w) - \mu_r A_r = 0$, the following equations can be deduced: $p(w) = \frac{\mu_r \phi + (\beta\mu_r - b^2)w}{2\beta\mu_r - b^2}$ and $A_r(w) = \frac{b(\phi - \beta w)}{2\beta\mu_r - b^2}$. Substituting $p(w)$ and $A_r(w)$ into the demand function, we have $D = \frac{\beta\mu_r(\phi - \beta w)}{2\beta\mu_r - b^2}$. Substituting it into $\prod_M$, the Hessian matrix of $\prod_M$ on variables w and τ is given as follows:

$$H_2 = \begin{pmatrix} \frac{\partial^2 \Pi_M}{\partial w^2} & \frac{\partial^2 \Pi_M}{\partial w \partial \tau} \\ \frac{\partial^2 \Pi_M}{\partial \tau \partial w} & \frac{\partial^2 \Pi_M}{\partial \tau^2} \end{pmatrix} = \begin{pmatrix} \frac{-2\beta^2\mu_r}{2\beta\mu_r - b^2} & \frac{\beta^2\mu_r(re_1 - \Delta)}{2\beta\mu_r - b^2} \\ \frac{\beta^2\mu_r(re_1 - \Delta)}{2\beta\mu_r - b^2} & -c_l \end{pmatrix}$$

According to the assumption, we can conclude that H_2 is negative definite. Solving $\frac{\partial \Pi_M}{\partial w} = \frac{\beta\mu_r[\phi - 2\beta w + \beta(c_m - \Delta\tau) + \beta r(e_0 + \tau e_1)]}{2\beta\mu_r - b^2} = 0$ and $\frac{\partial \Pi_M}{\partial \tau} = \frac{\beta\mu_r(\phi - \beta w)(\Delta - re_1)}{2\beta\mu_r - b^2} - \tau c_l = 0$, we can obtain w^{sym*} and τ^{sym*}. After substituting them into $p(w)$ and $A_r(w)$, we get the expression of p^{sym*} and A_r^{sym*}, and Theorem 1 is proved. □

Proof of Theorem 2. By solving the Hessian matrix of $E(\prod_M^{asy})$, it is found that when $h(\varepsilon) < \frac{2c_l}{\beta(\Delta - re_1)^2}$, $E(\prod_M^{asy})$ is a concave function of w and τ. Solving the equations

$$\frac{\partial E(\prod_M^{asy})}{\partial \tau} = h(\varepsilon)(\phi - \beta w)(\Delta - re_1) - \tau c_l = 0$$

and

$$\frac{\partial E(\prod_M^{asy})}{\partial w} = h(\varepsilon)[\phi - 2\beta w + \beta(c_m - \Delta\tau) + \beta r(e_0 + \tau e_1)] = 0,$$

We obtain w^{asy*} and τ^{asy*}. Substituting them into $p(w)$ and $A_r(w)$, analytical expressions of p^{asy*} and A_r^{asy*} are obtained. Theorem 2 is proved. □

Proof of Corollary 1. Let $g(\varepsilon) = \frac{4\beta\varepsilon(2\beta\overline{\mu_r}-b^2)}{[2\beta(\overline{\mu_r}+\varepsilon)-b^2][2\beta(\overline{\mu_r}-\varepsilon)-b^2]} - \ln\frac{2\beta(\overline{\mu_r}+\varepsilon)-b^2}{2\beta(\overline{\mu_r}-\varepsilon)-b^2}$, then $g(0) = 0$ and $g\prime(\varepsilon) = \frac{32\beta^3\varepsilon^2(2\beta\overline{\mu_r}-b^2)}{[2\beta(\overline{\mu_r}+\varepsilon)-b^2]^2[2\beta(\overline{\mu_r}-\varepsilon)-b^2]^2} > 0$ hold. Because $0 < \varepsilon < \overline{\mu_r}$, we have $g(\varepsilon) > 0$ and $h'(\varepsilon) = \frac{b^2}{8\beta\varepsilon^2}\cdot g(\varepsilon) > 0$. Furthermore, we obtain $\frac{\partial w^{asy*}}{\partial\varepsilon} = -\frac{c_l h'(\varepsilon)(\Delta-re_1)^2[\phi-\beta(c_m+re_0)]}{[2c_l-\beta h(\varepsilon)(\Delta-re_1)^2]^2} < 0$ and $\frac{\partial\tau^{asy*}}{\partial\varepsilon} = \frac{2c_l h'(\varepsilon)(\Delta-re_1)[\phi-\beta(c_m+re_0)]}{[2c_l-\beta h(\varepsilon)(\Delta-re_1)^2]^2} > 0$.

After taking the first derivatives of w^{asy*} and τ^{asy*} with respect to r and G, it is found that

$$\frac{\partial w^{asy*}}{\partial r} = \frac{c_l e_0}{2c_l - \beta h(\varepsilon)(\Delta - re_1)^2} + \frac{2e_1 c_l h(\varepsilon)(\Delta - re_1)[\phi - \beta(c_m + re_0)]}{[2c_l - \beta h(\varepsilon)(\Delta - re_1)^2]^2} > 0,$$

$$\frac{\partial\tau^{asy*}}{\partial r} = -\frac{\beta e_0 h(\varepsilon)(\Delta - re_1)}{2c_l - \beta h(\varepsilon)(\Delta - re_1)^2} - \frac{e_1 h(\varepsilon)[\phi - \beta(c_m + re_0)][2c_l + \beta h(\varepsilon)(\Delta - re_1)^2]}{[2c_l - \beta h(\varepsilon)(\Delta - re_1)^2]^2} < 0,$$

$$\frac{\partial w^{asy*}}{\partial G} = -\frac{2c_l h(\varepsilon)(\Delta - re_1)[\phi - \beta(c_m + re_0)]}{[2c_l - \beta h(\varepsilon)(\Delta - re_1)^2]^2} < 0,$$

$$\frac{\partial\tau^{asy*}}{\partial G} = \frac{h(\varepsilon)[\phi - \beta(c_m + re_0)][2c_l + \beta h(\varepsilon)(\Delta - re_1)^2]}{[2c_l - \beta h(\varepsilon)(\Delta - re_1)^2]^2} > 0.$$

Proof completed. □

Proof of Corollary 2. Based on Equations (7) and (12), we have

$$\begin{aligned}\Delta E(\Pi_M) &= \Pi_M^{sym*} - E(\Pi_M^{asy*}) \\ &= \frac{\mu_r c_l[\phi-\beta(c_m+re_0)]^2}{2[2c_l(2\beta\mu_r-b^2)-\beta^2\mu_r(\Delta-re_1)^2]} - \frac{c_l[\phi-\beta(c_m+re_0)]^2[2\mu_r c_l-h^2(\varepsilon)(\Delta-re_1)^2(2\beta\mu_r-b^2)]}{2(2\beta\mu_r-b^2)[2c_l-\beta h(\varepsilon)(\Delta-re_1)^2]^2}.\end{aligned}$$

When $\frac{\beta\mu_r}{2\beta\mu_r-b^2} < h(\varepsilon) < \frac{2c_l}{\beta(\Delta-re_1)^2}$, we have $\beta\mu_r - h(\varepsilon)(2\beta\mu_r - b^2) < 0$. Hence, $\frac{\partial E(\Pi_M^{asy*})}{\partial\varepsilon} = \frac{2c_l^2 h'(\varepsilon)(\Delta-re_1)^2[\phi-\beta(c_m+re_0)]^2[\beta\mu_r-h(\varepsilon)(2\beta\mu_r-b^2)]}{(2\beta\mu_r-b^2)[2c_l-\beta h(\varepsilon)(\Delta-re_1)^2]^3} < 0$ and $\frac{\partial\Delta E(\Pi_M)}{\partial\varepsilon} = -\frac{2c_l^2 h'(\varepsilon)(\Delta-re_1)^2[\phi-\beta(c_m+re_0)]^2[\beta\mu_r-h(\varepsilon)(2\beta\mu_r-b^2)]}{(2\beta\mu_r-b^2)[2c_l-\beta h(\varepsilon)(\Delta-re_1)^2]^3} > 0$ hold. That is, as ε increases, $E(\Pi_M^{asy*})$ decreases while $\Delta E(\Pi_M)$ increases. Similarly, it can be proved that when $\frac{1}{2} < h(\varepsilon) < \frac{\beta\mu_r}{2\beta\mu_r-b^2}$, $\frac{\partial E(\Pi_M^{asy*})}{\partial\varepsilon} > 0$ and $\frac{\partial\Delta E(\Pi_M)}{\partial\varepsilon} < 0$ hold. That is, when the value of ε increases, $E(\Pi_M^{asy*})$ increases while $\Delta E(\Pi_M)$ decreases. Proof completed. □

Proof of Corollary 3. Based on Equations (6) and (11), we have

$$\begin{aligned}\Delta E(\Pi_R) &= E(\Pi_R^{asy*}) - \Pi_R^{sym*} \\ &= \frac{\mu_r c_l^2[\phi-\beta(c_m+re_0)]^2}{2(2\beta\mu_r-b^2)[2c_l-\beta h(\varepsilon)(\Delta-re_1)^2]^2} - \frac{\mu_r c_l^2[\phi-\beta(c_m+re_0)]^2(2\beta\mu_r-b^2)}{2[2c_l(2\beta\mu_r-b^2)-\beta^2\mu_r(\Delta-re_1)^2]^2}.\end{aligned}$$

Recall that $h'(\varepsilon) > 0$ holds. So we have $\frac{\partial E(\Pi_R^{asy*})}{\partial\varepsilon} = \frac{\beta h'(\varepsilon)\mu_r c_l^2(\Delta-re_1)^2[\phi-\beta(c_m+re_0)]^2}{(2\beta\mu_r-b^2)[2c_l-\beta h(\varepsilon)(\Delta-re_1)^2]^3} > 0$ and $\frac{\partial\Delta E(\Pi_R)}{\partial\varepsilon} = \frac{\beta h'(\varepsilon)\mu_r c_l^2(\Delta-re_1)^2[\phi-\beta(c_m+re_0)]^2}{(2\beta\mu_r-b^2)[2c_l-\beta h(\varepsilon)(\Delta-re_1)^2]^3} > 0$. That is, when the value of ε increases, both $E(\Pi_R^{asy*})$ and $\Delta E(\Pi_R)$ increase. Proof completed. □

Proof of Corollary 4. When $\frac{1}{2} < h(\varepsilon) < \frac{\beta\mu_r}{2\beta\mu_r - b^2}$, we have $\beta\mu_r - h(\varepsilon)(2\beta\mu_r - b^2) > 0$ and $\frac{\partial SW^{asy*}}{\partial \varepsilon} = \frac{h'(\varepsilon)c_l^2(\Delta - re_1)^2[\phi - \beta(c_m + re_0)]^2}{(2\beta\mu_r - b^2)[2c_l - \beta h(\varepsilon)(\Delta - re_1)^2]^3}\{2[\beta\mu_r - h(\varepsilon)(2\beta\mu_r - b^2)] + \beta\mu_r\} + \frac{\beta h'(\varepsilon)\mu_r c_l(\Delta - re_1)[\phi - \beta(c_m + re_0)]^2}{(2\beta\mu_r - b^2)^2[2c_l - \beta h(\varepsilon)(\Delta - re_1)^2]^3}\left\{\beta\mu_r c_l(\Delta - re_1) + g(2\beta\mu_r - b^2)[2c_l + \beta h(\varepsilon)(\Delta - re_1)^2]\right\} > 0$. Together with Corollaries 2 and 3 and the fact that consumer surplus and total subsidies increase with ε, we conclude that the social welfare increases with ε. Proof completed. □

Proof of Corollary 5. Based on Equation (13), we obtain that

$$\frac{\partial J^{asy*}}{\partial \varepsilon} = \frac{\beta\mu_r c_l h'(\varepsilon)(\Delta - re_1)[\phi - \beta(c_m + re_0)]}{(2\beta\mu_r - b^2)[2c_l - \beta h(\varepsilon)(\Delta - re_1)^2]^2}\left\{e_0\beta(\Delta - re_1) + \frac{e_1[\phi - \beta(c_m + re_0)][2c_l + \beta h(\varepsilon)(\Delta - re_1)^2]}{2c_l - \beta h(\varepsilon)(\Delta - re_1)^2}\right\}$$

$$\frac{\partial J^{asy*}}{\partial r} = -\frac{\beta^2\mu_r c_l e_0}{(2\beta\mu_r - b^2)[2c_l - \beta h(\varepsilon)(\Delta - re_1)^2]} - \frac{4\mu_r c_l \beta^2 e_1^2 h^2(\varepsilon)(\Delta - re_1)^2[\phi - \beta(c_m + re_0)]^2}{(2\beta\mu_r - b^2)[2c_l - \beta h(\varepsilon)(\Delta - re_1)^2]^3} - \frac{\beta\mu_r c_l e_1 h(\varepsilon)[\phi - \beta(c_m + re_0)]\{e_1[\phi - \beta(c_m + re_0)] + 2\beta e_0(\Delta - re_1)\}}{(2\beta\mu_r - b^2)[2c_l - \beta h(\varepsilon)(\Delta - re_1)^2]^2}.$$

$$\frac{\partial J^{asy*}}{\partial G} = \frac{\beta\mu_r c_l e h1(\varepsilon)[\phi - \beta(c_m + re_0)]^3[2c_l + 3\beta h(\varepsilon)(\Delta - re_1)^2]}{(2\beta\mu_r - b^2)[2c_l - \beta h(\varepsilon)(\Delta - re_1)^2]^3} + \frac{2\beta^2\mu_r c_l e_0 h(\varepsilon)(\Delta - re_1)[\phi - \beta(c_m + re_0)]}{(2\beta\mu_r - b^2)[2c_l - \beta h(\varepsilon)(\Delta - re_1)^2]^2}.$$

According to the assumptions, we have $\frac{\partial J^{asy*}}{\partial \varepsilon} > 0, \frac{\partial J^{asy*}}{\partial G} > 0$ and $\frac{\partial J^{asy*}}{\partial r} < 0$. That is, when the basic market scale is large, the carbon emissions increase with ε and G but decrease with carbon tax. □

Proof of Corollary 6. According to Theorems 1 and 2, we have

$$\tau^{sym*} - \tau^{asy*} = \frac{2c_l(\Delta - re_1)[\phi - \beta(c_m + re_0)][\beta\mu_r - h(\varepsilon)(2\beta\mu_r - b^2)]}{[2c_l(2\beta\mu_r - b^2) - \beta^2\mu_r(\Delta - re_1)^2][2c_l - \beta h(\varepsilon)(\Delta - re_1)^2]},$$

$$A_r^{\text{sym}*} - A_r^{\text{asy}*} = \frac{\beta b c_l(\Delta - re_1)^2[\phi - \beta(c_m + re_0)][\beta\mu_r - h(\varepsilon)(2\beta\mu_r - b^2)]}{(2\beta\mu_r - b^2)[2c_l(2\beta\mu_r - b^2) - \beta^2\mu_r(\Delta - re_1)^2][2c_l - \beta h(\varepsilon)(\Delta - re_1)^2]},$$

$$w^{sym*} - w^{asy*} = \frac{c_l(2\beta\mu_r - b^2)[\phi + \beta(c_m + re_0)] - \beta^2\mu_r\phi(\Delta - re_1)^2}{2\beta c_l(2\beta\mu_r - b^2) - \beta^3\mu_r(\Delta - re_1)^2} - \frac{c_l[\phi + \beta(c_m + re_0)] - \beta\phi h(\varepsilon)(\Delta - re_1)^2}{\beta[2c_l - \beta h(\varepsilon)(\Delta - re_1)^2]}$$

$$p^{sym*} - p^{asy*} = \frac{\beta\mu_r - b^2}{2\beta\mu_r - b^2} \cdot \left\{\frac{c_l(2\beta\mu_r - b^2)[\phi + \beta(c_m + re_0)] - \beta^2\mu_r\phi(\Delta - re_1)^2}{2\beta c_l(2\beta\mu_r - b^2) - \beta^3\mu_r(\Delta - re_1)^2} - \frac{c_l[\phi + \beta(c_m + re_0)] - \beta\phi h(\varepsilon)(\Delta - re_1)^2}{\beta[2c_l - \beta h(\varepsilon)(\Delta - re_1)^2]}\right\}.$$

It can be seen that if $h(\varepsilon) < \frac{\beta\mu_r}{2\beta\mu_r - b^2}$, there are $\tau^{sym*} > \tau^{asy*}$ and $A_r^{\text{sym}*} > A_r^{\text{asy}*}$. If $\frac{2c_l[1 - (2\beta\mu_r - b^2)]}{\beta(\Delta - re_1)^2} + \beta\mu_r < h(\varepsilon) < \frac{c_l[\phi + \beta(c_m + re_0)][1 - (2\beta\mu_r - b^2)]}{\beta\phi(\Delta - re_1)^2} + \beta\mu_r$, $w^{sym*} < w^{asy*}$ and $p^{sym*} < p^{asy*}$ hold. Proof completed. □

References

1. Guide, V.D.R.; Van Wassenhove, L.N. The evolution of closed-loop supply chain research. *Oper. Res.* **2009**, *57*, 10–18. [CrossRef]
2. Liu, Z.; Anderson, T.D.; Cruz, J.M. Consumer environmental awareness and competition in two-stage supply chain. *Eur. J. Oper. Res.* **2012**, *218*, 602–613. [CrossRef]
3. Qu, S.; Jiang, G.; Ji, Y.; Zhang, G.; Mohamed, N. Newsvendor's optimal decisions under stochastic demand and cap-and-trade regulation. *Environ. Dev. Sustain.* **2021**, *23*, 17764–17787. [CrossRef]
4. Qu, S.; Yang, H.; Ji, Y. Low-carbon supply chain optimization considering warranty period and carbon emission reduction level under cap-and-trade regulation. *Environ. Dev. Sustain.* **2021**, *23*, 18040–18067. [CrossRef]
5. Dou, G.; Cao, K. A joint analysis of environmental and economic performances of closed-loop supply chains under carbon tax regulation. *Comput. Ind. Eng.* **2020**, *146*, 106624. [CrossRef]
6. Konstantaras, I.; Skouri, K.; Benkherouf, L. Optimizing inventory decisions for a closed-loop supply chain model under a carbon tax regulatory mechanism. *Int. J. Prod. Econ.* **2021**, *239*, 108185. [CrossRef]
7. Jauhari, W.A.; Pujawan, I.N.; Suef, M. A closed-loop supply chain inventory model with stochastic demand, hybrid production, carbon emissions, and take-back incentives. *J. Clean. Prod.* **2021**, *320*, 128835. [CrossRef]
8. Yang, Y.X.; Goodarzi, S.; Bozorgi, A.; Fahimnia, B. Carbon cap-and-trade schemes in closed-loop supply chains: Why firms do not comply? *Transp. Res. E-Log.* **2021**, *156*, 102486. [CrossRef]
9. Guo, Y.; Wang, M.M.; Yang, F. Joint emission reduction strategy considering channel inconvenience under different recycling structures. *Comput. Ind. Eng.* **2022**, *169*, 108159. [CrossRef]
10. Alegoz, M.; Kaya, O.; Bayindir, Z.P. Closing the loop in supply chains: Economic and environmental effects. *Comput. Ind. Eng.* **2020**, *142*, 106366. [CrossRef]
11. Zhang, F.; Li, N. The impact of CSR on the performance of a dual-channel closed-loop supply chain under two carbon regulatory policies. *Sustainability* **2022**, *14*, 3021. [CrossRef]
12. Li, J.; Lai, K.K.; Li, Y.M. Remanufacturing and low-carbon investment strategies in a closed-loop supply chain under multiple carbon policies. *Int. J. Logist. Res. Appl.* **2022**. online. [CrossRef]
13. Bai, Q.; Xu, J.; Chauhan, S.S. Effects of sustainability investment and risk aversion on a two-stage supply chain coordination under a carbon tax policy. *Comput. Ind. Eng.* **2020**, *142*, 106324. [CrossRef]
14. Bai, Q.; Xu, J.; Gong, Y.; Chauhan, S.S. Robust decisions for regulated sustainable manufacturing with partial demand information: Mandatory emission capacity versus emission tax. *Eur. J. Oper. Res.* **2022**, *298*, 874–893. [CrossRef]
15. Dou, G.; Choi, T.M. Does implementing trade-in and green technology together benefit the environment? *Eur. J. Oper. Res.* **2021**, *295*, 517–533. [CrossRef]
16. Shang, M.; Li, H.; Wang, Y.P.; Qin, Y.Y.; Liu, Y.; Tan, Y. Optimal decisions in a closed-loop supply chain under different policies of government intervention. *Sustain. Energy Technol. Assess.* **2021**, *47*, 101283. [CrossRef]
17. Liao, H.; Li, L. Environmental sustainability EOQ model for closed-loop supply chain under market uncertainty: A case study of printer remanufacturing. *Comput. Ind. Eng.* **2021**, *151*, 106525. [CrossRef]
18. Golpira, H.; Javanmardan, A. Robust optimization of sustainable closed-loop supply chain considering carbon emission schemes. *Sustain. Prod. Consum.* **2022**, *30*, 640–656. [CrossRef]
19. Guo, J.; Wang, G.L.; Wang, G.; Wang, Z.; Liang, C.J.; Gen, M. Research on remanufacturing closed loop supply chain based on incentive-compatibility theory under uncertainty. *Ann. Oper. Res.* 2022, *online*. [CrossRef]
20. Xu, Z.; Pokharel, S.; Elomri, A. An eco-friendly closed-loop supply chain facing demand and carbon price uncertainty. *Ann. Oper. Res.* 2022, *online*. [CrossRef]
21. Shu, T.; Peng, Z.Z.; Chen, S.; Wang, S.; Lai, K.K.; Yang, H. Government subsidy for remanufacturing or carbon tax rebate: Which is better for firms and a low-carbon economy. *Sustainability* **2017**, *9*, 156. [CrossRef]
22. Bai, Q.; Xu, J.; Zhang, Y. Emission reduction decision and coordination of a make-to-order supply chain with two products under cap-and-trade regulation. *Comput. Ind. Eng.* **2018**, *119*, 131–145. [CrossRef]
23. Qu, S.; Zhou, Y.; Zhang, Y.; Wahab, M.I.M.; Zhang, G.; Ye, Y. Optimal strategy for a green supply chain considering shipping policy and default risk. *Comput. Ind. Eng.* **2019**, *131*, 172–186. [CrossRef]
24. Bazan, E.; Jaber, M.Y.; El Saadany, A.M.A. Carbon emissions and energy effects on manufacturing-remanufacturing inventory models. *Comput. Ind. Eng.* **2015**, *88*, 307–316. [CrossRef]
25. Wang, Y.; Chen, W.; Liu, B. Manufacturing/remanufacturing decisions for a capital-constrained manufacturer considering carbon emission cap and trade. *J. Clean. Prod.* **2017**, *140*, 1118–1128. [CrossRef]
26. Chen, Y.; Li, B.; Zhang, G.; Bai, Q. Quantity and collection decisions of the remanufacturing enterprise under both the take-back and carbon emission capacity regulations. *Transp. Res. E-Log.* **2020**, *141*, 102032. [CrossRef]
27. Bai, Q.; Xu, J.; Zhang, Y. The distributionallly robust optimization model for a remanufacturing system under cap-and-trade policy: A newsvendor approach. *Ann. Oper. Res.* **2022**, *309*, 731–760. [CrossRef]
28. He, R.; Xiong, Y.; Lin, Z. Carbon emissions in a dual channel closed loop supply chain: The impact of consumer free riding behavior. *J. Clean. Prod.* **2016**, *134*, 384–394. [CrossRef]
29. Bazan, E.; Jaber, M.Y.; Zanoni, S. Carbon emissions and energy effects on a two-level manufacturer-retailer closed-loop supply chain model with remanufacturing subject to different coordination mechanisms. *Int. J. Prod. Econ.* **2017**, *183*, 394–408. [CrossRef]

30. Yang, L.; Hu, Y.; Huang, L. Collecting model selection in a remanufacturing supply chain under cap-and-trade regulation. *Eur. J. Oper. Res.* **2020**, *287*, 480–496. [CrossRef]
31. Jauhari, W.A.; Adam, N.A.F.P.; Rosyidi, C.N.; Pujawan, I.N.; Shah, N.H. A closed-loop supply chain model with rework, waste disposal, and carbon emissions. *Oper. Res. Perspect.* **2020**, *7*, 100155. [CrossRef]
32. Shekarian, E.; Marandi, A.; Majava, J. Dual-channel remanufacturing closed-loop supply chains under carbon footprint and collection competition. *Sustain. Prod. Consum.* **2021**, *28*, 1050–1075. [CrossRef]
33. Wang, Z.; Wu, Q. Carbon emission reduction and product collection decisions in the closed-loop supply chain with cap-and-trade regulation. *Int. J. Prod. Res.* **2021**, *59*, 4359–4383. [CrossRef]
34. Kim, S.H.; Netessine, S. Collaborative cost reduction and component procurement under information asymmetry. *Manag. Sci.* **2013**, *59*, 189–206. [CrossRef]
35. Ma, P.; Shang, J.; Wang, H.Y. Enhancing corporate social responsibility: Contract design under information asymmetry. *Omega-Int. J. Manag. S* **2017**, *67*, 19–30. [CrossRef]
36. Guan, X.; Mantrala, M.; Bian, Y.W. Strategic information management in a distribution channel. *J. Retail.* **2019**, *95*, 42–56. [CrossRef]
37. Cao, H.; Guan, X.; Fan, T.J.; Zhou, L. The acquisition of quality information in a supply chain with voluntary vs. mandatory disclosure. *Prod. Oper. Manag.* **2020**, *29*, 595–616. [CrossRef]
38. Guan, X.; Wang, Y.; Yi, Z.; Chen, Y.J. Inducing consumer online reviews via disclosure. *Prod. Oper. Manag.* **2020**, *29*, 1956–1971. [CrossRef]
39. Zhang, P.; Xiong, Y.; Xiong, Z.K.; Yan, W. Designing contracts for a closed-loop supply chain under information asymmetry. *Oper. Res. Lett.* **2014**, *42*, 150–155. [CrossRef]
40. Wei, J.; Govindan, K.; Li, Y.J.; Zhao, J. Pricing and collecting decisions in a closed-loop supply chain with symmetric and asymmetric information. *Comput. Oper. Res.* **2015**, *54*, 257–265. [CrossRef]
41. De Giovanni, P. Closed-loop supply chain coordination through incentives with asymmetric information. *Ann. Oper. Res.* **2017**, *253*, 133–167. [CrossRef]
42. Gao, J.; Liang, Z.; Shang, J.; Xu, Z.S. Remanufacturing with patented technique royalty under asymmetric information and uncertain markets. *Technol. Econ. Dev. Econ.* **2020**, *26*, 599–620. [CrossRef]
43. Wang, W.B.; Zhou, S.Y.; Zhang, M.; Sun, H.; He, L. A closed-loop supply chain with competitive dual collection channel under asymmetric information and reward–penalty mechanism. *Sustainability* **2018**, *10*, 2131. [CrossRef]
44. Chen, Z.Y.; Huang, L.Z. Digital twins for information-sharing in remanufacturing supply chain: A review. *Energy* **2021**, *220*, 119712. [CrossRef]
45. Wu, Q.; Xu, X.; Lin, R. Government incentive mechanism of closed-loop supply chain based on information asymmetry. *Rairo-Oper. Res.* **2021**, *55*, 3359–3378. [CrossRef]
46. Wang, Q.; Chen, K.B.; Wang, S.B.; Cao, X.G. Optimal decisions in a closed-loop supply chain: Fairness concerns, corporate social responsibility and information value. *Ann. Oper. Res.* **2022**, *309*, 277–304. [CrossRef]
47. Du, S.; Liu, H.; Song, M. Production optimization considering environmental performance and preference in the cap-and-trade system. *J. Clean Prod.* **2016**, *112*, 1600–1607. [CrossRef]
48. Xu, J.; Qi, Q.; Bai, Q. Coordinating a dual-channel supply chain with price discount contracts under carbon emission capacity regulation. *Appl. Math. Model.* **2018**, *56*, 449–468. [CrossRef]

sustainability

Article

Analysis of $M/M/1/N$ Stochastic Queueing—Inventory System with Discretionary Priority Service and Retrial Facility

K. Jeganathan [1], S. Vidhya [2], R. Hemavathy [2], N. Anbazhagan [3], Gyanendra Prasad Joshi [4,*], Chanku Kang [5] and Changho Seo [5,*]

1 Ramanujan Institute for Advanced Study in Mathematics University of Madras, Chepauk, Chennai 600005, India; kjeganathan@unom.ac.in
2 Department of Mathematics, Queen Mary's College, Chennai 600005, India; seetharamanvidhya@gmail.com (S.V.); hemaths@gmail.com (R.H.)
3 Department of Mathematics, Alagappa University, Karaikudi 630003, India; anbazhagann@alagappauniversity.ac.in
4 Department of Computer Science and Engineering, Sejong University, Seoul 05006, Korea
5 Department of Convergence Science, Kongju National University, Gongju 32588, Korea; giallar@naver.com
* Correspondence: joshi@sejong.ac.kr (G.P.J.); chseo@kongju.ac.kr (C.S.)

Citation: Jeganathan, K.; Vidhya, S.; Hemavathy, R.; Anbazhagan, N.; Joshi, G.P.; Kang, C.; Seo, C. Analysis of $M/M/1/N$ Stochastic Queueing—Inventory System with Discretionary Priority Service and Retrial Facility. *Sustainability* **2022**, *14*, 6370. https://doi.org/10.3390/su14106370

Academic Editors: Shaojian Qu, Qingguo Bai, Ying Ji and Congjun Rao

Received: 19 April 2022
Accepted: 13 May 2022
Published: 23 May 2022

Abstract: In this paper, we analyze a queueing–inventory system with two classes of customers, high priority (HP) and low priority (LP), under the discretionary priority discipline. The LP customers are served in two stages: preliminary service in stage-I and main service in stage-II. In contrast, HP customers require only the main service. Whenever the inventory level is less than the threshold level during the stage-I service of an LP customer, an arriving HP customer is allowed to interrupt the service of an LP customer by adopting the mixed-priority discipline. Otherwise, non-preemptive priority discipline is used in both stages. The interrupted LP customer moves to orbit and retries for the service whenever the server is free. The waiting hall of finite capacity is afforded for the HP customer only. The orbital search is provided for LP customers in orbit. The inventory is replenished following the (s, Q) ordering policy, with the lifetimes of the items being exponentially distributed. An expression for the stability condition is determined explicitly, and system performance measures are evaluated. Numerical examples are formulated for different sets of input values of the parameters.

Keywords: discretionary priority; mixed priority; preemptive priority; non-preemptive priority; infinite orbit

MSC: 60K25

1. Introduction

The model developed in this paper was motivated by the author's personal experience at a fertilizer company located in Thiruvannamalai, a town in the state of Tamilnadu. The customers of the company were farmers who came to buy fertilizers for their agricultural land. To do so, the company made it mandatory to register their cultivation land details beforehand, to buy the required amounts of fertilizer using cash. The farmers who had registered earlier were considered registered farmers, and those who were new to this process were considered unregistered farmers. Therefore, the registered farmer directly went to buy the product, whereas an unregistered farmer had to go through the registration process first and then buy the product. Consider the registration of land with the survey number to be a stage-I service, and buying the product to be a stage-II service. When an unregistered farmer is in stage-I, and in the meantime, a registered farmer arrives, the seller pauses the service for the unregistered farmer to assist the registered farmer. The farmer whose service has been paused goes around the town to make other purchases or waits nearby until the registered farmer's order is completed. This situation motivated the author to focus on such stochastic modelling.

The service rule applied in the above real-life situation illustrates the discretionary priority service process. It describes an intermediate priority between preemptive and non-preemptive priorities. This type of service system is widely used in telecommunication systems, hospitals, industrial administration, railway network services, flight landings in an airport, signal processing, military fields, etc. When observing a railway network, the administration gives a green signal to super-fast trains while pausing local trains in cases of insufficient track facilities at a particular junction. Once a local train has crossed the junction, the super-fast train cannot be allowed to interrupt the service of the local train. Therefore, providing preemptive priority in one stage and non-preemptive priority in the next stage of the service process is called discretionary service discipline.

Literature Review

In a queueing–inventory model, every customer is provided with an item from the inventory at the end of the service. Melikov and Molchanov [1] discussed the optimal policies for reordering stocks in an inventory system with queueing of order requests. Sigman and Simichi [2] introduced a service facility in an $M/G/1$ queueing–inventory model with limited inventory under light traffic. Berman and Kim [3] formulated two cases of a queueing–inventory model with a service facility. The queueing capacity of one was infinite, and the other was finite. Further, Berman and Sapna [4] developed the aforementioned models with arbitrarily distributed service time. Recently, Sangeetha and Sivakumar [5] discussed a perishable inventory system with Markovian arrivals and an exponentially distributed service rate. Jeganathan et al. [6] modeled an $M/M/1/N$ queueing–inventory system with a retrial facility. The reader can refer to the recent papers on retrial systems in [7–9]. For more details, the references [10–22] are essential for understanding queueing–inventory systems with service facilities.

With a single server, customers of multiple classes can be served following various service disciplines in many real-life situations. For instance, there have been many COVID-19 vaccination clinics opened in the country recently; in all such clinics, two types of individuals arrive and are categorized: individuals who booked their slots beforehand and those who come without bookings. Both individuals are administered by a single server. The server makes sure that all individuals who booked ahead are served completely before moving serving individuals without bookings. Additionally, suppose a "booked" individual arrives during the provision of the service to an "unbooked" one. In that case, the former has to wait until the service's completion, and their vaccination will start immediately after the latter's departure. This way of providing services is known as non-preemptive priority discipline, a service carried out without any interruption.

Falin et al. [23] studied in-depth a more general model with two classes of customers, using different service distributions for both types of customers, to examine the stochastic decomposition and asymptotic behavior of the stationary characteristics. Jeganathan et al. [24] developed a queueing–inventory model with two types of customers based on their priority levels and rendered services through non-preemptive priority discipline. Chakravarthy [25] studied a non-preemptive priority queue with two classes of customers and introduced a new dynamic rule to provide services to lower-priority customers in the presence of higher-priority customers. Korenevskaya et al. [26] considered a retrial queueing system with Poisson arrival following non-preemptive priority. Recently, Krishnamoorthy and Divya [27] analyzed a queue with non-preemptive priority under a marked Markovian arrival process with a distinct phase-type distribution of services.

Non-preemptive priority discipline cannot be applied in all priority queues, because some situations need preemptive priority discipline. For instance, in a retail shop, two types of customers approach: those of one type have used buy online and pick up in-store (BOPIS), and those of the other come for in-store purchases. The services to in-store purchasers are interrupted on the arrival of BOPIS customers, as the service times required for BOPIS customers are low compared with those for shop-in-store customers. A service that is carried out with interruption of a high-priority (HP) customer in favor of a low-priority (LP) customer is known as preemptive priority discipline. Krishna Kumar et al. [28] discussed the $M/G/1$ queue under preemptive resume priority with a service in two phases. Tarabia [29] discussed a queueing system with two priority classes under preemptive discipline, where an interrupted service of a low priority is resumed when there is no high priority need in the system. Jeganathan et al. [30] discussed a single-server inventory system with an interruption and a finite waiting room. Gao and Wang [31] analyzed a single-server retrial queue with Poisson arrival. The customers are served through preemptive priority with orbital search.

The preemptive and non-preemptive service disciplines are both labored in some situations and are known as mixed-priority discipline. Like in situations as explained for preemptive priority, a BOPIS customer may opt to wait until the service completion of an ongoing in-store purchase. Thus, they adopt a non-preemptive priority. However, not all BOPIS customers would opt to wait, and some may interrupt and avail service following preemptive priority. Hence, a service discipline that focuses on both preemptive and non-preemptive priority customer under the Bernoulli's schedule is known as mixed-priority discipline.

Adiri and Domb [32] used this service discipline in their model that serves m priority classes of customers. If the positive value d is fixed, and if the variation between the class indices is outreach d, then preemptive priority discipline is allowed. Otherwise, non-preemptive priority is followed. Cho and Un [33] analyzed the $M/G/1$ queue combining preemptive and non-preemptive discipline and considered three schemes: the elapsed service time, the ratio of elapsed to total service time, and the remaining service time as the discretion for preemption, where each is based on the parameters of the low-priority job. Fajardo and Drekic [34] categorized two distinct types of priority classes: (i) urgent class and (ii) non-urgent class among N priority classes. Within each category, non-preemptive priority is adopted, and between the categories, preemptive priority is followed.

A priority mechanism is a method of scheduling that allows customers of different classes to receive different levels of service from a single server. It is used in communication systems, healthcare systems, and inventory systems. The priority discipline in queueing systems can be preemptive or non-preemptive. However, both disciplines have flaws. Under preemptive discipline, a low-priority (LP) job whose service is almost complete may be preempted. Even if a low-priority (LP) job with a long service time has just entered service, a high-priority (HP) job may wait under non-preemptive discipline. Allowing the server to decide whether or not to continue or stop the LP customer's service can help avoid these unpleasant situations. In reality, especially in the areas of telecommunication, communication networks, and the chemical industry, among the entire service procedure, some service stages can be interrupted while some service stages cannot be interrupted.

From all these priorities, a new priority rule known as discretionary priority was formulated and was first studied by Avi-itzhak et al. [35]. This priority discipline is widely utilized in telecommunication and communication networks. Melkonian and Kaiser [36] examined discretionary priority with general service time distribution. Lian and Zhao [37] modeled an $M/G/1$ queue with service in two stages, following discretionary priority with preemptive priority in stage-I and non-preemptive priority in stage-II. Zhao and Lian [38] formulated a two-stage $M/G/1$ queue with discretionary priority and constructed an embedded Markov model for specific time points on the time axis.

Lian and Zhao [39] investigated the two-stage service process for two classes of customers. They are called LP and HP customers. They assumed that each class of customer would have a two-stage service. However, the crucial assumption of their model is that the HP customer can be allowed to interrupt the service of an LP customer while they are being serviced in stage-I only. In stage-II, interruption is not allowed. Ning Zhao and Yaya Guo [40] investigated the same two-stage service for two classes of customers whose arrival occurs according to the Markovian arrival process and whose service process follows the phase-type distribution. Jeganathan et al. [11] analyzed the threshold-based priority services in the queueing inventory system.

To the best of our knowledge, so far, the models following discretionary priority discipline are discussed only in queueing systems. There is no queueing system attached to an inventory, which is considered under discretionary priority discipline, which creates a research gap in the queueing–inventory system. Of course, the discretionary priority discipline applied in [37–40] mostly assumed that the interruption of HP customers would happen only at stage-I service but not at stage-II without any other restriction. Even though the service discipline involving discretionary priority rules is not applied in the QIS, the concept of providing discretionary priority service methodology based on threshold level is not yet discussed even in queueing theory. As per the literature review analysis on the QIS, the discretionary priority service discipline is not applied. This is to be considered a research gap in the QIS. This model is specifically designed to fill such a gap. However, the novelty of this proposed model is that it introduces the discretionary priority service methodology with the threshold-based inventory level. Additionally, the research work is an attempt to weigh between the priority disciplines and find the optimal priority discipline that helps improve a business setup. Together with discretionary priority, we combined the concepts of orbital search, retrial facility, and interruption, based on Bernoulli's schedule.

The paper is organized as follows: In Section 2, the description of the model and notations are defined. Section 3 deals with analysis, the steady-state solution of the model, the system performance measures, and cost analysis. In Section 4, some numerical examples pertaining to the real-time usage of the model are explained. Finally, Section 5 holds the conclusion of the paper.

2. Model Description

2.1. *Definitions*

2.1.1. Non-Preemptive Priority (Head-of-The-Line Priority Discipline)

During the service of an LP customer, the interruption of an HP customer is strictly not allowed, and this service discipline is called non-preemptive priority.

2.1.2. Preemptive Priority Discipline

When the server is busy with an LP customer, the interruption of an HP customer is always allowed, and this service discipline is called preemptive priority.

2.1.3. Mixed-Priority Discipline

In mixed-priority service, both preemptive and non-preemptive priority discipline can be adopted. If an HP customer arrives while an LP customer is being served, the arriving HP customer may or may not interrupt the LP customer's service based on the Bernoulli's schedule.

2.1.4. Discretionary Priority Discipline

Discretionary priority service discipline is defined as in which the server allows preemption of HP customer when the server is busy with LP customer at stage-I and preemption of HP customer is not allowed if the server is busy with LP customer in stage-II (see [38]).

2.2. *Notations*

The following standard notations are used in this paper.

A_{ij}	: The matrix A as a element at (i,j)th position
0	: Zero matrix
I	: Identitiy matrix
δ_{ij}	: $\begin{cases} 1 & i = j \\ 0 & i \neq j \end{cases}$
$H(y)$	: $\begin{cases} 1 & y \geq 0 \\ 0 & y < 0 \end{cases}$
e	: Column vector with 1 in each entry
W	:The Set of all Whole Numbers
V_1	:$\{1, 2, \ldots, S\}$
V_2	:$\{0, 1, 2, \ldots, N\}$

2.3. Explanation of the Model

This model explores the discretionary priority service discipline in a single-server, perishable queueing–inventory system (SSPQIS) with two classes of customers. The inventory system is made up of a single commodity with the size S, a finite waiting hall with the size N, and an infinite orbit. This system provides the service facility for two types of customers, say HP and LP, served in two distinct stages. Every customer of the system has to go through two stages of service, namely, preliminary service and main service. In preliminary service, a customer undergoes registration with the server, whereas the customer purchases the product in the main service. Every customer in the system goes through preliminary service only once. Once the registration process has been completed successfully, the customer can directly advance to stage-II service in the current and upcoming purchases. Depending on this service scheme, the customer who has already completed the preliminary service and approaches the system only to avail of the main service is initialized as an HP customer, and the customer who is new to the system and accordingly has to go through the preliminary service before moving to obtain the main service is termed as an LP customer.

The service discipline utilized in the system depends upon the threshold level. The threshold, L, is a predetermined level on a certain amount of inventory. Suppose the inventory level is greater than this threshold level. In that case, the server follows non-preemptive priority discipline in stage-I service. That is, if an LP customer is being served with preliminary service, then the arriving HP customer cannot be allowed to interrupt the service. On the other hand, if the quantity of inventory is less than or equal to the threshold, then discretionary priority service discipline is adopted, allowing the interruption of HP customers during the service of LP customers in stage-I under Bernoulli's schedule. The interrupted LP customer moves into orbit. During the stage-II service, the server always follows non-preemptive priority discipline.

While the server follows mixed-priority service discipline, the arriving HP customer may interrupt the service or decide to wait until the service of the LP customer is completed in stage-I. When inventory is less than or equal to L, the arriving HP customer may either interrupt the service of an LP customer with probability r or move to the waiting hall with probability $(1 - r)$. If an HP customer decides to wait, then the upcoming HP customers also have to wait in the queue. However, if the HP customer interrupts, then the LP customer moves to orbit with probability, p, and tries for service when the system is void of any HP customers. If $r = 0$, then there is no HP customer interrupts an LP customer. Thus, the server always follows non-discretionary priority discipline. If $r \in (0, 1]$, then the discussion completely comes under the discretionary priority discipline. When $r = 1$, the preemption of HP customers towards the LP customers' service will certainly occur. Thus, the server adopts a preemptive priority discipline. Additionally, if $r \in (0, 1)$, then the priority discipline is called mixed-priority discipline.

The parameters λ_H and λ_L denote the arrival rate of HP and LP customers, and both independently follow the Poisson process. The waiting hall is used only for HP customers. An arriving HP customer moves to the waiting hall in any one of the following circumstances:

- The HP customer is at service;
- The LP customer is receiving stage-II service;
- The LP customer is at stage-I service, when the inventory is greater than L;
- The HP customer decides to wait with probability $(1-r)$;
- With zero inventory level.

Similarly, an LP customer will move to an orbit with probability p in any one of the following circumstances:

- When the inventory is zero;
- On interruption from an HP customer at stage-I;
- If the server is busy with HP or LP customers.

Otherwise, they are considered lost with probability $(1-p)$. The retrial customer approaches the system under the classical retrial policy, and the subsequent retrial process follows an exponential distribution. The service rate of HP customer in stage-II is μ_H, and LP customer in stage-I is μ_{L_1} and stage-II is μ_{L_1}. All the service rates follow the exponential distribution independently. Here, we assume $\mu_{L_1} < \mu_{L_2}$. Inventory decreases one by one after each service is completed. Whenever the server is free with positive inventory, they search for an LP customer from the orbit with probability q, or does not search with probability $(1-q)$. The items in the inventory are subject to perishables except for the items in the service process. The perishable rate is defined by γ. If the inventory reaches s, $Q(=S-s)$ items are ordered using the (s,Q) ordering policy. The parameter β indicates the replenishment rate of the reorder process. The lifetime and replenishment time follow the exponential distribution.

3. Analysis

From the suppositions made in the system, we see that the stochastic process $\{A(t), t \geq 0\} = \{(A_1(t), A_2(t), A_3(t), A_4(t)), t \geq 0\}$ is a continuous time Markov chain with state space E, where

$$
\begin{aligned}
A_1(t) &: \text{Number of Customers in the Orbit at time } t\\
A_2(t) &: \text{Inventory level at time } t\\
A_3(t) &: \text{The Status of Server at time } t\\
A_4(t) &: \text{Number of HP in Waiting Hall at time } t\\
A_3(t) &: \begin{cases} 0 & \text{the server is free.}\\ 1 & \text{the server is busy with LP customer in stage-I.}\\ 2 & \text{the server is busy with LP customer in stage-II.}\\ 3 & \text{the server is busy with HP customer in stage-II.} \end{cases}\\
E_1 &: \{(i, j=0, k=0, l) : i \in W, l \in V_2\}\\
E_2 &: \{(i, j, k=0, l=0) : i \in W, j \in V_1\}\\
E_3 &: \{(i, j, k, l) : i \in W, j \in V_1, k \in \{1,2,3\}, l \in V_2\}\\
E &: E_1 \cup E_2 \cup E_3.
\end{aligned}
$$

The transition of infinitesimal generator matrix is of the following structure:

$$
K = \begin{array}{c} \\ (0)\\ (1)\\ (2)\\ \vdots\\ (M-1)\\ (M)\\ (M+1)\\ \vdots \end{array}
\begin{array}{c}
\begin{array}{cccccccc} (0) & (1) & (2) & \cdots & (M-1) & M & (M+1) & \cdots \end{array}\\
\begin{pmatrix}
\mathbb{K}_{00} & \mathbb{K}_{01} & \mathbf{0} & \cdots & \mathbf{0} & \mathbf{0} & \mathbf{0} & \cdots\\
\mathbb{K}_{10} & \mathbb{K}_{11} & \mathbb{K}_{01} & \cdots & \mathbf{0} & \mathbf{0} & \mathbf{0} & \cdots\\
\mathbf{0} & \mathbb{K}_{21} & \mathbb{K}_{22} & \cdots & \mathbf{0} & \mathbf{0} & \mathbf{0} & \cdots\\
\vdots & \vdots & \vdots & \vdots & \vdots & \ddots & \vdots & \ddots\\
\mathbf{0} & \mathbf{0} & \mathbf{0} & \cdots & \mathbb{K}_{M-1,M-1} & \mathbb{K}_{01} & \mathbf{0} & \cdots\\
\mathbf{0} & \mathbf{0} & \mathbf{0} & \cdots & \mathbb{K}_{M,M-1} & \mathbb{K}_{M,M} & \mathbb{K}_{01} & \cdots\\
\mathbf{0} & \mathbf{0} & \mathbf{0} & \cdots & \mathbf{0} & \mathbb{K}_{M+1,M} & \mathbb{K}_{M+1,M+1} & \cdots\\
\vdots & \vdots & \vdots & \ddots & \vdots & \vdots & \vdots & \ddots
\end{pmatrix}
\end{array}
\quad (1)
$$

Here, we assume that i_1, i_2 represent the number of customer in the orbit, j_1, j_2 represent the present stock level of the system, k_1, k_2 denote the server status and l_1, l_2 indicate the number of HP customer in the waiting hall to explore the transition of states of the system, where $K_{i_1i_2}, \quad i_2 = i_1 + 1, \quad i_1 = 0,1,2,\ldots$

$$\begin{aligned}
K_{i_1i_2} &= \begin{cases} B_0, & j_2 = j_1, \quad j_1 = 0 \\ B_1, & j_2 = j_1, \quad j_1 = 1,2,\ldots,L \\ B_2, & j_2 = j_1, \quad j_1 = L+1, L+2,\ldots,S \\ \mathbf{0} & otherwise. \end{cases} \\
B_0 &= \begin{cases} p\lambda_L, & k_2 = k_1, \quad k_1 = 0 \\ & l_2 = l_1, \quad l_1 = 0,1,\ldots,N \\ 0, & otherwise. \end{cases} \\
B_1 &= \begin{cases} p\lambda_L, & k_2 = k_1, \quad k_1 = 1,2,3. \\ & l_2 = l_1, \quad l_1 = 0,1,\ldots,N \\ r\lambda_H, & k_2 = 3, \quad k_1 = 1 \\ & l_2 = l_1, \quad l_1 = 0 \\ 0, & otherwise. \end{cases} \\
B_2 &= \begin{cases} p\lambda_L, & k_2 = k_1, \quad k_1 = 1,2,3. \\ & l_2 = l_1, \quad l_1 = 0,1,\ldots,N \\ 0, & otherwise. \end{cases}
\end{aligned} \tag{2}$$

$K_{i_1i_2}, \quad i_2 = i_1 - 1, \quad i_1 = 1,2,\ldots$

$$K_{i_1i_2} = \begin{cases} G_1 & j_2 = j_1, \quad j_1 = 1,2,\ldots,S \\ H & j_2 = j_1 - 1, \quad j_1 = 2,3,\ldots,s \\ \mathbf{0}, & otherwise. \end{cases}$$

$$G_1 = \begin{cases} i_1\lambda_r, & k_2 = 1, \quad k_1 = 0 \\ & l_2 = 0, \quad l_1 = 0 \\ 0, & otherwise. \end{cases} \tag{3}$$

$$H = \begin{cases} q\mu_{L_2}, & k_2 = 1, \quad k_1 = 2 \\ & l_2 = 0, \quad l_1 = 0 \\ q\mu_H, & k_2 = 1, \quad k_1 = 3 \\ & l_2 = 0, \quad l_1 = 0 \\ 0, & otherwise. \end{cases}$$

$K_{i_1i_2}, \quad i_2 = i_1$ and $i_1 = 0,1,2,\ldots$

$$K_{00} = \begin{cases} F_{j_1} & j_2 = j_1, \quad j_1 = 0,1,\ldots,S \\ C_{j_1} & j_2 = j_1 - 1, \quad j_1 = 1,2,\ldots,S \\ D_0 & j_2 = Q, \quad j_1 = 0 \\ D_1 & j_2 = Q + j_1, \quad j_1 = 1,\ldots,s \\ \mathbf{0}, & otherwise. \end{cases} \tag{4}$$

and

$$K_{i_1 i_2} = \begin{cases} F_{j_1} & j_2 = j_1, \quad j_1 = 0,1,\ldots,S \\ C_{j_1} & j_2 = j_1 - 1, \quad j_1 = 1,2,\ldots,S \\ D_0 & j_2 = Q, \quad j_1 = 0 \\ D_1 & j_2 = Q + j_1, \quad j_1 = 1,\ldots,s \\ \mathbf{0}, & otherwise. \end{cases} \tag{5}$$

where

$$F_0 = \begin{cases} \lambda_H + \beta & k_2 = k_1, \quad k_1 = 0 \\ & l_2 = l_1 + 1, \quad l_1 = 0,1,\ldots,N \\ -(\lambda_H + \beta + p\lambda_L) & k_2 = k_1, \quad k_1 = 0 \\ & l_2 = l_1, \quad l_1 = 0,1,\ldots,N-1 \\ -(\beta + p\lambda_L) & k_2 = k_1, \quad k_1 = 0 \\ & l_2 = l_1, \quad l_1 = N \\ 0, & otherwise. \end{cases}$$

$$F_{j_1} = \begin{cases} \lambda_H & k_2 = 3, \quad k_1 = 1 \\ & l_2 = l_1 = 0 \\ \mu_{L_1} & k_2 = k_1 + 1, \quad k_1 = 1 \\ & l_2 = l_1, \quad l_1 = 0,1,\ldots,N \\ \lambda_H & k_2 = k_1, \quad k_1 = 2,3 \\ & l_2 = l_1 + 1, \quad l_1 = 0,1,\ldots,N \\ (1-r)\lambda_H & k_2 = k_1, \quad k_1 = 1 \\ & l_2 = l_1 + 1, \quad l_1 = 0 \\ \lambda_H & k_2 = k_1, \quad k_1 = 1 \\ & l_2 = l_1 + 1, \quad l_1 = 1,2,\ldots,N-1 \\ -((j_1 - 1)\gamma + \lambda_H + H(s - j_1)\beta + \bar{\delta}_{0i_1} i_1 \lambda_r) & k_2 = k_1, \quad k_1 = 0 \\ & l_2 = l_1, \quad l_1 = 0 \\ -((j_1 - 1)\gamma + \lambda_H + \mu_{L_1} + p\lambda_L + H(s - j_1)\beta) & k_2 = k_1, \quad k_1 = 1 \\ & l_2 = l_1, \quad l_1 = 0,1,\ldots,N-1 \\ -((j_1 - 1)\gamma + \mu_{L_1} + p\lambda_L + H(s - j_1)\beta) & k_2 = k_1, \quad k_1 = 1 \\ & l_2 = l_1, \quad l_1 = N \\ -((j_1 - 1)\gamma + \mu_{L_2} + \lambda_H + p\lambda_L + H(s - j_1)\beta) & k_2 = k_1, \quad k_1 = 2 \\ & l_2 = l_1, \quad l_1 = 0,1,\ldots,N-1 \\ -((j_1 - 1)\gamma + \mu_{L_2} + p\lambda_L + H(s - j_1)\beta) & k_2 = k_1, \quad k_1 = 2 \\ & l_2 = l_1, \quad l_1 = N \\ -((j_1 - 1)\gamma + \mu_H + \lambda_H + p\lambda_L + H(s - j_1)\beta) & k_2 = k_1, \quad k_1 = 3 \\ & l_2 = l_1, \quad l_1 = 0,1,\ldots,N-1 \\ -((j_1 - 1)\gamma + \mu_H + p\lambda_L + H(s - j_1)\beta) & k_2 = k_1, \quad k_1 = 3 \\ & l_2 = l_1, \quad l_1 = N \\ 0, & otherwise. \end{cases}$$

$$C_1 = \begin{cases} \mu_{L_2}, & k_2 = 0, \quad k_1 = 2 \\ & l_2 = l_1, \quad l_1 = 0,1,\ldots,N \\ \mu_H, & k_2 = 0, \quad k_1 = 3 \\ & l_2 = l_1, \quad l_1 = 0,1,\ldots,N \\ 0, & otherwise. \end{cases}$$

$$C_{j_1} = \begin{cases} (j-1)\gamma, & k_2 = k_1 \quad k_1 = 0,1,2,3 \\ & l_2 = l_1, \quad l_1 = 0,1,\ldots,N \\ \mu_{L_2}, & k_2 = 0 \quad k_1 = 2 \\ & l_2 = 0 \quad l_1 = 0 \\ \mu_{L_2}, & k_2 = 3, \quad k_1 = 2 \\ & l_2 = l_1 - 1, \quad l_1 = 1,2,\ldots,N \\ \mu_H, & k_2 = 0, \quad k_1 = 3 \\ & l_2 = 0, \quad l_1 = 0 \\ \mu_H, & k_2 = 3 \quad k_1 = 3 \\ & l_2 = l_1 - 1 \quad l_1 = 1,2,\ldots,N \\ 0, & otherwise. \end{cases}$$

$$D_0 = \begin{cases} \beta & k_2 = 0, \quad k_1 = 0 \\ & l_2 = l_1, \quad l_1 = 0,1,\ldots,N \\ 0, & otherwise. \end{cases}$$

$$D_1 = \begin{cases} \beta, & k_2 = k_1, \quad k_1 = 0,1,2,3 \\ & l_2 = l_1, \quad l_1 = 0,1,2,\ldots,N \\ 0, & otherwise. \end{cases}$$

$$C_1 = \begin{cases} (1-q)\mu_{L_2}, & k_2 = 0, \quad k_1 = 2 \\ & l_2 = l_1, \quad l_1 = 0,1,\ldots,N \\ (1-q)\mu_H, & k_2 = 0, \quad k_1 = 3 \\ & l_2 = l_1, \quad l_1 = 0,1,\ldots,N \\ 0, & otherwise. \end{cases}$$

$$C_{j_1} = \begin{cases} (j-1)\gamma, & k_2 = k_1 \quad k_1 = 0,1,2,3 \\ & l_2 = l_1, \quad l_1 = 0,1,\ldots,N \\ (1-q)\mu_{L_2}, & k_2 = 0 \quad k_1 = 2 \\ & l_2 = 0 \quad l_1 = 0 \\ (1-q)\mu_{L_2}, & k_2 = 3, \quad k_1 = 2 \\ & l_2 = l_1 - 1, \quad l_1 = 1,2,\ldots,N \\ (1-q)\mu_H, & k_2 = 0, \quad k_1 = 3 \\ & l_2 = 0, \quad l_1 = 0 \\ (1-q)\mu_H, & k_2 = 3 \quad k_1 = 3 \\ & l_2 = l_1 - 1 \quad l_1 = 1,2,\ldots,N \\ 0, & otherwise. \end{cases}$$

3.1. Steady-State Analysis

According to the Neuts [41] matrix geometric method, the level-dependent quasi-birth-and-death process is truncated at the point $M \geq 1$, where M denotes the number of customers in orbit. After this truncation point, the LDQBD process changed into a level-independent QBD process. That is, the retrial customers follow the constant retrial policy to approach the system. To proceed further, we need to find the stability condition of the system at this truncation point. First, we require the stationary probability vector to the rate matrix K_M is formed by $K_M = K_{M(M-1)} + K_{MM} + K_{01}$, which is

$$
K_M = \begin{array}{c} \\ (0) \\ (1) \\ (2) \\ \vdots \\ (s-1) \\ (s) \\ \vdots \\ (Q) \\ (Q+1) \\ \vdots \\ (S-1) \\ (S) \end{array}
\begin{array}{c}
\begin{array}{cccccccccccc} (0) & (1) & (2) & \cdots & (s-1) & (s) & \cdots & (Q) & (Q+1) & \cdots & (S-1) & (S) \end{array} \\
\left(\begin{array}{cccccccccccc}
F_0 & & & & & & & D_0 & & & & \\
C_1 & F_1 & & & & & & & D_1 & & & \\
& C_2 & F_2 & & & & & & & & & \\
& & \ddots & \ddots & & & & & & \ddots & & \\
& & & & F_{s-1} & & & & & & D_1 & \\
& & & & C_s & F_s & & & & & & D_1 \\
& & & & & \ddots & \ddots & & & & & \\
& & & & & & & F_Q & & & & \\
& & & & & & & C_Q & F_{Q+1} & & & \\
& & & & & & & & \ddots & \ddots & & \\
& & & & & & & & & & F_{S-1} & \\
& & & & & & & & & & C_S & F_S
\end{array} \right)
\end{array}
$$

Theorem 1. *The steady-state probability vector $\pi = (\pi^{(0)}, \pi^{(1)}, \pi^{(2)}, \dots, \pi^{(S)})$, related to the rate matrix K_M, is given by*

$$\pi^{(l)} = \pi^{(Q)}\Omega_l, \quad l = 0,1,2,\dots S. \tag{6}$$

where

$$\Omega_l = \begin{cases} (-1)^{Q-l} C_Q F_{Q-1}^{-1} C_{Q-1} F_{Q-2}^{-1} \cdots C_{l+1} F_l^{-1}, & l = 0,1,2,\dots,Q-1, \\ I, & l = Q, \\ (-1)^{2Q+1-l} \sum_{i=0}^{S-l} [(C_Q F_{Q-1}^{-1} C_{Q-1} F_{Q-2}^{-1} \cdots C_{s-i+1} F_{s-i}^{-1}) & \\ \quad (D_1 F_{S-i})(C_{S-i} F_{S-i-1}^{-1} \cdots C_{l+1} F_l^{-1})] & l = Q+1,\dots,S \end{cases}$$

and $\pi^{(Q)}$ can be obtained by solving

$$\begin{aligned} \pi^{(Q)} \Big[(-1)^Q C_Q F_{Q-1}^{-1} C_{Q-1} F_{Q-2}^{-1} \cdots C_1 F_0^{-1} D_0 + F_Q + (-1)^Q \Big[\sum_{i=0}^{s-1} (C_Q F_{Q-1}^{-1} C_{Q-1} F_{Q-2}^{-1} \cdots \\ C_{s-i+1} F_{s-i}^{-1})(D_1 F_{S-i})(C_{S-i} F_{S-i-1}^{-1} \cdots C_{Q+2} F_{Q+1}^{-1}) \Big] C_{Q+1} \Big] = \mathbf{0} \end{aligned} \tag{7}$$

and

$$\begin{aligned} \pi^{(Q)} \Big[\sum_{i=0}^{Q-1} ((-1)^{Q-l} C_Q F_{Q-1}^{-1} C_{Q-1} F_{Q-2}^{-1} \cdots C_{l+1} F_l^{-1}) + I + \sum_{i=0}^{2Q+1-l} ((-1)^{2Q+1-l} (C_Q F_{Q-1}^{-1} \\ C_{Q-1} F_{Q-2}^{-1} \cdots C_{s-i+1} F_{s-i}^{-1})(D_1 F_{S-i})(C_{S-i} F_{S-i-1}^{-1} \cdots C_{l+1} F_l^{-1})) \Big] \mathbf{e} = 1 \end{aligned} \tag{8}$$

Proof. We know that $\pi K_M = \mathbf{0}$ and $\pi \mathbf{e} = \mathbf{0}$. The first equality of the above produces the set of equations as follows:

$$\pi^{(l)} F_l + \pi^{(l+1)} C_{l+1} = \mathbf{0} \quad l = 0,1,2,\dots,Q-1 \tag{9}$$
$$\pi^{(l-Q)} D_0 + \pi^{(l)} F_l + \pi^{(l+1)} C_{l+1} = \mathbf{0} \quad l = Q \tag{10}$$
$$\pi^{(l-Q)} D_1 + \pi^{(l)} F_l + \pi^{(l+1)} C_{l+1} = \mathbf{0} \quad l = Q+1, Q+2, \dots, S-1 \tag{11}$$
$$\pi^{(l-Q)} D_1 + \pi^{(l)} F_l = \mathbf{0} \quad l = S \tag{12}$$

On solving the above equations recursively, we obtain Equation (6). Additionally, on substituting the value of Ω_l in Equation (6) and in the normalizing condition, we yield $\pi^{(Q)}$. □

After finding the stationary probability vector, π, to the generator matrix, K_M, the stability condition of the system can be derived easily. To derive such condition, we require the Neuts [41] inequality, which is formed at the truncation point, M. So using Theorem 1, we derive the stability condition of the system.

Theorem 2. *The stability condition at the truncation point, M, of the proposed system is given by*

$$\sum_{j=0}^{N}\pi^{(0,0,j)}p\lambda_L+\sum_{i=1}^{L}\sum_{K=1}^{3}\sum_{j=0}^{N}\pi^{(i,k,j)}p\lambda_L+\sum_{i=L+1}^{S}\sum_{K=1}^{3}\sum_{j=0}^{N}\pi^{(i,k,j)}r\lambda_H<\sum_{i=1}^{S}\pi^{(i,0,0)}M\lambda_r+ \sum_{i=2}^{S}\pi^{(i,2,0)}q\mu_{L_2}+\sum_{i=2}^{S}\pi^{(i,3,0)}q\mu_H \tag{13}$$

Proof. From the Neuts [41] matrix-geometric approach, we obtain the following constrain to derive a stability condition:

$$\pi K_{01}\mathbf{e}<\pi K_{M(M-1)}\mathbf{e}$$

$$[\pi^{(0)}B_0,\pi^{(1)}B_1,\ldots,\pi^{(L)}B_1,\pi^{(L+1)}B_2,\ldots,\pi^{(S)}B_2]\mathbf{e}<[\pi^{(1)}G_1+\pi^{(2)}H,\ldots,\pi^{(S-1)}G_1 +\pi^{(S)}H,\pi^{(S)}G_1]\mathbf{e}.$$

On computing the above inequality explicitly, we obtain
LHS:

$$\pi K_{01}\mathbf{e}=[\theta_0+(\theta_1+\Psi_1)+(\theta_2+\Psi_2)+\ldots+(\theta_L+\Psi_L)+\theta_{(L+1)}+\ldots+\theta_S]$$

where
$\theta_0=\sum_{j=0}^{N}\pi^{(0,0,j)}p\lambda_L$,
$\theta_i=\sum_{k=1}^{3}\sum_{j=0}^{N}\pi^{(i,k,j)}p\lambda_L \quad i=1,2,\ldots,S \quad \textit{and} \quad \Psi_i=\pi^{(i,0,0)}r\lambda_H \quad i=1,2,\ldots,S$
simplifying further, we obtain

$$\sum_{j=0}^{N}\pi^{(0,0,j)}p\lambda_L+\sum_{i=1}^{L}\sum_{k=1}^{3}\sum_{j=0}^{N}\pi^{(i,k,j)}p\lambda_L+\sum_{i=L+1}^{S}\sum_{k=1}^{3}\sum_{j=0}^{N}\pi^{(i,k,j)}r\lambda_H \tag{14}$$

RHS:

$$\begin{aligned}\pi K_{M(M-1)}\mathbf{e}&=\pi^{(1,0,0)}M\lambda_r+\pi^{(2,2,0)}q\mu_{L_2}+\pi^{(2,3,0)}q\mu_H+pi^{(2,0,0)}M\lambda_r+\\ &\pi^{(3,2,0)}q\mu_{L_2}+\pi^{(3,3,0)}q\mu_H+\ldots+\pi^{(S-1,0,0)}M\lambda_r+\pi^{(S,2,0)}q\mu_{L_2}+\pi^{(S,3,0)}q\mu_H+\pi^{(S,0,0)}M\lambda_r\\ &=\sum_{i=1}^{S}\pi^{(i,0,0)}M\lambda_r+\sum_{i=2}^{S}\pi^{(i,2,0)}q\mu_{L_2}+\sum_{i=2}^{S}\pi^{(i,3,0)}q\mu_H.\end{aligned} \tag{15}$$

Therefore, from (14) and (15), the result holds. □

3.2. Computation of R-Matrix

After the truncation point M, the stationary probability vector u to the infinitesimal generator matrix K is dependent on the matrix R, where R is the minimal non-negative solution of the matrix quadratic equation $R^2K_{M,M-1}+RK_{MM}+K_{01}=\mathbf{0}$.

Theorem 3. *Due to the specific structure of K, R can be determined by*

$$R^2K_{M,M-1}+RK_{MM}+K_{01}=\mathbf{0} \tag{16}$$

where R is the minimal non-negative solution of the quadratic Equation (16). The square matrix R is of dimension $S[(3N+4)+(N+1)]$ and is defined by

$$R = \begin{pmatrix} R_{(0,0)} & R_{(0,1)} & R_{(0,2)} & \cdots & R_{(0,S)} \\ R_{(1,0)} & R_{(1,1)} & R_{(1,2)} & \cdots & R_{(1,S)} \\ R_{(2,0)} & R_{(2,1)} & R_{(2,2)} & \cdots & R_{(2,S)} \\ \vdots & \vdots & \vdots & \ddots & \vdots \\ R_{(S,S)} & R_{(S,1)} & R_{(S,2)} & \cdots & R_{(S,S)} \end{pmatrix}. \tag{17}$$

where the block $R_{(0,0)}$ has dimension (N+1)(N+1) and of the form

$$R_{(0,0)} = \begin{pmatrix} r_{00} & r_{01} & r_{02} & \cdots & r_{0N} \\ r_{10} & r_{11} & r_{12} & \cdots & r_{1N} \\ \vdots & \vdots & \vdots & \ddots & \vdots \\ r_{N0} & r_{N1} & r_{N2} & \cdots & r_{NN} \end{pmatrix}.$$

The consecutive blocks in the first row with dimension (N+1)(3N+4) will be

$$R_{(0,j)} = \begin{bmatrix} R_{0j}^{(0,0)} & R_{0j}^{(0,1)} & R_{0j}^{(0,2)} & R_{0j}^{(0,3)} \end{bmatrix}, j = 1,2,\ldots,S.$$

where

$$R_{0j}^{(0,0)} = \begin{bmatrix} r_{00}^{(00,j0)} \\ r_{10}^{(00,j0)} \\ \vdots \\ r_{N0}^{(00,j0)} \end{bmatrix} \text{ and}$$

$$R_{0j}^{(0,k)} = \begin{bmatrix} r_{00}^{(00,jk)} & r_{01}^{(00,jk)} & r_{02}^{(00,jk)} & \cdots & r_{0N}^{(00,jk)} \\ r_{10}^{(00,jk)} & r_{11}^{(00,jk)} & r_{12}^{(00,jk)} & \cdots & r_{1N}^{(00,jk)} \\ \vdots & \vdots & \vdots & \ddots & \vdots \\ r_{N0}^{(00,jk)} & r_{N1}^{(00,jk)} & r_{N2}^{(00,jk)} & \cdots & r_{NN}^{(00,jk)} \end{bmatrix}, j = 1,2,\ldots,S., \quad k = 1,2,3.$$

similarly, for the blocks in the first column, $R_{(i,0)}, i = 1,2,\ldots,S$.

$$R_{(i,0)} = \begin{bmatrix} R_{i0}^{(0,0)} \\ R_{i0}^{(1,0)} \\ R_{i0}^{(2,0)} \\ R_{i0}^{(3,0)} \end{bmatrix}, i = 1,2,\ldots,S.$$

where

$$R_0^{(0,0)} = \begin{bmatrix} 0 & 0 & 0 & \cdots & 0 \end{bmatrix} \text{ and}$$

$$R_{i0}^{(k,0)} = \begin{bmatrix} r_{00}^{(ik,00)} & r_{01}^{(ik,00)} & r_{02}^{(ik,00)} & \cdots & r_{0N}^{(ik,00)} \\ r_{10}^{(ik,00)} & r_{11}^{(ik,00)} & r_{12}^{(ik,00)} & \cdots & r_{1N}^{(ik,00)} \\ \vdots & \vdots & \vdots & \ddots & \vdots \\ r_{N0}^{(ik,00)} & r_{N1}^{(ik,00)} & r_{N2}^{(ik,00)} & \cdots & r_{NN}^{(ik,00)} \end{bmatrix}, i = 1,2,\ldots,S, \quad k = 1,2,3.$$

Considering the remaining blocks $R_{(i,j)}, i = 1,2,\ldots,S$ *and* $j = 1,2,\ldots,S$

$$R_{(i,j)} = \begin{bmatrix} R_{ij}^{(0,0)} & R_{ij}^{(0,1)} & R_{ij}^{(0,2)} & R_{ij}^{(0,3)} \\ R_{ij}^{(1,0)} & R_{ij}^{(1,1)} & R_{ij}^{(1,2)} & R_{ij}^{(1,3)} \\ R_{ij}^{(2,0)} & R_{ij}^{(2,1)} & R_{ij}^{(2,2)} & R_{ij}^{(2,3)} \\ R_{ij}^{(3,0)} & R_{ij}^{(3,1)} & R_{ij}^{(3,2)} & R_{ij}^{(3,3)} \end{bmatrix}$$

where

$$R_{ij}^{(k,0)} = \begin{bmatrix} r_{00}^{(ik,j0)} \\ r_{01}^{(ik,j0)} \\ r_{02}^{(ik,j0)} \\ \vdots \\ r_{0N}^{(ik,j0)} \end{bmatrix}$$

$$R_{ij}^{(k,l)} = \begin{bmatrix} r_{00}^{(ik,jl)} & r_{01}^{(ik,jl)} & r_{02}^{(ik,jl)} & \cdots & r_{0N}^{(ik,jl)} \\ r_{10}^{(ik,jl)} & r_{11}^{(ik,jl)} & r_{12}^{(ik,jl)} & \cdots & r_{1N}^{(ik,jl)} \\ \vdots & \vdots & \vdots & \ddots & \vdots \\ r_{N0}^{(ik,jl)} & r_{N1}^{(ik,jl)} & r_{N2}^{(ik,jl)} & \cdots & r_{NN}^{(ik,jl)} \end{bmatrix}$$

$$R_{ij}^{(0,l)} = \begin{bmatrix} 0 & 0 & 0 & \ldots & 0 \end{bmatrix} \textit{ and}$$

$$R_{ij}^{(0,0)} = [0],\ i = 1,2,\ldots,S.,\ j = 1,2,\ldots,S.,\ k = 1,2,3. \textit{ and } l = 1,2,3.$$

Proof. The structure of R matrix has $S[(3N+3)+(N+1)]$ non-zero rows and can be formed depending on the special structure of K_{01} matrix .

$$K_{01} = \begin{array}{c} \\ (0) \\ (1) \\ (2) \\ \vdots \\ (L) \\ (L+1) \\ \vdots \\ (S) \end{array} \begin{array}{c} \begin{array}{cccccccc} (0) & (1) & (2) & \cdots & (L) & (L+1) & \cdots & (S) \end{array} \\ \begin{pmatrix} B_0 & & & & & & & \\ & B_1 & & & & & & \\ & & B_1 & & & & & \\ & & & \ddots & & & & \\ & & & & B_1 & & & \\ & & & & & B_2 & & \\ & & & & & & \ddots & \\ & & & & & & & B_2 \end{pmatrix} \end{array}$$

where the blocks B_0 has all positive rows and B_1, B_2 has one zero row in each. Therefore, the R-matrix has $S[(3N+3)+(N+1)]$ non-zero rows. Let us assume the R matrix is as in (17). With this assumption, exploiting the structure of the matrices $K_{M,M-1}, K_{MM}$ and R in Equation (16), we obtain the following set of nonlinear equations:

$$R_{i0}B_0 + R_{i1}C_1 + \delta_{i0}B_0 = \mathbf{0}, \quad i = 0,1,2,\ldots,S. \tag{18}$$

$$G_1\left\{\sum_{j=0}^{S} R_{ij}R_{jk}\right\} + H\left\{\sum_{j=0}^{S} R_{ij}R_{j(k+1)}\right\} + R_{ik}F_k + R_{i(k+1)}C_{k+1} + \delta_{ik}B_1 = \mathbf{0} \quad (19)$$
$$i = 0,1,2,\ldots,S, \quad k = 1,2,\ldots,L.$$

$$G_1\left\{\sum_{j=0}^{S} R_{ij}R_{jk}\right\} + H\left\{\sum_{j=0}^{S} R_{ij}R_{j(k+1)}\right\} + R_{ik}F_k + R_{i(k+1)}C_{k+1} + \delta_{ik}B_2 = \mathbf{0} \quad (20)$$
$$i = 0,1,2,\ldots,S, \quad k = L+1, L+2,\ldots,Q-1.$$

$$G_1\left\{\sum_{j=0}^{S} R_{ij}R_{jQ}\right\} + H\left\{\sum_{j=0}^{S} R_{ij}R_{j(Q+1)}\right\} + R_{iQ}F_Q + R_{i0}D_0 + R_{i(Q+1)}C_{Q+1} + B_2 = \mathbf{0} \quad (21)$$
$$i = Q.$$

$$G_1\left\{\sum_{j=0}^{S} R_{ij}R_{jk}\right\} + H\left\{\sum_{j=0}^{S} R_{ij}R_{j(k+1)}\right\} + R_{ik}F_k + R_{i0}D_0 + R_{i(k+1)}C_{k+1} + \delta_{ik}B_2 = \mathbf{0} \quad (22)$$
$$i = 0,1,2,\ldots,S, \quad k = Q+1, Q+2,\ldots,S-1.$$

$$G_1\left\{\sum_{j=0}^{S} R_{ij}R_{jS}\right\} + R_{i(S-Q)}D_1 + R_{iS}F_S + \delta_{iS}B_2 = \mathbf{0} \quad (23)$$
$$i = 0,1,2,\ldots,S.$$

After solving all Equations (18)–(23) by the Gauss–Seidel iterative method, we obtain the entries of the R-matrix. □

Now, we derive the stationary probability vector, $u = (u(0), u(1), u(2), \ldots)$, to the infinitesimal generator matrix, K, as given below.

Theorem 4. *Due to the special structure of K, the stationary probability vector* $u = (u(0), u(1), u(2), \ldots)$ *can be expressed as*

$$u(i+M-1) = u(M-1)R^i; \qquad i \geq 0 \quad (24)$$

where the matrix R is the unique non-negative solution with spectral radius less than 1, of the following equation:

$$R^2 K_{M,M-1} + RK_{MM} + K_{01} = \mathbf{0} \quad (25)$$

yields the solution to the vectors $(u(0), u(1), \ldots)$ *as*

$$u(i) = \begin{cases} \delta v(0) \prod_{j=i}^{M} K_{j,j-1}(-K_{j-1})^{-1} & 0 \leq i \leq M-1 \\ \delta v(0) R^{i-M} & i \geq M \end{cases} \quad (26)$$

where $v(0)$ *is the unique solution of the system*

$$\begin{aligned} v(0)(K_{MM} + RK_{M+1,M}) &= \mathbf{0} \\ v(0)(I-R)^{-1}\mathbf{e} &= 1 \end{aligned} \quad (27)$$

and

$$\delta = [1 + v(0) \sum_{i=0}^{M-1} \prod_{j=i}^{M} K_{j,j-1}(-K_{j-1})^{-1}\mathbf{e}]^{-1}. \tag{28}$$

Proof. The sub-vector, $(u(0), \ldots, u(M-1))$, can be obtained by solving

$$u(0)K_{00} + u(1)K_{21} = \mathbf{0}, \tag{29}$$

$$u(i-1)K_{01} + u(i)K_{ii} + u(i+1)K_{i+1,i} = \mathbf{0}, \quad 1 \le i \le M-1. \tag{30}$$

Now, simplifying (29) and (30) over the range $1 \le i \le M-1$, we obtain

$$u(i) = u(i+1)K_{i+1,i}(-K_i)^{-1} \quad 1 \le i \le M-1$$

where

$$K_i' = \begin{cases} K_{00} & i = 0 \\ K_{ii} + K_{i,i-1}(-K_{i-1}')^{-1} & 1 \le i \le M. \end{cases}$$

Now, by applying block Gaussian elimination, the partitioned sub-vector (u(M), u(M+1), ⋯) corresponding to non-boundary states, satisfies the relation.

$$\begin{bmatrix} u(M) & u(M+1) & \cdots \end{bmatrix} \begin{bmatrix} K_M' & K_{01} & 0 & 0 & \cdots \\ K_{M+1,M} & K_{M+1,M+1} & K_0 & 0 & \cdots \\ 0 & K_{M+2,M+1} & K_{M+2,M+2} & K_{01} & \vdots \\ \vdots & 0 & \ddots & \ddots & \ddots \end{bmatrix} = \mathbf{0} \tag{31}$$

Let

$$\delta = \sum_{i=M}^{\infty} u(i)\mathbf{e} \tag{32}$$

$$\text{and} \quad v(i) = \delta^{-1}u(M+i), \quad i \ge 0 \tag{33}$$

From (31), we obtain

$$u(M)K_M' + u(M+1)K_{M+1,M} = \mathbf{0}$$
$$u(M+i) = u(M+i-1)R, \quad i \ge 1$$

Using (25), we obtain

$$u(M)K_M' + u(M)RK_{M+1,M} = \mathbf{0}$$
$$v(0)K_M' + v(1)K_{M+1,M} = \mathbf{0} \tag{34}$$

Since $\sum_{i=0}^{\infty} v(i)\mathbf{e} = 1$, we obtain

$$v(0)(I-R)^{-1}\mathbf{e} = 1 \tag{35}$$

Additionally,

$$u(i) = \delta v(0)R^{i-M}, \quad i \ge M. \tag{36}$$

From (32), we obtain

$$u(i) = \delta v(0) \prod_{j=i}^{M} K_{j,j-1}(-K'_{j-1})^{-1}, \quad 0 \le i \le M-1 \tag{37}$$

From Equations (34) and (35), we obtain (27)

where $v(0)$ is the unique solution of the system (28). Additionally, from (36) and (37), we obtain (28)

$$\delta = [1 + v(0) \sum_{i=0}^{M-1} \prod_{j=i}^{M} K_{j,j-1}(-K'_{j-1})^{-1}\mathbf{e}]^{-1}$$

□

3.3. System Performance Measures

In this section, we put forward some important performance measures in order to construct the mean total cost of the proposed model.

1. Expected inventory level:

$$E_I = \sum_{i_1=0}^{\infty} \sum_{j_1=1}^{S} j_1 \left\{ u(i_1, j_1, 0, 0) + \sum_{k_1=1}^{3} \sum_{l_1=0}^{N} u(i_1, j_1, k_1, l_1) \right\}$$

2. Expected perishable rate:

$$E_P = \sum_{i_1=0}^{\infty} \left\{ \sum_{j_1=1}^{S} j_1 \gamma u(i_1, j_1, 0, 0) + \sum_{j_1=1}^{S} \sum_{k_1=1}^{3} \sum_{l_1=0}^{N} (j_1 - 1)\gamma u(i_1, j_1, k_1, l_1) \right\}$$

3. Expected reorder rate:

$$\begin{aligned} E_R = & \sum_{i_1=0}^{\infty} (s+1)\gamma u(i_1, (s+1), 0, 0) + \sum_{i_1=0}^{\infty} \sum_{k_1=1}^{3} \sum_{l_1=0}^{N} s\gamma u(i_1, (s+1), k_1, l_1) + \\ & \sum_{i_1=0}^{\infty} \mu_{L_2} u(i_1, (s+1), 2, 0) + \sum_{i_1=0}^{\infty} \sum_{l_1=0}^{N} \mu_H \, u(i_1, (s+1), 3, l_1). \end{aligned}$$

4. Expected number of HP customers in the waiting hall:

$$E_{WH} = \sum_{i_1=0}^{\infty} \sum_{l_1=0}^{N} l_1 \left\{ u(i_1, 0, 0, l_1) + \sum_{j_1=1}^{S} \sum_{k_1=1}^{3} u(i_1, j_1, k_1, l_1) \right\}$$

5. Expectation of HP customer entering into the waiting hall:

$$E_{EWH} = \sum_{i_1=0}^{\infty} \left\{ \sum_{l_1=0}^{N-1} \lambda_H \, u(i_1, 0, 0, l_1) + \sum_{j_1=1}^{S} \lambda_H u(i_1, j_1, 0, 0) + \sum_{j_1=1}^{S} \sum_{k_1=1}^{3} \sum_{l_1=0}^{N-1} \lambda_H \, u(i_1, j_1, k_1, l_1) \right\}$$

6. Expected waiting time of an HP:

$$E_{WHP} = E_{WH} / E_{EWH}$$

7. Expected number of LP customers in the orbit:

$$E_O = \sum_{i_1=0}^{\infty} i_1 \, u(i_1)\mathbf{e}$$

8. Expected number of LP customers lost:

$$E_{LPL} = \sum_{i_1=0}^{\infty} \sum_{l_1=0}^{N} (1-p)\lambda_L \, u(i_1,0,0,l_1) + \sum_{i_1=0}^{\infty} \sum_{1_1=1}^{S} \sum_{k_1=1}^{3} \sum_{l_1=0}^{N} (1-p)\lambda_L \, u(i_1,j_1,k_1,l_1)$$
$$+ \sum_{i_1=0}^{\infty} \sum_{j_1=1}^{L} \sum_{l_1=0}^{N} r\lambda_H \, u(i_1,j_1,0,l_1)$$

9. Expected number of HP customers lost:

$$E_{HPL} = \sum_{i_1=0}^{\infty} \left\{ \lambda_H \, u(i_1,0,0,N) + \sum_{j_1=1}^{S} \sum_{k_1=1}^{3} \lambda_H \, u(i_1,j_1,k_1,N) \right\}$$

10. Expected number of times an HP customer interrupts an LP customer:

$$E_{HPIL} = \sum_{i_1=0}^{\infty} \sum_{j_1=1}^{L} r\lambda_H \, u(i_1,j_1,1,0)$$

11. Probability of server being idle:

$$E_{SI} = \sum_{i_1=0}^{\infty} \left\{ \sum_{l_1=0}^{N} u(i_1,0,0,l_1) + \sum_{j_1=1}^{S} u(i_1,j_1,0,0) \right\}$$

12. Probability of server being busy:

$$E_{SB} = \sum_{i_1=0}^{\infty} \sum_{j_1=1}^{S} \sum_{k_1=1}^{3} \sum_{l_1=0}^{N} u(i_1,j_1,k_1,l_1)$$

13. Expected number of times a server carries out the orbital search:

$$E_{OS} = \sum_{i_1=1}^{\infty} \sum_{j_1=1}^{S} \{ q\mu_{L_2} \, u(i_1,j_1,2,0) + q\mu_H \, u(i_1,j_1,3,0) \}$$

14. Overall rate of retrial:

$$E_{ORR} = \sum_{i_1=1}^{\infty} i_1\lambda_r \, u(i_1)\mathbf{e}$$

15. Successful rate of retrial:

$$E_{SRR} = \sum_{i_1=1}^{\infty} \sum_{j_1=1}^{S} i_1\lambda_r \, u(i_1,j_1,0,0)$$

16. Fraction of successful rate of retrial:

$$E_{FSRR} = E_{SRR}/E_{ORR}$$

3.4. Cost Analysis

The different costs we used are as follows:

c_h = cost of carrying the inventory per unit item per unit time;
c_r = cost of placing an order;
c_p = perishable cost of an item per unit time;
c_{wh} = cost due to waiting time of a HP customer per unit time;
c_{wo} = cost due to waiting time of a LP customer per unit time;
c_{ll} = cost due to loss of an LP customer per unit time;
c_{hl} = cost due to loss of an HP customer per unit time;

The expected total cost (TC) is given by:

$$TC(S,s) = c_h * E_I + c_p * E_P + c_r * E_R + c_{wh} * E_{WHP} + c_{wo} * E_O + c_{ll} * E_{LPL} + c_{hl} * E_{HPL}$$

4. Numerical Discussions

In this section, we discuss significant examples that illustrate the effectiveness of the proposed queueing–inventory model more practically. Through numerical discussion, we are provided with a broader view of the model in a more practical situation. Additionally, to enhance understanding and to find solutions to real-life problems, the results discussed would be more useful. To do so, we first fix the values for parameters. The optimal set of values for the parameters and costs are $L = 6$; $N = 8$; $Q = S - s$; $\lambda_H = 0.90$; $\lambda_L = 0.47$; $\lambda_r = 0.64$; $\mu_{L_1} = 2.45$; $\mu_{L_2} = 2.63$; $\mu_H = 2.79$; $p = 0.6$; $q = 0.67$; $c_h = 0.006$; $c_r = 2.4$; $c_p = 0.1$; $c_{wh} = 1.4$; $c_{wo} = 0.5$; $c_{ll} = 2$; $c_{hl} = 1$.

Example 1. *Optimal Analysis: To determine the optimal value for S, s, and r, we vary S and s under a finite set of values and r from 0 to 1 as it is the probability of interruption. Therefore, the total cost as a function of S and s for different values of r is tabulated in Tables 1–3.*

Table 1. Convex with $r = 0$.

S | s	1	2	3	4	5	6	7
28	3.187610	3.132378	3.114067	**3.102593**	3.110016	3.132837	3.164871
29	3.170042	3.116397	3.108295	**3.093581**	3.099043	3.120513	3.151056
30	3.157677	3.103865	3.097418	**3.088793**	3.092117	3.111897	3.140980
31	3.153721	3.101571	3.095904	**3.089001**	3.095888	3.114345	3.142301
32	3.162313	3.113600	3.111644	**3.106273**	3.114821	3.135832	3.163509

Table 2. Convex with $r = 0.23$.

S | s	1	2	3	4	5	6	7
28	3.156901	3.107505	3.094919	**3.087682**	3.098520	3.119306	3.155629
29	3.140263	3.092402	3.089834	**3.079195**	3.088026	3.107578	3.142287
30	3.128573	3.080375	3.079441	**3.074810**	3.081410	3.099400	3.132570
31	3.125040	3.078352	3.078033	**3.075131**	3.085328	3.102140	3.134130
32	3.133964	3.090639	3.094001	**3.092575**	3.104508	3.124117	3.155706

In Tables 1–3, the values shown in bold are the least in the respective rows and the values underlined are the least in that respective column. Therefore, the value that is bold and underlined is the optimal value of total cost.

Case (i): Non-discretionary priority discipline.

From Table 1, we see that for $r = 0$ the optimum value of total cost is obtained at $S^* = 30$, $s^* = 4$ and $TC^* = 3.088793$. This provides the optimum result when the interruption of an HP customer is not allowed.

Table 3. Convex with $r = 1$.

S | s	1	2	3	4	5	6	7
28	3.027256	**3.005787**	3.018169	3.027169	3.050287	3.068396	3.120758
29	3.014473	**2.994246**	3.015881	3.020835	3.041772	3.058894	3.109163
30	3.005563	**2.984233**	3.007418	3.018146	3.036500	3.052381	3.100798
31	3.003744	**2.983274**	3.006425	3.018997	3.041181	3.056248	3.103275
32	3.013988	**2.996584**	3.023343	3.037225	3.061479	3.080069	3.126179
33	3.022344	**3.005740**	3.032852	3.047970	3.072497	3.092072	3.138950

Case (ii): Discretionary priority discipline.

- *Mixed priority: At $r = 0.23$, the optimum value is obtained similarly at $S^* = 30$, $s^* = 4$ and $TC^* = 3.074810$. as shown in Table 2. This TC^* shows the impact of interruption, based on the discretionary priority. On comparing TC^* with case (i), as we expected, the discretionary priority service process gives the minimum optimal total cost rather than the usual priority service patterns.*
- *Preemptive priority discipline: At $r = 1$, the optimum total cost is obtained at $S^* = 31$, $s^* = 2$, and $TC^* = 2.983274$, as shown in Table 3. Even though the TC^* of this case is minimum than that of the case (i), we conclude that case (ii) provides the best service discipline. This is because, when a company allows preemptive priority, it may lose its LP customers. The growth of the company obviously depends upon all types of customers. So, the company must satisfy them by providing their best service. In such a way, discretionary priority discipline is considered the best one.*

Example 2. *In this example we study the influence of β, γ, λ_H, λ_r, and μ_H under various interruption rates, r, on the total cost. The results are tabulated in Table 4 and observations made are the following: with the increase in perishable rate as expected, the total cost increases and, on increased reorder rate, the total cost decreases. Additionally, the TC^* increases when the arrival rate increases and service rate decreases.*

Case (i): Non-discretionary priority discipline.

For $r = 0$, the optimum total cost $TC^* = 2.925533$ is obtained at $\gamma = 0.05$ and $\beta = 1.8$. For higher values of either arrival rate or perishable rate, the TC^* increases.

Case (ii): Discretionary priority discipline.

- *Mixed priority: At r = 0.5, the total cost TC^* follows the same pattern of increment and decrement as in case (i), for all the parameters considered in Table 4, respectively. The optimum total cost obtained is $TC^* = 2.890126$ at $\gamma = 0.05$ and $\beta = 1.8$. Thus, the optimum value obtained in discretionary priority is less than that of non-preemptive priority.*
- *Preemptive priority discipline: At $r = 1$, for the same value of the parameter as in case (i), we obtain the optimum total cost to be $TC^* = 2.48149$, which is less than those of non-preemptive and discretionary priority disciplines. However, depending on this TC^*, we cannot make the inference that preemptive priority discipline is economical. Although it seems like preemptive priority discipline is more profitable than the other priority discipline, it leads to the loss of new patrons for the business. A new LP customer who is ignored or made to wait for a long duration may leave the system before trying the product even once. Hence, discretionary priority is best suited to increase the customer base.*

Table 4. The effect of the parameters γ, β, λ_H, λ_r, μ_H, and r.

γ	λ_H	λ_r	μ_H	$r=0$		$r=0.5$		$r=1$	
				$\beta=1.7$	$\beta=1.8$	$\beta=1.7$	$\beta=1.8$	$\beta=1.7$	$\beta=1.8$
0.05	0.85	0.60	2.75	2.979013	2.973860	2.942602	2.938585	2.899332	2.896864
			2.80	2.954551	2.948898	2.918055	2.913559	2.874628	2.871711
			2.85	2.931692	2.925533	2.895107	2.890126	2.851522	2.848149
		0.70	2.75	2.990586	2.985467	2.954113	2.950135	2.910731	2.908310
			2.80	2.966412	2.960791	2.929858	2.925397	2.886322	2.883448
			2.85	2.943827	2.937697	2.907188	2.902239	2.863496	2.860163
	0.95	0.60	2.75	2.991737	2.987089	2.957554	2.953994	2.916601	2.914532
			2.80	2.966064	2.960974	2.931800	2.927819	2.890698	2.888237
			2.85	2.941959	2.936423	2.907617	2.903210	2.866367	2.863511
		0.70	2.75	2.991737	2.996258	2.957554	2.963084	2.916601	2.923494
			2.80	2.966064	2.970501	2.931800	2.937274	2.890698	2.897572
			2.85	2.941959	2.946289	2.907617	2.913009	2.866367	2.873197
0.1	0.85	0.60	2.75	3.106973	3.099336	3.064676	3.058380	3.014288	3.009778
			2.80	3.083714	3.075479	3.041333	3.034460	2.990770	2.985716
			2.85	3.062043	3.053205	3.019577	3.012126	2.968839	2.963238
		0.70	2.75	3.118530	3.110927	3.076163	3.069904	3.025633	3.021167
			2.80	3.095568	3.087364	3.053121	3.046281	3.002421	2.997407
			2.85	3.074181	3.065369	3.031652	3.024231	2.980781	2.975217
	0.95	0.60	2.75	3.118081	3.111224	3.078620	3.073032	3.031223	3.027343
			2.80	3.093557	3.086174	3.054006	3.047916	3.006432	3.002082
			2.85	3.070585	3.062675	3.030949	3.024355	2.983203	2.978382
		0.70	2.75	3.127136	3.120334	3.087575	3.082047	3.040010	3.036197
			2.80	3.102982	3.095650	3.063340	3.057305	3.015606	3.011320
			2.85	3.080360	3.072498	3.040641	3.034099	2.992743	2.987982
0.15	0.85	0.60	2.75	3.227871	3.124522	3.180782	3.172613	3.124522	3.118335
			2.80	3.205585	3.101958	3.158412	3.149586	3.101958	3.095151
			2.85	3.184858	3.080966	3.137602	3.128126	3.080966	3.073538
		0.70	2.75	3.239407	3.229767	3.239407	3.184112	3.135819	3.129678
			2.80	3.217425	3.207099	3.217425	3.161390	3.113568	3.106804
			2.85	3.196987	3.185980	3.196987	3.140222	3.092876	3.085485
	0.95	0.60	2.75	3.237823	3.229153	3.194032	3.186784	3.141275	3.135916
			2.80	3.214209	3.204947	3.170322	3.162508	3.117363	3.111471
			2.85	3.192134	3.182281	3.148160	3.139778	3.095006	3.088582
		0.70	2.75	3.246812	3.238204	3.202911	3.195730	3.149959	3.144675
			2.80	3.223575	3.214372	3.179589	3.171837	3.126443	3.120623
			2.85	3.201857	3.192058	3.157792	3.149469	3.104462	3.098105

Example 3. *Waiting Time Analysis: In this example, we investigate the waiting times of HP customers and LP customers for changes in arrival rate and service rate. The data are presented in graphs in Figures 1–6, with the following observations.*

Case (i): Non-discretionary priority discipline.

At $r=0$, the measure E_{WHP} increases with an increase in arrival rate and decreases with an increase in service rate. Similarly, the waiting time of an LP customer, E_{WLP}, is directly proportional to the arrival rate and inversely proportional to the service rates, μ_{L_1} and μ_{L_2}. Additionally, with an increase in retrial rate, E_{WLP} decreases.

Case (ii): Discretionary priority discipline.

- *Mixed priority: At* $r = 0.5$*, the measures* E_{WHP} *and* E_{WLP} *show the same variations as in case (i) for the respective parameters, but the difference is that the waiting time of HP customers is less, and the waiting time of LP customers is higher, compared with those in case (i).*
- *Preemptive Priority discipline: At* $r = 1$*, similar to case (i), the waiting times remain with the same variation with respect to the parameters considered in Figures 1–6, and the measure* E_{WHP} *is considerably less and* E_{WLP} *is much higher compared with those in case (i).*

With these observations, we can understand that the waiting time of HP customers will reach the peak with non-preemptive priority and that the waiting time of LP customers will reach the peak with preemptive priority. However, discretionary priority discipline is preferable to the other disciplines for maintaining an optimal waiting time for each type of customer.

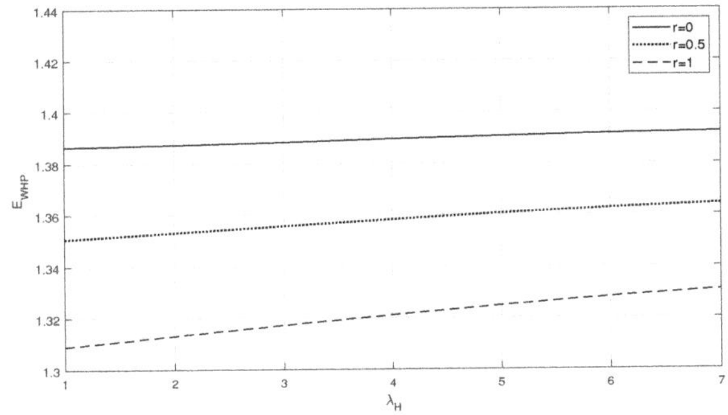

Figure 1. r vs. λ_H on E_{WHP}.

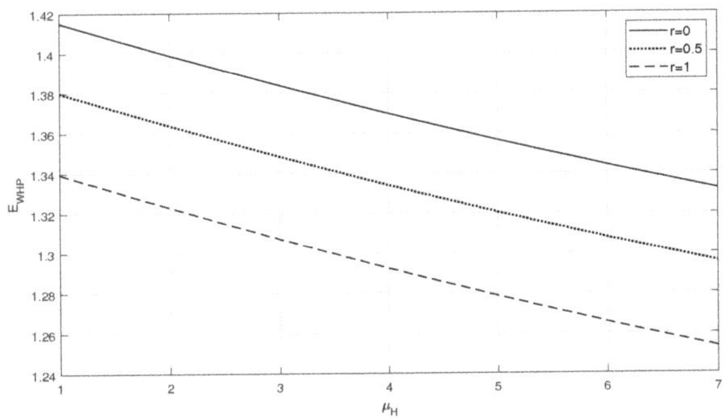

Figure 2. r vs. μ_H on E_{WHP}.

Example 4. *This example examines the expected number of HP customers and LP customers lost. The results of* E_{HPL} *and* E_{LPL} *are tabulated in Tables 5 and 6, and the observations are as follows:*

Case (i): Non-discretionary Priority discipline.

In terms of $r = 0$*,* E_{HPL}*, and* E_{LPL}*, they show an increase in the arrival rate of HP customers. However, with the HP customers, a gradual decline in the loss of customers can be noticed with an increase in waiting for hall capacity and reorder rate. For* E_{LPL}*, the loss slightly reduces with an increase in service rates.*

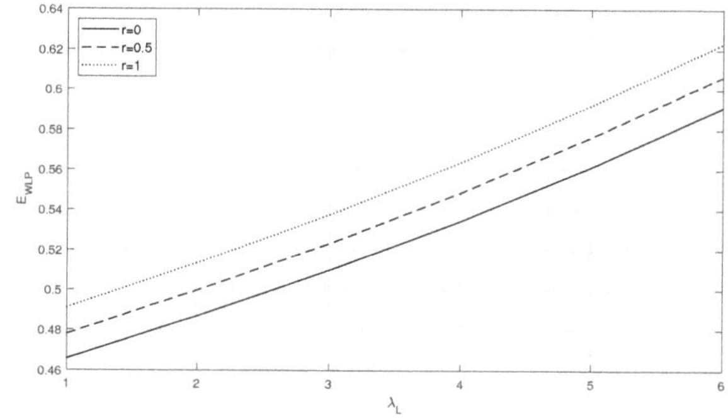

Figure 3. r vs. λ_L on E_{WLP}.

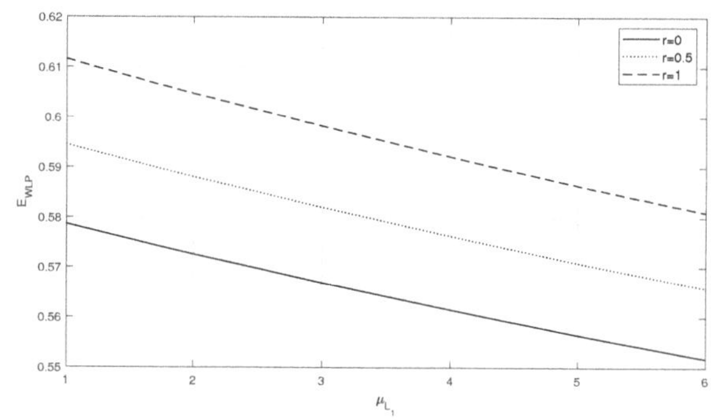

Figure 4. r vs. μ_{L_1} on E_{WLP}.

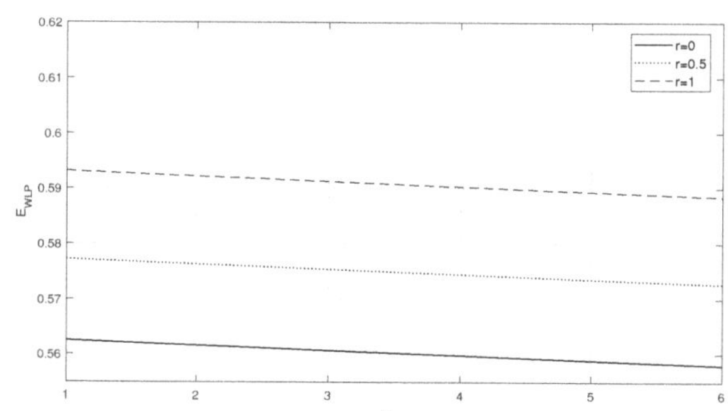

Figure 5. r vs. μ_{L_2} on E_{WLP}.

Case (ii): Discretionary priority discipline.

- *Mixed priority: At* $r = 0.5$*, the loss of HP customers encountered by following discretionary priority is less compared with non-preemptive priority. In the same way, the loss of LP customers by following discretionary priority is greater compared with non-preemptive priority.*
- *Preemptive Priority: As seen previously, the measure* E_{HPL} *is a little less compared with case (i), and* E_{LPL} *is slightly greater than mixed priority. Thus, the loss of an HP customer heightens with non-discretionary priority, and an LP customer rises with preemptive priority. However, the intensified loss of any one type of customer will affect the reputation of the organization, which in turn decreases the growth of the business. Hence, to maintain an admissible loss of any type of customer, the following discretionary priority is advisable.*

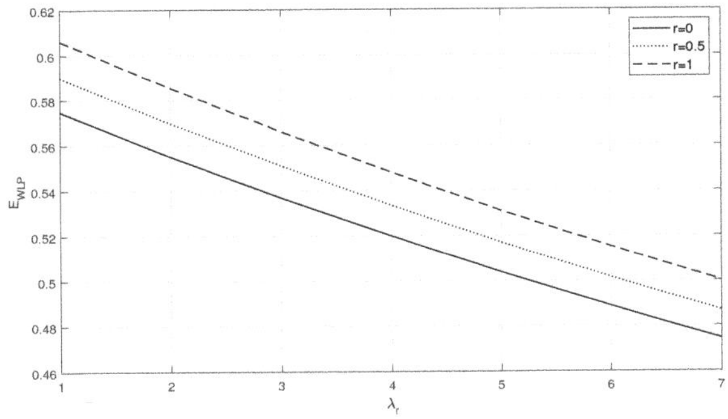

Figure 6. r vs. λ_r on E_{WLP}.

Table 5. Expected number of HP customers lost.

N	β	μ_H	$r=0$		$r=0.5$		$r=1$	
			$\lambda_H=0.85$	$\lambda_H=0.95$	$\lambda_H=0.85$	$\lambda_H=0.95$	$\lambda_H=0.85$	$\lambda_H=0.95$
7	1.6	2.75	0.003336	0.005413	0.003179	0.005173	0.002990	0.004882
		2.80	0.003239	0.005245	0.003083	0.005006	0.002894	0.004717
		2.85	0.003156	0.005097	0.002999	0.004859	0.002811	0.004570
	1.7	2.75	0.003173	0.005191	0.003024	0.004963	0.002845	0.004687
		2.80	0.003068	0.005012	0.002920	0.004786	0.002742	0.004511
		2.85	0.002977	0.004854	0.002829	0.004628	0.002652	0.004355
	1.8	2.75	0.003054	0.005025	0.002911	0.004806	0.002741	0.004542
		2.80	0.002943	0.004839	0.002802	0.004622	0.002633	0.004360
		2.85	0.002846	0.004673	0.002706	0.004457	0.002537	0.004197
8	1.6	2.75	0.001648	0.002828	0.001570	0.002702	0.001476	0.002549
		2.80	0.001591	0.002723	0.001514	0.002598	0.001421	0.002447
		2.85	0.001543	0.002631	0.001466	0.002507	0.001373	0.002358
	1.7	2.75	0.001547	0.002683	0.001474	0.002565	0.001387	0.002422
		2.80	0.001486	0.002572	0.001414	0.002456	0.001327	0.002314
		2.85	0.001432	0.002474	0.001361	0.002359	0.001275	0.002219
	1.8	2.75	0.001474	0.002576	0.001405	0.002464	0.001323	0.002328
		2.80	0.001409	0.002461	0.001341	0.002350	0.001260	0.002217
		2.85	0.001353	0.002358	0.001286	0.002249	0.001205	0.002117
9	1.6	2.75	0.000821	0.001491	0.000782	0.001424	0.000735	0.001344
		2.80	0.000789	0.001426	0.000750	0.001361	0.000704	0.001282
		2.85	0.000761	0.001371	0.000723	0.001306	0.000677	0.001228
	1.7	2.75	0.000760	0.001399	0.000724	0.001337	0.000681	0.001262
		2.80	0.000725	0.001331	0.000690	0.001271	0.000647	0.001197
		2.85	0.000695	0.001272	0.000660	0.001212	0.000618	0.001140
	1.8	2.75	0.000716	0.001331	0.000683	0.001273	0.000642	0.001203
		2.80	0.000679	0.001261	0.000646	0.001205	0.000607	0.001136
		2.85	0.000647	0.001199	0.000615	0.001144	0.000576	0.001076

Table 6. Expected number of LP customers lost.

μ_H	μ_{L_2}	μ_{L_1}	$r = 0$		$r = 0.5$		$r = 1$	
			$\lambda_H = 0.85$	$\lambda_H = 0.95$	$\lambda_H = 0.85$	$\lambda_H = 0.95$	$\lambda_H = 0.85$	$\lambda_H = 0.95$
2.75	2.60	2.40	0.223480	0.223851	0.223730	0.224161	0.224073	0.224585
		2.45	0.223453	0.223827	0.223700	0.224133	0.224038	0.224551
		2.50	0.223427	0.223803	0.223671	0.224107	0.224004	0.224519
	2.65	2.40	0.223458	0.223831	0.223708	0.224141	0.224051	0.224566
		2.45	0.223431	0.223807	0.223678	0.224113	0.224016	0.224532
		2.50	0.223405	0.223783	0.223649	0.224087	0.223983	0.224500
	2.70	2.40	0.223436	0.223812	0.223686	0.224122	0.224030	0.224547
		2.45	0.223409	0.223787	0.223656	0.224094	0.223995	0.224513
		2.50	0.223383	0.223764	0.223628	0.224068	0.223962	0.224481
2.80	2.60	2.40	0.223403	0.223775	0.223653	0.224086	0.223998	0.224514
		2.45	0.223374	0.223750	0.223622	0.224057	0.223962	0.224479
		2.50	0.223347	0.223726	0.223592	0.224030	0.223927	0.224445
	2.65	2.40	0.223379	0.223754	0.223630	0.224065	0.223975	0.224493
		2.45	0.223351	0.223729	0.223598	0.224037	0.223939	0.224458
		2.50	0.223324	0.223705	0.223569	0.224009	0.223904	0.224425
	2.70	2.40	0.223357	0.223734	0.223608	0.224046	0.223953	0.224474
		2.45	0.223329	0.223709	0.223576	0.224017	0.223917	0.224439
		2.50	0.223302	0.223685	0.223547	0.223989	0.223883	0.224406
2.85	2.60	2.40	0.223327	0.223701	0.223577	0.224013	0.223924	0.224443
		2.45	0.223297	0.223674	0.223545	0.223983	0.223887	0.224407
		2.50	0.223269	0.223649	0.223514	0.223954	0.223851	0.224373
	2.65	2.40	0.223302	0.223679	0.223553	0.223991	0.223900	0.224422
		2.45	0.223273	0.223653	0.223521	0.223961	0.223863	0.224386
		2.50	0.223245	0.223628	0.223490	0.223933	0.223827	0.224351
	2.70	2.40	0.223279	0.223658	0.223530	0.223970	0.223878	0.224402
		2.45	0.223250	0.223632	0.223498	0.223941	0.223840	0.224366
		2.50	0.223222	0.223607	0.223467	0.223912	0.223804	0.224331

Example 5. *System performances with respect to r, μ_H, μ_{L_1}, and λ_H: In this example, we discuss the other system performances with respect to r, μ_H, μ_{L_1}, and λ_H. The corresponding values are tabulated in Table 7 and the observations are as follows:*

1. *The E_{HPIL} increases with μ_H despite the increment in μ_{L_1} on positive interruption. Since an increase in service rates allows the server to serve all customers in less time, the measure E_{SB} decreases with an increase in service rate and increases with arrival rate.*
2. *It is clear from the table that, as the service rate, μ_H, increases, the server may complete the service to all HP customers, and the LP customers obtain a successful chance of retrial.*

Example 6. *In this example, we deal with the expected number of HP customers in the waiting hall (E_{WH}) and of LP customers in the orbit (E_O). The results are tabulated in Tables 8 and 9. The observations are as follows.*

Case (i): Non-preemptive priority discipline.

At $r = 0$, E_{WH} increase with N, λ_H and γ and decreases with increase in μ_H and r. Additionally, E_O increases with r and λ_H, whereas it decreases with an increase in service rates μ_H, μ_{L_1}, and μ_{L_2}.

Table 7. System performances with respect to r, μ_H, μ_{L_1}, and λ_H.

r	μ_H	μ_{L_1}	E_{HPIL}		E_{SB}		E_{SRR}	
			$\lambda_H = 0.85$	$\lambda_H = 0.95$	$\lambda_H = 0.85$	$\lambda_H = 0.95$	$\lambda_H = 0.85$	$\lambda_H = 0.95$
0.5	2.75	2.40	0.001919	0.002247	0.971408	0.972786	0.015359	0.014580
		2.45	0.001921	0.002249	0.971198	0.972595	0.015423	0.014632
		2.50	0.001923	0.002251	0.970997	0.972412	0.015484	0.014682
	2.80	2.40	0.001921	0.002251	0.970795	0.972169	0.015546	0.014750
		2.45	0.001923	0.002253	0.970575	0.971969	0.015615	0.014806
		2.50	0.001925	0.002255	0.970364	0.971778	0.015679	0.014859
	2.85	2.40	0.001922	0.002255	0.970186	0.971555	0.015732	0.015732
		2.45	0.001923	0.002257	0.969956	0.971346	0.015804	0.015804
		2.50	0.001925	0.002259	0.969736	0.971146	0.015873	0.015873
1	2.75	2.40	0.003858	0.004527	0.970324	0.971718	0.015939	0.015181
		2.45	0.003863	0.004532	0.970121	0.971534	0.015995	0.015224
		2.50	0.003868	0.004536	0.969926	0.971358	0.016048	0.015266
	2.80	2.40	0.003860	0.004536	0.969673	0.971062	0.016135	0.015358
		2.45	0.003865	0.004540	0.969460	0.970869	0.016196	0.015406
		2.50	0.003870	0.004544	0.969255	0.970684	0.016252	0.015451
	2.85	2.40	0.003860	0.004543	0.969027	0.970408	0.016329	0.015536
		2.45	0.003865	0.004547	0.968804	0.970206	0.016394	0.015588
		2.50	0.003870	0.004551	0.968589	0.970012	0.016454	0.015636

Table 8. Expected number of HP customers in the waiting hall.

N	γ	μ_H	$r = 0$		$r = 0.5$		$r = 1$	
			$\lambda_H = 0.85$	$\lambda_H = 0.95$	$\lambda_H = 0.85$	$\lambda_H = 0.95$	$\lambda_H = 0.85$	$\lambda_H = 0.95$
7	0.05	2.75	1.166812	1.294542	1.133508	1.259561	1.092839	1.216619
		2.80	1.155714	1.281839	1.122121	1.246556	1.081005	1.203148
		2.85	1.145370	1.269916	1.111490	1.234334	1.069927	1.190461
	0.1	2.75	1.176179	1.304286	1.138123	1.264415	1.092326	1.216121
		2.80	1.165538	1.292048	1.127166	1.251849	1.080899	1.203065
		2.85	1.155644	1.280587	1.116958	1.240061	1.070223	1.190789
	0.15	2.75	1.185531	1.314007	1.143894	1.270412	1.094334	1.218132
		2.80	1.175302	1.302194	1.133339	1.258259	1.083308	1.205488
		2.85	1.165815	1.291153	1.123528	1.246881	1.073028	1.193620
8	0.05	2.75	1.196611	1.333320	1.162571	1.297436	1.120983	1.253362
		2.80	1.184390	1.319239	1.150062	1.283055	1.108028	1.238515
		2.85	1.173006	1.306032	1.138391	1.269550	1.095910	1.224544
	0.1	2.75	1.207098	1.344284	1.168216	1.303401	1.121370	1.253814
		2.80	1.195381	1.330726	1.156191	1.289522	1.108882	1.239453
		2.85	1.184499	1.318037	1.145001	1.276513	1.097230	1.225963
	0.15	2.75	1.217523	1.355182	1.174969	1.310463	1.124237	1.256745
		2.80	1.206273	1.342103	1.163403	1.297056	1.112209	1.242855
		2.85	1.195845	1.329887	1.152660	1.284515	1.101009	1.229833

Table 8. *Cont.*

N	γ	μ_H	$r=0$		$r=0.5$		$r=1$	
			$\lambda_H=0.85$	$\lambda_H=0.95$	$\lambda_H=0.85$	$\lambda_H=0.95$	$\lambda_H=0.85$	$\lambda_H=0.95$
9	0.05	2.75	1.217496	1.361326	1.182931	1.324789	1.140678	1.279882
		2.80	1.204370	1.346108	1.169522	1.309276	1.126827	1.263907
		2.85	1.192158	1.331848	1.157027	1.294722	1.113888	1.248891
	0.1	2.75	1.228956	1.373376	1.189474	1.331744	1.141857	1.281196
		2.80	1.216383	1.358733	1.176600	1.316790	1.128530	1.265767
		2.85	1.204714	1.345043	1.164633	1.302789	1.116108	1.251293
	0.15	2.75	1.240316	1.385311	1.197085	1.339751	1.145488	1.284957
		2.80	1.228252	1.371198	1.184714	1.325320	1.132663	1.270051
		2.85	1.217078	1.358028	1.173237	1.311835	1.120736	1.256093

Table 9. Expected number of LP Customers in the orbit.

μ_H	μ_{L_2}	μ_{L_1}	$r=0$		$r=0.5$		$r=1$	
			$\lambda_H=0.85$	$\lambda_H=0.95$	$\lambda_H=0.85$	$\lambda_H=0.95$	$\lambda_H=0.85$	$\lambda_H=0.95$
2.75	2.60	2.40	0.168374	0.205167	0.172715	0.210847	0.177406	0.217027
		2.45	0.166878	0.203129	0.171119	0.208675	0.175696	0.214703
		2.50	0.165467	0.201211	0.169613	0.206630	0.174082	0.212513
	2.65	2.40	0.167100	0.203432	0.171412	0.209068	0.176071	0.215202
		2.45	0.165626	0.201427	0.169838	0.206931	0.174385	0.212914
		2.50	0.164236	0.199539	0.168354	0.204918	0.172793	0.210758
	2.70	2.40	0.165892	0.201789	0.170176	0.207385	0.174805	0.213474
		2.45	0.164439	0.199815	0.168624	0.205280	0.173141	0.211219
		2.50	0.163068	0.197957	0.167160	0.203297	0.171571	0.209095
2.80	2.60	2.40	0.162540	0.197251	0.166733	0.202716	0.171263	0.208663
		2.45	0.161115	0.195324	0.165212	0.200662	0.169633	0.206463
		2.50	0.159772	0.193510	0.163777	0.198726	0.168094	0.204388
	2.65	2.40	0.161327	0.195611	0.165492	0.201035	0.169992	0.206938
		2.45	0.159923	0.193714	0.163992	0.199012	0.168384	0.204770
		2.50	0.158599	0.191929	0.162577	0.197107	0.166866	0.202727
	2.70	2.40	0.160177	0.194057	0.164315	0.199443	0.168786	0.205303
		2.45	0.158792	0.192190	0.162836	0.197450	0.167199	0.203167
		2.50	0.157487	0.190431	0.161440	0.195573	0.165701	0.201154
2.85	2.60	2.40	0.157094	0.189915	0.161148	0.195181	0.165528	0.200911
		2.45	0.155734	0.188089	0.159696	0.193232	0.163970	0.198822
		2.50	0.154452	0.186369	0.158325	0.191396	0.162500	0.196853
	2.65	2.40	0.155937	0.188360	0.159964	0.193587	0.164315	0.199275
		2.45	0.154597	0.186563	0.158532	0.191669	0.162779	0.197218
		2.50	0.153333	0.184870	0.157181	0.189861	0.161329	0.195278
	2.70	2.40	0.154839	0.186888	0.158841	0.192078	0.163164	0.197726
		2.45	0.153518	0.185118	0.157428	0.190188	0.161648	0.195698
		2.50	0.152272	0.183450	0.156095	0.188406	0.160217	0.193786

Case (ii): Discretionary priority discipline.

- *Mixed priority: In contrast to case (i), we obtain the same increment and reduction in the case $r = 0.5$ for the parameters, respectively. However, with discretionary priority, E_{WH} decreases and E_O increases.*
- *Preemptive priority discipline: As discussed in case (i), the changes with respect to other parameters at $r = 1$ are in the same order, but E_{WH} decreases, and E_O increases slightly higher compared with case (i). With this, we conclude that the waiting hall accumulates non-preemptive priority, and the customers in orbit increase with preemptive priority. This also affects the business environment and once again shows the importance of discretionary priority discipline.*

5. Conclusions

The queueing–inventory model studied here considered two classes of customers who require two stages of service for LP customers and a single stage of service for HP customers. In this model, the case $r = 0$ provides the result of a non-discretionary priority service process. In contrast, the case $r \in (0, 1]$ provides the result of a discretionary priority service process. In this paper, a comparison between non-discretionary and discretionary priority service discipline is carried out. We conclude that each of them shows a significant result for discretionary priority rather than non-discretionary priority from the numerical illustrations. A priority discipline, either preemptive or non-preemptive, adopted individually might reflect negatively on the growth of the customer base, the organization's reputation, and the company's profit. Therefore, to overcome such economic issues, the organization has to plan wisely when it comes to service provision. Our model recommends that through discretionary priority discipline, the above mentioned issues can be overcome to a certain extent. This work is an attempt to utilize discretionary priority in the queueing–inventory system. The model can be extended by changing the arrival rate from the Poisson process to the Markovian process. This model deals with the discretion rule based on the predetermined inventory level. Instead, a discretion rule based on service time can be formulated, which is another extension.

Author Contributions: Conceptualization, K.J.; data curation, K.J. and R.H.; formal analysis, K.J., R.H., and N.A.; funding acquisition, G.P.J. and C.S.; investigation, N.A. and C.K.; methodology, S.V. and N.A.; project administration, G.P.J.; resources, G.P.J. and C.S.; software, S.V.; supervision, C.S.; validation, C.K.; visualization, R.H. and C.K.; writing—original draft preparation, K.J.; writing—review and editing, G.P.J. All authors have read and agreed to the published version of the manuscript.

Funding: This work was supported by the Institute of Information and Communications Technology Planning and Evaluation (IITP) Grant by the Korean Government through MSIT (development of user identity certification and management technology for self-sovereign identity applications.) under Grant 2021-0-00565.

Institutional Review Board Statement: Not applicable.

Informed Consent Statement: Not applicable.

Data Availability Statement: Not applicable.

Acknowledgments: Anbazhagan would like to thank RUSA Phase 2.0 (F 24-51/2014-U), DST-FIST (SR/FIST/MS-I/2018/17), DST-PURSE 2nd Phase programme (SR/PURSE Phase 2/38), Govt. of India.

Conflicts of Interest: The authors declare no conflict of interest.

References

1. Melikov, A.Z.; Molchanov, A.A. Stock optimization in transport/storage. *Cybern. Syst. Anal.* **1992**, *28*, 484–487. [CrossRef]
2. Sigman, K.; Simichi-Levi, D. Light traffic heuristic for an $M/G/1$ queue with limited inventory. *Ann. Oper. Res.* **1992**, *40*, 371–380. [CrossRef]
3. Berman,O.; Kim, E. Stochastic models for inventory management at service facility. *Commun. Stat. Stoch. Model.* **1999**, *15*, 695–718. [CrossRef]

4. Berman, O.; Sapna, K.P. Inventory Management for service facility for systems with arbitrarily distributed service time. *Commun. Stat. Stoch. Model.* **2000**, *16*, 343–360. [CrossRef]
5. Sangeetha, N.; Sivakumar, B. Optimal service rates of a perishable inventory system with service facility. *Int. J. Math. Oper. Res.* **2020**, *16*, 515–550. [CrossRef]
6. Jeganathan, K.; Selvakumar, S.; Anbazhagan, N.; Amutha, S.; Porpattama, H. Stochastic modeling on $M/M/1/N$ inventory system with queue-dependent service rate and retrial facility. *Aims Math.* **2021**, *6*, 7386–7420. [CrossRef]
7. Jacob, J.; Shajin, D.; Krishnamoorthy, A.; Vishnevsky, V.; Kozyrev, D. Queueing-Inventory with One Essential and m Optional Items with Environment Change Process Forming Correlated Renewal Process (MEP). *Mathematics* **2022**, *10*, 104. [CrossRef]
8. Sugapriya, C.; Nithya, M.; Jeganathan, K.; Anbazhagan, N.; Joshi, G.P.; Yang, E.; Seo, S. Analysis of Stock-Dependent Arrival Process in a Retrial Stochastic Inventory System with Server Vacation. *Processes* **2022**, *10*, 176. [CrossRef]
9. Klimenok, V.I.; Dudin, A.N.; Vishnevsky, V.M.; Semenova, O.V. Retrial $BMAP/PH/N$ Queueing System with a Threshold-Dependent Inter-Retrial Time Distribution. *Mathematics* **2022**, *10*, 269. [CrossRef]
10. Amirthakodi, M.; Sivakumar, B. An inventory system with service facility and feedback customers. *Int. J. Ind. Syst. Eng.* **2019**, *33*, 374–411.
11. Jeganathan, K.; Anbazhagan, N.; Amutha, S.; Vinitha, V. A queueing inventory system with threshold based priority service and a flexible server. *J. Stat. Manag. Syst.* **2020**, *24*, 663–684. [CrossRef]
12. Jeganathan, K.; Selvakumar, S.; Saravanan, S.; Anbazhagan, N.; Amutha, S.; Cho, W.; Joshi, G.P.; Ryoo, J. Performance of Stochastic Inventory System with a Fresh Item, Returned Item, Refurbished Item, and Multi-Class Customers. *Mathematics* **2022**, *10*, 1137. [CrossRef]
13. Jeganathan, K.; Reiyas, M.A.; Selvakumar, S.; Anbazhagan, N. Analysis of Retrial Queueing-Inventory System with Stock Dependent Demand Rate: (s,S) Versus (s,Q) Ordering Policies. *Int. J. Appl. Comput. Math.* **2020**, *5*, 1–27. [CrossRef]
14. Wang, J.; Qiu, Q.; Wang, H. Joint optimization of condition-based and age-based replacement policy and inventory policy for a two-unit series system. *Reliab. Eng. Syst. Saf.* **2021**, *205*, 107251. [CrossRef]
15. Zhang, J.; Zhao, X.; Song, Y.; Qiu, Q. Joint optimization of condition-based maintenance and spares inventory for a series–parallel system with two failure modes, *Comput. Ind. Eng.* **2022**, *168*, 108094. [CrossRef]
16. Krishnamoorthy, A.; Joshua, A.N.; Kozyrev, D. Analysis of a Batch Arrival, Batch Service Queuing-Inventory System with Processing of Inventory While on Vacation. *Mathematics* **2021**, *9*, 419. [CrossRef]
17. Maqbali, K.A.K.A.; Joshua, V.C., Krishnamoorthy A. On a Single Server Queueing Inventory System with Common Life Time for Inventoried Items. *Commun. Comput. Inf. Sci.* **2021**, *1391*, 186–197.
18. Reshmi, P.S.; Jose, K.P. A queueing–inventory system with perishable items and retrial of customers. *Malaya J. Math.* **2019**, *7*, 165–170. [CrossRef]
19. Sivakumar, B.; Arivarignan, A. A modified lost sales inventory system with two types of customers. *Qual. Technol. Quant. Manag.* **2008**, *5*, 339–349. [CrossRef]
20. Uzunoglu Kocer, U.; Yalçin, B. Continuous review (s, Q) inventory system with random lifetime and two demand classes. *Opsearch* **2019**, *57*, 104–118. [CrossRef]
21. Yadavalli, V.S.S.; Anbazhagan, N.; Jeganathan, K. A retrial inventory system with impatient customers. *Int. J. Appl. Math. Inf. Sci.* **2015**, *9*, 637–650.
22. Zhang, J.; Zhao, X.; Song, Y.; Qiu, Q. Joint optimization of maintenance and spares ordering policy for a use-oriented product-service system with multiple failure modes. *Appl. Stoch. Model. Bus Ind.* **2021**, *37*, 1123– 1142. [CrossRef]
23. Falin, G.I.; Artalejo, J.R.; Martin, M. On the single server retrial queue with priority customers. *Queueing Syst.* **1993**, *14*, 439–455. [CrossRef]
24. Jeganathan, K.; Anbazhagan, N.; Kathiresan, J. A Retrial Inventory System with Non-preemptive Priority Service. *Int. J. Inf. Manag. Sci.* **2013**, *24*, 57–77.
25. Chakravarthy, S.R. A Dynamic Non-preemptive Priority Queueing Model with Two Types of Customers. In Proceedings of the Mathematics and Computing, ICMC, Varanasi, India, 9–11 January 2018; Springer: Singapore, 2018; Volume 253, pp. 23–42
26. Korenevskaya, M.; Zayats, O.; Ilyashenko, A.; Muliukha, V. Retrial Queuing System with Randomized Push-Out Mechanism and Non-Preemptive Priority. *Procedia Comput. Sci.* **2019**, *150*, 716–725. [CrossRef]
27. Krishnamoorthy, A.; Divya, V. (M, MAP)/(PH, PH)/1 Queue with Non-preemptive Priority and Working Vacation under N-Policy. *J. Indian Soc. Probab. Stat.* **2020**, *21*, 69–122. [CrossRef]
28. Krishna, Kumar, B.; Pavai, Madheswari, S.; Arivudainambi, D. 2005. On the Busy Period of an $M/G/1$ Retrial Queueing System with Two-Phase Service and Preemptive Resume. *Stoch. Model. Appl.* **2005**, *8*, 18–34.
29. Tarabia, A.M.K. Analysis of $M/M/1$ Queueing System with two Priority Classes. *OPSEARCH* **2007**, *44*, 346–365. [CrossRef]
30. Jeganathan, K.; Kathiresan, J.; Anbazhagan, N. A retrial inventory system with priority customers and second optional service. *Opsearch* **2016**, *53*, 808–834 [CrossRef]
31. Gao, S.; Wang, J. Stochastic analysis of a preemptive retrial queue with orbital search and multiple vacations. *Rairo-Oper. Res.* **2018**, *54*, 231–249. [CrossRef]
32. Adiri, I.; Domb, I. A Single Server Queueing System Working under Mixed Priority Discipline. *Oper. Res.* **1982**, *30*, 97–115. [CrossRef]

33. Cho, Y.; Un, C. Analysis of the M/G/1 Queue under a combined Preemptive/Nonpreemptive Priority Discipline. *IEEE Trans. Commun.* **1993**, *41*, 132–141. [CrossRef]
34. Fajardo, V. A.; Drekic, S. On a General Mixed Priority Queue with Server Discretion. *Stoch. Model.* **2016**, *32*, 1–31. [CrossRef]
35. Avi-Itzhak, B.; Brosh, I.; Naor, P. On discretionary priority queueing. *Zamm-J. Appl. Math. Mech. Fiir Angew. Math. Mech.* **1964**, *44*, 235–242. [CrossRef]
36. Melkonian, V.; Kaiser, M. Discretionary Priority Discipline: A Reasonable Compromise between Preemptive and Nonpreemtive Disciplines. *Appl. Math. Lett.* **1996**, *9*, 91–94. [CrossRef]
37. Lian, Z.; Zhao, N. A two-stage M/G/1 queue with discretionary priority. In Proceedings of the 2011 IEEE International Conference on Industrial Engineering and Engineering Management, Singapore, 6–9 December 2011; pp. 1402–1406.
38. Zhao, N.; Guo, Y.; Lian, Z.; Wang, M. Analysis of a $MAP/PH/1$ queue with discretionary priority. In Proceedings of the 2015 Winter Simulation Conference (WSC), Huntington Beach, CA, USA, 6–9 December 2015; pp. 2872–2883.
39. Lian, Z.; Zhao, N. A Two-Stage Discretionary Priority Service System with Markovian Arrival Inputs. In Proceedings of the 2010 IEEE International Conference on Industrial Engineering and Engineering Management, Macao, China, 7–10 December 2010; pp. 438–442.
40. Zhao, N.; Lian, Z.; Wu, K. Analysis of a $MAP/PH/1$ Queue with Discretionary Priority Based on Service Stages. *Asia-Pac. J. Oper. Res.* **2015**, *32*, 1550042. [CrossRef]
41. Neuts, M. *Matrix Geometric Solutions in Stochastic Models: An Algorithmic Approach*; John Hopkins University Press: Baltimore, MD, USA, 1981.

sustainability

Article

Research on the Relationship between Digital Transformation and Performance of SMEs

Xiaoyan Teng [1], Zhong Wu [2,*] and Feng Yang [1]

[1] Business School, University of Shanghai for Science and Technology, Shanghai 200093, China; 201670074@st.usst.edu.cn (X.T.); 211420092@st.usst.edu.cn (F.Y.)
[2] School of Management, Shanghai University of International Business and Economics, Shanghai 201620, China
* Correspondence: wuzhong_1968@163.com

Abstract: Objective: Through an empirical analysis of the performance of SMEs undergoing digital transformation, this study attempts to identify the influencing factors that determine their sustainable development to provide reference for academic researchers and industrial decision makers. Method: This study first uses an interview method to investigate the impact of SMEs' three main resources on digital transformation: digital technology, employee digital skills, and digital transformation strategy. Second, we assess the impact of digital transformation on financial performance. Using the structural equation model, 335 valid questionnaires were recovered through the questionnaire method, and the key factors were identified using SPSS and SPSSAU tools. Results: In the Chinese context, digital transformation affects SME performance, and the three resources mentioned above are positively correlated with SMEs' digital transformation. Digital transformation is positively correlated with performance, and it is the mediator of the impact of digital transformation strategies on performance. Conclusion: For SMEs, focusing on investing in digital technologies, employee digital skills, and digital transformation strategies are three key factors that are beneficial for digital transformation, thus helping to improve performance and maintain their sustainable development.

Keywords: resource-based view; digital transformation; small- and medium-sized enterprises; performance; sustainable development

Citation: Teng, X.; Wu, Z.; Yang, F. Research on the Relationship between Digital Transformation and Performance of SMEs. *Sustainability* **2022**, *14*, 6012. https://doi.org/10.3390/su14106012

Academic Editor: Nicola Saccani

Received: 21 April 2022
Accepted: 11 May 2022
Published: 16 May 2022

Publisher's Note: MDPI stays neutral with regard to jurisdictional claims in published maps and institutional affiliations.

1. Introduction

The fourth industrial revolution, marked mainly by digital transformation, is booming. The essence of digital transformation is to use the digital characteristics of copying, linking, simulation, and feedback to quantify all aspects of the business. With clear quantitative indicators and data, targeted analysis and optimization can be carried out, and management can be refined to every detail, thereby improving the overall operating level of the enterprise. Small- and medium-sized enterprises (SMEs) should adapt to the new round of scientific and technological revolution and the trend of industrial transformation; grasp the dividends of digital technology; and improve their abilities of rapid perception, agile response, and intelligent decision making in the digital age to enhance their ability to deal with risks and sustainable development. Judging from the successful cases of many traditional industries, it is normal for enterprises to increase the efficiency of each link by 30–50% after completing the digital transformation, and the entire enterprise can improve its operating efficiency by 8–10 times.

The coronavirus disease 2019 (COVID-19) pandemic has further accelerated the process of digital transformation of SMEs [1], as well as the development strategy of China's new infrastructure; thus, there are more SMEs as research objects. Therefore, in this study, we focus on digital transformation. However, digital transformation is a long-term complex system engineering that consists of four consecutive stages of digital transformation to support the enterprise's digital system development [2]. Moreover, digital transformation is "the most common and complex stage" [3]. Most SMEs, with their limited resources, have difficulties in dealing with this complex situation. We exploratively survey SMEs that have implemented digital transformation, trying to find the key factors (resources) that affect digital transformation to solve the difficulties and challenges of the digital transformation of SMEs.

We believe that the purpose of digital transformation is innovation. Through digital transformation, SMEs have found a new paradigm for development [4]. Compared with large enterprises, SMEs have the advantage of flexibility and can accept innovation. Therefore, as a method of organizational change, it is relatively easy for SMEs to carry out digital transformation. Some scholars have found from the perspective of VUCA that during the COVID-19 pandemic and blockade period, digitally immature organizations are vulnerable, while organizations with a high degree of digitalization are usually more flexible [5].

Although numerous academic studies on digital transformation have appeared recently, it should be pointed out that these studies are mostly aimed at digital native enterprises, platform enterprises, and large enterprises. Owing to limited resources, SMEs have slowed the digital transformation process. Therefore, the empirical research on the digital transformation of SMEs is insufficient. Relevant empirical research focuses on the success factors of the digital transformation of SMEs [6], but there is a lack of empirical research on the relationship between different stages of digital transformation and performance [7]. Therefore, our research goal is to investigate the current stage of digital transformation of small- and medium-sized enterprises, find out the key influencing factors of digital transformation from the perspective of the enterprise, and verify the impact mechanism of digital transformation.

The research object of this paper is small- and medium-sized enterprises in China. Based on previous research, a resource-based digital transformation performance framework is adopted. Three influencing factors of digital transformation of small- and medium-sized enterprises are identified, and the influence mechanism is empirically analyzed using the structural equation model. The study found that the success of digital transformation requires a combination of three factors: digital technology, digital skills, and digital transformation strategy. Digital technology is the foundation of digital transformation, digital skills are the key to digital transformation, and digital transformation strategy is the primary task of digital transformation. This study enriches and expands the research in the field of digital transformation, helps to deepen the knowledge and understanding of digital transformation of SMEs, and promotes the success of digital transformation by investing in key resources. In addition, it provides a path, method, and reference for the management practice of SMEs.

2. Literature Review

2.1. Digital Transformation

Recent research helps us to understand the process of digital transformation of enterprises, and the perspective of the existing literature helps to clarify the concept and definition of digital transformation. After reviewing 23 typical definitions, scholar Vial believes that digital transformation refers to the process of triggering organizations to make strategic responses through digital technologies such as information, computing, and communication, changing their structure, boundaries and even value generation paths, and then realizing the process of enterprise entity evolution [8]. Some scholars also believe that digital transformation is a high-level transformation that is based on digitization and digitalization, further touches the company's core business, and aims to create a new business model. Digital transformation entails the development of digital technology and support capabilities to create a dynamic digital business model [9]. Other scholars believe that enterprise digital transformation refers to the process of triggering major changes in enterprise organizational characteristics and reconstructing the organizational structure, behavior, and operating system through the combined application of information technology (IT), computing, communication, and connection technologies [10].

Based on previous research, we believe that digital transformation is anchored in digital technologies, including artificial intelligence, big data, cloud computing, and blockchain, to empower enterprises and vigorously develop new technologies, new products, new models, and new formats, such that enterprises can obtain a sustainable development model with diversified efficiency. This concept embodies the four aspects that enterprises need to grasp systematically to carry out digital transformation. First, digital transformation is a systemic change triggered by digital technology. Second, the main task of digital transformation is the reconstruction and innovation of value systems. Third, the core path of digital transformation is the ability to form new kinetic energy, continuously create new value, and achieve sustainable development. Ultimately, the key elements of digital transformation are technology, people (skills), and strategies that are appropriate to the stage of the business. Therefore, in the digital transformation process, enterprises need a clear strategy (path), correct method (technology), suitable talent, and so on to form an effective digital transformation system, systematically promote transformation and change, and accelerate the progress of the innovation stage of continuous development.

2.2. Key Factors for the Digital Transformation of SMEs

The digital transformation of SMEs is a path of differentiation that is inherently constrained by industry, scale, stage, technology, and resources. Meanwhile, an enterprise's digital transformation is a complex systematic project. It entails not only the use of online payment and sales tools, but also transformation model selection, business model innovation, organizational structure optimization, and asset management changes. Systematic digital empowerment is carried out through R&D, production, sales, warehousing, logistics, marketing, and other links, and finally, the industrial closed loop is realized, involving all aspects of intelligent production, digital logistics, data risk, and other issues.

Some scholars have found that SMEs face many difficulties in digital transformation [11]. First, SMEs lack adequate knowledge about themselves. Digital transformation is not only a technological update, but also an all-round change in business philosophy, strategy, organization, and operation, which requires overall planning. Most SMEs have a strong desire for digital transformation, but they generally lack clear strategic goals and practical paths, and they face challenges in technology, business capacity building, and talent training. Second, the application of digital technology in my country's SMEs is not high at present, and more than 70% of SMEs have not yet undergone large-scale digital transformation. The application level of digital technology is not high, which is reflected not only in the application depth of digital technology, but also in the application breadth of digital technology. Finally, digital transformation requires the support of financial, material, human, and other resources. Insufficient strength or following the current trend of transformation will not contribute to the sustainable development of SMEs.

In recent years, several studies have been conducted on the factors that influence the success of digital transformation. Some studies report that the six dimensions of an enterprise's strategic vision, consistency of vision and digital transformation investment, suitability of innovation culture, possession of sufficient intellectual property assets and know-how, strength of digital capabilities, and use of digital technology can be improved. This enables a business to achieve a successful competitive position through digital transformation [12]. Some studies have also found that there are four factors for successful digital transformation: customer centricity, governance, innovation, and resource acquisition [13]. "In addition to technology adoption, important factors for successful digital transformation are the organization's ability to change and its operational excellence in integrating external digital services with internal IT support" [14]. Support from the top management team is also key to digital transformation [15]. Digital transformation can be successful for SMEs through planning, aligning organizational interests, consistent and regular communication, providing resources and tools, mobilizing people, and establishing accountability and timelines for deliverables [16]. A unified strategic objective is also critical for successful and rapid digital transformation [17]. There are 11 determinants of success that are critical to SMEs' digital transformation efforts, with external support for digital being the first step and operational technology readiness being the most challenging determinant of success [18]. Cooperation is reflected in various stages of SME digital transformation projects, such as business needs adjustment, project portfolio creation, technology solution selection, and later stages [19]. Studies have also found that three major resources—IT, human resources, and business strategy—have a positive impact on the digital transformation of SMEs; however, these factors are barriers to the digital transformation of these enterprises. Furthermore, digital transformation has a positive impact on the business outcomes of SMEs [20]. From a holistic perspective, six key factors for the digital transformation of SMEs are identified from the three dimensions of technology, organization, and environment: government support, partnership, top management, digital strategy, IT infrastructure, and IT management capabilities [6]. Employee skills are the moderating factor. Digital transformation is a long-term and arduous task, facing challenges in technology, business capacity building, talent training, and so on, but it is not unattainable. Only by focusing on key elements and by adding external assistance according to their own strength can SMEs iteratively innovate (Table 1).

Table 1. Summary of the literature reviewed.

Reference	Method	Context	Key Factors
[12]	Literature review	Media company, lodging company, retail company, pharmaceutical company	Strategic vision, culture of innovation, know-how and intellectual property, digital capability, strategic alignment, technology assets
[13]	Qualitative research	Large organization	Customer centricity, governance, innovation, and resource attainment.
[14]	Multiple case studies	Three companies from different industries that are in different stages of digital transformation	Technology adoption, the ability of an organization to change and operational excellence in the integration of external digital services with internal IT support
[15]	In-depth interview	Large German firms	Top managers: understanding digitalization, setting the formal context for digitalization, and leading change
[16]	Exploratory factor analysis	Education	Planning, aligned organizational interests, consistent and regular communication, provision of resources and tools, engaging faculty and creating accountability and timelines with deliverables
[17]	Case study	Public health system	A unified purpose is essential to successful, rapid digital transformation
[18]	Exploratory factor analysis	Small and medium enterprises	Eleven success determinants, namely: business partner digital maturity, cybersecurity maturity, change management competency, digitalization readiness preassessment, external support for digitalization, information and digital technology expertise, information and digital technology readiness, management competency for digital transformation, manufacturing digitalization strategic road mapping, operations technology readiness and resource availability
[19]	Longitudinal case studies	Two manufacturing SMEs	The collaboration manifests itself at various stages of the transformation projects, such as the business needs alignment, project portfolio creation, technology solution selection, and post-mortem phase
[20]	Qualitative research	Small and medium enterprises (SMEs)	Three main resources (IT, human resources, and business strategy) have a positive impact on the digital transformation of SMEs
[7]	Qualitative research	Small and medium enterprises (SMEs)	Technological and environmental factors have a positive impact on organizational capabilities, and then promote the success of DT of SMEs. Organizational capabilities play an intermediary role in the influence of technological and environmental factors on DT. In addition, employee skills positively moderate the relationship between organizational capabilities and the success of DT

3. Materials and Methods

3.1. Resource-Based View

The American economist Penrose was one of the earliest scholars to advocate for the concept of enterprise resources. Penrose defines resources as "the physical objects purchased, leased, or manufactured by an enterprise for its own use and the labor force employed on certain terms." She also emphasizes that the heterogeneity (rather than homogeneity) of productive or potentially productive services produced by firms from their own resources gives each firm its unique characteristics [21]. The publication of Wernerfelt's "Resource-based Theory of Enterprises" in 1984 signified the birth of resource-based theory [22]. Barney's research suggests that there may be a kind of heterogeneity or difference between companies and that some companies maintain a competitive advantage [23]. Therefore, resource-based theory emphasizes strategic choices and believes that the strategic task of corporate management is to find, develop, and allocate this part of the distinctive key resources in order to maximize business returns. According to Peteraf's point of view, the resources controlled by an enterprise must satisfy four conditions (i.e., valuable, rare, inimitable, non-substitutable) to generate sustainable competitive advantage for the enterprise [24]. Adopting the resource-based view and the inheritance and development of this theory, Teece et al. defined dynamic capability as the ability of a company to integrate, construct, and reconfigure internal and external capabilities to cope with a rapidly changing environment [25]; from here, dynamic capability theory gradually formed and then developed rapidly.

At different stages of digital transformation, enterprises have different requirements for organizational structure, organizational culture, growth strategy, and digital and other related resources or capabilities, and their coordination and adaptation will help to promote enterprise digital transformation [7]. However, some studies have found that SMEs have difficulty starting their digital journey because of a lack of resources and expertise. As an enabling mechanism, dynamic capabilities can facilitate digital transformation [26]. Additionally, perception ability, learning ability, integration ability, digital leaders, and key resources have an important impact on the digital transformation of enterprises, among which perception ability and learning ability are also important "triggers" for enterprise digital transformation [27]. While mediating the relationship between digital transformation and enterprise performance, dynamic capabilities also mediate the joint effect of digital transformation and individual forgetting on innovation performance, and the joint effect of digital transformation and entrepreneurial orientation on innovation performance. The three interactive effects of entrepreneurial orientation also play a mediating role in the relationship between performance [28].

By embracing resource-based theory (including dynamic capability theory), we reviewed the relevant literature to better understand our assumptions about digital transformation and performance and the three key resources of SMEs—digital technology, employee digital skills, and digital transformation strategy.

3.2. Proposed Conceptual Model and Hypothesis Formulation

Resource-based theory posits that an enterprise is a collection of various resources. Owing to various reasons, the resources owned by enterprises are different and heterogeneous, and this heterogeneity determines the differences in the competitiveness of enterprises. However, resources can only be used as a basis for competitive advantage if they conform to the VRIN framework. Specifically, VRIN refers to valuable resources (which are the basis for companies to conceive and execute corporate strategies, as well as improve efficiency and effectiveness); rare (that is, scarce) resources (because even if the resource is valuable, once owned by most companies, it cannot bring competitive advantage or sustainable competitive advantage); imperfectly imitable resources (that is, resources that cannot be imitated, which generally need to have the following three characteristics simultaneously: unique historical conditions, vague causes, and social complexity); and non-substitutable resources (that is, irreplaceable resources, which do not have a substitute that is both replicable and not scarce). This is because enterprises have different tangible and intangible resources, which can be transformed into unique capabilities, and resources are immobile and difficult to replicate among enterprises; these unique resources and capabilities are the source of lasting competitive advantages for enterprises [22]. Compared with the external conditions of the enterprise, the internal conditions of the enterprise play a decisive role in obtaining competitive advantage in the market. Therefore, our research focused on the organization itself and identified three key resources (elements) that SMEs in digital transformation need: digital technology, digital skills, and digital transformation strategy. We use structural equation modeling to verify the impact of digital technology (x_1), digital skills (x_2), and digital transformation strategy (x_3) on digital transformation (Y_1) and performance (Y_2). See Figure 1.

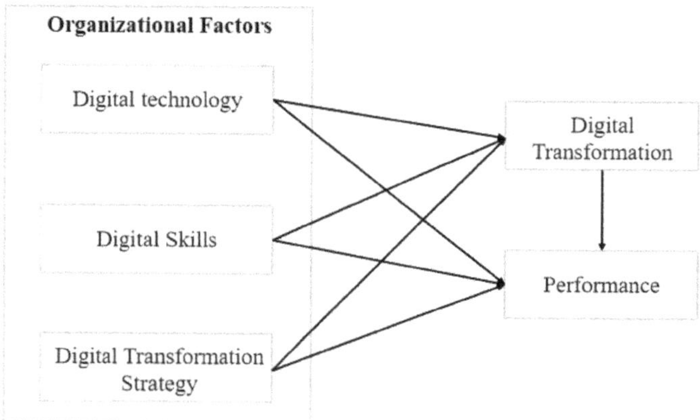

Figure 1. Conceptual model and research hypothesis.

The model parameters are as follows:

$$y_i = a + By_i + \Gamma x_i + \zeta_i \ \mathrm{var}(\zeta_i) = \Psi$$

Given p-endogenous and q-exogenous variables:

y_i is a p × 1 vector of observed endogenous variables
x_i is a q × 1 vector of observed exogenous variables
a is a p × 1 vector of regression intercepts
B is a p × p matrix of regression slopes
Γ is a p × q matrix of regression slopes
ζ is a p × 1 vector of disturbances (i.e., residuals)
Ψ is a p × p covariance matrix of disturbances

3.2.1. Digital Technology

The basic idea behind the implementation of digital transformation by SMEs is to change work and business processes based on digital-technology-driven improvements; in this way, business operations can be made more efficient. Artificial intelligence has been proposed for decades, but it has only experienced explosive growth in the past two years. The reason lies in the increasing maturity of technologies such as cloud computing, the Internet of things (IoT), and big data. Cloud computing provides an open platform for artificial intelligence, the IoT ensures real-time data sharing, and big data provides unlimited resources for deep learning. In addition, the development of blockchain technology in 2018 can make up for the shortcomings of artificial intelligence in data security and data factors, and provide a reliable guarantee for the application of artificial intelligence in different scenarios. Some studies have found that digital transformation can be achieved through artificial intelligence [29]. The IoT plays an important role in digital transformation [30,31]. Blockchain technology affects digital transformation [32,33]. Notably, 5G networks also play a role in digital transformation [34]. The use of social media in management affects the overall performance of enterprises [35]. In summary, driven by the digital technology represented by cloud computing, big data, artificial intelligence, and blockchain, the value chain is breaking and reshaping, new ecosystems are emerging, and industry boundaries are becoming increasingly blurred. An increasing number of traditional enterprises are trying to calmly restructure their business models to cope with industry changes and expand their businesses across traditional industry boundaries.

Hypothesis 1 (H1). *Digital technology is positively correlated with the digital transformation of SMEs.*

3.2.2. Digital Skills

In the digital transformation process, an increasing number of companies are beginning to realize the importance of talent team building, because the implementation of digital strategies requires the support of digital talents with business capabilities, overall views, as well as digital concepts and skills. Kane believes that people are the real key to digital transformation [36]. Therefore, the self-management team is key to starting the digital transformation of enterprise management [37]. The rapid development of employees' individual cognitive and process capabilities promotes the digital transformation process of enterprises [38]. Bessonova et al. discussed the digital literacy of employees as readiness for digital transformation [39]. Studies have found that employees' digital mindset can affect their participation in or exit from a company's digital transformation program [40]. Personal digital ability is related to SME growth and innovation performance [41]. However, some companies have not yet started digital transformation, which is also due to the lack of human capital and the digital skills of employees [42]. Therefore, organizations must develop virtual human resource development and use learning resources to prepare for the future [43]. The empowerment of the organization is effective only when the right people are selected, and everything we do in the future will be multiplied with half the effort. Therefore, regarding digital transformation, the biggest challenge of the organization is not the specific business, but the people.

Hypothesis 2 (H2). *Employee digital skills are positively correlated with SME digital transformation.*

3.2.3. Digital Transformation Strategy

A digital transformation strategy is a prerequisite for successful digital transformation. By developing an effective, clear, and sound digital transformation strategy, one can ensure that one's digital transformation is as seamless as possible. A digital transformation strategy is like a personalized map that can bring great value in business transformation. The formulation and implementation of a digital transformation strategy has become a key concern for organizations prior to digital transformation across many traditional industries [44]. Four common types of digital transformation strategies are formed via two dimensions: the use of digital technologies and the preparation of business models for digital operations [45]. Studies have found that SMEs are strengthening their digital transformation strategies through innovative technologies and new values that restructure business models and processes [46]. Presently, many enterprises have completed the strategy formulation stage of digital transformation and are in the strategy implementation stage [47]. A new strategic implementation framework can be adopted in digital transformation, which is divided into three phases: planning, implementation, and review [48]. In summary, digital transformation is the best strategy for an enterprise, and it must be reflected in the whole process execution of business, operation, and performance appraisal. It is no longer sufficient to stay at the technical level; it is necessary to digitalize all aspects of decision making, work, and cooperation, and provide the best experience for customers.

Hypothesis 3 (H3). *Digital transformation strategies are positively correlated with the digital transformation of SMEs.*

3.2.4. Digital Transformation and Performance

Existing research finds that digital transformation has an impact on corporate financial performance [49]. Enterprise digital transformation has a positive effect on enterprise performance [50]. The positive impact of digital transformation on enterprise performance is more evident in large enterprises, state-owned enterprises, mature enterprises, and non-manufacturing (service) enterprises. The positive impact of digital transformation on enterprise performance and supply chain integration is not significant in small- and medium-sized enterprises; supply chain integration cannot play an intermediary role in the relationship between the digital transformation of manufacturing enterprises and enterprise performance [51]. Three major resources (IT, human resources, and business strategy) have a positive impact on the digital transformation of SMEs, and digital transformation has a positive impact on the business results of SMEs [20]. Digital transformation has a positive effect on the dynamic capabilities of enterprises; the higher the intensity of individual forgetting and the stronger the entrepreneurial orientation, the stronger the positive impact of digital transformation on the dynamic capabilities and innovation performance of enterprises [28]. Some of the benefits of digital transformation for enterprises are the transformation and optimization of traditional stock businesses through digital technology; improving the level of large-scale production and trading of traditional products; and realizing value benefits, such as efficiency improvement, cost reduction, and quality improvement.

Hypothesis 4 (H4). *Digital transformation is positively correlated with financial performance.*

3.2.5. Complementarity of Resources

Digital technology, employee digital skills, and digital transformation strategies are the three core elements for SMEs to carry out digital transformation from point to face, local to global, and in stages. All three are indispensable in the ongoing digital transformation process. Previous research has found that, in addition to technology adoption, important factors for successful digital transformation are the organization's ability to change and its operational excellence in integrating external digital services with internal IT support [14]. Another study reports that digital transformation at the enterprise level requires attention to the alignment of strategy, vision, and digital transformation investment; the suitability of innovation culture; intellectual property and know-how; the strength of digital capabilities; and the use of digital technology [12]. Digital transformation strategies have a positive impact on short- and long-term financial performance [52]. There is a significant correlation between big data, IoT, blockchain, and performance [53]. Other studies suggest that the variables that have the greatest impact on SME performance are management, technology, and marketing and innovation capabilities [54]. In short, the digital transformation of an enterprise is a dynamic process. Continuous iteration, the development of digital products and services based on data, and the realization of intelligent operations through digital technology are essential. When digital technology penetrates into every corner of an enterprise, continuous and rapid innovation capabilities will become the core driving force for enterprises to maintain competitiveness, thereby helping enterprises to make steady progress on the road of digital transformation. Enterprises that achieve intelligent operations can make real-time and correct decisions; fully integrate talent, data, and intelligent technology; drive process transformation and incorporate agility and rapid response capabilities; continuously improve user experience; and achieve breakthrough business results. Sustainable digital transformation can only be achieved through sustainable technological innovation and intelligent operations.

Hypothesis 5 (H5). *The relationship between (a) digital technology, (b) employee digital skills, and (c) digital transformation strategy and financial performance is mediated through digital transformation.*

4. Results

4.1. Measurement and Data Collection

We tested five hypotheses among SMEs (1–299 employees) from various provinces within China, and the questionnaire was designed after on-site interviews with SMEs. After the review of relevant experts and the pre-test, we comprehensively considered the analysis results of 30 questionnaires and China's national conditions; adjusted the evaluation of questions, wording, and consistency; and improved the reliability and effectiveness of the questionnaire. The survey targets are executives of SMEs, who are familiar with the specific operation of the company and can ensure that the data truly reflect the actual application of digital technology in SMEs. The quantitative data were collected through an online survey platform that collected 335 complete and correctly filled questionnaires by the end of 2021. See Appendix A.

Independent variable: The design of the independent variables was conducted in advance through corporate interviews. Employee digital skills (DS) uses a seven-point Likert scale ranging from 1 ("strongly disagree") to 7 ("strongly agree"). Digital transformation strategy is evaluated using a seven-point Likert scale ranging from 1 ("strongly disagree") to 7 ("strongly agree"). Digital technology (DT) uses a seven-point Likert scale ranging from 1 ("very low") to 7 ("very high").

Dependent variable: Except for listed companies, Chinese SMEs are not obliged to publish financial data. The Entrepreneur Orientation Scale adopted by Lumpkin and Dess [55] was used. Financial performance variables (FP) were measured using a seven-point Likert scale ranging from 1 ("strongly disagree") to 7 ("strongly agree"). The scale measures how SME executives feel about the financial performance of their companies after digital transformation.

Mediating variable: The mediating variable, digital transformation (DX), is an assessment of an enterprise's digital transformation maturity, using the seven-point Likert scale adopted by Bley et al. [56], ranging from 1 ("very low") to 7 ("very high").

Control variable: Financial performance may be affected by other factors, and control variables can prevent test failures for the dependent variables [57,58]. The control variables are the number of employees, gender of executives, and readiness for digital transformation. The readiness for digital transformation is measured by Xiaodong Ma [59].

4.2. Measurement Model

The SEMs approach used in this study is a method for establishing, estimating, and testing a causal relationship model. The model contains both observable explicit variables and latent variables that cannot be directly observed. Table 2 shows the descriptive statistics of the data sample: medium-sized enterprises account for 77% of the total. The information and communication technology industry, media, and finance accounted for 34%. Levels 2 and 3 of digital transformation are the most common, accounting for 33%.

SPSS 21 software is used to test reliability and validity, and the results are shown in Table 3, evaluated using mean, SD, Cronbach's alpha, CR, and AVE. All α and CR values exceeded the threshold of 0.70, and all AVE values were above the recommended threshold of 0.50; thus, all indicators showed good reliability. We used factor loadings and square root of AVE to measure convergent and discriminant validity, respectively. The results showed that all factor loadings were greater than 0.7 at the significance level of $p < 0.01$, indicating good convergent validity. The correlation coefficient of each factor with other factors confirmed the discriminant validity.

Table 2. Sample demographic information.

Measure	Items	Size	Percentage
Firm scale (people)	Micro enterprises (<10)	1	0.30%
	Small enterprise (10–99)	77	22.99%
	Medium enterprise (100–299)	257	76.72%
Industry	Information and communication technology industry, media, and finance	114	34.03%
	Consumer industry (entertainment and leisure, retail trade, etc.)	68	20.30%
	Service industries (utilities, healthcare, education, etc.)	63	18.81%
	Capital-intensive industries (high-end manufacturing, oil and gas, basic product manufacturing, chemical and pharmaceutical, etc.)	65	19.40%
	Localization industries (agriculture, personal and local services, hospitality, construction, etc.)	25	7.46%
Levels of digital transformation	Level 0: Data not applied (no data analysis tools, lack of data awareness at decision-making level, decision making relying on leadership "whims", etc.)	5	1.49%
	Level 1: Scattered use of data (simple data, used for statistical reports, infrequent use, unable to handle large-scale data sets, etc.)	45	13.43%
	Level 2: Technology center assists decision making (with certain data thinking, used by major technical departments, unable to respond in real time, etc.)	109	32.54%
	Level 3: Systematic operation of the technology center (complex data applications, technology struggling to meet simple needs, low ability to capture business opportunities, etc.)	109	32.54%
	Level 4: Business-centered data-based operations (data analysis, freeing up technical time and manpower, business-oriented data assets, strong ability to capture business opportunities, etc.)	62	18.51%
	Level 5: Data competitiveness formation (innovative business models; data ecology formation; deep accumulation of data, models, and application assets, with data operation and practical experience; etc.)	5	1.49%

Table 3. Descriptive statistics and correlations.

Items	Mean	SD	Cronbach's Alpha	CR	AVE	DX	DS	DTS	DT	FP
DX	4.5711	0.98601	0.751	0.876	0.772	1.00				
DS	4.7373	0.98504	0.732	0.849	0.751	0.625 **	1.00			
DTS	5.4145	0.78292	0.792	0.919	0.854	0.182 **	0.247 **	1.00		
DT	4.8280	0.91577	0.718	0.831	0.754	0.631 **	0.633 **	0.302 **	1.00	
FP	5.1602	0.86940	0.767	0.882	0.834	0.270 **	0.345 **	0.683 **	0.393 **	1.00

Note(s): ** $p < 0.01$, n = 335. Digital transformation (DX); employee digital skills (DS); digital transformation strategy (DTS); digital technology (DT); financial performance (FP).

4.3. Regression Analysis

After factor analysis and reliability tests, we perform multiple regression analysis to test hypotheses H1–H4, and Table 4 shows the results of the multiple regression analysis.

Table 4. Results regression analysis.

Variables	DX	DX	DX	FP
DT	0.000 **			
DS		0.000 **		
DTS			0.001 **	
DX				0.000 **
R2	0.398	0.391	0.033	0.073

Note(s): ** $p < 0.01$, Digital transformation (DX); employee digital skills (DS); digital transformation strategy (DTS); digital technology (DT); financial performance (FP).

4.4. Structural Model

To ensure the consistency of the estimates, we use the "path analysis" in the web version of SPSSAU to estimate the coefficients and significance of each path. Figure 2 shows the path coefficients in the test results obtained using the structural equation model. It can be seen from Figure 2 that when digital transformation has an impact on financial performance, this path does not appear significant (z = 0.158, p = 0.875 > 0.05), thus indicating that digital transformation has no impact on financial performance. When digital skills have an impact on financial performance, this path does not appear significant (z = 1.872, p = 0.061 > 0.05), thus indicating that employees' digital skills have no impact on financial performance. When the digital transformation strategy has an impact on financial performance, the standardized path coefficient value is 0.615 > 0, and this path shows a significant level of 0.01 (z = 15.292, p = 0.000 < 0.01), thus indicating that the digital transformation strategy will have an impact on financial performance (significant positive impact relationship). When digital technology has an impact on financial performance, the standardized path coefficient value is 0.138 > 0, and this path shows a significant level of 0.05 (z = 2.523, p = 0.012 < 0.05), thus indicating that digital technology will have a significant impact on financial performance (positive influence relationship). When employees' digital skills have an impact on digital transformation, the standardized path coefficient value is 0.380 > 0, and this path shows a significant level of 0.01 (z = 7.466, p = 0.000 < 0.01), thus indicating that employees' digital skills will have an impact on digital transformation (significant positive impact relationship). When the digital transformation strategy has an impact on the digital transformation, this path does not appear significant (z = −0.786, p = 0.432 > 0.05), thus indicating that the digital transformation strategy has no influence on the digital transformation. When digital technology has an impact on digital transformation, the standardized path coefficient value is 0.401 > 0, and this path shows a significant level of 0.01 (z = 7.749, p = 0.000 < 0.01), which indicates that digital technology will have a significant impact on digital transformation (positive influence relationship).

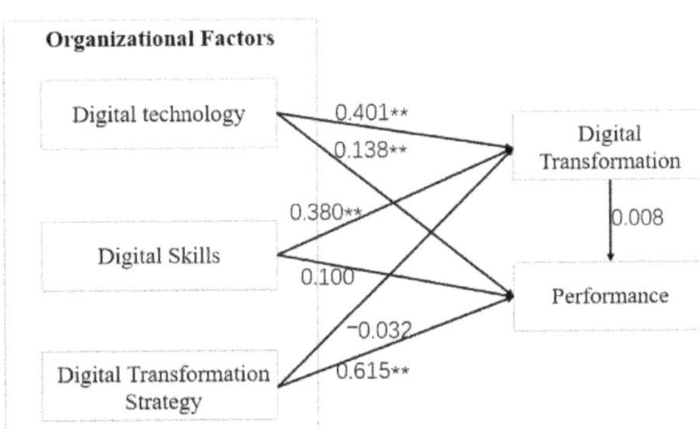

Figure 2. Results of structure model analysis. Note(s): ** $p < 0.01$.

We use structural equation modeling to verify the impact of digital technology (x_1), digital skills (x_2), and digital transformation strategy (x_3) on digital transformation (y_1) and performance (y_2). The results are shown in the following equation: $y_1 = 0.401^*x_1 + 0.38^*x_2 - 0.032^*x_3$, $y_2 = 0.008^*y_1 + 0.138^*x_1 + 0.100^*x_2 + 0.615^*x_3$.

4.5. Mediating Effect

We use the process plug-in in SPSS and the bootstrap method to test the mediation effect. That is, we measure the direct and indirect effects between the mediator variable DX and the dependent variable FP, and the effects of the three independent variables, DT, DS, DTS, and DX. Table 5 shows the results of the mediation analysis.

Table 5. Results of mediation analysis.

Hypothesis	Effects	Pathways	P	Coeff.	95%CI		Mediation Existence
DT- > DX- > FP	Direct Effect	c′	0.000	0.350	0.229	0.471	Mediating effect is not significant
	Indirect Effect	a*b	0.000	0.022	−0.064	0.115	
	Total Effect	c	0.000	0.373 **	0.279	0.466	
DS- > DX- > FP	Direct Effect	c′	0.000	0.255	0.141	0.368	Mediating effect is not significant
	Indirect Effect	a*b	0.000	0.050	−0.032	0.144	
	Total Effect	c	0.000	0.304 **	0.215	0.393	
DTS- > DX- > FP	Direct Effect	c′	0.000	0.728	0.641	0.815	Partial
	Indirect Effect	a*b	0.001	0.030	0.006	0.055	
	Total Effect	c	0.000	0.759 **	0.671	0.846	

** $p < 0.01$.

In terms of the mediation path DT > DX > FP, findings showed that no clear mediating role between digital technology and financial performance. Its structural equation model results are as follows: $y_2 = 3.361 + 0.373^*x_1$, $y_1 = 1.290 + 0.680^*x_1$, $y_2 = 3.319 + 0.350^*x_1 + 0.033^*y_1$.

For the mediation path DS > DX > FP, findings showed that no clear mediating role between digital skills and financial performance. Its structural equation model results are as follows: $y_2 = 3.719 + 0.304^*x_2$, $y_1 = 1.606 + 0.626^*x_2$, $y_2 = 3.592 + 0.255^*x_2 + 0.079^*y_1$.

Regarding the mediation path DTS > DX > FP, findings showed that digital transformation partially mediates the relationship between digital transformation strategy and financial performance. Its structural equation model results are as follows: $y_2 = 1.053 + 0.759^*x_3$, $y_1 = 3.327 + 0.230^*x_3$, $y_2 = 0.611 + 0.728^*x_3 + 0.133^*y_1$.

5. Discussion

This paper contributes to a better understanding of the factors influencing the digital transformation of SMEs and their relationship to performance. The empirical examination examines the relationship between three key resources of SMEs (digital technology, employee digital skills, and digital transformation strategy) and digital transformation, and the impact of these three resources and digital transformation on financial performance. These results suggest some important findings.

First, digital technology has a positive impact on the digital transformation of SMEs (Hypothesis 1/H1).This shows that digital technologies (resources including artificial intelligence, blockchain, cloud technology, big data, 5G, IoT, social media, and cyber-security technologies) contribute to the digital transformation of SMEs. This shows that digital technology is the foundation of digital transformation, and SMEs should consider the different stages they are in and choose the digital technology that suits them, in order to contribute to digital transformation, rather than blindly choosing digital technologies without analysis, or simply using digital technologies to gain various benefits from digital transformation [6].

This result is consistent with RBV's complementary view that digital technology needs to be deeply integrated with the business and management of the enterprise, combined with the enterprise's strategy and the digital skills of its people, in order to form a strong dynamic capability to realize the value of digital technology [8].This can also explain why many SMEs have invested in digital technology, but digital transformation has not achieved much. Therefore, SMEs must understand and correctly use different digital technologies to give full play to their value in enabling digital transformation.

Second, employee digital skills have a positive impact on SMEs' digital transformation (Hypothesis 2/H2). This shows that digital transformation depends on employees with digital skills. Digital technology does not function independently, nor does it always work well. To successfully master these digital methods and technologies, SMEs will rely on a large number of employees who are familiar with and adept at digital tools [6]. For SMEs to successfully implement digital transformation, people are as important as technology, and everything from corporate vision and strategy to execution involves people. Employee digital skills are key to digital transformation.

Third, the digital transformation strategy has a relevant but not significant impact on the digital transformation of SMEs (Hypothesis 3/H3). To carry out digital transformation, we must first carry out top-level design. Previous studies have found that in order to successfully achieve digital transformation, companies must first develop a comprehensive digital transformation strategy [60]. It should be taken as an important part of the development strategy to root the data-driven concepts, methods, and mechanisms in the overall development strategy, and systematically design digital transformation around the vision, goals, business ecological blueprint, and other major strategic directions proposed by the company's overall development strategy. The digital transformation strategy of small- and medium-sized enterprises must be guided by value innovation and completely return to the essence of the business, so that it is possible to carry out digital transformation effectively, quickly, and comprehensively.

The results of intermediary analysis (h5a to h5c) show a more complex situation. The relationship between (a) digital technology, (b) employee digital skills, (c) digital transformation strategy, and financial performance is mediated through digital transformation. Regarding transformation, on the direct and indirect paths (H5c), digital transformation strategies had a positive mediating effect on financial performance, but digital technology (H5a) and employee digital skills (H5b) had no significant mediating effects on financial performance. We did multiple rounds of testing for the mediation effect, but the results were not perfect. We consider the following possibilities. One is that the digital transformation of SMEs has just started, and digital transformation is a long-term and complex system engineering [2], so the effect is not obvious, and the financial performance is not ideal. Second, at the moment of the epidemic, the cash flow of SMEs is very tight, and the price of digital technology is currently relatively expensive, so it is difficult for SMEs to bear the cost of digital transformation. Third, large enterprises, Internet companies, and information technology companies are the first to start digital transformation, which can attract talents with digital skills that are currently scarce at high salaries. SMEs are relatively less attractive, which leads to high recruitment costs and affects the process of digital transformation. Finally, it is relatively easy to formulate a digital transformation strategy. Referring to the digital transformation strategy of a benchmark enterprise and combining it with the actual situation of the enterprise can formulate a digital transformation strategy suitable for the enterprise.

6. Conclusions

This research helps us to understand the resources that SMEs need to deploy to succeed in their digital transformation. Digital technologies, employees digital skills, and digital transformation strategies can help to drive digital transformation, which in turn can improve the financial performance of SMEs. The digital transformation strategy is identified as the key factor affecting financial performance. In previous research, Rogers pointed out that "digital transformation is not fundamentally a technology, but a strategy" [61]. Small- and medium-sized enterprises generally have problems such as a lack of digital transformation thinking, a weak digital foundation, and large obstacles to digital transformation [11]. Digital technology is a major obstacle to the digital transformation of SMEs. The digital skills of employees is another major obstacle to the digital transformation of SMEs [62]. The limited capital of SMEs, coupled with various obstacles, complicates the path of digital transformation. Various types of SMEs are significantly different, which increases the difficulty of digital transformation. The sooner SMEs respond to these challenges, the sooner they can capitalize on the impact to better cope with competition.

This study has several limitations. First, the data were collected in mainland China, and regional restrictions may limit the generalization of the findings. Replicating this study in other parts of the world may improve the reliability and validity of the measurement models. Second, in China, except for listed companies, ordinary SMEs are not obliged to publish financial data. Therefore, the financial performance indicators in this study are based on self-assessed data. Studies have shown that self-assessment data correlate with objective performance measures [63].

We identify some research questions that can advance the digital transformation agenda of SMEs. First, how does digital technology affect the financial performance of SMEs? How different is the degree of impact of each of the different digital technologies? Second, how do employee digital skills affect the financial performance of SMEs? How are the digital skills assessment models? Third, what are the specific indicators of the five stages of digital transformation? How does each affect the financial performance of SMEs? Fourth, are the digital transformation paths of SMEs in different industries different? Fifth, how do different application scenarios of digital transformation of SMEs affect financial performance? Finally, what dynamic capabilities will be formed by digital technologies, employees' digital skills, and digital transformation strategies to influence the financial performance of SMEs?

Author Contributions: Conceptualization, X.T. and Z.W.; methodology, X.T. and F.Y.; software, X.T.; validation, X.T. and F.Y.; formal analysis, X.T.; investigation, X.T.; resources, X.T.; data curation, X.T. and Z.W.; writing—original draft preparation, X.T.; writing—review and editing, X.T. and Z.W.; visualization, X.T.; supervision, Z.W. All authors have read and agreed to the published version of the manuscript.

Funding: This research received no external funding.

Institutional Review Board Statement: Not applicable.

Informed Consent Statement: Not applicable.

Data Availability Statement: The dataset generated and analyzed in this study is not publicly available. Dataset is available from the corresponding author on reasonable request.

Conflicts of Interest: The authors declare no conflict of interest.

Appendix A

Table A1. Research questionnaire.

Item	Observed Variables and Latent Constructs	Mean	Std. Deviation
General Information	Q1. The size of SMEs	2.76	0.432
	Q2. Gender of executives	2.46	1.330
	Q3. Executive jobs	3.58	1.032
Digital Transformation (DX)	DX1.Assess your organization's digital transformation maturity compared to peers	4.50	1.089
	DX2.Assessment of the use of digital technology	4.61	1.228
	DX3.Assess how widely your own digital technology is used	4.60	1.330

Table A1. *Cont.*

Item	Observed Variables and Latent Constructs	Mean	Std. Deviation
Employee Digital Skills (DS)	DS1.We advance continuous learning in digital technologies	5.08	1.268
	DS2.A balance between general digital skills and specialized digital roles is adequate	4.69	1.322
	DS3.We can assemble teams with the right mix of skills for each digital project	4.69	1.484
	DS4.Employees are compound talents who understand both business and digitalization	4.28	1.615
	DS5.My organization provides employees with resources or opportunities to acquire the right digital skills for digital transformation	4.95	1.460
Digital Transformation Strategy (DTS)	DTS1.Your company's digital transformation strategy can increase sales	5.65	1.219
	DTS2.Your company's digital transformation strategy can improve competitiveness	5.19	1.502
	DTS3.Your company's digital transformation strategy can fundamentally change business processes	4.93	1.434
	DTS4.Your company's digital transformation strategy can improve customer experience and satisfaction	5.60	1.194
	DTS5.Your company's digital transformation strategy can improve innovation capabilities	5.64	1.247
	DTS6.Your company's digital transformation strategy can improve business decisions	5.12	1.266
	DTS7.Your company's digital transformation strategy can improve efficiency	5.78	1.178
Digital Technology (DT)	DT1.To what extent your company uses artificial intelligence	4.31	1.463
	DT2.To what extent your company uses blockchain technology	4.33	1.406
	DT3.To what extent your company uses cloud technologies (cloud computing, edge algorithms, cloud-edge collaboration)	4.64	1.482
	DT4.To what extent your company uses big data and data analytics	5.06	1.375
	DT5.To what extent your company is using mobile technology 5G	4.84	1.516
	DT6.To what extent is your company using IoT	4.70	1.429
	DT7.To what extent your company uses social media (collaboration technology)	5.39	1.322
	DT8.To what extent your company uses cybersecurity technologies	5.36	1.346
Financial Performance (FP)	FP1.Digital transformation of your business can help increase sales	5.30	1.206
	FP2.Digital transformation of your business can help return on sales	5.14	1.208
	FP3.Digital transformation of your business can help increase gross profit	5.14	1.255
	FP4.Your company's digital transformation can help increase net profit	5.14	1.283
	FP5.Digital transformation of your business can help return on equity	5.03	1.234
	FP6.Digital transformation of your business can help return on investment	5.23	1.256

References

1. Soto-Acosta, P. COVID-19 pandemic: Shifting digital transformation to a high-speed gear. *Inf. Syst. Manag.* **2020**, *37*, 260–266. [CrossRef]
2. Kääriäinen, J.; Kuusisto, O.; Pussinen, P.; Saarela, M.; Saari, L.; Hänninen, K. Applying the positioning phase of the digital transformation model in practice for smes: Toward systematic development of digitalization. *Int. J. Inf. Syst. Proj. Manag.* **2020**, *8*, 24–43. [CrossRef]
3. Furjan, M.T.; Tomičić-Pupek, K.; Pihir, I. Understanding digital transformation initiatives: Case studies analysis. *Bus. Syst. Res.* **2020**, *11*, 125–141. [CrossRef]
4. Zhu, H.R. Promote digital transformation of enterprises with integrated innovation. *Chinese Inf. Ind.* **2019**, 38–40.
5. Fletcher, G.; Griffiths, M. Digital transformation during a lockdown. *Int. J. Inf. Manag.* **2020**, *55*, 102185. [CrossRef]
6. Zhang, X.; Xu, Y.; Ma, L. Research on successful factors and influencing mechanism of the digital transformation in SMEs. *Sustainability* **2022**, *14*, 2549. [CrossRef]
7. Verhoef, P.C.; Broekhuizen, T.; Bart, Y.; Bhattacharya, A.; Qi Dong, J.; Fabian, N.; Haenlein, M. Digital transformation: A multidisciplinary reflection and research agenda. *J. Bus. Res.* **2021**, *122*, 889–901. [CrossRef]
8. Vial, G. Understanding digital transformation: A review and a research agenda. *J. Strateg. Inf. Syst.* **2019**, *28*, 118–144. [CrossRef]
9. Chen, J.; Yang, W.C.; Yu, F. Ecological collaborative innovation strategy in digital transformation—Based on the strategy discussion of Huawei Enterprise Business Group (EBG) China. *Tsinghua Manag. Rev.* **2019**, 22–26.

10. Huang, L.H.; Zhu, H.L.; Liu, W.H.; Dou, Y.F.; Wang, J.Z.; Cai, L.; Chen, Y.B.; Liao, P.W.; Wu, X.B.; Xie, K.; et al. Enterprise digital transformation and management: Research framework and outlook. *J. Manag. Sci.* **2021**, *24*, 26–35. [CrossRef]
11. Zhang, X.H. Obstacles, driving factors and path dependence of digital transformation of small and medium-sized enterprises: Based on a survey of 377 small and medium-sized enterprises in the tertiary industry. *China Circ. Econ.* **2020**, *34*, 72–82. [CrossRef]
12. Gurbaxani, V.; Dunkle, D. Gearing up for successful digital transformation. *MIS Q. Exec.* **2019**, *18*, 209–220. [CrossRef]
13. Mhlungu, N.S.M.; Chen, J.Y.J.; Alkema, P. The underlying factors of a successful organisational digital transformation. *SA J. Inf. Manag.* **2019**, *21*, 1–10. [CrossRef]
14. Ivančić, L.; Vukšić, V.B.; Spremić, M. Mastering the digital transformation process: Business practices and lessons learned. *Technol. Innov. Manag. Rev.* **2019**, *9*, 36–51. [CrossRef]
15. Wrede, M.; Velamuri, V.K.; Dauth, T. Top managers in the digital age: Exploring the role and practices of top managers in firms' digital transformation. *Manag. Dec. Econ.* **2020**, *41*, 1549–1567. [CrossRef]
16. Mirzaian, E.; Franson, K.L. Leading a digital transformation in pharmacy education with a pandemic as the accelerant. *Pharmacy* **2021**, *9*, 19. [CrossRef]
17. Sullivan, C.; Wong, I.; Adams, E.; Fahim, M.; Fraser, J.; Ranatunga, G.; Busato, M.; McNeil, K. Moving Faster than the COVID-19 Pandemic: The Rapid, Digital Transformation of a Public Health System. *Appl. Clin. Inform.* **2021**, *12*, 229–236. [CrossRef]
18. Ghobakhloo, M.; Iranmanesh, M. Digital transformation success under Industry 4.0: A strategic guideline for manufacturing SMEs. *J. Manuf. Technol. Manag.* **2021**, *32*, 1533–1556. [CrossRef]
19. Brodeur, J.; Pellerin, R.; Deschamps, I. Collaborative approach to digital transformation (CADT) model for manufacturing SMEs. *J. Manuf. Technol. Manag.* **2022**, *33*, 61–83. [CrossRef]
20. Hai, N.T. Digital transformation barriers for small and medium enterprises in Vietnam today. *LAPLAGE EM Rev.* **2021**, *7*, 416–426. [CrossRef]
21. Penrose, E.T. *The Theory of the Growth of the Firm*; Oxford University Press: Oxford, UK, 1995.
22. Wernerfelt, B. A resource-based view of the firm. *Strat. Manag. J.* **1984**, *5*, 171–180. [CrossRef]
23. Barney, J. Firm resources and sustained competitive advantage. *J. Manag.* **1991**, *17*, 99–120. [CrossRef]
24. Peteraf, M.A. The cornerstones of competitive advantage: A resource-based view. *Strateg. Manag. J.* **1993**, *14*, 179–191. [CrossRef]
25. Teece, D.J.; Pisano, G.; Shuen, A. Dynamic capabilities and strategic management. *Strateg. Manag. J.* **1997**, *18*, 509–533. [CrossRef]
26. Fischer, M.; Imgrund, F.; Janiesch, C.; Winkelmann, A. Strategy archetypes for digital transformation: Defining meta objectives using business process management. *Inf. Manag.* **2020**, *57*, 103262. [CrossRef]
27. Matarazzo, M.; Penco, L.; Profumo, G.; Quaglia, R. Digital transformation and customer value creation in Made in Italy SMEs: A dynamic capabilities perspective. *J. Bus. Res.* **2021**, *123*, 642–656. [CrossRef]
28. Wang, C. Research on the effect mechanism of digital transformation on enterprise innovation performance. *Contemp. Econ. Manag.* **2021**, *43*, 34–42. [CrossRef]
29. Magistretti, S.; Dell'Era, C.; Messeni Petruzzelli, A. How intelligent is Watson? Enabling digital transformation through artificial intelligence. *Bus. Horiz.* **2019**, *62*, 819–829. [CrossRef]
30. Gopal, G.; Suter-Crazzolara, C.; Toldo, L.; Eberhardt, W. Digital transformation in healthcare—Architectures of present and future information technologies. *Clin. Chem. Lab. Med.* **2019**, *57*, 328–335. [CrossRef]
31. Salvini, G.; Hofstede, G.J.; Verdouw, C.N.; Rijswijk, K.; Klerkx, L. Enhancing digital transformation towards virtual supply chains: A simulation game for Dutch floriculture. *Prod. Plan. Control.* **2020**, *2020*, 1–18. [CrossRef]
32. Hartley, J.L.; Sawaya, W.J. Tortoise, not the hare: Digital transformation of supply chain business processes. *Bus. Horiz.* **2019**, *62*, 707–715. [CrossRef]
33. Evangelatos, N.; Upadya, S.P.; Venne, J.; Satyamoorthy, K.; Brand, H.; Ramashesha, C.S.; Brand, A. Digital transformation and governance innovation for public biobanks and free/libre open source software using a blockchain technology. *Omi. A J. Integr. Biol.* **2020**, *24*, 278–285. [CrossRef]
34. Attaran, M.; Attaran, S. Digital transformation and economic contributions of 5G networks. *Int. J. Enterp. Inf. Syst.* **2020**, *16*, 58–79. [CrossRef]
35. Vardarlier, P.; Ozsahin, M. Digital transformation of human resource management: Social media's performance effect. *Int. J. Innov. Technol. Manag.* **2021**, *18*, 2150005. [CrossRef]
36. Kane, G. The technology fallacy: People are the real key to digital transformation. *Res. Technol. Manag.* **2019**, *62*, 44–49. [CrossRef]
37. Grab, B.; Olaru, M.; Gavril, R. Self-managed teams as a key to unlocking digital transformation in business management. *Qual. Success* **2019**, *20*, 280–286.
38. Butschan, J.; Heidenreich, S.; Weber, B.; Kraemer, T. Tackling hurdles to digital transformation—The role of competencies for successful industrial internet of things (IIoT) implementation. *Int. J. Innov. Manag.* **2019**, *23*, 1950036. [CrossRef]
39. Bessonova, T.V.; Goryacheva, O.N. Human resources management in the conditions of digital transformation. *Laplage Rev.* **2020**, *6*, 12–18. [CrossRef]
40. Solberg, E.; Traavik, L.E.M.; Wong, S.I. Digital mindsets: Recognizing and leveraging individual beliefs for digital transformation. *Calif. Manag. Rev.* **2020**, *62*, 105–124. [CrossRef]
41. Scuotto, V.; Nicotra, M.; Del Giudice, M.; Krueger, N.; Gregori, G.L. A microfoundational perspective on SMEs' growth in the digital transformation era. *J. Bus. Res.* **2021**, *129*, 382–392. [CrossRef]

42. Bikse, V.; Lusena-Ezera, I.; Rivza, P.; Rivza, B. The development of digital transformation and relevant competencies for employees in the context of the impact of the COVID-19 pandemic in latvia. *Sustainability* **2021**, *13*, 9233. [CrossRef]
43. Bennett, E.E.; McWhorter, R.R. Virtual HRD's role in crisis and the post COVID-19 professional lifeworld: Accelerating skills for digital transformation. *Adv. Dev. Hum. Resour.* **2021**, *23*, 5–25. [CrossRef]
44. Chanias, S.; Myers, M.D.; Hess, T. Digital transformation strategy making in pre-digital organizations: The case of a financial services provider. *J. Strateg. Inf. Syst.* **2019**, *28*, 17–33. [CrossRef]
45. Tekic, Z.; Koroteev, D. From disruptively digital to proudly analog: A holistic typology of digital transformation strategies. *Bus. Horiz.* **2019**, *62*, 683–693. [CrossRef]
46. Min, S.A.; Kim, B.Y. SMEs' digital transformation competencies on platform empowerment: A case study in South Korea. *J. Asian Financ. Econ. Bus.* **2021**, *8*, 897–907. [CrossRef]
47. Lichtenthaler, U. Building blocks of successful digital transformation: Complementing technology and market issues. *Int. J. Innov. Technol. Manag.* **2020**, *17*, 2050004. [CrossRef]
48. Barrane, F.Z.; Ndubisi, N.O.; Kamble, S.; Karuranga, G.E.; Poulin, D. Building trust in multi-stakeholder collaborations for new product development in the digital transformation era. *Benchmarking* **2021**, *28*, 205–228. [CrossRef]
49. Mubarak, M.F.; Shaikh, F.A.; Mubarik, M.; Samo, K.A.; Mastoi, S. The impact of digital transformation on business performance: A study of Pakistani SMEs. *Eng. Technol. Appl. Sci. Res.* **2019**, *9*, 5056–5061. [CrossRef]
50. Hu, Q. The mechanism and performance of enterprise digital transformation. *Zhejiang Acad. J.* **2020**, *2020*, 146–154. [CrossRef]
51. Li, Q.; Liu, L.G.; Shao, J.B. Digital transformation, supply chain integration and corporate performance: The moderating effect of entrepreneurship. *Econ. Manag.* **2021**, *43*, 5–23. [CrossRef]
52. Wang, H.; Feng, J.; Zhang, H.; Li, X. The effect of digital transformation strategy on performance: The moderating role of cognitive conflict. *Int. J. Confl. Manag.* **2020**, *31*, 441–462. [CrossRef]
53. Bhatti, A.; Malik, H.; Kamal, A.Z.; Aamir, A.; Alaali, L.A.; Ullah, Z. Much-needed business digital transformation through big data, internet of things and blockchain capabilities: Implications for strategic performance in telecommunication sector. *Bus. Process Manag. J.* **2021**, *27*, 1854–1873. [CrossRef]
54. Kim, S.S. Sustainable growth variables by industry sectors and their influence on changes in business models of smes in the era of digital transformation. *Sustainability* **2021**, *13*, 7114. [CrossRef]
55. Lumpkin, G.T.; Dess, G.G. Linking two dimensions of entrepreneurial orientation to firm performance: The moderating role of environment and industry life cycle. *J. Bus. Ventur.* **2001**, *16*, 429–451. [CrossRef]
56. Bley, K.; Leyh, C.; Schäffer, T. Digitization of German enterprises in the production sector-Do they know how "digitized" they are? In Proceedings of the Twenty second Americas Conference on Information Systems, San Diego, CA, USA, 11–14 August 2016; pp. 1–10.
57. Creswell, J.W. *Research Design: Qualitative, Quantitative, and Mixed Methods Approaches*; Sage Publications: Thousand Oaks, CA, USA, 2014.
58. Field, A. *Discovering Statistics Using IBM SPSS Statistics and Sex and Drugs and Rock'n'Roll*; Sage Publications: London, UK, 2013.
59. Ma, X.D. *Digital Transformation Methodology Landing Path and Data Center*; Machinery Industry Press: Beijing, China, 2021.
60. Chen, X.P. *One Book to Understand Digital Transformation*; Machinery Industry Press: Beijing, China, 2021.
61. Rogers, D. *The Digital Transformation Playbook: Rethink Your Business for the Digital Age*; Columbia University Press: New York, NY, USA, 2016.
62. Liu, T.; Zhang, X.H. The current situation, problems and countermeasures of digital transformation of small and medium-sized enterprises in my country. *Guizhou Soc. Sci.* **2021**, *2021*, 148–155. [CrossRef]
63. Brush, C.G.; Vanderwerf, P.A. A comparison of methods and sources for obtaining estimates of new venture performance. *J. Bus. Ventur.* **1992**, *7*, 157–170. [CrossRef]

sustainability

Article

Decisions on Pricing, Sustainability Effort, and Carbon Cap under Wholesale Price and Cost-Sharing Contracts

Doo-Ho Lee and Jong-Chul Yoon *

Department of Artificial Intelligence and Software, Kangwon National University, 346 Joongang-ro, Samcheok-si 29513, Gangwon-do, Korea; enjdhlee@kangwon.ac.kr
* Correspondence: media19@kangwon.ac.kr

Abstract: Rapid economic growth and industrialization have brought material abundance and convenience, but also social and environmental problems such as global warming, climate change, and ozone depletion. For this reason, the public and governments have continued to make efforts to reduce carbon oxide emissions worldwide over the past few decades. To achieve this mission, cap-and-trade regulations have been introduced as one of the most effective market-based mechanisms to control carbon emissions. Accordingly, sustainability efforts, including the development of green products and innovating manufacturing technologies, are being made by companies in supply chains to reduce their carbon emissions. In the context of sustainability innovations and carbon emission constraints, this article investigates pricing decisions, the degree of sustainability efforts, and carbon caps under two different supply chain contracts—in this case, wholesale price contract and cost-sharing contract. This article establishes a Stackelberg game model under each of the supply chain contract types and presents the equilibrium decisions made by players of the game. Major findings of this article reveal that (i) the performance of the supply chain is considerably affected by the presence of a carbon cap; (ii) the higher the carbon cap set by a government is, the more sustainability innovation efforts the supply chain makes; and (iii) the supply chain can improve its profitability and its sustainability under a cost-sharing contract.

Keywords: pricing; sustainability effort; carbon cap; wholesale price contract; cost-sharing contract

Citation: Lee, D.-H.; Yoon, J.-C. Decisions on Pricing, Sustainability Effort, and Carbon Cap under Wholesale Price and Cost-Sharing Contracts. *Sustainability* **2022**, *14*, 4863. https://doi.org/10.3390/su14084863

Academic Editors: Shaojian Qu, Qingguo Bai, Ying Ji and Congjun Rao

Received: 14 March 2022
Accepted: 15 April 2022
Published: 18 April 2022

1. Introduction

Economic growth, driven by rapid industrialization, has significantly improved the well-being of human societies. The global rush for economic development has led to unprecedented material affluence and convenience, but these are not without cost. In particular, environmental degradation has been a major concern in both developed and developing countries. With the growing consensus that economic development cannot be sustained without environmental protection, sustainability has become a mainstream imperative with regard to the running of a global economy. When attempting to enhance sustainability, numerous challenges exist, but global warming—a direct result of the rapid rise in carbon dioxide and other greenhouse gases—has become a primary concern for a global society, given its widespread impact on our planet. It is widely recognized that climate change threatens our ecosystem, affecting all aspects of life on the Earth. Apart from the long-term adverse impact on biodiversity, many people around the world have already fallen victim to the effects of global warming. An example of this is desertification expedited by rising global temperatures, which has displaced half a billion people while also destabilizing the food supply, particularly in Africa, where the majority of people live in poverty. The rising sea level also threatens the lives of people living on islands [1]. According to earlier research in this area, by 2100, 48 islands may disappear, and up to 187 million people may lose their homes to the rising ocean [2,3]. Given the increasing number of droughts, floods, and extreme weather events triggered by climate change, the United Nations forecasts that the number of climate migrants may reach one billion worldwide by 2050 [4].

In recognition of the fact that climate change is one of the most serious threats faced by humanity, the world has made efforts to mitigate the harmful effects of global warming. In 1997, developed countries agreed on a legally binding treaty to reduce greenhouse gas emissions, known as the Kyoto Protocol, along with the introduction of mechanisms to curb the release of greenhouse gases. The three market-based mechanisms adopted under the Kyoto Protocol—the clean development mechanism, joint implementation, and emission trading—are considered effective ways to control greenhouse gas emissions [5]. The Paris Agreement, adopted in 2015, further strengthened global efforts to address climate change; it sets a specific target for limiting global warming—1.5~2 °C above preindustrial levels—and requires nearly all countries to reduce their national emissions. Climate change is also included as an important agenda item in the Sustainable Development Goals, which were agreed upon by all United Nations member states in 2015.

Private companies have also joined such sustainability efforts either voluntarily or involuntarily by implementing environmentally friendly practices. The growing awareness of 'responsible consumption' has called for companies increasingly to care about eco-friendly business practices, from manufacturing and packaging to marketing. Over the past decade, the private sector has actively recognized the issue of climate change as part of its 'Corporate Social Responsibility' (CSR). In Velazquez-Cazares et al. [6], CSR is defined as "the responsibility of the enterprises for their impact on the society", highlighting their role in the sustainability, competitiveness, and innovation of the business practices. Globalization, technology, inequalities, crisis, and climate change have pressured companies toward CSR. At the World Economic Forum in 2015, for instance, the CEOs of 79 companies and 20 economic sectors announced their commitment to sharing the responsibility to reduce carbon emissions. On the other hand, other companies are forced to comply with regulations to cut carbon emissions by means of, for instance, cap-and-trade systems. Under a cap-and-trade system, the government puts a cap on the carbon emissions allowed by each company but allows them to trade emissions allowances. This market-based approach is accepted as an effective measure by which to control carbon emissions as it gives companies more flexibility to control their emissions as they operate their business. Studies undertaken to date have shown that the emission cap has a direct impact on a carbon trading price [7–9]. For example, in 2006, companies were allocated 10% more than the emission cap they actually needed to cover their 2005 emissions. This caused carbon trading prices to plummet by 60% in a week in carbon trading markets [10,11]. Therefore, there may be an inverse relationship between the cap and the emission trading price.

A wholesale price contract is a contractual agreement that is widely adopted in many supply chains. With a wholesale price contract, a manufacturer charges a retailer for each product without collaboration (cooperation). However, various strategies to induce collaboration among supply chain participants have been proposed to achieve sustainable supply chain operation. Among them, cost sharing is considered one of the most desirable cooperative contracts, in which supply chain participants can enhance not only their profitability but also their sustainability by sharing their costs incurred during production, marketing, or R&D [12–16]. Thus, this paper focuses on supply chains with wholesale price contracts and cost-sharing contracts that are governed by a cap-and-trade system. As mentioned, because the emission trading price is directly influenced by the emission cap, the supply chain participants' decisions will be affected by the cap. In addition, the collaboration among supply chain participants also has a large effect on their production decisions. To the best of the authors' knowledge, there has been no study on setting the carbon cap under a wholesale price contract and a cost-sharing contract in consideration of the inverse relationship between the carbon cap and the emission trading price. Thus, the main research questions addressed in this article are as follows:

- Under wholesale price and cost-sharing contracts, how does the carbon cap affect manufacturers' sustainability efforts and production decisions?

- Under both contracts, how does the carbon cap affect the profits of supply chain participants?
- In view of the profitability and sustainability of the supply chain, which is better: a wholesale price contract or a cost-sharing contract?
- Is it possible for a government to determine an optimal carbon cap in order to maximize social welfare under both contracts?

The rest of this article is organized as follows. In Section 2, a review of the relevant literature is presented. Section 3 reviews the notations and the assumptions in this article. In Section 4, the main results and various numerical examples are presented. The last section provides a summary and some directions for future research.

2. Literature Review

This section reviews the relevant literature considering three different streams of research on sustainable operations: sustainability innovation, collaboration decisions, and operational management under cap-and-trade regulations.

2.1. Sustainability Innovation in Supply Chains

The topic of sustainable operations is a growing concern for supply chain management [17,18]. Hassini et al. [19] defined sustainable supply chain management as the "management of supply chain operations, resources, information, and funds in order to maximize the supply chain profitability while at the same time minimizing the environmental impacts and maximizing social well-being". For the purpose of this article, this subsection focuses on sustainability innovations in supply chains.

Product design is a crucial tool for those involved in sustainable development. Krikke et al. [20] suggested a model that integrates product design with a closed-loop supply chain design. By combining costs and environmental impacts into an objective function, they evaluated three potential designs of a refrigerator in different supply chain configurations. Subramanian et al. [21] examined optimal product-design decisions and demonstrated how financial aid could be used as a lever to encourage firms to design environmentally friendly products by considering two key decisions related to product design: performance and remanufacturability. Chen [22] developed a quality-based model that considers the green features of products as a quality attribute and analyzed strategic decision issues related to the development of sustainable products. Shen et al. [23] discussed optimal product line design for green and nongreen products in terms of quality differentiation. They argued that only when the quality difference between green and nongreen products is large enough does designing and selling green products benefit consumers and generate high social welfare. Zhu and He [24] studied green product design in competitive supply chains and derived equilibrium decisions on the greenness degrees of products. Thies et al. [25] conducted a valuable survey of sustainable product designs with extensive bibliographical references

Firms are increasingly paying more attention to sustainable technology innovations in recent years because of stricter regulations and greater consumer concern over corporate social responsibility and environmental protection. The introduction of new product designs and technology/process innovations requires firms to evaluate the complex cost and environmental tradeoffs. Stuart et al. [26] developed a technology selection model for a firm that considers yield, reliability, and environmental impact tradeoffs. They modeled several constraints for environmental effects, including material/energy consumption and technology/process waste generation. Debo et al. [27] solved the issue of joint pricing and production technology selections by manufacturers in the market where new and remanufactured products coexist. They showed that investing in remanufacturing technology is more profitable when there are more environmentally conscious customers in the market. Drake et al. [9] studied the impact of the carbon tax and cap-and-trade regulations on firms' technology selections and capacity investment decisions. They revealed that a firm's profits are greater and carbon emissions are lower under governmental cap-and-trade regulations, while production quantities are greater under a carbon tax regulation. Raz et al. [28] considered a newsvendor problem for decisions affecting production quantity and environmentally focused process design efforts. They determined a firm's sustainable technology innovation efforts at each stage of the product lifecycle.

2.2. Collaboration Decisions in Supply Chain

This work is also related to the literature on supply chain collaboration between manufacturers and their upstream suppliers or downstream retailers. Klassen and Vachon [29] discovered that supply chain collaboration plays a key role in determining both the form and level of environmental technology investments. Zhu et al. [30] and Green et al. [31] also support the view that collaboration between supply chain members on sustainability innovation efforts enhances both environmental and financial performance outcomes and that collaborative innovation is critical for the success of a circular economy initiative.

As pointed out by Ge et al. [32], most studies in this line of research deal with horizontal R&D cooperation in supply chains [33–35]. Few papers investigate collaboration among firms with vertical relationships. Talluri et al. [36] considered a manufacturer allocating funds to its suppliers to improve performance and found conditions in which the manufacturer benefits from collaborating with its suppliers. They presented a Markowitz portfolio investment model that can be used to help a manufacturer optimally allocate its available dollars while maintaining expected benefits. Under production cost uncertainty, Kim and Netessine [37] investigated how information asymmetry and procurement contracting strategies affect supply chain members' incentives to collaborate. According to their results, a manufacturer prefers an expected margin commitment contract if collaboration results in a large reduction in the unit production cost and demand fluctuations are low. Ghosh and Shah [14] explored collaboration issues in a green supply chain and studied the influence of cost-sharing contracts on the key decisions by supply chain members undertaking green initiatives. Two models of cost-sharing contracts were considered—one where a retailer drives a cost-sharing contract and the other where supply chain members negotiate it.

More recently, Ji et al. [38] investigated the collaboration between manufacturers and retailers in online and offline shops and analyzed how a governmental cap-and-trade policy affects supply chain profits and social welfare by adding supply chain members' efforts to reduce their carbon emissions. Yenipazarli [39] studied the impact of collaboration via supply chain contracts on suppliers' investments in carbon emission reductions during their manufacturing processes. Further, Dai et al. [15] devised game models to investigate two collaboration mechanisms between supply chain members in relation to green R&D investments: cartelization and cost-sharing contracts. Hong and Lee [16] analyzed two popular contract types, revenue- and cost-sharing contracts, considering the development time uncertainties caused by the R&D capabilities of suppliers. They argued that a contract mechanism to determine the fraction of the revenue or cost that a manufacturer shares with its supplier affects supply chain profits and the expected time required for new product development. Ji et al. [11] derived manufacturer equilibrium production decisions under wholesale pricing and revenue sharing contracts and determined optimal governmental carbon caps. They showed that the equilibrium production quantities increase with a carbon cap, whereas manufacturer profit can decrease with stronger carbon cap regulations. They also found that the overall supply chain profit under a revenue-sharing contract is greater than that under a wholesale price contract. Cachon [40] provided an extensive review of the supply chain literature, focusing on the management of incentive conflicts with contracts, where various contracts in supply chains are identified, and their benefits and drawbacks are illustrated.

2.3. Operations Management under Cap-and-Trade Regulation

Given the increasing severity of global warming, many studies of supply chains with cap-and-trade regulations have been actively conducted over the past few decades. The literature on supply chain management under cap-and-trade regulations includes numerous topics, including pricing, production, inventory, carbon emissions reduction, and carbon cap determination.

Xu et al. [41] studied joint production and the pricing problems of a manufacturer who produces multiple products under cap-and-trade and carbon tax regulations. They presented comparative results on the impacts of the two types of regulation on total carbon emissions, the profits of the firm, and social welfare. Xu et al. [42] extended the work of Xu et al. [41] to study production and pricing issues in a make-to-order (MTO) supply chain with cap-and-trade regulations. Xu et al. [42] analyzed the impact of emission trading prices on the optimal production decisions and the profits of the supply chain members. Bai et al. [43] also considered an MTO supply chain and showed that the profit of a decentralized supply chain can be improved and that carbon emissions can be significantly reduced through supply chain coordination efforts. Yang et al. [44] modeled chain-to-chain competition under cap-and-trade regulations. In their model, each supply chain can either be vertically or horizontally integrated. Yang et al. [44] suggested that a retailer-driven revenue-sharing contract may encourage manufacturers to give up horizontal cooperation, which not only helps all supply chain members achieve a win–win situation but also benefits the environment via a greater reduction in the rate of emissions. Wang et al. [45] examined manufacturing/remanufacturing planning decisions by manufacturers under cap-and-trade regulations. They showed that capital constraints force manufacturers to remanufacture used products at a high level of quality and can considerably reduce carbon emissions. Xu et al. [46] analyzed the equilibrium decisions and coordination mechanisms of a two-echelon supply chain under cap-and-trade regulations and revealed that a two-part tariff contract outperforms a revenue-sharing contract in terms of the profit of a supply chain and the reduction in carbon emissions. Shu et al. [47] dealt with a closed-loop supply chain considering social responsibility and explored the effects of carbon caps and CSR on retailers' recycling decision behavior. They showed that manufacturers' CSR has a positive impact on recycling rates and that CSR dramatically leads to a decrease in carbon emissions. From an inventory management perspective, Hua et al. [7] discussed how firms could control carbon emissions as part of their inventory management effort under cap-and-trade regulations. They demonstrated that the carbon cap and emission trading price have a considerable impact on order decisions, carbon emissions, and total costs. Ji et al. [48] analyzed decisions on carbon emissions reductions and inventory replenishment efforts given low-carbon preferences held by consumers. They suggested that a joint carbon tax and cap-and-trade policy work best for emissions reduction and that a government must tighten the carbon cap to realize a clean industry. Shen et al. [49] studied an inventory model for perishable items under a carbon tax policy with collaborative preservation technology investments. They identified optimal decisions on production, delivery, ordering, and investments to maximize the total supply chain profits considering a carbon tax policy. Bai et al. [50] discussed the influence of reducing carbon emissions on the coordination of the supply chain with a vendor-managed inventory system. They showed that a revenue-sharing contract could be introduced to improve profits and reduce carbon emissions in a decentralized supply chain.

Determination of the carbon cap is a crucial decision by a government as part of its work to maximize social welfare. Du et al. [51] considered a supply chain in which an emission-dependent manufacturer and an emission permit supplier interact with each other under cap-and-trade regulations. They determined the optimal allocation of the carbon cap of a manufacturer through consideration of what was termed a fairness distribution in social welfare. Xu et al. [41] found that as the unit environmental damage of a product increases, the government's optimal carbon cap decreases or remains constant. He et al. [52] studied the optimal production of a self-pricing manufacturer and the optimal carbon cap of a government. They found that the optimal carbon cap and the total carbon emissions initially increase and then decrease as the emissions intensity increases. Ji et al. [11] also argued that if the environmental impact of carbon emissions is low or high, the government should keep the optimal carbon cap unchanged; otherwise, the government may decrease the cap.

3. Research Methodology

As discussed, there are many in-depth studies of sustainable innovations and collaborative decisions in supply chains with carbon emission regulations. The literature review in Section 2 shows that governmental cap-and-trade regulations represent a key factor in the profitability and sustainability of supply chains. However, most studies overlook the relationship between cap-and-trade regulations and manufacturers' sustainability innovation efforts. Because a significant relationship is found between a carbon cap and decision making in supply chains, it is natural that governmental cap-and-trade regulations affect the sustainability innovation efforts of a supply chain. This article aims to identify the relationship between such a cap and sustainability efforts in a supply chain. Moreover, most studies assumed that carbon trading prices are permanently constant for analytical simplicity, but this article assumes an inverse relationship between a carbon cap and the emission trading price. According to Section 1, the assumption of such an inverse relationship is reasonable and helps supply chain participants gain meaningful managerial insights.

This article provides Stackelberg game models under both wholesale price and cost-sharing contracts and presents the equilibrium decisions made by players of each game. In each game, supply chain participants want to maximize or achieve their goals, and their decisions influence each other. The sequence of each Stackelberg game is detailed in Section 4.

3.1. Notations

This article uses the notations in Table 1.

Table 1. Notations.

Parameters	Descriptions
a	Maximal price of emission credit
b	Price sensitivity of carbon cap
β	Change in production cost as a result of the sustainability effort
η	Unit environmental damage of carbon emission
λ	Carbon emission intensity
Decision Variables	**Descriptions**
e	Sustainability effort
p	Retail price
w	Wholesale price
P	Carbon cap
δ	Collaboration level
Functions	**Descriptions**
d	Demand for the product
M	Total carbon emission
π_m	Manufacturer's profit
π_r	Retailer's profit
$\pi_{sc} = \pi_m + \pi_r$	Supply chain profit
SW	Social welfare

3.2. Assumptions

This article considers a simple supply chain composed of a downstream retailer and an upstream manufacturer that develops a sustainable product. The supply chain distributes only one type of sustainable product. The retailer purchases the manufacturer's products at the wholesale price w. The retailer provides the product to the consumers in the market at a retail price p. In the supply chain considered here, the manufacturer makes a sustainability effort e, which can boost product demand d. This assumption is intuitive because consumers may have higher utility by consuming the sustainable product, meaning that their willingness-to-pay increases, leading to greater demand. Hence, the demand function for the product is formulated as shown below

$$d = 1 - p + e \tag{1}$$

As indicated in Equation (1), the potential market base is assumed to be scaled to 1. Note that without a loss of generality, we assume that e has a positive coefficient of one (i.e., effort sensitivity of demand $\gamma = 1$). This assumption is justified because the results for an arbitrary γ can be obtained by scaling the equilibrium effort $e = \gamma e'$.

To capture the decreasing marginal effect of the sustainability efforts, assume that the investment in developing the sustainable product is an increasing and convex function of the efforts. Hence, the effort cost can be expressed as $\theta e^2/2$, where θ is the cost coefficient of the effort and assume that $\theta = 1$ for analytical convenience. In addition, assume that the sustainability effort affects the unit production cost $c = c_0 + \beta e$ with the base cost c_0 normalized to 0. β represents the scenario where the sustainability effort results in an increase in the unit production cost. For example, in the coffee industry, environment-friendly manufacturing processes increase the unit production cost by about 30%. [53]. Additionally, assume that $\beta \in [0, 1)$ so that the manufacturer exerts positive sustainability efforts.

Based on an MTO supply chain, this article deals with two types of contracts between the manufacturer and the retailer, specifically a wholesale price contract and a cost-sharing contract. In such a supply chain, the manufacturer and the retailer do not have to focus on the inventory cost or the salvage cost for unsold products. The purpose of this paper is to specify not only the decisions of the supply chain members but also the optimal carbon cap of the government under a cap-and-trade regulation taking into consideration the two types of contracts. Consider the wholesale price contract (cost-sharing contract) as a noncooperative contract (cooperative contract). Under both contracts, the government is assumed to play the role of a Stackelberg game leader. A detailed description of the Stackelberg game under each contract is presented in Section 4. According to the cap-and-trade regulation, the manufacturer is initially given a free quantity of carbon emission credits (also known as carbon emission permits) from a central government. The manufacturer is obligated to hold the credits in an amount larger than or equal to its carbon emissions. The manufacturer can purchase or sell these credits through an outside emissions trading system (ETS). The manufacturer produces the product depending on both the retailer's order and the allocated cap. In this situation, the government wants to determine the optimal carbon cap which maximizes social welfare.

Let λ be the carbon emissions incurred when producing one unit of product. Therefore, the total carbon emissions associated with production is $M = \lambda d$. A similar expression of M can be found in Ji et al. [11] and Gong and Zhou [54]. If the total carbon emission exceeds the given cap (i.e., $M > P$), the manufacturer must purchase the emission credits via the ETS to avoid a large penalty. Otherwise, the manufacturer can sell the excess credits via the ETS to obtain economic incentives. Similar to Hua et al. [7], Benjaafar et al. [8], and Ji et al. [11], assume that the carbon trading price is a linear function of the carbon cap, i.e., $b_0 = a - bP \geq 0$ (or equivalently $P \leq a/b$), where a is the maximal price of the emission credits and b is the price sensitivity of the cap. The retailer also faces a challenge to lessen the amount of carbon emissions it creates during marketing and selling activities, but the retailer's carbon emissions are neglected because the amount is relatively small.

4. Analysis and Interpretation of Results

With the notations and assumptions in Section 3, the main results and numerical examples of this article are presented in Propositions and Corollaries. Note that all numerical experiments are carried out based on artificially simulated parameter settings. All proofs in Section 4 are given in Appendix A.

4.1. Equilibrium Analysis with the Given Carbon Cap

This subsection presents the equilibrium decisions of the manufacturer and the retailer with the given carbon cap P. Hereafter we utilize the superscript $l \in \{W, C\}$, where W and C represent the wholesale price contract and the cost-sharing contract, respectively.

4.1.1. Wholesale Price Contract

First, consider the wholesale price contract between the manufacturer and the retailer. The profit functions of the manufacturer and the retailer are expressed, respectively, as follows

$$\pi_m = (w - \beta e)d - \frac{e^2}{2} - (a - bP)(M - P) \text{ and } \pi_r = (p - w)d, \tag{2}$$

where π_m and π_r denote the profits of the manufacturer and the retailer, respectively. The sequence of the two-stage Stackelberg game under the wholesale price contract is as follows. In the first stage of the game, the manufacturer announces the wholesale price w and the sustainability efforts e. After specifying the values of w and e, the retailer determines the retail price for the product p. Applying backward induction, the equilibrium values of the decision variable and the profit for each player can be obtained in Proposition 1.

Proposition 1. *Given the carbon cap P, the equilibrium values in the wholesale price contract game are determined as follows*

$$\begin{aligned} w^W &= \tfrac{2-\beta+\lambda(a-bP)}{3-\beta},\ p^W = \tfrac{3+\beta(1-\beta+\lambda(a-bP))}{(1+\beta)(3-\beta)},\ e^W = \tfrac{(1-\beta)(1-\lambda(a-bP))}{(1+\beta)(3-\beta)},\ d^W = \tfrac{1-\lambda(a-bP)}{(1+\beta)(3-\beta)}, \\ M^W &= \tfrac{\lambda(1-\lambda(a-bP))}{(1+\beta)(3-\beta)},\ \pi_m^W = \tfrac{(1-\lambda(a-bP))^2}{2(1+\beta)(3-\beta)} + P(a-bP),\ \textit{and}\ \pi_r^W = \left(\tfrac{1-\lambda(a-bP)}{(1+\beta)(3-\beta)}\right)^2. \end{aligned} \tag{3}$$

When the government initially sets the carbon cap to zero, the inequality $a\lambda < 1$ should be assumed to avoid negative demand for the product. From Proposition 1, Corollaries 1–3 can be derived as shown below.

Corollary 1. *Under the wholesale price contract, the equilibrium decisions have the following relationships:*

$$\frac{\partial w^W}{\partial P} < 0,\ \frac{\partial p^W}{\partial P} < 0,\ \frac{\partial e^W}{\partial P} > 0,\ \frac{\partial d^W}{\partial P} > 0,\ \textit{and}\ \frac{\partial M^W}{\partial P} > 0. \tag{4}$$

Corollary 1 presents the important fact that the equilibrium decision of each supply chain member depends on the carbon cap. The equilibrium wholesale price decreases with the carbon cap, which encourages the retailer to decrease its retail price. As a result, the demand for the product inevitably increases. This occurs because as the allocated carbon cap increases, the carbon trading price decreases, which leads to a higher marginal profit per unit of the product. Another important fact is that the manufacturer's sustainability effort increases as the carbon cap increases. The manufacturer can obtain higher profits as it produces more products. Therefore, the manufacturer may try to increase the demand for the product, which requires a greater sustainability effort. As the carbon cap increases, sustainability efforts increase, and the manufacturer can thus obtain higher profits. As the cap increases, the production quantity also increases, resulting in more carbon emissions.

Corollary 2. *Under the wholesale price contract, the equilibrium profits have the following relationships:*

$$\frac{\partial \pi_r^W}{\partial P} > 0 \textit{ and } \frac{\partial \pi_m^W}{\partial P} = \begin{cases} > 0,\ \textit{if}\ 0 < b < K_1 \textit{ and } 0 < P < K_2, \\ > 0,\ \textit{if}\ b > K_1, \\ < 0,\ \textit{if}\ 0 < b < K_1 \textit{ and } P > K_2, \end{cases} \tag{5}$$

where $K_1 = \frac{2(1+\beta)(3-\beta)}{\lambda^2}$ *and* $K_2 = \frac{a(1+\beta)(3-\beta)+b\lambda(1-a\lambda)}{b(2(1+\beta)(3-\beta)-b\lambda^2)}$.

Corollary 2 expresses the relationship between the equilibrium profit and the carbon cap. As indicated in Equation (5), the retailer's equilibrium profit increases with respect to the cap. This conclusion is rather intuitive because the higher the carbon cap, the higher the retailer's marginal profit and the higher the demand for the product and the retailer's profit will increase as well. When the price sensitivity of the carbon cap b is lower than the threshold K_1, the trend of the manufacturer's profit with respect to the carbon cap differs based on the range of the cap. That is, if the cap is under the threshold K_2, the manufacturer's profit increases along with the cap. Otherwise, the manufacturer's profit decreases. This means that the manufacturer's profit reaches its maximum when the government sets the carbon cap to K_2. In contrast, when b exceeds the threshold K_1, the manufacturer's profit always increases with the cap. Figure 1 illustrates the relationship between the carbon cap and the manufacturer's profit with the following parameter settings: $\lambda = 0.8$, $\beta = 0.05$, $a = 1$, $K_1 = 9.68$ (a) $b = 0.5 < K_1$, and (b) $b = 10 > K_1$, which confirms that when $0 < b < K_1$, the maximal manufacturer's profit is obtained at $K_2 = 1.0817$, and also confirms that when $b > K_1$, the manufacturer's profit increases with the cap. In previous studies, such as Xu et al. [41], Gong and Zhou [54], Zhang and Xu [55], Du et al. [56], Xu et al. [57], and Zhang et al. [58], it is argued that the carbon cap always has a positive impact on the manufacturer's profit. However, our result shows that the manufacturer's profit is either positively or negatively affected by the carbon cap; thus, determining the appropriate carbon cap is crucial for the manufacturer.

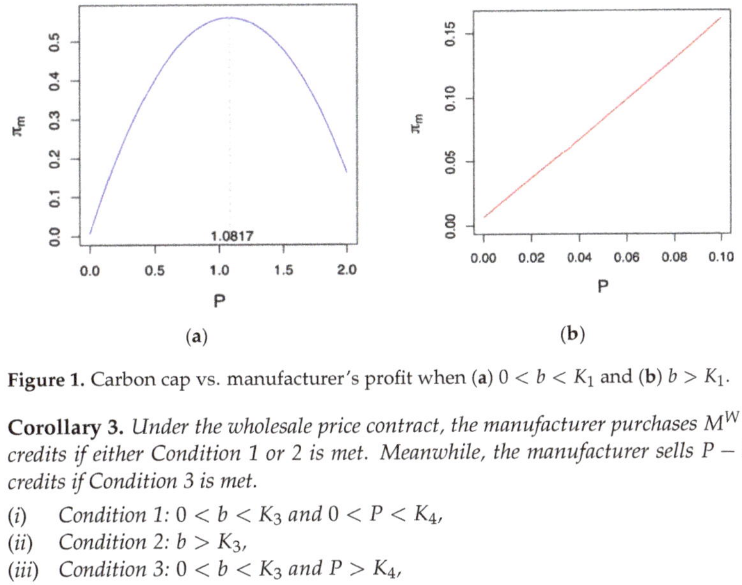

Figure 1. Carbon cap vs. manufacturer's profit when (**a**) $0 < b < K_1$ and (**b**) $b > K_1$.

Corollary 3. *Under the wholesale price contract, the manufacturer purchases* $M^W - P$ *emission credits if either Condition 1 or 2 is met. Meanwhile, the manufacturer sells* $P - M^W$ *emission credits if Condition 3 is met.*

(*i*) *Condition 1:* $0 < b < K_3$ *and* $0 < P < K_4$,
(*ii*) *Condition 2:* $b > K_3$,
(*iii*) *Condition 3:* $0 < b < K_3$ *and* $P > K_4$,

where $K_3 = \frac{(1+\beta)(3-\beta)}{\lambda^2}$ *and* $K_4 = \frac{\lambda(1-a\lambda)}{(1+\beta)(3-\beta)-b\lambda^2}$.

Corollary 3 shows the manufacturer's equilibrium behavior for purchasing/selling emission credits. When the price sensitivity of the carbon cap b is under the threshold K_3, the manufacturer's purchasing/selling behavior differs according to the range of the carbon cap. If the cap is lower than the threshold K_4, the manufacturer must purchase additional emission credits to avoid the penalty and enhance the cap level. Otherwise, the manufacturer is willing to sell excess emission credits to achieve a financial incentive. In contrast, when b exceeds the threshold K_3, the manufacturer always purchases the emission credits. In summary, there is a threshold for purchasing and selling the emission credits because of the linearly inverse relationship between the carbon cap and the emission trading price. Figure 2 depicts the manufacturer's decision on purchasing/selling the emission credits with the following parameter settings: $\lambda = 0.5$, $\beta = 0.05$, $a = 0.4$, $K_3 = 12.39$, $K_4 = 0.1469$, (a) $b = 1.5 < K_3$, and (b) $b = 13 > K_3$. Figure 2a shows that the manufacturer purchases (sells) $M^W - P$ $(P - M^W)$ emission credits if $P < K_4$ $(P > K_4)$. If $P = K_4$, the manufacturer neither purchases nor sells the emission credits because $M^W = P$. Meanwhile, as shown in Figure 2b, the manufacturer always purchases $M^W - P$ emission credits if $b > K_3$.

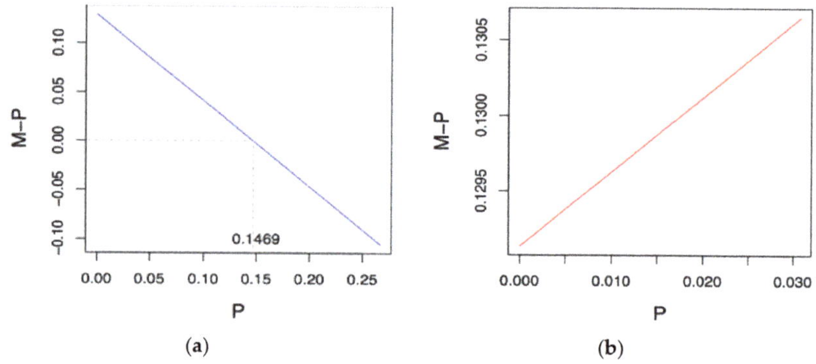

Figure 2. Carbon cap vs. emission credit when (**a**) $0 < b < K_3$ and (**b**) $b > K_3$.

4.1.2. Cost-Sharing Contract

Next, consider the collaboration model. Under a cost-sharing contract between supply chain members, the retailer shares δ fraction of the manufacturer's cost related to the production and sustainability innovation efforts. The profit functions of the manufacturer and the retailer are then expressed, respectively, as shown below.

$$\pi_m = (w - (1-\delta)\beta e)d - \frac{(1-\delta)e^2}{2} - (a - bP)(M - P) \text{ and } \pi_r = (p - w - \delta\beta e)d - \frac{\delta e^2}{2}. \quad (6)$$

Under the cost-sharing contract, the sequence of the three-stage Stackelberg game is as follows. In the first stage, the retailer determines the collaboration level δ that maximizes its profit. In the second stage, the manufacturer announces the wholesale price w and the sustainability effort e. After specifying the values of w and e, the retailer determines the retail price for the product p in the last stage. Applying backward induction, the equilibrium values of the decision variable and the profit for each player are obtained in Proposition 2.

Proposition 2. *Given the carbon cap P, the equilibrium values in the cost-sharing contract game are determined as follows:*

$$\begin{gathered}
\delta^C = \frac{(1-\beta)^2}{8},\ w^C = \frac{(1+\beta)(2-\beta)(7+2\beta-\beta^2)+\lambda(a-bP)(3-\beta)(2+5\beta+\beta^3)}{4(5+6\beta-3\beta^2)},\\
p^C = \frac{21+14\beta-11\beta^2-\lambda(a-bP)(1-10\beta+\beta^2)}{4(5+6\beta-3\beta^2)},\ e^C = \frac{2(1-\beta)(1-\lambda(a-bP))}{5+6\beta-3\beta^2},\\
d^C = \frac{(7+2\beta-\beta^2)(1-\lambda(a-bP))}{4(5+6\beta-3\beta^2)},\ M^C = \frac{\lambda(7+2\beta-\beta^2)(1-\lambda(a-bP))}{4(5+6\beta-3\beta^2)},\\
\pi_m^C = \frac{(7+2\beta-\beta^2)(1-\lambda(a-bP))^2}{8(5+6\beta-3\beta^2)} + P(a-bP),\ \text{and}\ \pi_r^C = \frac{(9-2\beta+\beta^2)(1-\lambda(a-bP))^2}{16(5+6\beta-3\beta^2)}.
\end{gathered} \quad (7)$$

Under the cost-sharing contract, we assume that $a\lambda < 1$ to avoid the negative demand for the product. Note that the equilibrium retailer's collaboration level must be lower than

0.125 because δ^C is a decreasing function when $\beta \in (0,\ 1)$. From Proposition 2, Corollaries 4–6 are derived as shown below.

Corollary 4. *Under the cost-sharing contract, the equilibrium decisions have the following relationships:*

$$\frac{\partial w^C}{\partial P} < 0,\ \frac{\partial e^C}{\partial P} > 0,\ \frac{\partial d^C}{\partial P} > 0,\ \frac{\partial M^C}{\partial P} > 0,\ and\ \frac{\partial p^C}{\partial P} = \begin{cases} > 0,\ if\ 0 < \beta < 5 - 2\sqrt{6}, \\ < 0,\ if\ 5 - 2\sqrt{6} < \beta < 1. \end{cases} \tag{8}$$

Under the cost-sharing contract, if $0 < \beta < 5 - 2\sqrt{6}$ $(5 - 2\sqrt{6} < \beta < 1)$, the retail price increases (decreases) with respect to the carbon cap. Regarding the wholesale price, the sustainability effort, and the demand for the product, relationships similar to Corollary 1 are found. Accordingly, further statements are omitted.

Corollary 5. *Under the cost-sharing contract, the equilibrium profits have the following relationships:*

$$\frac{\partial \pi_r^C}{\partial P} > 0\ and\ \frac{\partial \pi_m^C}{\partial P} = \begin{cases} > 0,\ if\ 0 < b < K_5\ and\ 0 < P < K_6, \\ > 0,\ if\ b > K_5, \\ < 0,\ if\ 0 < b < K_5\ and\ P > K_6, \end{cases} \tag{9}$$

where $K_5 = \frac{8(5+6\beta-3\beta^2)}{\lambda^2(7+2\beta-\beta^2)}$ *and* $K_6 = \frac{4a(5+6\beta-3\beta^2)+b\lambda(1-a\lambda)(7+2\beta-\beta^2)}{b(8(5+6\beta-3\beta^2)-b\lambda^2(7+2\beta-\beta^2))}$.

Corollary 5 shows results similar to those of Corollary 2. Even under the cost-sharing contract, the retailer's profit increases as the carbon cap increases. When the price sensitivity of the carbon cap b is under the threshold K_5, the trend of the manufacturer's profit differs according to the range of the cap. That is, if the cap is under (exceeds) threshold K_6, the manufacturer's profit increases (decreases) along with the cap. Hence, the manufacturer obtains its maximal profit at $P = K_6$. On the other hand, when b exceeds the threshold K_5, the manufacturer's profit always increases. Comparing the critical cap shown in Corollaries 2 and 5, the following relation is found that

$$K_6 - K_2 = \frac{\lambda(2 - a\lambda)(1 - \beta^4)}{(2(3-\beta)(1+\beta) - b\lambda^2)(8(5 + 6\beta - 3\beta^2) - b\lambda^2(7 + 2\beta - \beta^2))} > 0, \tag{10}$$

which means that by adopting the cost-sharing contract, the manufacturer maintains its profit growth with a relatively higher carbon cap. This argument is depicted in Figure 3 with the following parameter settings: $\lambda = 0.8$, $\beta = 0.05$, $a = 1$, $b = 0.5$, $K_2 = 1.0817$, and $K_6 = 1.085$.

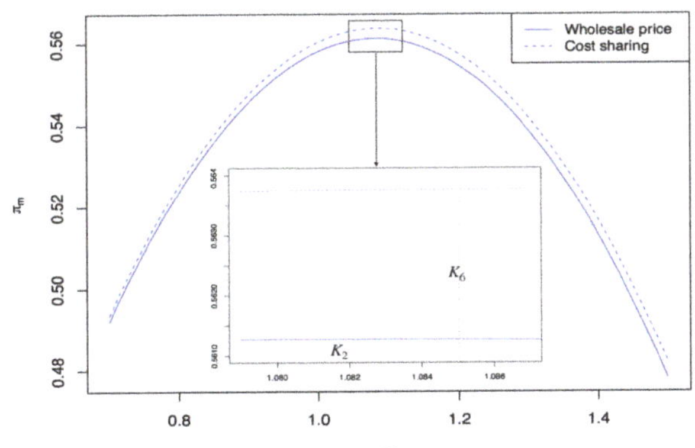

Figure 3. Comparison of the critical carbon cap between two contracts with respect to the manufacturer's profit.

Corollary 6. *Under the cost-sharing contract, the manufacturer purchases $M^C - P$ emission credits if either Condition 4 or 5 is met. Meanwhile, the manufacturer sells $P - M^C$ emission credits if Condition 6 is met.*

(i) *Condition 4: $0 < b < K_7$ and $0 < P < K_8$,*
(ii) *Condition 5: $b > K_7$,*
(iii) *Condition 6: $0 < b < K_7$ and $P > K_8$,*

where $K_7 = \frac{4(5+6\beta-3\beta^2)}{\lambda^2(7+2\beta-\beta^2)}$ and $K_8 = \frac{\lambda(1-a\lambda)(7+2\beta-\beta^2)}{4(5+6\beta-3\beta^2)-b\lambda^2(7+2\beta-\beta^2)}$.

Corollary 6 shows the manufacturer's equilibrium decision with regard to purchasing/selling the emission credits under the cost-sharing contract. Corollary 6 is similar to Corollary 3; thus, redundant statements are omitted here. Comparing the critical cap shown in Corollaries 3 and 6, the following relation is found that

$$K_8 - K_4 = \frac{\lambda(1-a\lambda)(1-\beta^4)}{((3-\beta)(1+\beta)-b\lambda^2)(4(5+6\beta-3\beta^2)-b\lambda^2(7+2\beta-\beta^2))} > 0, \quad (11)$$

which means that the threshold of purchasing/selling the emission credits with the cost-sharing contract is higher than that under the wholesale price contract. Figure 4 shows the difference in the manufacturer's carbon trading decision between the wholesale price contract and the cost-sharing contract with the following parameter settings: $\lambda = 0.5$, $\beta = 0.05$, $a = 0.4$, $b = 1.5$, $K_4 = 0.1469$, and $K_8 = 0.1534$. Figure 4 confirms the following: (i) when $P < K_4$, the manufacturer must purchase additional emission credits under both contracts; (ii) when $P > K_8$, the manufacturer can sell excess emission credits under both contracts; and (iii) when $K_4 < P < K_8$, the manufacturer purchases emission credits under the cost-sharing contract, whereas it sells them under the wholesale price contract.

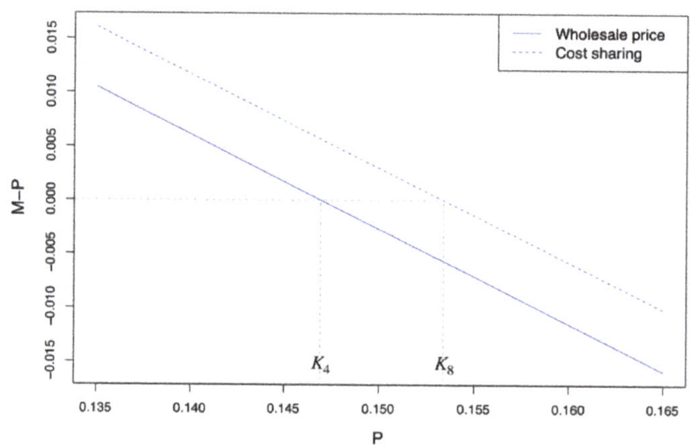

Figure 4. Comparison of the critical carbon cap between two contracts with respect to the emission credits.

4.1.3. Comparison

One may be curious about the differences in the equilibrium values between two contracts. This subsection conducts a comparative analysis by comparing Propositions 1 and 2. The result is as follows.

Corollary 7. *The following relation is found that*

$$e^W < e^C,\ d^W < d^C,\ w^W < w^C,\ p^W < p^C,\ M^W < M^C,\ \pi_m^W < \pi_m^C,\ \pi_r^W < \pi_r^C,\ \text{and}\ \pi_{sc}^W < \pi_{sc}^C. \quad (12)$$

Corollary 7 shows that a cost-sharing contract outperforms a wholesale price contract in terms of the profitability and sustainability of a supply chain. It must be first noted that the sustainability effort under the cost-sharing contract is greater than that under the wholesale price contract. Cost sharing with the retailer can help the manufacturer reduce the burden of product development and reduce production costs, ultimately leading to greater sustainability efforts. This makes it possible for the product to become more sustainable, which increases consumer utility as well as the demand for the product. Let m^W and m^C denote the marginal cost under the wholesale price and cost-sharing contracts, respectively. Following Equations (2) and (6), the relation $m^W = \beta e^W < (1-\delta^C)\beta e^C = m^C$ is obtained. That is, the higher marginal cost under the cost-sharing contract forces the manufacturer to raise the wholesale price, which results in an increase in the retail price. Corollary 7 also confirms that the profit of each supply chain member as well as the supply chain profit will increase if the retailer adopts a cost-sharing contract. Under the cost-sharing contract, the demand and prices increase at the same time, resulting in a natural increase in the profit of each supply chain member and leading to an increase in the overall supply chain profits. Finally, more carbon emissions are inevitable owing to the increased production of the product under the cost-sharing contract.

4.2. Government's Optimal Decision on Carbon Cap

Thus far, this article has regarded the carbon cap P as an exogenous parameter initially set by the government. However, henceforth, we consider P as an endogenous decision variable. In reality, the government may have a goal when implementing cap-and-trade regulations. Assume that the following possible objective is considered by the government: the government wants to maximize social welfare. By determining the optimal carbon cap, the government will achieve this objective.

Once again, this section analyzes the government's optimal decisions regarding the carbon cap under both contracts. To do this, we initially formulate the revenue and cost terms separately. We then integrate them to derive the objective function of the government problem. Social welfare SW is a function of the cap and takes the form of

$$SW(P) = \int_{p}^{p_{\max}} (1 - p + e)dp + \pi_{sc} - \eta M, \tag{13}$$

where $p_{\max}$ is the maximum price of the product. The first term $\int_{p}^{p_{\max}} (1-p+e)dp$ represents the value of the consumer surplus, which, in economics, is defined as the gap between the maximum price consumers are willing to pay and the actual price they pay. From simple algebra, it is obtained that $\int_{p}^{p_{\max}} (1-p+e)dp = d^2/2$. The second term is the supply chain profit, which is defined as the sum of the manufacturer's and retailer's profits. The government wants to minimize environmental degradation as the manufacturer produces the product. Let η be the unit environmental impact of the product. The coefficient η measures the degree of environmental damage due to the manufacturer's production process and the associated pollutants in monetary terms. Thus, the total environmental damage costs caused by production are expressed as ηM. It is important to note that the carbon trading cost (revenue) is not included in the social welfare function in Equation (13) because it is a transaction between the manufacturer and the government.

Assume that the government is an overall game leader under both a wholesale price contract and a cost-sharing contract. Let P^l denote the equilibrium carbon cap. Under each contract, P^l is obtained by solving the following maximization problem:

$$P^l = \underset{0 \le P \le a/b}{\operatorname{argmax}}\, SW^l(P), \; for\, l \in \{W,\, C\}. \tag{14}$$

The main results of this subsection are presented in Propositions 3 and 4.

Proposition 3. *Under the wholesale price contract, the government's optimal decision on the carbon cap is given by:*

(i) *If* $b < K_9$, $P^W = \frac{a\left((1+\beta)^2(3-\beta)^2 - b\lambda^2(6+2\beta-\beta^2)\right) + b\lambda(6-3\eta\lambda+\beta(2-\beta)(1-\eta\lambda))}{b\left(2(1+\beta)^2(3-\beta)^2 - b\lambda^2(6+2\beta-\beta^2)\right)}$,

(ii) *If* $b = K_9$ *and* $\eta > K_{10}$, $P^W = 0$,

(iii) *If* $b = K_9$ *and* $\eta < K_{10}$, $P^W = \frac{a}{b}$,

(iv) *If* $b > K_9$, $P^W = \frac{a}{b}$,

where $K_9 = \frac{2(1+\beta)^2(3-\beta)^2}{\lambda^2(6+2\beta-\beta^2)}$ *and* $K_{10} = \frac{(6+2\beta-\beta^2)(2-a\lambda)}{2\lambda(1+\beta)(3-\beta)}$.

Proposition 3 shows that the government's optimal decision about the carbon cap differs based on the price sensitivity of the cap b and the unit environmental impact cost η. If $b < K_9$, the social welfare function is strictly concave with respect to the cap; therefore, solving the first-order condition determines the global maximizer of the government's problem in Equation (14). If $b = K_9$ and $\eta > K_{10}$ ($\eta < K_{10}$), social welfare is a linearly decreasing (increasing) function of the carbon cap; thus, the optimal carbon cap equals zero (a/b). Lastly, social welfare is a convex and increasing function of the cap if $b > K_9$. Accordingly, the optimal carbon cap equals a/b. Figure 5 depicts Proposition 3 with the following parameter settings: $\lambda = 0.8$, $\beta = 0.1$, $a = 1$, $K_9 = 5.1374$, and $K_{10} = 1.4553$. In Figure 5a, the set $b = 2 < K_9$, and the social welfare is maximized at $P^C = 0.3248$. Figure 5b–d correspond to Proposition 3(ii), 3(iii) and 3(iv), respectively.

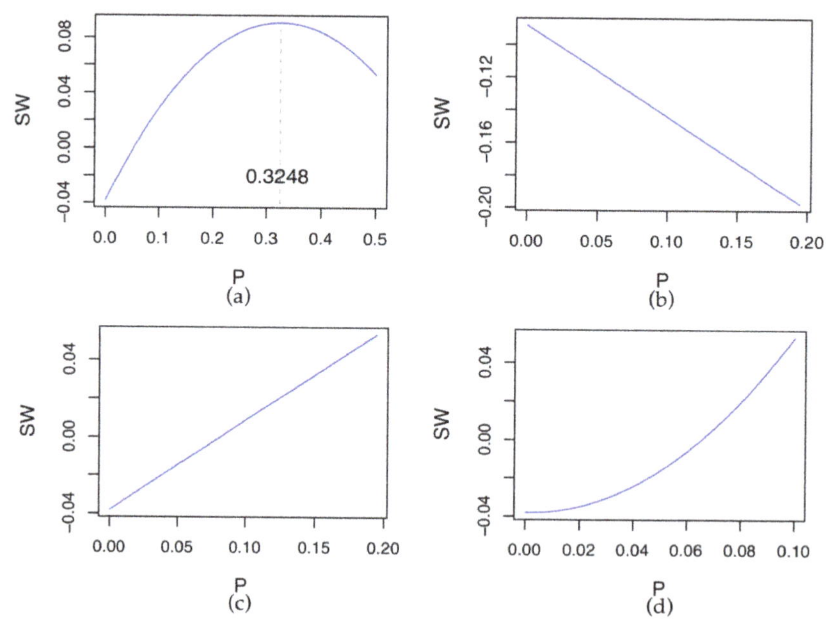

Figure 5. Carbon cap vs. social welfare when (**a**) $b = 2 < K_9$, (**b**) $b = K_9$ and $\eta = 2 > K_{10}$, (**c**) $b = K_9$ and $\eta = 1 < K_{10}$, (**d**) $b = 10 > K_9$.

Proposition 4. *Under the cost-sharing contract, the government's optimal decision on the carbon cap is given by*

$$\text{(i)}\quad \text{If } b < K_{11},\ P^C = \frac{a\left(16(5+6\beta-3\beta^2)^2 - b\lambda^2(279+324\beta-134\beta^2-28\beta^3+7\beta^4)\right) + b\lambda(279+324\beta-134\beta^2-28\beta^3+7\beta^4-4\eta\lambda(5+6\beta-3\beta^2)(7+2\beta-\beta^2))}{b\left(32(5+6\beta-3\beta^2)^2 - b\lambda^2(279+324\beta-134\beta^2-28\beta^3+7\beta^4)\right)},$$

(ii) *If* $b = K_{11}$ *and* $\eta > K_{12}$, $P^C = 0$,
(iii) *If* $b = K_{11}$ *and* $\eta < K_{12}$, $P^C = \frac{a}{b}$,
(iv) *If* $b > K_{11}$, $P^C = \frac{a}{b}$,

$$\text{where } K_{11} = \frac{32(5+6\beta-3\beta^2)^2}{\lambda^2(279+324\beta-134\beta^2-28\beta^3+7\beta^4)} \text{ and } K_{12} = \frac{4(5+6\beta-3\beta^2)(2-a\lambda)}{b\lambda^3(7+2\beta-\beta^2)}.$$

Proposition 4 presents the government's optimal decision regarding the carbon cap under the cost-sharing contract. Because Proposition 4 is similar to Proposition 3, we skip any redundant explanations. Following Propositions 3 and 4, Corollaries 8 and 9 are obtained as follows:

Corollary 8. *Assuming* $b < \min\{K_9, K_{11}\}$, *it is found that*

$$\frac{\partial P^l}{\partial \eta} < 0, \text{ for } l \in \{W, C\}. \tag{15}$$

Corollary 8 confirms that the carbon cap decreases as the unit environmental impact cost η increases. As indicated in Equation (13), η has a negative influence on social welfare. When η increases, the total environmental damage cost also increases; thus, the government wants to reduce the environmental damage by reducing the production quantity of the product. In our model, the only way for the government to reduce the manufacturer's production quantity is to lower the carbon cap. Note that from Corollaries 1 and 4, the demand for the product decreases as the carbon cap decreases.

Corollary 9. *Assuming* $b < \min\{K_9, K_{11}\}$, *it is found that*

$$P^W - P^C = \begin{cases} \leq 0, \text{ if } \eta \leq K_{13} = \frac{(2-a\lambda)(111+148\beta-46\beta^2-28\beta^3+7\beta^4)}{8\lambda(1+\beta)(3-\beta)(5+6\beta-3\beta^2)-b\lambda^3(3+10\beta-5\beta^2)}, \\ > 0, \text{ otherwise.} \end{cases} \tag{16}$$

Corollary 9 shows a comparison of the carbon caps under each contract. If η is low, the government sets a higher carbon cap under the cost-sharing contract. By setting a higher carbon cap when η is relatively low, the government wants to maintain the cooperation between the manufacturer and the retailer for greater supply chain profit and greater consumer surplus. On the other hand, if η is high, the government sets a higher carbon cap under the wholesale price contract. By doing so, the government encourages the manufacturer to increase its sustainable innovation efforts, which boosts the demand for the product. The increased demand raises consumer surplus as well as the supply chain profit, which compensates for the total environmental impact cost. Figure 6 shows a comparison between the optimal carbon caps under each type of contract with the following parameter settings: $\lambda = 0.8$, $\beta = 0.1$, $a = 1$, $b = 2$, $\eta = 0.5$, $K_{13} = 1.3711$, $P^W = 0.4069$, and $P^C = 0.4137$. As illustrated in Figure 6, the optimal carbon cap under the cost-sharing contract is greater than that under the wholesale price contract when η is under the threshold K_{13}.

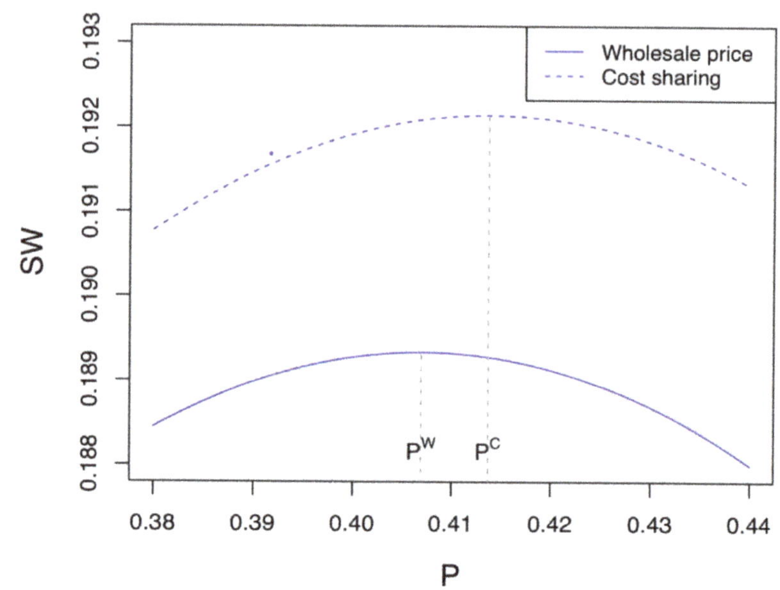

Figure 6. Comparison of the optimal carbon caps when $\eta < K_{13}$.

5. Conclusions

This article discussed equilibrium decisions about pricing and sustainability efforts in a supply chain under wholesale price and cost-sharing contracts. The major findings of this article are summarized below:

- Under wholesale price and cost-sharing contracts, the carbon cap has a positive impact on the manufacturer's sustainability efforts, which leads to an increase in the demand for the product. Because of the increased demand, the wholesale and retail prices fall at the same time;
- Under both contracts, the retailer's profit increases with respect to the carbon cap. Therefore, the retailer would like the government to set a higher carbon cap. However, the manufacturer's profit does not increase consistently with a higher carbon cap. This implies that proper cap setting has a significant impact on the manufacturers' profit;
- The cost-sharing contract outperforms the wholesale price contract in terms of product demand, prices, and the profitability of supply chain members. This fact reveals that the cost-sharing contract is a win–win strategy for both the manufacturer and the retailer. Under the cost-sharing contract, the profitability of the supply chain can improve, which has a positive impact on the sustainability of the supply chain. Because of the increasing manufacturers' profits, they can invest in more sustainable product development and manufacturing processes. This makes the supply chain more sustainable. Therefore, policy makers must provide subsidies and legal systems related to supply chain cooperation to provide a business environment in which supply chain members can collaborate more easily and efficiently;
- When the government wants to determine the carbon cap under both contracts, there is an optimal cap that maximizes social welfare. This optimal carbon cap decreases when the environmental damage caused by the production of the product increases. It is also found that when this type of environmental damage is relatively low, the government sets a higher cap under the cost-sharing contract so that the supply chain members maintain their cooperation.

This article provides several recommendations for firms that undertake sustainability efforts and face carbon emission constraints. However, there are also limitations that can be expanded upon and improved in future research: (1) Newsvendor model: the assumption of uncertain demand for the product is more complex but more realistic; (2) retailer's sustainability effort: an extended model can assume that the retailer can participate in sustainability activities, such as green retailing and green marketing; (3) other contracts: a comparative analysis of other supply chain cooperation contracts will provide more meaningful insights; and (4) multiple participants: in reality, multiple manufacturers and retailers either collaborate or compete in various supply chains. It would be worthwhile to analyze a supply chain with more complex but realistic conditions.

Author Contributions: Conceptualization, D.-H.L. and J.-C.Y.; Methodology, D.-H.L.; Software, D.-H.L.; validation, J.-C.Y.; Formal analysis, D.-H.L. and J.-C.Y.; Investigation, D.-H.L. and J.-C.Y.; Data curation, J.-C.Y.; writing—original draft preparation, D.-H.L. and J.-C.Y.; writing—review and editing, D.-H.L. and J.-C.Y.; visualization, D.-H.L. and J.-C.Y.; supervision, D.-H.L. and J.-C.Y.; funding acquisition, J.-C.Y. All authors have read and agreed to the published version of the manuscript.

Funding: This research was supported by Basic Science Research through the National Research Foundation of Korea (NRF), funded by the Ministry of Education (NRF-2019R1I1A3A01061157).

Institutional Review Board Statement: Not applicable.

Informed Consent Statement: Not applicable.

Data Availability Statement: Data are contained within the article.

Conflicts of Interest: The authors declare no conflict of interest.

Appendix A

Proof of Proposition 1. Because $\partial^2\pi_r/\partial p^2 = -2 < 0$, π_r is strictly concave with respect to p. Thus, solving the first-order condition (FOC) of the retailer's problem yields $p = (1+e+w)/2$. Integrating p into the manufacturer's problem, the following Hessian matrix is obtained:

$$\mathbf{H}^W = \begin{pmatrix} \frac{\partial^2\pi_m}{\partial w^2} & \frac{\partial^2\pi_m}{\partial w\partial e} \\ \frac{\partial^2\pi_m}{\partial e\partial w} & \frac{\partial^2\pi_m}{\partial e^2} \end{pmatrix} = \begin{pmatrix} -1 & \frac{1+\beta}{2} \\ \frac{1+\beta}{2} & -1-\beta \end{pmatrix}. \quad \text{(A1)}$$

We define Δ_k^l as the leading principal minor of order k in $\mathbf{H}^l$ for $l \in \{W, C\}$. We then find that $\Delta_1^W = -1 < 0$ and $\Delta_2^W = (1+\beta)(3-\beta)/4 > 0$, which implies that the manufacturer's profit is strictly concave with respect to w and e. By solving the FOCs of the manufacturer's problem, the equilibrium decisions and profits under the wholesale price contract are determined, as presented in Equation (3). □

Proof of Corollary 1. Under the wholesale price contract, the first-order derivatives of the equilibrium decisions with respect to the carbon cap P are given by

$$\begin{gathered} \frac{\partial w^W}{\partial P} = -\frac{b\lambda}{3-\beta} < 0,\ \frac{\partial p^W}{\partial P} = -\frac{b\beta\lambda}{(1+\beta)(3-\beta)} < 0,\ \frac{\partial e^W}{\partial P} = \frac{b\lambda(1-\beta)}{(1+\beta)(3-\beta)} > 0, \\ \frac{\partial d^W}{\partial P} = \frac{b\lambda}{(1+\beta)(3-\beta)} > 0, \text{ and } \frac{\partial M^W}{\partial P} = \frac{b\lambda^2}{(1+\beta)(3-\beta)} > 0. \end{gathered} \quad \text{(A2)}$$

Equation (A2) confirms Corollary 1. □

Proof of Corollary 2. Under the wholesale price contract, the first-order derivatives of the equilibrium profits with respect to the carbon cap P are given by

$$\frac{\partial\pi_r^W}{\partial P} = \frac{2b\lambda(1-\lambda(a-bP))}{(1+\beta)^2(3-\beta)^2} > 0 \text{ and } \frac{\partial\pi_m^W}{\partial P} = \frac{b\lambda(1-\lambda(a-bP))}{(1+\beta)(3-\beta)} + a - 2bP. \quad \text{(A3)}$$

When $b > K_1$, $\partial\pi_m^W/\partial P > 0$ regardless of P. When $0 < b < K_1$, $\partial\pi_m^W/\partial P > 0$ (< 0) if $0 < P < K_2$ $(P > K_2)$. □

Proof of Corollary 3. Under the wholesale price contract, the following equation holds:

$$M^W - P = \frac{\lambda(1-a\lambda) - P((1+\beta)(3-\beta) - b\lambda^2)}{(1+\beta)(3-\beta)}. \tag{A4}$$

When either Condition 1 or 2 in Corollary 3 is satisfied, $M^W > P$, which means that the carbon emissions caused by production exceed the carbon cap. Therefore, the manufacturer should purchase additional emission credits. On the other hand, when Condition 3 in Corollary 3 is satisfied, $M^W < P$, which means that the carbon emission levels are under the cap. Accordingly, the manufacturer can sell excess emission credits. □

Proof of Proposition 2. Because $\partial^2\pi_r/\partial p^2 = -2 < 0$, π_r is strictly concave with respect to p. Thus, solving the FOC of the retailer's problem yields $p = (1+w+e(1+\beta\delta))/2$. Integrating p into the manufacturer's problem, the following Hessian matrix is obtained:

$$\mathbf{H}^C = \begin{pmatrix} \frac{\partial^2\pi_m}{\partial w^2} & \frac{\partial^2\pi_m}{\partial w\partial e} \\ \frac{\partial^2\pi_m}{\partial e\partial w} & \frac{\partial^2\pi_m}{\partial e^2} \end{pmatrix} = \begin{pmatrix} -1 & \frac{1+\beta(1-2\delta)}{2} \\ \frac{1+\beta(1-2\delta)}{2} & -(1-\delta)(1+\beta(1-\beta\delta)) \end{pmatrix}. \tag{A5}$$

We then find that $\Delta_1^C = -1 < 0$ and $\Delta_2^C = (1+\beta)(3-\beta)/4 - \delta$. If the condition $\delta < (1+\beta)(3-\beta)/4$ is met, $\Delta_2^C > 0$, implying that the manufacturer's profit is strictly concave with respect to w and e. By solving the FOCs of the manufacturer's problem, we have

$$w = \frac{(1+\beta)(2-\beta)(1-\delta) + \lambda(a-bP)\left(1+\beta-\delta(2-\beta+\beta^2)\right)}{(1+\beta)(3-\beta) - 4\delta} \text{ and } e = \frac{(1-\beta)(1-\lambda(a-bP))}{(1+\beta)(3-\beta)-4\delta}. \tag{A6}$$

To determine the retailer's optimal collaboration level, substitute w and e in Equation (A6) into the retailer's objective function and the second-order condition of the retailer's problem then yields

$$\frac{\partial^2\pi_r}{\partial\delta^2} = -\frac{2(1-\beta)^2(1-\lambda(a-bP))^2(3+10\beta-5\beta^2+16\delta)}{((1+\beta)(3-\beta)-4\delta)^4} < 0. \tag{A7}$$

Hence, by solving the FOC of the retailer's problem, we have $\delta = (1-\beta)^2/8$. Finally, we calculate the equilibrium decisions and profits under the cost-sharing contract, as presented in Equation (7). □

Proof of Corollary 4. Under the cost-sharing contract, the first-order derivatives of the equilibrium decisions with respect to the carbon cap P are given by

$$\begin{gathered} \frac{\partial w^C}{\partial P} = -\frac{b\lambda(3-\beta)(2+5\beta+\beta^3)}{4(5+6\beta-3\beta^2)} < 0,\ \frac{\partial e^C}{\partial P} = \frac{2b\lambda(1-\beta)}{5+6\beta-3\beta^2} > 0,\ \frac{\partial d^C}{\partial P} = \frac{b\lambda(7+2\beta-\beta^2)}{4(5+6\beta-3\beta^2)} > 0, \\ \frac{\partial M^C}{\partial P} = \frac{b\lambda^2(7+2\beta-\beta^2)}{4(5+6\beta-3\beta^2)} > 0, \text{ and } \frac{\partial p^C}{\partial P} = \frac{b\lambda(1-10\beta+\beta^2)}{4(5+6\beta-3\beta^2)}. \end{gathered} \tag{A8}$$

If $0 < \beta < 5-2\sqrt{6}$, then $\partial p^C/\partial P > 0$; otherwise, $\partial p^C/\partial P < 0$. □

Proof of Corollary 5. Under the cost-sharing contract, the first-order derivatives of the equilibrium profits with respect to the carbon cap P are given by

$$\frac{\partial\pi_r^C}{\partial P} = \frac{b\lambda(9-2\beta+\beta^2)(1-\lambda(a-bP))}{8(5+6\beta-3\beta^2)} > 0 \text{ and } \frac{\partial\pi_m^C}{\partial P} = \frac{b\lambda(7+2\beta-\beta^2)(1-\lambda(a-bP))}{4(5+6\beta-3\beta^2)} + a - 2bP. \tag{A9}$$

When $b > K_5$, $\partial\pi_m^C/\partial P > 0$ regardless of P. When $0 < b < K_5$, $\partial\pi_m^C/\partial P > 0$ (< 0) if $0 < P < K_6$ ($P > K_6$). This completes the proof. □

Proof of Corollary 6. Under the cost-sharing contract, the following equation holds:

$$M^C - P = \frac{\lambda(1-a\lambda)(7+2\beta-\beta^2) - P\big(4(5+6\beta-3\beta^2) - b\lambda^2(7+2\beta-\beta^2)\big)}{4(5+6\beta-3\beta^2)}. \quad \text{(A10)}$$

The rest of the proof of Corollary 6 is quite similar to that of Corollary 3. Consequently, the details are omitted. □

Proof of Corollary 7. From Equations (3) and (7), the following relationships can be obtained:

$$\begin{aligned} e^W - e^C &= -\frac{(1-\beta)^3(1-\lambda(a-bP))}{(1+\beta)(3-\beta)(5+6\beta-3\beta^2)} < 0,\ d^W - d^C = -\frac{(1-\beta)^4(1-\lambda(a-bP))}{4(1+\beta)(3-\beta)(5+6\beta-3\beta^2)} < 0,\\ w^W - w^C &= -\frac{(1-\beta)^4(2-\beta)(1-\lambda(a-bP))}{4(3-\beta)(5+6\beta-3\beta^2)} < 0,\ p^W - p^C = -\frac{(1-\beta)^3(3+\beta)(1-\lambda(a-bP))}{4(1+\beta)(3-\beta)(5+6\beta-3\beta^2)} < 0,\\ M^W - M^C &= -\frac{\lambda(1-\beta)^4(1-\lambda(a-bP))}{4(1+\beta)(3-\beta)(5+6\beta-3\beta^2)} < 0,\ \pi_m^W - \pi_m^C = -\frac{(1-\beta)^4(1-\lambda(a-bP))^2}{8(1+\beta)(3-\beta)(5+6\beta-3\beta^2)} < 0,\\ &\text{and } \pi_r^W - \pi_r^C = -\frac{(1-\beta)^6(1-\lambda(a-bP))^2}{16(1+\beta)^2(3-\beta)^2(5+6\beta-3\beta^2)} < 0. \end{aligned} \quad \text{(A11)}$$

Because $\pi_{sc}^l = \pi_m^l + \pi_r^l$ for $l \in \{W,\ C\}$, it is found that $\pi_{sc}^W - \pi_{sc}^C < 0$. □

Proof of Proposition 3. Proposition 3 describes the optimal government decision on the carbon cap under the wholesale price contract. Integrating the equilibrium values in Equation (3) into the social welfare function, the SOC of the government's problem is given by

$$\frac{\partial^2 SW^W}{\partial P^2} = b\left(-2 + \frac{b\lambda^2(6+2\beta-\beta^2)}{(1+\beta)^2(3-\beta)^2}\right). \quad \text{(A12)}$$

If $b < K_9$, the social welfare function SW^W is strictly concave with respect to the carbon cap P and solving the FOC of the government's problem yields the optimal carbon cap P^W. If $b = K_9$ and $\eta > K_{10}$ ($\eta < K_{10}$), SW^W is a linearly decreasing (increasing) function of P within the interval $P \in [0,\ a/b]$. Therefore, P^W should be equal to the minimum (maximum) value of P. Lastly, if $b > K_9$, SW^W is a convex and increasing function of P. Thus, P^W should be equal to the maximum value of P. □

Proof of Proposition 4. The proof of Proposition 4 is quite similar to that of Proposition 3. As above, the details are omitted. □

Proof of Corollary 8. Under each supply chain contract, the following are obtained:

$$\frac{\partial P^W}{\partial \eta} = -\frac{\lambda^2(1+\beta)(3-\beta)}{2(1+\beta)^2(3-\beta)^2 - b\lambda^2(6+2\beta-\beta^2)} \text{ and } \frac{\partial P^C}{\partial \eta} = -\frac{4\lambda^2(5+6\beta-3\beta^2)(7+2\beta-\beta^2)}{32(5+6\beta-3\beta^2)^2 - b\lambda^2(279+324\beta-134\beta^2-28\beta^3+7\beta^4)}. \quad \text{(A13)}$$

In Equation (A13), the condition $b < \min\{K_9,\ K_{11}\}$ guarantees that $\partial P^l/\partial\eta < 0$ for $l \in \{W,\ C\}$. □

Proof of Corollary 9. Assuming that $b < \min\{K_9,\ K_{11}\}$ and following Propositions 3 and 4, we have Equation (A14).

$$P^W - P^C = \frac{\lambda(1-\beta)^4\begin{pmatrix} \eta(8\lambda(1+\beta)(3-\beta)(5+6\beta-3\beta^2) - b\lambda^3(3+10\beta-5\beta^2)) \\ -(111+148\beta-46\beta^2-28\beta^3+7\beta^4)(2-a\lambda) \end{pmatrix}}{\left(2(1+\beta)^2(3-\beta)^2 - b\lambda^2(6+2\beta-\beta^2)\right)\left(32(5+6\beta-3\beta^2)^2 - b\lambda^2(279+324\beta-134\beta^2-28\beta^3+7\beta^4)\right)}. \quad \text{(A14)}$$

If $\eta \le K_{13}$ in Equation (A14), $P^W \le P^C$; otherwise, $P^W > P^C$. □

References

1. One day We'll Disappear: Tuvalu's Sinking Islands. Available online: http://www.theguardian.com/global-development/2019/may/16/one-day-disappear-tuvalu-sinking-islands-rising-seas-climate-change (accessed on 2 February 2022).
2. Disappearing Island Nations are the Sinking Reality of Climate Change. Available online: http://qrius.com/disappearing-island-nations-are-the-sinking-reality-of-climate-change/ (accessed on 4 February 2022).
3. It's Official: Sea Level Rise Could Soon Displace Up to 187 Million People. Available online: http://www.sciencealert.com/sea-level-rise-could-displace-187-million-climate-refugees (accessed on 4 February 2022).
4. Climate Migrants Might Reach One Billion by 2050. Available online: http://reliefweb.int/report/world/climate-migrants-might-reach-one-billion-2050 (accessed on 4 February 2022).
5. Mechanisms under the Kyoto Protocol. Available online: http://unfccc.int/process/the-kyoto-protocol/mechanisms (accessed on 5 February 2022).
6. Velazquez-Cazares, M.G.; Leon-Castro, E.; Blanco-Mesa, F.; Alvarado-Altamirano, S. The ordered weighted average corporate social responsibility. *Kybernetes* **2021**, *50*, 203–220. [CrossRef]
7. Hua, G.; Cheng, T.C.E.; Wang, S. Managing carbon footprints in inventory management. *Int. J. Prod. Econ.* **2011**, *132*, 178–185. [CrossRef]
8. Benjaafar, S.; Li, Y.; Daskin, M. Carbon Footprint and the Management of Supply Chains: Insights from Simple Models. *IEEE Trans. Autom. Sci. Eng.* **2013**, *10*, 99–116. [CrossRef]
9. Drake, D.F.; Kleindorfer, P.R.; van Wassenhove, L.N. Technology choice and capacity portfolios under emissions regulation. *Prod. Oper. Manag.* **2016**, *25*, 1006–1025. [CrossRef]
10. Benz, E.; Trück, S. Modeling the Price Dynamics of CO2 Emission Allowances. *Energy Econ.* **2009**, *31*, 4–15. [CrossRef]
11. Ji, T.; Xu, X.; Yan, X.; Yu, Y. The production decisions and cap setting with wholesale price and revenue sharing contracts under cap-and-trade regulation. *Int. J. Prod. Res.* **2020**, *58*, 128–147. [CrossRef]
12. Chao, G.H.; Iravani, S.M.; Savaskan, R.C. Quality improvement incentives and product recall cost sharing contracts. *Manag. Sci.* **2009**, *55*, 1122–1138. [CrossRef]
13. Leng, M.; Parlar, M. Game-theoretic analyses of decentralized assembly supply chains: Non-cooperative equilibria vs. coordination with cost-sharing contracts. *Eur. J. Oper. Res.* **2010**, *204*, 96–104. [CrossRef]
14. Ghosh, D.; Shah, J. Supply chain analysis under green sensitive consumer demand and cost sharing contract. *Int. J. Prod. Econ.* **2015**, *164*, 319–329. [CrossRef]
15. Dai, R.; Zhang, J.; Tang, W. Cartelization or Cost-sharing? Comparison of cooperation modes in a green supply chain. *J. Clean. Prod.* **2017**, *156*, 159–173. [CrossRef]
16. Hong, J.; Lee, P. Supply chain contracts under new product development uncertainty. *Sustainability* **2019**, *11*, 6858. [CrossRef]
17. Drake, D.F.; Spinler, S. Sustainable operations management: An enduring stream or a passing fancy. *Manuf. Serv. Oper. Manag.* **2013**, *15*, 689–700. [CrossRef]
18. Beltagui, A.; Kunz, N.; Gold, S. The role of 3D printing and open design on adoption of socially sustainable supply chain innovation. *Int. J. Prod. Econ.* **2020**, *221*, 107462. [CrossRef]
19. Hassini, E.; Surti, C.; Searcy, C. A literature review and a case study of sustainable supply chains with a focus on metrics. *Int. J. Prod. Econ.* **2012**, *140*, 69–82. [CrossRef]
20. Krikke, H.; Bloemhof-Ruwaard, J.; van Wassenhove, L.N. Concurrent product and closed-loop supply chain design with an application to refrigerator. *Int. J. Prod. Res.* **2003**, *42*, 3689–3719. [CrossRef]
21. Subramanian, R.; Gupta, S.; Talbot, B. Product design and supply coordination under extended producer responsibility. *Prod. Oper. Manag.* **2009**, *18*, 259–277. [CrossRef]
22. Chen, C. Design for the environment: A quality-based model for green product development. *Manag. Sci.* **2001**, *47*, 250–263. [CrossRef]
23. Shen, B.; Cao, Y.; Xu, X. Product line design and quality differentiation for green and non-green products in a supply chain. *Int. J. Oper. Res.* **2020**, *58*, 148–164. [CrossRef]
24. Zhu, W.; He, Y. Green product design in supply chains under competition. *Eur. J. Oper. Res.* **2017**, *258*, 165–180. [CrossRef]
25. Thies, C.; Kieckhäfer, K.; Spengler, T.S.; Sodhi, M.S. Operations Research for Sustainability Assessment of Products: A Review. *Eur. J. Oper. Res.* **2019**, *274*, 1–21. [CrossRef]
26. Stuart, J.A.; Ammons, J.C.; Turbini, L.J. A product and process selection model with multidisciplinary environmental considerations. *Oper. Res.* **1999**, *47*, 221–234. [CrossRef]
27. Debo, L.G.; Toktay, L.B.; van Wassenhove, L.N. Market segmentation and product technology selection for remanufacturing. *Manag. Sci.* **2005**, *47*, 881–893.
28. Raz, G.; Druehl, C.T.; Blass, V. Design for the environment: Life-cycle approach using a newsvendor model. *Prod. Oper. Manag.* **2013**, *22*, 940–957. [CrossRef]
29. Klassen, R.; Vachon, S. Collaboration and evaluation in the supply chain: The impact on plant-level environmental investment. *Prod. Oper. Manag.* **2003**, *12*, 336–352. [CrossRef]
30. Zhu, Q.; Geng, Y.; Lai, K. Circular economy practices among Chinese manufacturers varying in environmental-oriented supply chain cooperation and the performance implications. *J. Environ. Manag.* **2010**, *91*, 1324–1331. [CrossRef]

31. Green, K.W., Jr.; Zelbst, P.J.; Bhadauria, V.S.; Meacham, J. Do environmental collaboration and monitoring enhance organizational performance? *Ind. Manag. Data Syst.* **2012**, *112*, 186–205. [CrossRef]
32. Ge, Z.; Hu, Q.; Xia, Y. Firms' R&D cooperation behavior in a supply chain. *Prod. Oper. Manag.* **2014**, *23*, 599–609.
33. Bhaskaran, S.R.; Krishnan, V. Effort, revenue, and cost sharing mechanisms for collaborative new product development. *Manag. Sci.* **2009**, *55*, 1152–1169. [CrossRef]
34. Ge, Z.; Hu, Q. Collaboration in R&D activities: Firm-specific decisions. *Eur. J. Oper. Res.* **2008**, *185*, 864–883.
35. Veugelers, R. Collaboration in R&D: An assessment of theoretical and empirical findings. *Economist* **1998**, *146*, 419–443.
36. Talluri, S.; Narasimhan, R.; Chung, W. Manufacturer cooperation in supplier development under risk. *Eur. J. Oper. Res.* **2010**, *207*, 165–173. [CrossRef]
37. Kim, S.-H.; Netessine, S. Collaborative cost reduction and component procurement under information asymmetry. *Manag. Sci.* **2013**, *59*, 189–206. [CrossRef]
38. Ji, J.; Zhang, Z.; Yang, L. Comparisons of initial carbon allowance allocation rules in an O2O retail supply chain with the cap-and-trade regulation. *Int. J. Prod. Econ.* **2017**, *187*, 68–84. [CrossRef]
39. Yenipazarli, A. To collaborate or not to collaborate: Prompting upstream eco-efficient innovation in a supply chain. *Eur. J. Oper. Res.* **2017**, *260*, 571–587. [CrossRef]
40. Cachon, G.P. *Handbooks in Operations Research and Management Science: Supply Chain Management: Design, Coordination and Operations*; Elsevier: Amsterdam, The Netherlands, 2003; pp. 227–339.
41. Xu, X.; Xu, X.; He, P. Joint production and pricing decisions for multiple products with cap-and-trade and carbon tax regulations. *J. Clean. Prod.* **2016**, *112*, 4093–4106. [CrossRef]
42. Xu, X.; Zhang, W.; He, P.; Xu, X. Production and pricing problems in make-to-order supply chain with cap-and-trade regulation. *Omega* **2017**, *66*, 248–257. [CrossRef]
43. Bai, Q.; Xu, J.; Zhang, Y. Emission reduction decision and coordination of a make-to-order supply chain with two products under cap-and-trade regulation. *Comput. Ind. Eng.* **2018**, *119*, 131–145. [CrossRef]
44. Yang, L.; Zhang, Q.; Ji, J. Pricing and carbon emission reduction decisions in supply chains with vertical and horizontal cooperation. *Int. J. Prod. Econ.* **2017**, *191*, 286–297. [CrossRef]
45. Wang, Y.; Chen, W.; Liu, B. Manufacturing/remanufacturing decisions for a capital-constrained manufacturer considering carbon emission cap and trade. *J. Clean. Prod.* **2017**, *140*, 1118–1128. [CrossRef]
46. Xu, J.; Chen, Y.; Bai, Q. A two-echelon sustainable supply chain coordination under cap-and-trade regulation. *J. Clean. Prod.* **2016**, *112*, 42–56. [CrossRef]
47. Shu, T.; Liu, Q.; Chen, S.; Wang, S.; Lai, K.K. Pricing decisions of CSR closed-loop supply chains with carbon emission constraints. *Sustainability* **2018**, *10*, 4430. [CrossRef]
48. Ji, S.; Zhao, D.; Peng, X. Joint decisions on emission reduction and inventory replenishment with overconfidence and low-carbon preference. *Sustainability* **2018**, *10*, 1119. [CrossRef]
49. Shen, Y.; Shen, K.; Yang, C. A production inventory model for deteriorating items with collaborative preservation technology investment under carbon tax. *Sustainability* **2019**, *11*, 5027. [CrossRef]
50. Bai, Q.; Gong, Y.; Jin, M.; Xu, X. Effects of carbon emission reduction on supply chain coordination with vendor-managed deteriorating product inventory. *Int. J. Prod. Econ.* **2019**, *208*, 83–99. [CrossRef]
51. Du, S.; Zhu, L.; Liang, L.; Ma, F. Emission-dependent supply chain and environment-policy-making in the 'cap-and-trade' system. *Energy Policy* **2013**, *57*, 61–67. [CrossRef]
52. He, P.; Dou, G.; Zhang, W. Optimal Production Planning and Cap Setting Under Cap-and-Trade Regulation. *J. Oper. Res. Soc.* **2017**, *68*, 1094–1105. [CrossRef]
53. The Economics of Sustainable Coffee Production. Available online: http://www.triplepundit.com/story/2014/economics-sustainable-coffee-production/39121 (accessed on 13 March 2022).
54. Gong, X.; Zhou, S.X. Optimal Production Planning with Emissions Trading. *Oper. Res.* **2013**, *61*, 908–924. [CrossRef]
55. Zhang, B.; Xu, L. Multi-item Production Planning with Carbon cap and Trade Mechanism. *Int. J. Prod. Econ.* **2013**, *144*, 118–127. [CrossRef]
56. Du, S.; Hu, L.; Song, M. Production Optimization Considering Environmental Performance and Preference in the Cap-and-Trade System. *J. Clean. Prod.* **2016**, *112*, 1600–1607. [CrossRef]
57. Xu, X.; He, P.; Xu, H.; Zhang, Q. Supply Chain Coordination with Green Technology under Cap-and-Trade Regulation. *Int. J. Prod. Econ.* **2017**, *183*, 433–442. [CrossRef]
58. Zhang, W.; Hua, Z.; Xia, Y.; Huo, B. Dynamic Multi-Technology Production-Inventory Problem with Emissions Trading. *IIE Trans.* **2016**, *48*, 110–119. [CrossRef]

sustainability

Article

Measurement of China's Building Energy Consumption from the Perspective of a Comprehensive Modified Life Cycle Assessment Statistics Method

Qiurui Liu [1], Juntian Huang [2], Ting Ni [3] and Lin Chen [1,*]

1 School of Management, Wuhan Institute of Technology, Wuhan 430205, China; qiurui.liu@hotmail.com
2 The Second Construction Co., Ltd., China Construction Third Engineering Bureau, Wuhan 430074, China; xingchenjt@outlook.com
3 College of Environmental & Civil Engineering, Chengdu University of Technology, Chengdu 610059, China; niting17@cdut.edu.cn
* Correspondence: linchen@wit.edu.cn

Abstract: This paper proposes a new life cycle assessment (LCA) statistics method to calculate the energy consumption of Chinese buildings from the perspective of LCA under the sustainable supply chain system. We divide the life cycle of buildings into the materialization stage, the construction stage, and the operation stage. Based on the new LCA statistics method, we obtain the following findings. First, the growth of total building energy consumption has slowed down since 2014, and its share of the Chinese total energy consumption levels off, remaining at about 40%. In 2018, the stages of materialization, construction, and operation account for about 34.02%, 4.65%, and 61.33% in total building energy consumption, respectively. Second, the materialization and operation stages are the main sources of energy consumption in the whole supply chain. Energy consumption in the materialization stage has been declining year by year since 2014, due to the impact of energy-saving policy. Moreover, we find that energy consumption in the operation and construction stages has been increasing year by year. Finally, in the life cycle of Chinese buildings, energy consumption in the operation stage plays a dominant role. This paper puts forward some managerial suggestions to relevant departments and provides some measures to optimize energy consumption in the Chinese building industry.

Keywords: sustainability; supply chain of construction industry; building energy consumption; life cycle assessment; building embodied sector

Citation: Liu, Q.; Huang, J.; Ni, T.; Chen, L. Measurement of China's Building Energy Consumption from the Perspective of a Comprehensive Modified Life Cycle Assessment Statistics Method. *Sustainability* **2022**, *14*, 4587. https://doi.org/10.3390/su14084587

Academic Editors: Shaojian Qu, Qingguo Bai, Ying Ji, Congjun Rao and Nicholas Chileshe

Received: 16 February 2022
Accepted: 7 April 2022
Published: 12 April 2022

1. Introduction

Nowadays, energy shortage and excessive emission of greenhouse gas have become hot topics in the world [1,2]. After more than 30 years of rapid economic development, China has become the world's largest carbon transfer site and has surpassed the United States as the world's largest carbon emitter since 2005 [3,4]. Economic development is inevitably accompanied by an increase in energy consumption [5]. In 2017, China's total energy consumption reached 4.49 billion tons of standard coal, 6.9 times higher than 1978, with an average annual growth rate of about 5.4% [6]. The Chinese Revolutionary Action Plan on Energy Production and Consumption has set a goal of limiting total energy consumption to 6 billion tons of coal equivalent, with non-fossil energy accounting for no less than 20% by 2030 [7]. As a result, China faces serious challenges in reducing energy consumption and coping with increasing environmental problems [8].

With the above attention to energy issues, the energy transformation of China has achieved great progress in recent years. The development of new energy and renewable energy is gradually improving [9–11]. However, energy problems such as excessive production, unreasonable energy consumption, energy shortage, and lack of effective regulation still exist [12,13]. In addition to energy transformation and the active promotion of new energy development, energy conservation and emission reduction are also very important to solve energy problems [14–16].

Specifically, the construction industry is one of the most energy and carbon-intensive industries, accounting for about 30% to 40% of the global total energy consumption and more than one-third of the global CO_2 emissions [16]. China has become the second-largest country in terms of building energy consumption. With the urbanization in China in recent years, total building energy consumption keeps increasing. As demonstrated in Guo et al. [17], building energy consumption in China has grown continually from 2001 to 2014. By the end of 2015, China's overall floorage had exceeded 60 billion square meters. According to the new Urbanization Plan (2014–2020), China's overall floorage will exceed 70 billion square meters by 2030, and the government sector will face a more severe situation regarding energy conservation and emission reduction [18,19].

To achieve the goal of energy conservation and emission reduction in the building sector, it is necessary to have an accurate baseline evaluation of building energy consumption, and a clear goal of building energy consumption control. Evaluation and accurate data of building energy consumption are also essential to provide a benchmark for policy-making of the construction sector, which can serve as an important reference for evaluating the measures of energy conservation and emission reduction.

In previous literature, several energy conversion methods have been designed with changes in statistical range to calculate building energy consumption [20,21]. This microscopic view provides detailed data for the energy consumption planning of a specific building. However, building energy consumption is inseparable from other industries as all industries cannot exist independently of buildings. The heterogeneity of the existing literature hinders the government's efforts in energy planning and policy assessment from a macrolevel. A more comprehensive perspective should be considered when measuring and evaluating building energy consumption [22]. Some researchers have calculated the energy consumption generated by the whole supply chain of the Chinese construction industry and then determined the energy consumption of the Chinese buildings [23,24]. Therefore, it is worth using the LCA theory from the national level of the supply chain of the construction industry to evaluate the energy consumption of the Chinese buildings.

LCA has been applied to the construction industry for about 10 years [25,26]. As it takes a comprehensive and systematic approach to environmental assessment; LCA methods have been incorporated into the construction industry to select eco-friendly products [27], evaluate performance [1], and optimize construction processes [28]. In the specific LCA steps, a building can be subdivided into the materialization stage, transportation stage, construction stage, operation stage, and demolition stage [28–30]. However, most studies focus on civil building and operation energy consumption, few of them cover the construction and demolition stages.

Moreover, there are some studies about the building energy consumption at the national level [6,23]. By tracing sources along the supply chains, Li et al. [6] find that the effects of China's building construction at the national and global scales are calculated as 55.21% and 5.67% of energy consumption, respectively. Some studies have found that energy consumption in the operation stage of buildings accounts for about 70% of the building's total energy consumption [23]; however, for the remaining 30%, which also has the potential to be optimized, there is a lack of attention. In other words, the calculation method of building energy consumption is fragmented [31]. Some researchers choose to accept or renounce all of the indexes and calculation methods from industrial enterprises when incorporating them into building energy consumption. Others mainly compute the energy consumption of building materials in the construction field but ignore the building energy consumption of industrial enterprises.

Based on the above discussions, this paper focuses on the following questions: First, how should the Chinese building energy consumption be calculated more comprehensively and accurately? Second, what are the differences among the materialization stage, construction stage, and operation stage of Chinese building energy consumption during recent years? Third, what policies should be kept or strengthened to control building energy consumption effectively?

To answer the aforementioned questions, we consider the building energy consumption within related industries, such as mineral mining, building materials manufacturing, and construction-related transportation, and incorporate them all into the calculation of the materialization, construction, and the operation stages of building energy consumption, based on LCA. A calculation model, which includes the materialization, construction, and the operation stages, is proposed to calculate the total energy consumption in various industries, from the perspective of the whole life cycle.

Based on the new LCA statistics method, several novel findings are obtained, as follows. Our first finding is that Chinese building energy consumption continues to grow. Since 2014, Chinese building energy consumption at each stage has been stable. The growth of total building energy consumption has slowed down, and its share of China's total energy consumption has leveled off, remaining at about 40%. Second, we find that the materialization and operation stages are the main sources of energy consumption in the whole supply chain. Energy consumption at the materialization stage is affected by energy-saving policy, and has been decreasing year by year since 2014. Third, the energy consumption in the operation and the construction stages has been increasing year by year. In the life cycle of Chinese buildings, energy consumption in the operation stage plays a dominant role.

This paper contributes to the extant studies in the following three aspects. First, we propose a novel calculation model to calculate the total energy consumption of different building stages in various industries from the perspective of the whole life cycle. Second, this paper focuses on building energy consumption within related industries and takes them all into consideration. The related industries include mineral mining, building materials manufacturing, construction-related transportation, construction, and so on. Third, we propose a new LCA statistics method to calculate the energy consumption of the Chinese buildings under the sustainable supply chain system.

Some managerial implications of this paper can be summarized as the following four aspects. First, we need to pay more attention to the recovery and reuse of construction waste and greatly encourage the use of construction waste recycling in prefabricated buildings. Second, it is hopeful to use BIM technology to promote the development of the construction industry and pave the way for energy conservation and emission reduction in the whole life cycle of buildings. Third, the application of hybrid renewable energy systems in ultra-low and near-zero energy consumption buildings should be vigorously promoted. Finally, improving people's consciousness of energy conservation is one of the most effective ways to reduce energy consumption in the building operation stage.

The remainder of the paper is structured as follows. In Section 2, compared with the existing literature on building energy consumption and life-cycle methods, this paper uses LCA statistics method to measure the building energy consumption, with a comprehensive consideration related to the construction industry. To be more in line with actual practice, the LCA-based evaluation model is constructed in Section 3 to cover the materialization, construction, and operation stages. Results and analysis of Chinese building energy consumption are provided in Section 4. Strategies to reduce energy consumption and carbon emissions are proposed for each stage in Section 5. In Section 6, we provide several theoretical contributions and managerial implications and outline some future research directions.

2. Literature Review

2.1. Scopes and Methods of Building Energy Consumption

There are different explanations of building energy consumption in academic literature, such as materialization, construction, operation, and energy consumption of the whole life cycle of buildings [27]. Generally, construction energy consumption refers to the energy consumption in the construction stage of the building [32], and materialization energy consumption refers to the total energy consumption involved in mining, processing, and manufacturing and transportation from the mining of raw materials to building materials, which are transported to the construction site [33]. Building operation energy consumption refers to the energy consumption in the operation stage [34–36]. Some scholars collectively refer to the sum of materialization and construction energy consumption as building energy consumption [37].

In the existing literature, there are many papers about different stages of building energy consumption. The materialization energy consumption of Chinese buildings calculated by Hong et al. [24] includes direct and indirect energy input in the whole supply chain of the construction industry. Zhang et al. [31] estimate the total energy consumption of the materialization, construction, and operation stages. Zhang and Wang [23] divide the building life cycle into three stages, namely, the construction, operation, and disposal stages. Zhang et al. [33] study the energy consumption of the construction stage, and that of the extraction, manufacturing, and transportation of building materials in China. Huo et al. [38] define Chinese building energy consumption as the operation energy consumption of domestic buildings, which excludes construction and industrial buildings. In general, energy consumption in the construction and operation stages is rarely studied by existing works.

Firstly, for the macro data, the basic idea is to separate the energy consumption associated with the building from the statistical data and then add them together [31]. Specifically, Zhang and Wang [23] calculate the direct energy consumption of Chinese buildings at different stages, the indirect energy consumption of electricity and heat, and the energy consumption generated by the supply chain, based on domestic macro statistics. Huo et al. [38] calculate the operation energy consumption of Chinese buildings by cutting the energy balance sheet in the Chinese Energy Statistics Yearbook. Li et al. [6] systematically quantify the embodied energy consumption of case buildings in Beijing based on the multi-scale intensity databases. Secondly, based on the comprehensive energy intensity method, the energy intensity of various buildings is investigated and calculated according to the energy dissipation characteristics and building area [39]. Thirdly, a model-based approach can be used to calculate or simulate energy consumption, and include top-down and bottom-up models based on the different levels of available input data. For instance, based on previous studies, Zhang et al. [33] discuss the energy consumption of the construction industry by adopting the modeling method of energy consumption calculation of extraction, manufacturing, and transportation. Hong et al. [24] calculate the materialization energy consumption of Chinese buildings by combining relevant economic and energy consumption data in the statistical yearbook. Recently, Wenninger et al. [40] put forward a new method, namely the QLattice method, to predict the annual energy consumption of German residential buildings. It should be noticed that the energy consumption of the construction industry has been recorded as energy consumption in the construction stage, without considering the transportation part in the energy statistical yearbook.

Our paper differs from the above papers in the following aspects. First, most of the data used in the previous papers come from the China Energy Statistical Yearbook, but only some parts of the data are selected for one stage. This ignores the potential energy consumption of other related industries. Therefore, we propose a novel calculation model to calculate the total energy consumption of different stages in various industries from the perspective of the whole life cycle. Second, most of existing studies mainly consider the energy consumption of materials or the operation stage of the building, there is little research on the construction stage. Different from the previous studies, this paper focuses on building energy consumption within related industries and takes them all into consideration. The related industries include mineral mining, building materials manufacturing, construction-related transportation, construction, and so on.

2.2. Energy Consumption Analysis Based on the LCA

Life cycle energy consumption of buildings refers to the total energy consumption in the whole life cycle of a building, including energy consumption in building material production, construction, building operation, and building demolition [28,29]. Therefore, life-cycle building energy consumption is more in line with the actual situation.

Many attempts have been made by scholars to evaluate the relationship between materials, components, systems of buildings, and the environment from the LCA perspective. Ramesh et al. [41] conduct a rigorous review of the analysis of building life-cycle energy generated by 73 cases in 13 countries and find that energy consumption in the operation and the materialization stages is the most important in the building life cycle. Hong et al. [42] develop an integrated framework for the embodied energy quantification of Chinese buildings from a multi-regional perspective. In the study of Chen et al. [43], building energy consumption refers to the terminal energy consumption in the transportation, construction, and operation stages of building materials. Motalebi et al. [44] propose a new framework combining mathematical optimization, building information modeling (BIM), and LCA to improve the energy efficiency of existing buildings through the application of energy retrofitting measures.

Three approaches of carrying LCA are mainly used. The first one is the Process-LCA. Based on the process analysis, different subsystems of the study subjects are divided, and the input of energy and resources, the output of emissions in each subsystem, and environmental impact are quantitatively analyzed. Each process is analyzed "from top to bottom" throughout the overall target system [45]. Felmer et al. [46] use LCA to evaluate the carbon footprint of a medium-rise, low-energy residential building in central Chile by using a large amount of wood products. The second approach is the Economic Input-Output based LCA (EIO-LCA), which is a combination of LCA and the input-output analysis method of environmental quantitative assessment, by adding environmental impact factors to the economic input-output model. EIO-LCA is often used to estimate the environmental impact of material manufacturing in infrastructure construction with a national average public data set. Chen et al. [47] use the EIO-LCA approach to quantify the net transfer of energy consumption and identify the transfer of energy consumption pressure in various sectors of economic activities. The third approach is the Hybrid-LCA that synthesizes the process LCA and EIO-LCA method [6,42]. It focuses on the advantages of these two methods together and considers the comprehensiveness, integrity, and particularity of the whole system, making results more reasonable and reliable [48]. The research on building energy consumption with LCA has been mature in the world; however, there are few pieces of research on the application of LCA to building energy consumption in China. The Chinese statistical statement system is also conducive to the study of LCA from a macro perspective.

Different from the previous studies, which rarely discuss the form of mixed existence of building energy consumption with other industries, this paper considers the building energy consumption within related industries, such as the mineral mining, building materials manufacturing, and construction-related transportation, and takes them all into calculation of the materialization, construction, and operation stages of building energy consumption, based on LCA.

2.3. Summary of Literature Review

Sections 2.1 and 2.2 review and summarize the research on scopes and methods of building energy consumption and energy consumption analysis, based on LCA. We find that there are three deficiencies in existing research. First, the existing literature mainly consider the energy consumption of materials or the operation stages of the building. There is a lack of comprehensive studies about the three-stage framework for the life cycle of buildings (i.e., materialization, construction, and operation stages). Second, most of the aforementioned papers do not take the mixed existence of building energy consumption with other industries into consideration. Third, the previous studies rarely discuss the form of a mixed existence of building energy consumption with other industries.

Table 1 demonstrates a summary of the above-mentioned literature, which is classified based on the scope of building energy consumption, whether the calculation of energy consumption covers other related industries and methods of calculation for building energy consumption. Compared with relevant works, the most important contributions of this paper are in the following four aspects. First, we focus on investigating a new model to calculate the total energy consumption of different building stages. Second, unlike some prior studies assuming that the building energy consumption does not contain other related industries, we investigate building energy consumption from the perspective of the whole life cycle by considering the energy consumption of other related industries. Third, we propose a new LCA statistics method to calculate the energy consumption of Chinese buildings, from the perspective of life cycle assessment under the sustainable supply chain system. These unique features make the present work different from the existing literature.

Table 1. Comparison of previous literature with this paper.

Articles	Scopes of the Building Energy Consumption	Whether the Calculation Covers Related Industries	Methods of Building Energy Consumption	LCA-Based Calculation Method
Hong et al. [42]	Materialization	No	Model-based	EIO-LCA
Zhang et al. [33]	Construction	No	Model-based	Process-LCA
Huo et al. [38]	Operation	Yes	Macro data	Hybrid-LCA
Zhang et al. [31]	Materialization Construction Operation	Yes	Macro data	Process-LCA
Zhang and Wang [23]	Construction Operation	Yes	Macro data	Hybrid-LCA
This paper	Materialization Construction Operation	Yes	Macro data	Process-LCA

3. LCA-Based Calculation Model

3.1. Identification and Division

By summarizing the existing studies of LCA, Zhang et al. [31] and Chen et al. [43], the whole life cycle of buildings is divided into three stages: the materialization, construction, and operation stages. The first is the materialization stage, which usually means product life cycle (namely, raw material supply, mining, manufacturing, transportation to the production site, and manufacturing). It is mainly the mining and processing of building materials in the industry, the production energy consumption of component manufacturing, and transportation. The second stage is the construction stage, which mainly refers to the total energy consumption of the construction site for various construction processes, mechanical operation, and related production auxiliary facilities, transportation to the construction site and construction, and energy consumption in the construction process of building demolition and waste transportation. The third stage is the operation stage, which includes the energy consumption involved in the use of the building for maintenance, repair, replacement, heating, ventilation, air conditioning, lighting, household appliances, elevators, cooking, etc.

3.2. Data Collection

As there are many types of direct and indirect energy consumption involved in building energy consumption at the macro level, it is difficult to calculate them under the life cycle of the buildings. In terms of various data sources, the relevant data in the

Chinese Statistical Yearbook is comprehensive and easy to access. Therefore, this paper puts forward a calculation method of various related industries based on it.

In the Chinese Statistical Yearbook, the Chinese energy consumption industry is divided into seven categories: (1) agriculture, forestry, animal husbandry, and fishery, (2) industry, (3) construction industry, (4) transportation, warehousing, and postal services, (5) wholesale, retail, accommodation, and catering industries, (6) other industries, and (7) living expenses. The daily operation of these seven industries is closely related to building energy consumption; therefore, it can well reflect the building energy consumption by proper treatment of these data.

The primary data in this paper are obtained from the Chinese Statistical Yearbook, the Chinese Energy Statistical Yearbook, and the Chinese Urban Construction Yearbook. The latest data of the energy part of the three data sources are all from 2018.

3.3. Building Energy Consumption

3.3.1. Building Energy Consumption of the Construction Industry

According to the statistical yearbook, the energy consumption of the construction industry mainly refers to the energy consumption of construction units, construction installation, main projects, temporary projects, affiliated industrial production units, and part of the building operation stage. Here, the energy consumption of each project and the affiliated industrial production unit are counted into the construction stage. Energy consumption of the construction industry in the statistical yearbook is used to represent the energy consumption in the construction stage:

$$E_a = E_1 \tag{1}$$

where E_a is building energy consumption of the construction industry and E_1 is total energy consumption of the construction industry.

3.3.2. Building Energy Consumption of the Engineering Industry

Energy consumption of the industrial industry mainly includes building energy consumption in the materialization and the operation stage. It includes all kinds of energy consumption used by industrial enterprises for production and non-production, no matter whether the energy products are used as fuel or as raw materials. It can be divided into three parts: (1) energy consumption of mining of building and raw materials, processing of building materials, production of machinery, (2) energy use of ancillary production departments that serve the main production system, and (3) energy consumption of non-productive sectors of industry, such as affiliated research institutes, schools, hospitals, canteens, nurseries, and construction teams.

The above energy consumption should be included in the measurement model. However, in the existing available statistics, only the final energy consumption were counted due to a lack of detailed statistical data. In this paper, building energy consumption of the industrial part is stated as follows:

$$E_b = (E_{211} + E_{223})X + (E_{212} + E_{224})Y + (E_{213} + E_{222})Z \tag{2}$$

where

- E_b: energy consumption of industrial buildings,
- E_{211} denotes total energy consumption of the ferrous metal mining industry,
- E_{212} represents total energy consumption of the non-ferrous metal mining industry,
- E_{213} denotes total energy consumption of the non-metallic ore mining industry,
- E_{222} captures total energy consumption of the non-metallic mineral manufacturing industry,
- E_{223} represents total energy consumption of the metal smelting and rolling processing industries,

- E_{224} stands for total energy consumption of the non-ferrous metal smelting and rolling processing industries, and
- X, Y, Z represent the ferrous/non-ferrous/non-metallic industries related to the ratio of building consumption. The detailed values of X, Y, and Z over the years are summarized in Table 2.

Table 2. X, Y, and Z values over the year.

Year	2000	2001	2002	2004	2005	2006	2007	2008	2009
X	0.41	0.41	0.41	0.41	0.41	0.50	0.50	0.50	0.50
Y	0.25	0.25	0.25	0.25	0.25	0.28	0.28	0.28	0.28
Z	0.90	0.90	0.90	0.90	0.90	0.90	0.90	0.90	0.90
Year	**2010**	**2011**	**2012**	**2014**	**2015**	**2016**	**2017**	**2018**	
X	0.55	0.55	0.55	0.58	0.58	0.58	0.58	0.58	
Y	0.28	0.32	0.32	0.32	0.32	0.35	0.35	0.35	
Z	0.90	0.90	0.90	0.90	0.90	0.90	0.90	0.90	

3.3.3. Building Energy Consumption of the Transportation and Storage Industry

In the statistical yearbook, energy consumption is counted by type. In this paper, building energy consumption of the transportation and storage industry refers to the energy consumption of business premises, warehouses, and affiliated units of transportation enterprises. On the one hand, coal consumption in transportation and warehousing is mainly used for the heating of related buildings. Therefore, the energy consumption of transportation and warehousing should be classified as building energy consumption. On the other hand, electricity consumption in transportation and warehousing is concentrated in three areas: railways, pipeline transportation, and urban public transportation. In addition to these three electricity consumptions, the rest of the consumption can be designated as building electricity consumption. Hence, the building energy consumption of the transportation and storage industry can be described as the following:

$$E_c = E_{31}\alpha_1 + E_{32}M\alpha_2 \quad (3)$$

where

- E_c represents building energy consumption of transportation and storage,
- E_{31} denotes coal consumption of transportation and storage,
- E_{32} captures electricity consumption of transportation and storage, and
- M stands for building electricity consumption of transportation and warehousing. This paper takes 40% [38].
- α_1 denotes coal conversion coefficient and standard coal coefficient of raw coal. It is 0.7143 million tons of standard coal/ton (China Energy Statistics Yearbook).
- α_2 denotes electric power conversion coefficient, and the conversion coefficient of thermal power generation is 122.9 million tons/100 million kWh (China Energy Statistics Yearbook).

3.3.4. Building Energy Consumption in Wholesale, Retail, Accommodation, and Catering Industries

Energy consumption of wholesale, retail, accommodation, and catering industries includes building, transportation, and electrical energy consumption. Here, transportation energy consumption should be deducted when calculating building energy consumption in these industries:

$$E_d = E_4 - E_{41} \quad (4)$$

$$E_{41} = E_{411} \times 95\% \times \alpha_3 + E_{412} \times 35\% \times \alpha_4 \quad (5)$$

where

- E_d represents building energy consumption in the wholesale, retail, accommodation, and catering industries,
- E_4 captures total energy consumption of the wholesale, retail, accommodation, and catering industries,
- E_{41} denotes transportation energy consumption in the wholesale, retail, accommodation, and catering industries;
- E_{411} represents gasoline consumption in the wholesale, retail, accommodation, and catering industries;
- E_{412} stands for diesel consumption in the wholesale, retail, accommodation, and catering industries;
- α_3 is the standard coal coefficient of gasoline, which is 14,714 tons of standard coal/ton (China Energy Statistics Yearbook), and
- α_4 is the standard coal coefficient of diesel oil, which is 144.571 million tons of standard coal/10 thousand tons (China Energy Statistics Yearbook).

3.3.5. Building Energy Consumption in Other Related Industries

Energy consumption of other related industries is also mainly concentrated in building, transport, and electrical energy consumption. As electricity energy consumption is often difficult to divide from building energy consumption, especially in the case of such a wide range of industries, electric energy consumption is included in building energy consumption in the calculation method:

$$E_e = E_5 - E_{51} \tag{6}$$

$$E_{51} = E_{511} \times 95\% \times \alpha_3 + E_{512} \times 35\% \times \alpha_4 \tag{7}$$

where

- E_e represents other industries related to building energy consumption,
- E_5 stands for the total energy consumption of other industries,
- E_{51} denotes the transportation energy consumption in other industries,
- E_{511} represents the amount of gasoline consumed by other industries, and
- E_{512} denotes diesel consumption in other industries.

3.3.6. Building Energy Consumption of Residential Consumption Part

In the statistical yearbook, domestic energy consumption is composed of coal, coke, gasoline, kerosene, diesel, natural gas, and electricity, among which gasoline and diesel are mainly used for transportation. Therefore, except for those usages of transportation, the energy consumption of daily life can be counted as building energy consumption:

$$E_f = E_6 - E_{61}\alpha_3 - E_{62} \times 95\% \times \alpha_4 \tag{8}$$

where

- E_f represents the building energy consumption contained in living consumption,
- E_6 denotes the total consumption of domestic energy,
- E_{61} captures the gasoline consumption in daily consumption, and
- E_{62} denotes the consumption of diesel in daily consumption.

3.3.7. The Heating Energy Consumption

Heating energy consumption mainly refers to the urban central heating data in the China Urban Construction Yearbook:

$$E_g = (E_{71} + E_{72})\alpha_5 \tag{9}$$

where

- E_g represents the building energy consumption of heating,

- E_{71} denotes total steam heating,
- E_{72} denotes total hot water heating, and
- α_5 is the thermal standard coal coefficient, which is 0.03412 million tons of standard coal/10,000 (China Energy Statistical Yearbook).

3.3.8. Building Energy Consumption Calculation Model

Based on the above calculation equations of building energy consumption in various industries, the calculation model of building energy consumption can be stated by stages, as follows:

$$E_y = E_c + E_d + E_e + E_f + E_g \tag{10}$$

$$E_z = E_a + E_b + E_c + E_d + E_e + E_f + E_g \tag{11}$$

where E_y is building energy consumption in the operation stage and E_z is building energy consumption.

4. Results and Analysis

The conversion coefficient of standard coal for energy used in this paper is from the Chinese Energy Statistics Yearbook. The public data of the Chinese Energy Statistical Yearbook is dealt with by the thermal power plant equivalent method. Therefore, the coal-fired power plant equivalent method is adopted in this study as a method of converting electric power into the standard coal equivalent. Table 3 displays the estimated results of building energy consumption in China from 2000 to 2018.

Table 3. Calculation results of building energy in China Unit: 10,000 tons of standard coal (tce).

Year	Construction	Materialization	Operation	Building Energy Consumption	Proportion
2000	2207	24,739.92	30,731.35	57,678.27	0.39
2001	2255.02	26,024.87	30,788.39	59,068.28	0.38
2002	2409.57	25,836.12	33,223.23	61,468.92	0.36
2003	2720.66	32,329.27	38,393.56	73,443.49	0.37
2004	3114.60	37,076.32	42,684.25	82,875.16	0.36
2005	3486	52,521.05	52,050.82	108,057.87	0.41
2006	3760.73	49,550.34	51,397.95	104,709.02	0.37
2007	4127.52	53,809.25	55,482.27	113,419.04	0.36
2008	3812.53	55,619.24	59,733.9	119,165.67	0.37
2009	4712	65,618.92	65,821.35	136,152.27	0.41
2010	5533	68,178.96	69,611.29	143,323.25	0.40
2011	6052	73,602.14	75,241.48	154,895.62	0.40
2012	6337	73,455.26	80,740.18	160,532.44	0.40
2013	7017	67,756.98	86,405.02	161,179.00	0.39
2014	7377	69,499.48	90,008.44	166,884.92	0.39
2015	7545	65,770.57	95,133.35	168,448.92	0.39
2016	7847	64,958.22	101,485.70	174,290.95	0.39
2017	8243	63,576.50	111,994.30	183,813.82	0.40
2018	8685	63,576.94	114,636.60	186,898.54	0.40

4.1. Current Situation of Building Energy Consumption in China

The Chinese building energy consumptions of the materialization, construction, and the operation stages in 2018 are provided in Figure 1. In 2018, the Chinese building energy consumption is about 1.869 billion tce, accounting for 40% of the total energy consumption. Energy consumption in the materialization stage is about 635.77 million tce, accounting for about 34.02% of building total energy consumption. Energy consumption in the construction stage is about 86.85 million tce, accounting for about 4.65% of total building energy consumption. Energy consumption in the operation stage is about 1146.37 million tce, accounting for 61.34% of the total building energy consumption.

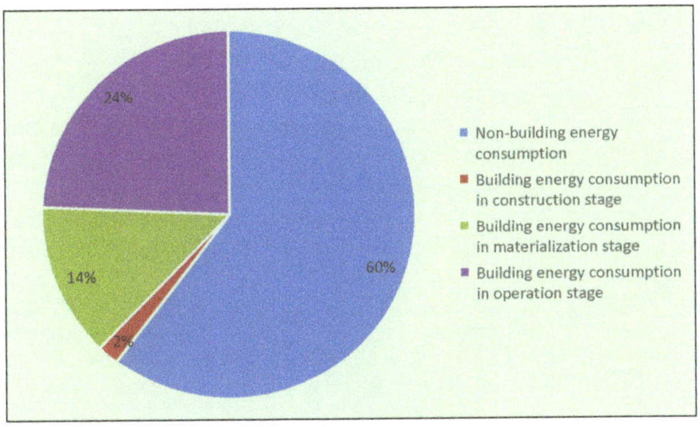

Figure 1. Building energy consumption in 2018.

From 2000 to 2018, the Chinese total building energy consumption keeps growing (see Figure 2). The total energy consumption of buildings has increased by 3.24 times, from 577 million tce in 2000 to 1.869 billion tce in 2018, with an average annual growth rate of 6.38%. Since 2014, its growth rate has slowed down to 2.29% annually, which is 26 million tce coal per year.

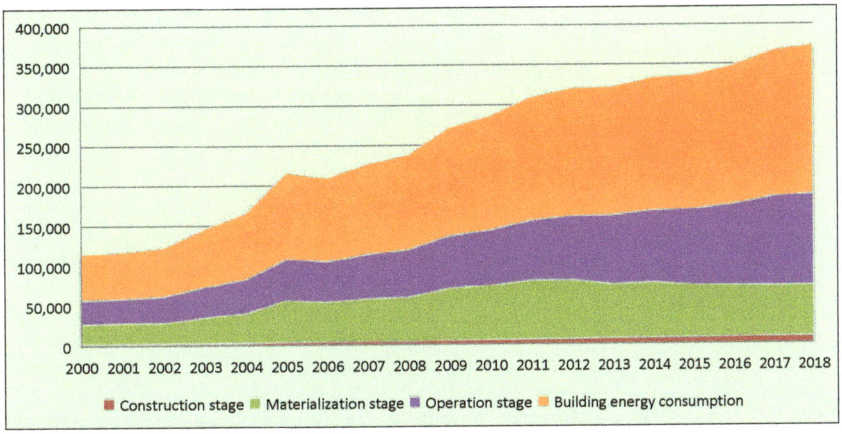

Figure 2. Total building energy consumption in China.

From 2000 to 2009, building energy consumption accounted for about 36% to 41% total energy consumption. In 2005 and 2009, the proportion of building energy consumption increased significantly, reaching 41%. From 2010 to 2018, the proportion of building energy consumption remained between 39% and 40% of total Chinese energy consumption. In general, the proportion of building energy consumption in total energy consumption fluctuated slightly between 36% and 41% from 2000 to 2009, and then tended to be stable around 39% and 40% from 2010 to 2018.

From 2000 to 2018, the average proportion of the materialization, construction, and operation stages in total building energy consumption is 4.02%, 42.77%, and 53.20%, respectively. Furthermore, the annual growth rates of building energy consumption in the materialization, construction, and operation stages are 5.09%, 7.48%, and 7.17%, respectively. Although building energy consumption of the construction stage accounts for about 4% of the building energy consumption, its annual growth rate is the highest in those 19 years. Therefore, building energy consumption in the construction stage should be also paid attention to, to control its growth.

4.2. Energy Consumption in the Materialization Stage

From 2000 to 2012, the Chinese building energy consumption in the materialization stage kept increasing, from 247.4 million tons of standard coal in 2000 to 734.5 million tce in 2012, an increase of 2.97 times, with an average annual growth rate of 8.73% (see Figure 3). Meanwhile, the energy consumption of the materialization stage has decreased from 695 million to 635.8 million tce per year since 2014, with a decrease of 11.84 million tce per year. During these 19 years, energy consumption in the materialization stage has been maintained within the range of 39.90% to 48.06% of total building energy consumption.

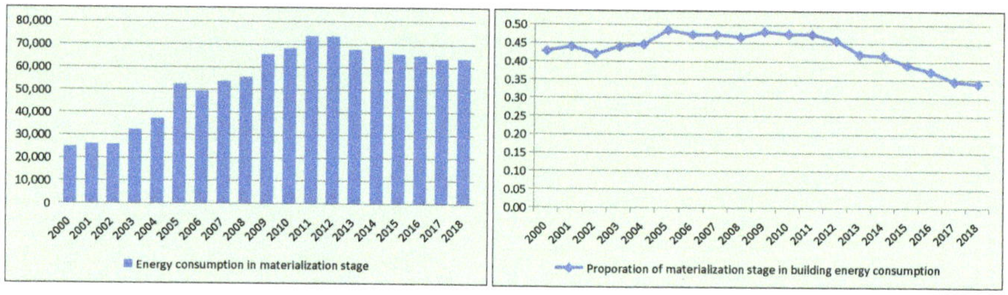

Figure 3. Chinese building energy consumption analysis of the materialization stage, from 2000 to 2018 (Unit: 10,000 tons of standard coal).

It is worth noticing that, since its peak in 2012 and 2014, energy consumption in the materialization stage as well as its proportion in building energy consumption have continued to decline speedily. This proves that the energy policy implemented in 2012 and 2014 has provided great help to energy conservation and emission reduction in the materialization stage of buildings. Some instructive policies related to the building industries are listed in the next section of strategies.

4.3. Energy Consumption in the Construction Stage

From 2000 to 2018, total Chinese building energy consumption in the construction stage continued to grow by four times, from 0.2 billion tons of standard coal in 2000 to 0.8 billion tce in 2018, with an increase about 0.03 billion tons of standard coal per year (see Figure 4). From 2000 to 2008, the energy consumption of the construction stage accounted for 1.19–1.48% of the total national energy consumption, and 3.40–4.34% of the total building energy consumption, illustrating a decreasing trend. On the other hand, from 2009 to 2018, the energy consumption of the construction stage accounted for 1.36% to 1.91% of national total energy consumption, and 3.80% to 4.71% of building energy consumption, indicating an increasing trend; therefore, its proportion in the total building energy consumption increases by 1.14% per year from 2000 to 2018.

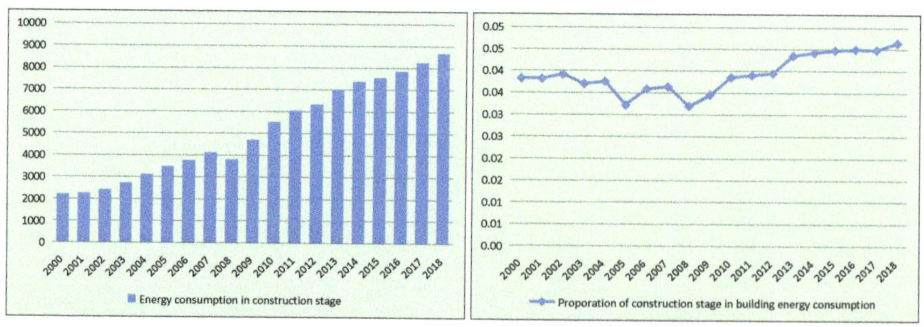

Figure 4. Analysis of Chinese building energy consumption in the construction period from 2000 to 2018 (Unit: 10,000 tons of standard coal).

Although energy consumption in the construction stage is relatively low, accounting for a relatively low proportion of total building energy consumption, it has been growing at a stable level, which is 7.48% annually. Energy consumption in the construction stage is greatly related to the national construction scale; therefore, its growth may be related to the rapid development of urbanization in China. Energy consumption of the construction stage is less affected by the outside world but is mainly related to the construction scale. Therefore, the best way to save energy during the construction stage is to improve construction technology and strengthen construction management.

4.4. Energy Consumption in the Operation Stage

From 2000 to 2018, the Chinese building energy consumption in the operation stage continued to increase from 307.3 million tons to 1146.36 million tons of standard coal. Building energy consumption has increased by 3.73 times, with an average annual growth rate of 7.17% (see Figure 5). From 2000 to 2011, the proportion of energy consumption of the building operation stage in building energy consumption fluctuated between 48.17% and 54.05%. Then, from 2011 to 2018, the proportion of building energy consumption of the operation stage in building energy consumption displayed an increasing trend, which increased significantly from 48.57% to 61.33%, with an annual growth rate of 3.29%, or an increase about 49.25 million tons of standard coal per year.

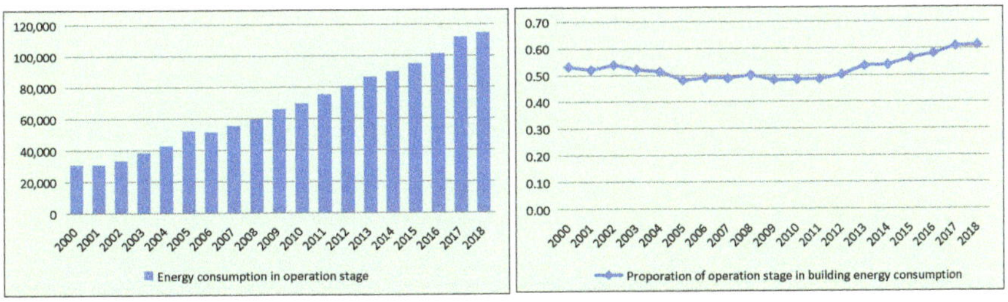

Figure 5. Energy consumption analysis of the Chinese building operation stage from 2000 to 2018 (Unit: 10,000 tons of standard coal).

Therefore, the change in building energy consumption is mainly caused by the energy consumption of the operation stage in recent years. The reduction of building energy consumption should start by restraining energy consumption in the operation stage.

5. Optimization Strategies

5.1. Optimization Strategies for the Materialization Stage

From the above analysis, energy consumption in the materialization stage demonstrates a downward trend while the energy consumption in other stages continues to increase, which proves that the production efficiency of building materials in China has been improving in recent years. This is owed to the Chinese government's adoption of many powerful command-controlled supervision measures during 2013 and 2014. For example, in the middle of 2013, the Chinese government implemented ten air pollution-related measures, including the implementation of a new mechanism of energy conservation and emission reductions, the strict control of newly increased energy consumption in high energy consumption industries, and so on [49]. Then, the Chinese government carried out the "2014–2015 action plan for energy conservation, emission reduction, and low carbon development". The plans are listed as follows:

(1) Industrial restructuring. Backward capacity should be eliminated, especially in the industries of iron making, steel making, cement (clinker and grinding), and flat glass.

(2) Transformation and innovation of energy-saving technology. Energy-saving technologies, such as waste heat and pressure utilization, energy system optimization,

and energy saving of motor systems, should be applied to reconstruct engineering equipment.

(3) Industrial energy saving. Energy efficiency benchmarking should be fully implemented in key energy-consuming industries, such as building materials, steel, electricity, and transportation. Energy management and control centers of industrial enterprises should also be established.

(4) Prefabricated industries should be supported to promote the modernization of the construction industry. Equivalent replacement or decrement replacement of energy consumption should be applied to the new capacity of the steel and building material industries.

However, there is still a gap between Chinese production efficiency and the international level. The extractive industry of raw materials and building materials manufacturing industry should continue to promote technological reform, develop new technology, and improve production efficiency. Research and development of new building materials should also be continued as the usage of new building materials can effectively save energy consumption.

Attention should also be paid to the recovery and reuse of construction wastes. The reuse of metal, wood, and available wastes can greatly reduce the input of raw materials and reduce energy consumption. At the same time, the amount of recycled steel in demolished buildings will keep increasing, which will have a significant impact on energy use and emissions in the construction sector.

5.2. Optimization Strategies for the Construction Stage

(1) Project management should be improved to reduce energy consumption. There still exists inefficiency or lack of regulation regarding the management of construction projects in China, resulting in accelerated mechanical aging, material damage, and unnecessary energy consumption. The immaturity of construction technology will not only lead to excessive energy consumption in the construction stage, but also lead to more energy consumption in the later operation stage.

(2) Selection of suppliers within a short distance between the product inventories and the construction site should be considered to reduce energy consumption in transportation.

(3) Adoption of green buildings. Full implementation of green building standards should be put into action for public welfare and large public buildings. Project capability and environmental impact assessment systems should be strictly implemented. For those regions that have not finished the energy-saving tasks, their new high energy consumption projects should be canceled or delayed.

5.3. Optimization Strategies for the Operation Stage

Energy consumption in the operation stage is the highest in building energy consumption, accounting for about 60%, and it continues to increase. Therefore, the operation stage is key to energy-saving in building energy consumption. The energy-saving strategy of the operation stage is listed as follows:

(1) Architectural design determines the energy consumption in the operation stage. Through controllable design, eco-friendly or new energy-saving building materials can be selected for the new buildings, which will be an effective method to reduce the consumption of resources and energy, from the perspective of the life cycle. For existing buildings, reasonable energy-saving renovation should also be carried out. The technology of building information modelling (BIM) has been considered an ideal digital tool to function as the data repository of all information relating to the building lifecycle [50]. The Ministry of Housing and Urban-Rural Development issued the Construction Industry Information Development Outline from 2011 to 2015 in May 2011 and the Architectural Engineering Design Information Model Mapping Standard in 2019, aiming to strengthen the management of projects and standardize building industry data.

(2) As energy consumption in the operation stage is to meet people's requirements for normal study, work, and life, the most effective way is to strengthen the management of energy conservation, to implement demand-side management mechanisms of power, and to enhance people's awareness of energy saving. Electricity and heat take most of the consumption energy in the building operation stage. For instance, reducing the usage of air conditioning is one of the most effective ways to save energy.
(3) As coal is still the main energy source of power and heat, the energy consumption structure should be adjusted to reduce inefficient and carbon-intensive coal use in electric power and thermal power enterprises. Meanwhile, strategies of energy efficiency are essential in energy policy, which could be created using the various approaches employed for energy savings in buildings [51]. The supply of non-fossil fuels and natural gas should be increased in residents' life or for the replacement of coal, with a strict replacement policy of equal or reduced coal consumption. Using the potentiality of electric vehicles toward the achievement of the zero-energy target extended to a buildings cluster level is also a measure to reduce coal burning, by exploiting renewable generation on and off-site [52].
(4) The government can give publicity to the catalog of energy-efficient air conditioners, refrigerators, and other products. They can also make policies to guide the production, purchase, and use of energy-efficient products.

6. Conclusions and Management Insights

6.1. Conclusions

This paper proposes a new statistical method of LCA, based on the Chinese Statistical Yearbook and analysis of Chinese building energy consumption with related industries from 2000 to 2018. The whole life cycle of buildings is divided into three stages: the materialization, construction, and operation stages. Different from other LCA methods, we consider the form of building energy consumption mixed with other industries. Our findings could be summarized as the following four aspects.

Firstly, this study demonstrates that the growth rate of Chinese building energy consumption has slowed down gently and its share in the Chinese total energy consumption is about 40%. This point is different form Zhang et al. [31], which obtained the conclusion that building energy accounted for about 43% of China's total energy consumption for 2011–2013. With our calculation method, building energy consumption takes about 39.53% of China's total energy consumption for 2011–2013, and this proportion decreased slightly to 39.43% in the recent five years (2014–2018). Therefore, we come to the conclusion that the share of building energy consumption in China's total energy consumption is nearly 40% for 2011–2018.

Secondly, with our proposed calculation method, which takes into account the building energy consumption in other related industries, Chinese building energy consumption increased with an annual rate of about 7.62% from 2001 to 2013. Comparing with Zhang et al. [31], which states that China's building energy consumption has increased about 7% annually since 2001 to 2013, the growth rate of China's building energy consumption obtained in our paper is higher than that in Zhang et al. [31]. Furthermore, we also calculate China's building energy consumption from 2013 to 2018. We find that the annual growth rate of China's building energy consumption is about 2.50% since 2013 to 2018. Compared with the growth rate from 2000 to 2013, the growth rate of China's building energy consumption has slowed down over 60%.

Thirdly, this paper discusses the differences among the different stages of building energy consumption. From 2000 to 2018, the stages of materialization, construction, and operation account for about 42.77%, 4.02%, and 53.20%, respectively, of total building energy consumption. Therefore, the materialization and operation stages are the main sources of building energy consumption. In these 19 years, the annual growth rates of building energy consumption in the materialization, construction, and operation stages are 5.09%, 7.48%, and 7.17%, respectively.

Lastly, another new discovery is that the growth of building energy consumption in recent years is mainly caused by the energy consumption of the operation stage. Energy consumption of the materialization stage has decreased, with 11.84 million tons of standard coal per year since 2014. On the contrary, energy consumption in the operation stage has increased, with 49.25 million tce per year, or with an annual growth rate of 5.40% of total building energy consumption from 2011 to 2018. Therefore, the variance of building energy consumption in 2014 to 2018 is mainly caused by the energy consumption in the operation stage. Controlling the growth of building energy consumption in the operation stage is more essential to energy conservation in the future.

6.2. Management Insights

Some managerial insights to control the building energy consumption are summarized as five aspects: policy orientation in China, how to recover and reuse construction wastes in the physicochemical stage, application of BIM to improve the informatization of the construction industry, emphasizing the application of renewable energy and energy conservation of the operation stage.

(1) This paper helps with the understanding of Chinese building energy consumption in 2000–2018. Nowadays, China is more eager than ever to save energy and reduce carbon emissions. During 2020 to 2021, the Chinese government has emphasized many times their determination to make efforts to save energy and reduce carbon emissions in international conferences. In 2021, they have put forward their 14th Five-Year Plan, which clearly demonstrates control of the growth of coal consumption. By 2030, China's civil construction area will exceed 70 billion square meters, their building industry will face a more severe situation of energy conservation. As building energy consumption still takes nearly 40% of their total energy consumption in recent years, the task of energy conservation in the building industry is quite urgent. In this case, the supervision intensity of the building industry has been enhanced. This forces building enterprises to put more attention into energy saving strategies.

(2) Attention should be paid to the recovery and reuse of construction wastes. The reuse of metal, wood, and available wastes can greatly reduce the input of raw materials and reduce energy consumption. The application of construction waste recycling in prefabricated buildings should be encouraged. Certain preferential tax policies could be applied to the projects that use construction waste recycling. Additionally, intelligent sorting standards should be developed in order to improve the sorting efficiency of construction waste recycling enterprises.

(3) A new finding is that although building energy consumption in the construction stage is only a small part of total building energy consumption and is often neglected, its growth is the fastest. Therefore, more attention should be paid to control its growth. Energy consumption in the construction stage is less affected by external factors. Improvement of construction technology, management, and design can be useful in energy-saving in the construction and operation stages. It is promising to use BIM to promote the construction industry, paving the way for energy saving and emission reduction in the whole building life cycle.

(4) The application of hybrid renewable energy systems in ultra-low energy and near-zero energy buildings should be promoted. For example, the hybrid geothermal-photovoltaic system has been used for heating and cooling. Companies should pay more attention to research and development in this field.

(5) The operation stage not only takes up most of a building's energy consumption, it also makes the greatest contribution to the growth of Chinese buildings' energy consumption from 2000 to 2018. Therefore, it is essential to control building energy consumption in the operation stage. Except for the design and application of new eco-friendly technology, an effective way to reduce building energy consumption in the operation stage may be to enhance people's awareness of energy conservation in their daily life.

6.3. Future Research Directions

There also exist some limitations of this study. First, the calculation method of building energy consumption in this paper is proposed based on the present situation, that there is a lack of well-directed and detailed statistical data on Chinese building energy consumption. Therefore, the first limitation is that the proposed method is more suitable for the macro analysis of building energy consumption. Second, the statistical range of energy of different industries is complex and hard to divide precisely [53]. Therefore, energy consumption in the building operation stage in the construction industry is divided roughly in this paper, which needs to be further studied. Third, there is a great difference in climate, geographical location, and economics among different provinces or regions. The building energy consumption of different Chinese regions such as urban areas is not discussed in this paper. Therefore, the building energy consumption of different provinces [17,49,54] also needs to be investigated in further studies. Last but not least, a future prediction of expected consumption should be further studied.

Author Contributions: Conceptualization, Q.L.; data curation, Q.L., J.H. and T.N.; formal analysis, J.H. and L.C.; investigation, Q.L., J.H. and T.N.; methodology, Q.L., J.H. and L.C.; resources, Q.L., J.H. and T.N.; software, Q.L., J.H. and L.C.; supervision, Q.L. and T.N.; validation, T.N.; visualization, L.C.; writing—original draft, Q.L., J.H. and L.C.; writing—review & editing, Q.L. and L.C. All authors have read and agreed to the published version of the manuscript.

Funding: This research was funded by the Scientific Research Program of Hubei Provincial Department of Education (Grant No. B2021076, Q20211507), the Internal Science Research Foundation of Wuhan Institute of Technology (Grant Nos. K2021050, K2021049), the National Natural Science Foundation of China (Grant No. 72102171), the Humanities and Social Sciences Youth Foundation, Ministry of Education of the People's Republic of China (Grant No. 21YJC630006), and the Youth Project of Philosophy and Social Science of Higher Education department of Hubei Province (Grant No. 21Q087).

Institutional Review Board Statement: Not applicable.

Informed Consent Statement: Not applicable.

Data Availability Statement: Not applicable.

Acknowledgments: The authors would like to thank the anonymous referees for their comments and suggestions.

Conflicts of Interest: The authors declare no conflict of interest.

References

1. Luo, X.; Oyedele, L.; Owolabi, H.; Bilal, M.; Ajayi, A.; Akinade, O. Life cycle assessment approach for renewable multi-energy system: A comprehensive analysis. *Energy Convers. Manag.* **2020**, *224*, 113354. [CrossRef]
2. Khan, H.R.; Awan, U.; Zaman, K.; Nassani, A.A.; Haffar, M.; Abro, M.M.Q. Assessing hybrid solar-wind potential for industrial decarbonization strategies: Global shift to green development. *Energies* **2021**, *14*, 7620. [CrossRef]
3. Wang, Y.; Wu, T.; Li, H.; Skitmore, M.; Su, B. A statistics-based method to quantify residential energy consumption and stock at the city level in China: The case of the Guangdong-Hong Kong-Macao Greater Bay Area cities. *J. Clean. Prod.* **2019**, *251*, 119637. [CrossRef]
4. Ikram, M.; Sroufe, R.; Awan, U.; Abid, N. Enabling progress in developing economies: A novel hybrid decision-making model for green technology planning. *Sustainability* **2022**, *14*, 258. [CrossRef]
5. Tabrizikahou, A.; Nowotarski, P. Mitigating the energy consumption and the carbon emission in the building structures by optimization of the construction processes. *Energies* **2021**, *14*, 3287. [CrossRef]
6. Li, Y.; Han, M.; Liu, S.; Chen, G. Energy consumption and greenhouse gas emissions by buildings: A multi-scale perspective. *Build. Environ.* **2019**, *151*, 240–250. [CrossRef]
7. Wen, Q.; Gu, J.; Hong, J.; Shen, G.; Yuan, M. Unfolding interregional energy flow structure of China's construction sector based on province-level data. *J. Environ. Manag.* **2019**, *253*, 109693. [CrossRef]
8. Chen, L.; Nan, G.; Li, M.; Feng, B.; Liu, Q. Manufacturer's online selling strategies under spillovers from online to offline sales. *J. Oper. Res. Soc.* **2022**. [CrossRef]
9. Yu, H.; Huang, Z.; Pan, Y.; Long, W. *Guidelines for Community Energy Planning*; Springer Science and Business Media LLC: Berlin, Germany, 2020.

10. Liu, B.; Wang, D.; Xu, Y.; Liu, C.; Luther, M. Vertical specialisation measurement of energy embodied in international trade of the construction industry. *Energy* **2018**, *165*, 689–700. [CrossRef]
11. Zhang, W.; Zhao, Y.; Huang, F.; Zhong, Y.; Zhou, J. Forecasting the energy and economic benefits of photovoltaic technology in China's rural areas. *Sustainability* **2021**, *13*, 8408. [CrossRef]
12. Wei, Y.; Liao, H. *Energy Economics: Energy Efficiency in China*; Springer Science and Business Media LLC: Berlin, Germany, 2016.
13. Xiao, Q.; Chen, L.; Xie, M.; Wang, C. Optimal contract design in sustainable supply chain: Interactive impacts of fairness concern and overconfidence. *J. Oper. Res. Soc.* **2021**, *72*, 1505–1524. [CrossRef]
14. Awan, U.; Khattak, A.; Rabbani, S.; Dhir, A. Buyer-driven knowledge transfer activities to enhance organizational sustainability of suppliers. *Sustainability* **2020**, *12*, 2993. [CrossRef]
15. Liu, X.; Chen, X.; Shahrestani, M. Optimization of insulation thickness of external walls of residential buildings in hot summer and cold winter zone of China. *Sustainability* **2020**, *12*, 1574. [CrossRef]
16. Li, P.; Rao, C.; Goh, M.; Yang, Z. Pricing strategies and profit coordination under a double echelon green supply chain. *J. Clean. Prod.* **2021**, *278*, 123694. [CrossRef]
17. Guo, Y.; Bart, D. Optimization of design parameters for office buildings with climatic adaptability based on energy demand and thermal comfort. *Sustainability* **2020**, *12*, 3540. [CrossRef]
18. Marchi, B.; Zanoni, S. Supply chain management for improved energy efficiency: Review and opportunities. *Energies* **2017**, *10*, 1618. [CrossRef]
19. Morella, P.; Lambán, M.P.; Royo, J.; Sánchez, J.C.; Corrales, L.C.N. Development of a new green indicator and its implementation in a cyber-physical system for a green supply chain. *Sustainability* **2020**, *12*, 8629. [CrossRef]
20. Sun, Y.; Haghighat, F.; Fung, B. A review of the-state-of-the-art in data-driven approaches for building energy prediction. *Energy Build.* **2020**, *221*, 110022. [CrossRef]
21. Begum, S.; Xia, E.; Ali, F.; Awan, U.; Ashfaq, M. Achieving green product and process innovation through green leadership and creative engagement in manufacturing. *J. Manuf. Technol. Manag.* **2021**. [CrossRef]
22. Yuan, X.; Sun, X.; Zhao, W.; Mi, Z.; Wang, B.; Wei, Y. Forecasting China's regional energy demand by 2030: A Bayesian approach. *Resour. Conserv. Recycl.* **2017**, *127*, 85–95. [CrossRef]
23. Zhang, X.; Wang, F. Hybrid input-output analysis for life-cycle energy consumption and carbon emissions of China's building sector. *Build. Environ.* **2016**, *104*, 188–197. [CrossRef]
24. Hong, J.; Shen, G.; Guo, S.; Xue, F.; Zheng, W. Energy use embodied in China's construction industry: A multi-regional input-output analysis. *Renew. Sustain. Energy Rev.* **2016**, *53*, 1303–1312. [CrossRef]
25. Buyle, M.; Braet, J.; Audenaert, A. Life cycle assessment in the construction sector: A review. *Renew. Sustain. Energy Rev.* **2013**, *26*, 379–388. [CrossRef]
26. Degen, F.; Schütte, M. Life cycle assessment of the energy consumption and GHG emissions of state-of-the-art automotive battery cell production. *J. Clean. Prod.* **2022**, *330*, 129798. [CrossRef]
27. Asdrubali, F.; Baldassarri, C.; Fthenakis, V. Life cycle analysis in the construction sector: Guiding the optimization of conventional Italian buildings. *Energy Build.* **2013**, *64*, 73–89. [CrossRef]
28. Chau, C.; Leung, T.; Ng, W. A review on life cycle assessment, life cycle energy assessment and life cycle carbon emissions assessment on buildings. *Appl. Energy* **2015**, *143*, 395–413. [CrossRef]
29. Sartori, I.; Sandberg, N.; Brattebø, H. Dynamic building stock modelling: General algorithm and exemplification for Norway. *Energy Build.* **2016**, *132*, 13–25. [CrossRef]
30. Ma, J.; Du, G.; Zhang, Z.; Wang, P.; Xie, B. Life cycle analysis of energy consumption and CO_2 emissions from a typical large office building in Tianjin, China. *Build. Environ.* **2017**, *117*, 36–48. [CrossRef]
31. Zhang, Y.; He, C.; Tang, B.; Wei, Y. China's energy consumption in the building sector: A life cycle approach. *Energy Build.* **2015**, *94*, 240–251. [CrossRef]
32. Cheng, Z.; Shahmir, N.; Lu, T. BIM-based investigation of total energy consumption in delivering building products. *Adv. Eng. Inform.* **2018**, *38*, 370–380.
33. Zhang, Y.; Yan, D.; Hu, S.; Guo, S. Modelling of energy consumption and carbon emission from the building construction sector in China, a process-based LCA approach. *Energy Policy* **2019**, *134*, 110949. [CrossRef]
34. Zhou, N.; Lin, J. The reality and future scenarios of commercial building energy consumption in China. *Energy Build.* **2008**, *40*, 2121–2127. [CrossRef]
35. Cai, W.; Wu, Y.; Zhong, Y.; Ren, H. China building energy consumption: Situation, challenges and corresponding measures. *Energy Policy* **2009**, *37*, 2054–2059. [CrossRef]
36. McNeil, M.; Feng, W.; Can, S.; Khanna, N.; Ke, J.; Zhou, N. Energy efficiency outlook in China's urban buildings sector through. *Energy Policy* **2016**, *97*, 532–539. [CrossRef]
37. Basbagill, J.; Flager, F.; Lepech, M.; Fischer, M. Application of life-cycle assessment to early stage building design for reduced embodied environmental impacts. *Build. Environ.* **2013**, *60*, 81–92. [CrossRef]
38. Huo, T.; Ren, H.; Zhang, X.; Cai, W.; Feng, W.; Zhou, N.; Wang, X. China's energy consumption in the building sector: A Statistical Yearbook-Energy Balance Sheet based splitting method. *J. Clean. Prod.* **2018**, *185*, 665–679. [CrossRef]
39. Abrahamse, W.; Steg, L.; Vlek, C.; Rothengatter, T. A review of intervention studies aimed at household energy conservation. *J. Environ. Psychol.* **2005**, *25*, 273–291. [CrossRef]

40. Wenninger, S.; Kaymakci, C.; Wiethe, C. Explainable long-term building energy consumption prediction using QLattice. *Appl. Energy* **2022**, *308*, 118300. [CrossRef]
41. Ramesh, T.; Prakash, R.; Shukla, K. Life cycle energy analysis of buildings: An overview. *Energy Build.* **2010**, *42*, 1592–1600. [CrossRef]
42. Hong, J.; Shen, G.; Li, C.; Liu, G.; Wu, Z.; Zhong, X. An integrated framework for embodied energy quantification of buildings in China: A multi-regional perspective. *Resour. Conserv. Recycl.* **2018**, *138*, 183–193. [CrossRef]
43. Chen, S.; Zhang, G.; Xia, X.; Setunge, S.; Shi, L. A review of internal and external influencing factors on energy efficiency design of buildings. *Energy Build.* **2020**, *216*, 109944. [CrossRef]
44. Motalebi, M.; Rashidi, A.; Nasiri, M.M. Optimization and BIM-based lifecycle assessment integration for energy efficiency retrofit of buildings. *J. Build. Eng.* **2022**, *49*, 104022. [CrossRef]
45. SäYnäJoki, A.; Heinonen, J.; Junnila, S.; Horvath, A. Can life-cycle assessment produce reliable policy guidelines in the building sector? *Environ. Res. Lett.* **2017**, *12*, 013001. [CrossRef]
46. Felmer, G.; Morales-Vera, R.; Astroza, R.; González, I.; Puettmann, M.; Wishnie, M. A lifecycle assessment of a low-energy mass-timber building and mainstream concrete alternative in central Chile. *Sustainability* **2022**, *14*, 1249. [CrossRef]
47. Chen, J.; Guo, Y.; Su, H.; Ma, X.; Zhang, Z.; Chang, B. Empirical analysis of energy consumption transfer in China's national economy from the perspective of production and demand. *Environ. Sci. Pollut. Res.* **2021**, *28*, 19202–19221. [CrossRef] [PubMed]
48. Kwok, K.; Kim, J.; Chong, W.; Ariaratnam, S. Structuring a comprehensive carbon-emission framework for the whole life cycle of building, operation, and construction. *J. Archit. Eng.* **2016**, *22*, 04016006. [CrossRef]
49. Yang, R.; Chen, W. Spatial correlation, influencing factors and environmental supervision mechanism construction of atmospheric pollution: An empirical study on SO_2 emissions in China. *Sustainability* **2019**, *11*, 1742. [CrossRef]
50. Marzouk, M.; Azab, S.; Metawie, M. BIM-based approach for optimizing life cycle costs of sustainable buildings. *J. Clean. Prod.* **2018**, *188*, 217–226. [CrossRef]
51. Aldhshan, S.R.S.; Abdul Maulud, K.N.; Wan Mohd Jaafar, W.S.; Karim, O.A.; Pradhan, B. Energy consumption and spatial assessment of renewable energy penetration and building energy efficiency in Malaysia: A review. *Sustainability* **2021**, *13*, 9244. [CrossRef]
52. Barone, G.; Buonomano, A.; Forzano, C.; Giuzio, G.F.; Palombo, A. Increasing self-consumption of renewable energy through the Building to Vehicle to Building approach applied to multiple users connected in a virtual micro-grid. *Renew. Energy* **2020**, *159*, 1165–1176. [CrossRef]
53. Gao, M.; Yang, H.; Xiao, Q.; Goh, M. A novel method for carbon emission forecasting based on Gompertz's law and fractional grey model: Evidence from American industrial sector. *Renew. Energy* **2022**, *181*, 803–819. [CrossRef]
54. Rao, C.; He, Y.; Wang, X. Comprehensive evaluation of non-waste cities based on two-tuple mixed correlation degree. *Int. J. Fuzzy Syst.* **2021**, *23*, 369–391. [CrossRef]

MDPI
St. Alban-Anlage 66
4052 Basel
Switzerland
Tel. +41 61 683 77 34
Fax +41 61 302 89 18
www.mdpi.com

Sustainability Editorial Office
E-mail: sustainability@mdpi.com
www.mdpi.com/journal/sustainability

www.ingramcontent.com/pod-product-compliance
Lightning Source LLC
Chambersburg PA
CBHW061553220926
51850CB00005B/326
9783036575810